The Ecology of Areas with Serpentinized Rocks

Geobotany 17

Series Editor

M.J.A. WERGER

a. The late professor H. Wild next to the endemic *Euphorbia wildii* on ultramafic soils on the Great Dyke, Zimbabwe. (Photograph by J. Proctor.)

b. Armerio arenarietum fontiqueri association at Carrazeda (Bragança), Portugal. Note the contrasting aspects of the foreground serpentinic habitat and the normal cultivated landscape on amphibolitic soil with large-leaved trees and rye field. (Photograph by E. M. de Sequeira.)

c. Cerastium arcticum spp. *edmondstonii*, a British endemic on the Keen of Hamar and an adjacent site on Unst, Shetland. (Photograph by J. Proctor.)

d. The Great Dyke near Mtoroshanga, Zimbabwe. (Photograph by J. Proctor.)

The Ecology of Areas with Serpentinized Rocks

A World View

edited by

B. A. ROBERTS AND J. PROCTOR

Springer Science+Business Media, B.V.

Library of Congress Cataloging-in-Publication Data

The Ecology of areas with serpentinized rocks : a world view / edited
by B.A. Roberts and J. Proctor.
  p.   cm – (Geobotany : 17)
  Includes bibliographical references and index.
  ISBN 978-94-010-5654-0   ISBN 978-94-011-3722-5 (eBook)
  DOI 10.1007/978-94-011-3722-5
  1. Serpentine plants–Ecology.   I. Roberts, Bruce A.
II. Proctor, J. III. Series.
OK938.S45E26   1991
581.5'222–dc20   90-44761

ISBN 978-94-010-5654-0

Printed on acid-free paper

(b)

(d)

(a)

(c)

a. Vertical cross-fibre (orange–brown colour) serpentinite (10–15 cm) cutting massive (yellow ochre colour serpentinized harzburgite with two small cushions of *Silene acaulis* in the lower right crevices. Bay of Islands complex. Newfoundland. Canada. (Photograph by J. Malpas.)

b. The strong contrast in the poorly vegetated, serpentinized Table Mountain area and the adjacent mixed volcanic and metamorphosed plutonic rocks. Bonne Bay, western Newfoundland. Humpback whales are seen in the foreground from Bonne Bay. (photograph by B. A. Roberts.)

c. The barren bluffs and screes on Red Mountain present a striking contrast with the subalpine *Nothfagus menziesi* forest and alpine tall tussock grassland on adjoining schist rock at Simonin Pass (1030 m). North West Otago. New Zealand. (Photograph by W. G. Lee.)

d. Dwarf shrubland at 300 m on the Anita Bay ultramafics, Lake Ronald area. New Zealand. Dominant species include *Leptospermum scoperium*, *Metrosideros umbellata* and *Nothofagus sonandri* var. *cliffortiodes*. (Photograph by W. G. Lee.)

Contents

X. The serpentine area of Ballachul‑ ....
by John Proctor

XI. The distribution and ecology of ....
by Wyllie ... Reba and Julia Stock

XII. The serpentine vegetation areas in Zürich ....
by Tobias ... areas and Klunstoradi Dale

XIII. The serpentine vegetation and ultramafic ....
by Donnasi ... V. Csele

SECTION I
INTRODUCTION

  I. Introduction
    by Bruce A. Roberts and John Proctor   1

SECTION II
GEOLOGY

  II. Serpentine and the geology of serpentinized rocks
    by John Malpas   7

SECTION III
NORTH AMERICA

  III. Plant life of western North American ultramafics
    by Art R. Kruckeberg   31
  IV. Ecology of serpentinized areas, Newfoundland, Canada
    by Bruce A. Roberts   75
  V. A phyto-ecological investigation of the Mount Albert serpentinized plateau
    by Luc Sirois and Miroslav M. Grandtner   115

SECTION IV
EUROPE

  VI. Chemical and ecological studies on the vegetation of ultramafic sites in Britain
    by John Proctor   135
  VII. The ecology of serpentinized areas of north-east Portugal
    by Eugenio Menezes de Sequeira and A. R. Pinto Da Silva   169
  VIII. Distribution of serpentinized massives on the Balkan peninsulas and their ecology
    by Budislav Tatic and V. Veljovic   199
  IX. The distribution and ecology of the vegetation of ultramafic soils in Italy
    by Ornella Vergnano Gambi   217

XIV. The serpentine flora and vegetation in the ....
by Javornik ... C. ...
XV. The serpentinized areas in New Zealand ....
by Wilfried G. Lee

XVI. Concluding remarks
by Bruce A. Roberts and John Proctor

Index

SECTION V
FAR EAST AND JAPAN

X. The vegetation over ultramafic rocks in the tropical Far East
by John Proctor   249

XI. The distribution and extent of serpentinized areas in Japan
by Naoharu Mizuno and Shiro Nosaka   271

SECTION VI
AFRICA

XII. The ecology of ultramafic areas in Zimbabwe
by John Proctor and Monica M. Cole   313

XIII. The vegetation over mafic and ultramafic rocks in the Transvaal Lowveld, South Africa
by Monica M. Cole   333

SECTION VII
AUSTRALASIA

XIV. The vegetation of the greenstone belts of Western Australia
by Monica M. Cole   343

XV. The serpentinized areas of New Zealand, their structure and ecology
by William G. Lee   375

SECTION VIII

XVI. Concluding remarks
by Bruce A. Roberts and John Proctor   419

Index   421

Introduction

B. A. ROBERTS & J. PROCTOR

The term 'serpentine' referring to rocks and minerals, can be traced back to the ancient Roman era of Dioscorides (A.D. 50) (Faust & Fahey 1962) and they suggest that the speckled colour of serpents and the fact that Dioscorides recommended it for the prevention of snakebite as two of many origins of the name. It is well known that the use of serpentinized rocks by sculptors and carvers since ancient times is a tradition that is still carried on today (Fig. 1). The differences in chemical composition of serpentinized rocks are many and specimens come in a wide range of attractive solid and mixed colours, e.g., red, green, blue and black. This combined with the ability to work easily with serpentinized rocks are the major reasons for their long use for ornamental purposes.

The mineral name serpentine is applied to a series of three polymorphic minerals: lizardite, antigorite, and chrysotile, the latter well known in its asbestos form in some localities (Fig. 2). All three polymorphs are produced by the hydration of the ferromagnesian minerals of ultramafic rocks (($MgFe)_2SiO_4$, approximate composition) of low temperature and pressure and paragenesis regimes favourable for the formation of each of the polymorphs. These conditions include temperatures of generally less than 530°C, and fluid pH's in excess of 10 with low pCO_2.

Serpentine protoliths are essentially ultramafic rocks which fall into a number of categories: (a) stratiform complexes, (b) concentrically zoned complexes, (c) ophiolite complexes and, (d) high-temperature peridotite complexes. The most commonly recognized are generally assigned to the ophiolite suite (Malpas, Ch. 2, this volume) and occur in most fold-mountain belts of the world. Serpentinized ultramafic rocks (($H_4(MgFe)_3Si_2O_9$, approximate composition) have many distinctive chemical and physical features which create an unusual environment for the plants which grow in the soils derived from them.

Serpentinized ultramafic rocks outcrop in many parts of the world and very often they are associated with an unusual flora with rare, endemic and different races of plants. These low-cover, sparse, botanically interesting areas have attracted world attention from pedologists and ecologists extensively since the last century but starting with the earliest published record of a plant restricted to serpentine by Caesalpino in the late sixteenth

(a)   (b)

Fig. 1. Attractive colours and the ability to work easily with serpentinized rocks are the major reasons for its long use for ornamental purposes. On the Lizard peninsula, England's southernmost point, more than 20 stone works have existed, some more than 50 years. (a) Serpentine stone works shop of R. H. Roberts (no relation to editor) has been a prominent shop at the Lizard turning out thousands of ornaments and souvenirs. (b) R. H. Roberts in his workshop where he has worked using local serpentine rock for more than 40 years.

century (1583) (Proctor & Woodell 1975; Werger, Wild & Drummond 1978; Vergnano-Gambi, Chapter IX, this volume). By the mid-1800s several accounts (e.g., Pančić 1859, Tatić & Veljović, Chapter VIII, this volume) and Caruel at the 1867 International Botanical Congress had conducted both botanical and zoological studies including the effects of man on the habitats of the old world serpentinized areas. Comprehensive reviews of the ecology of serpentinized areas since Proctor & Woodell (1975), include Kinzel (1982), and Brooks (1983, 1987).

Proctor & Woodell (1975) indicated that the only generalizations that can be made with confidence concerning the chemical composition of serpentine rocks are that iron (Fe) and magnesium (Mg) are always relatively high and silicon (Si) relatively low. Nevertheless, serpentines very commonly have other features of their chemical composition which may exert unusual influences on the soils and living organisms associated with them. Many have low calcium (Ca), potassium (K), phosphorus (P) and molybdenum (Mo) concentrations, for example, and also relatively high chromium (Cr), cobalt (Co) and nickel (Ni) concentrations.

Most current research on the ecology of serpentinized areas has focused on the major stress factors (Roberts 1986a, 1986b) including (1) low essential macro-nutrients, (2) absence of some micro-nutrients, (3) physical factors, cryoturbation (Roberts 1980a), drought (Carter, Proctor & Slingsby 1987), but especially (4) the form, type and extent of toxic concentrations of Co, Cr, Mg, Ni and other elements. The role of plants as accumulators, indicators and excluders (Baker 1981, Roberts 1980b, 1981) is now well researched.

(a)

(b)

Fig. 2. The term serpentine is properly used as a group name for the 3 polymorphic minerals lizardite, antigorite and crysotile. The most easily recognizable form – asbestos, represents well developed, cylindrical chrysotile fibres (a) open pit asbestos mining in the serpentinized rock at Baie Verte, Northeast Newfoundland, (b) close up of short asbestos fibres in serpentinized rock.

Plants with the ability to accumulate high concentrations of toxic elements such as Ni have been known (Minguzzi & Vergnano 1948) for many years. The discovery of the restriction of the species *Hybanthus floribundus* (Lindl.) F. Muell to Ni-toxic soils and of its status as an Ni accumulator in the mid-1960s (Cole 1973) in western Australia prompted an investigation of the species over its entire range. Brooks *et al.* (1977) applied the term hyperaccumulators to designate those species that accumulated at least 0.1% Ni in dry leaf concentration. By 1983, Brooks using mainly herbarium specimens had identified 44 species of the *Alyssum* genus to be hyperaccumulators of Ni. One of the highlights of more recent work has been the identification of the Ni complexes in the Ni-accumulating plants (Jaffré *et al.* 1976, Jaffré *et al.* 1979, Kersten *et al.* 1980, Brooks 1983) as nickel citrate, malic acid complexes, etc.

The potential methods for selective accumulation of Ni(II) ions by plants (Still & Williams 1980) have been studied and the inorganic element transfer through membranes (Williams 1981) are now better understood, but the additive protective and synergistic effects on plants with excess trace elements (Wallace 1982) are not well known. Some work (Johnston & Proctor 1984) for example, showed increases on the root surface phosphatase of a serpentine over a non-serpentine clone of *Festuca rubra*, but the ecological significance of the adapted phosphatase is unclear.

Interactions between Ca, Mg, and Ni are now known to be important. Proctor *et al.* (1981) showed that field capacity water extracts from serpentine soils provided a good method for judging the status of Ni and Mg toxicity. The protection from Ni toxicity afforded by Mg and Ca (Proctor & McGowan 1976, Robertson 1985) is important in the full understanding of the ecology of serpentinized areas. The fact that many factors acting interactively must be invoked to explain the cause of the unique flora and fauna of serpentinized areas was clearly outlined by Proctor & Woodell (1975).

The present volume includes details of much current research. Some important work could not be included, for example that in Switzerland by Juchler (1988). New studies of asbestos fibres and trace elements in earthworms and cattle as a result of a landslide of serpentinized rocks onto pasture land (Schreier & Taylor 1981, Schreier & Timmenga 1986, and Schreier *et al.* 1986) near the Canada–United States border at Everson, Washington present a new aspect of an old problem of landslides in serpentinized bedrock areas (Mizuno & Nosaka, Chapter II, this volume).

There is a need for more microbiological studies especially mycorrhizal (Hopkins 1987) research in serpentinized areas. Research on the establishment of plant communities on mine wastes (Winterhalder 1984, Moore & Zimmermann 1977) from serpentine asbestos areas and related conservation activities (Kruckeberg, Chapter III, this volume) should be continued.

This present volume draws together contributions from several of the world's leading authorities on the vegetation of serpentinized ultramafic soils and incorporates much previously unpublished information as well as references to literature which are little known. It has two editors, both of whom have been researching in the field for many years.

References

Baker, A. J. M. 1981. Accumulators and excluders – strategies in the response of plants to heavy metals. Journal of Plant Nutrition 3: 643–654.

Brooks, R. R., J. Lee, R. D. Reeves & T. Jaffré. 1977. Detection of nickeliferous rocks by analysis of herbarium specimens of indicator plants. Journal Geochem. Explor. 7: 49–57.

Brooks, R. R. 1983. Biological methods of prospecting for minerals. John Wiley & Sons, Toronto. 322 p.

Brooks, R. R. 1987. Serpentine and its vegetation, a multidisciplinary approach. Dioscorides Piess, Portland, Oregon. 454 p.

Carter, S. P., J. Proctor & D. R. Slingsby. 1987. Soil and vegetation of the Keen of Hamar serpentine, Shetland. Journal of Ecology. 75: 21–42.

Caruel, T. 1867. Sur la flore des gabbres de Toscane. Actes Congrès Intern. Botanique, Paris, pp. 58–63.

Cole, M. M. 1973. Geobotanical and biogeochemical investigations in the sclerophyllus woodland and scrub associations of the Eastern Goldfields of Western Australia, with particular reference to the role of *Hybanthus floribundus* (Lindl.) F. Muell. as nickel indicator and accumulator plant. J. Appl. Ecol. 10: 269–330.

Faust, G. T. & J. J. Fahey. 1962. The serpentine group minerals: Studies of the natural phases in the system MgO–SiO$_2$–H$_2$O and the systems containing the congeners of Magnesium. U.S. Dept. of the Interior. Geological Survey Professional Paper 384–A. 92 p.

Hopkins, N. A. 1987. Mycorrhizae in a California serpentine grassland community. Can. J. Bot. 65: 484–487.

Jaffré, T., R. R. Brooks, J. Lee and R. D. Reeves. 1976. *Sebertia acuminata*: a nickel-accumulating plant from New Caledonia. Science, N.Y. 193: 579–580.

Jaffré, T., W. Kersten, R. R. Brooks and R. D. Reeves. 1979. Nickel uptake by Flacourtiaceae of New Caledonia. Proc. R. Soc. Lond. 205: 385–394.

Johnston, W. R. & J. Proctor. 1984. The effects of magnesium, nickel, calcium and micronutrients on the root surface phosphatase activity of a serpentine and non-serpentine clone of *Festuca rubra* L. New Phytol. 96: 95–101.

Juchler, S. J. 1988. Die Boden auf serpentinit in der Subalpinen stufe Bei Daros 1. Bodenbildung. II Nikel-und Chromdynamik. ADAG Administration & Druck AG. Diss. Ethnr. 8716 Zurich 1988. 174 p.

Kersten, W. J., R. R. Brooks, R. D. Reeves & T. Jaffré. 1980. Nature of nickel complexes in *Psychotria douarrei* and other nickel-accumulating plants. Phytochemistry 19: 1963–1965.

Kinzel, H. 1982. Serpentin-pflanzen. In: Pflanzenökologie und Mineralstoffwechsel Verlag Eugen Ulmer, Stuttgart, pp. 381–410.

Minguzzi, C. & O. Vergnano. 1948. Il contenuto di nichel nelle ceneri di *Alyssum bertolonii*. Desv. Memorie Soc. tosc. Sci. nat. A55: 49–77.

Moore, T. R. & R. Zimmermann. 1977. Establishment of vegetation on serpentine asbestos mine wastes, southeastern Quebec, Canada. Journal Appl. Ecol. 14: 589–599.

Pančić, J. 1859. Die Flora der Serpentinberge in Mittelserbien. Verh. d. K. Zool. bot. Ges. Bd. IX, Wien.

Proctor, J. & S. R. J. Woodell. 1975. The ecology of serpentine soils. *In* Advances in Ecological Research 9: 256–347.

Proctor, J. & I. D. McGowan. 1976. Influence of magnesium on nickel toxicity. Nature, London. 260: 134.

Proctor, J., W. R. Johnston, D. A. Cottam & A. B. Wilson. 1981. Field-capacity water extracts from serpentine soils. Nature, London, 294: 245–246.

Roberts, B. A. 1980a. Some chemical and physical properties of serpentine soils from western Newfoundland. Can. J. Soil Sci. 60: 231–240.

Roberts, B. A. 1980b. Concentrations of micro-nutrients (trace elements) in native plants growing on serpentine soils from western Newfoundland. Paper presented at the Botany 80 Symposium at the University of British Columbia, Vancouver, 12–16 July 1980. Abstract published. Botanical Society of America, Miscellaneous Publication 158, p. 95.

Roberts, B. A. 1981. Ecology of serpentinized areas, west Newfoundland, Canada. Paper presented at the XIII International Botanical Congress, Sydney, Australia, August 1981. Abstract published, p. 328.

Roberts, B. A. 1986a. Stress in serpentinized areas. Invited paper presented at the stressed ecosystem, Canadian Botanical Association, 22nd Annual Meeting. Sudbury, June 1986. Abstract published #2, p. 21.

Roberts, B. A. 1986b. Soils of serpentinized areas, their ecology and distribution. Paper presented at the 13th Congress of the International Society of Soil Science, 13–20 August, Hamburg, Germany, Extended summary published, Commission V, p. 1253–1254.

Robertson, A. I. 1985. The poisoning of the roots of *Zea Mays* by nickel ions and the protection afforded by magnesium and calcium. New Phytol. 100: 173–189.

Schreier, H. & J. Taylor. 1981. Variations and mechanisms of asbestos fibre distribution in stream water. Technical Bulletin No. 118, Inland Waters Directorate, Pacific and Yukon Region, Water Quality Branch, Vancouver, B.C. pp. 1–17.

Schreier, H., J. A. Shelford & T. D. Nguyen. 1986. Asbestos fibers and trace metals in the blood of cattle grazing in fields inundated by asbestos-rich sediments. Environmental Research 41: 99–109.

Schreier, H. & H. J. Timmenga. 1986. Earthworm response to asbestos-rich serpentinitic sediments. Soil Biol. Biochem. 18: 85–89.

Still, E. R. & R. J. P. Williams. 1980. Potential methods for selective accumulation of nickel (II) ions by plants. Journal of Inorganic Biochemistry 13: 35–40.

Wallace, A. 1982. Additive, protective and synergistic effects on plants with excess trace elements. Soil Science 133: 319–323.

Werger, M. J. A., H. Wild & B. R. Drummond. 1978. Vegetation structure and substrate of the northern part of the Great Dyke, Rhodesia. Environment and Plant Communities. Vegetatio 37: 79–89.

Williams, R. J. P. 1981. Physicochemical aspects of inorganic element transfer through membranes. Phil. Trans. R. Soc. Lond. B294: 57–74.

Winterhalder, K. 1984. Environmental degradation and rehabilitation in the Sudbury area. Excerpt from Laurentian University Review, Northern Ontario. Environmental Perspectives, Vol. 16, No. 2. 47 p.

B. A. Roberts
Forestry Canada
P.O. Box 6028
St. John's
Newfoundland
Canada A1C 5X8

J. Proctor
Dept. of Biological and Molecular Sciences
University of Stirling
Stirling
Scotland FK9 4LA
U.K.

II

Serpentine and the geology of serpentinized rocks

J. MALPAS

Abstract. The mineral name 'serpentine' is applied to a series of three minerals, lizardite, antigorite, and chrysotile, the latter well known in its asbestos-form in some localities. All three are produced by the hydration of the ferromagnesian minerals of ultramafic rocks at low temperature and pressure under conditions favorable for the formation of each member of the group. These conditions include temperatures of generally less than 500°C, and fluid pH's in excess of 10 with low pCO_2. During serpentinization the only component mobilized to any extent is calcium which is found in the pyroxenes and plagioclases of some parent rocks. Isotopic evidence suggests that serpentinization occurs as a result of rock interaction with various waters, including sea-water, meteoric water and hydrothermal water, according to the environment in which alteration takes place.

Serpentine protoliths are essentially ultramafic rocks which fall into a number of categories: (a) stratiform complexes, (b) concentrically zoned complexes, (c) ophiolite complexes and, (d) high-temperature peridotite complexes. The most commonly recognized are generally assigned to the ophiolite suite and occur in most fold-mountain belts of the world.

B. A. Roberts and J. Proctor (eds), The ecology of areas with serpentinized rocks. A world view, 7–30.
© 1992 *Kluwer Academic Publishers.*

Introduction

Serpentine minerals are produced by the hydration of an assortment of minerals which commonly occur in ultramafic rock assemblages. These rocks contain iron and magnesium silicates such that they are dark in color or are said to have a high color index. The term 'serpentine' is properly used as a group name for the three minerals 'Lizardite', 'Chrysotile' and 'Antigorite', but in many cases is used more loosely as a general description of mineral phases when the individual mineralogy is not known. The most easily recognizable form – asbestos – represents well developed, cylindrical chrysotile fibres.

Hydration of the ultramafic protolith occurs at temperatures less than 500°C. These protoliths are either dunites, harzburgites, lherzolites, websterites or wehrlites, i.e., rocks containing the minerals olivine, orthopyroxene, clinopyroxene and an aluminium-bearing phase (Fig. 1). The relative rate of alteration of these minerals is variable but Wicks (1969) notes the general order olivine > orthopyroxene > clinopyroxene, reflecting the temperatures of their paragenesis, and the common hydrated equivalents of the ultramafic rocks contain the assemblage lizardite > chrysotile, brucite and magnetite. Small amounts of awaruite, native metals and low-sulfur sulfides are also common (Moody 1976). Brucite

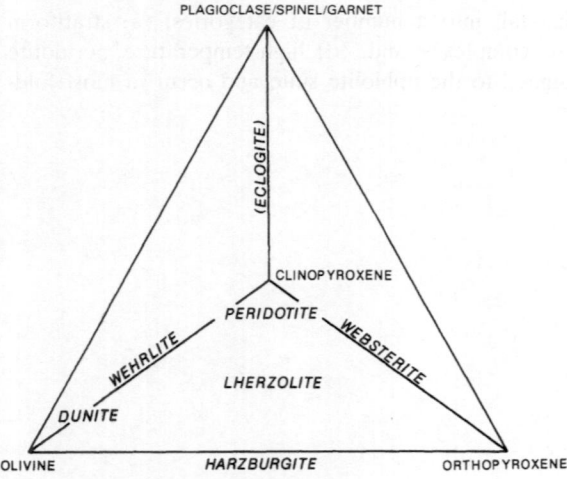

Fig. 1. Mineralogical composition of common ultramafic rocks. End member minerals are olivine, orthopyroxene, clinopyroxene and an aluminium-rich phase, plagioclase, spinel and garnet.

is often intimately intergrown with lizardite and/or chrysotile but may not be always optically discernible.

Serpentine paragenesis requires conditions favorable for the formation of each of the polymorphs. In most cases it appears that antigorite forms from chrysotile but the sequence of events that leads to the formation of asbestos fibres is less well understood.

Following a brief account of serpentine minerals, and the process of serpentinization and textures that result, this paper discusses world occurrences of ultramafic rocks with a view to understanding the geographic distribution of serpentinites.

Serpentine mineralogy

The principal minerals of the serpentine group, lizardite, chrysotile and antigorite, have the approximate composition $H_4 Mg_3 Si_2 O_9$, but there are minor and persistent variations in this chemistry, i.e.:

1. Antigorite has a relatively high weight percent SiO_2 and relatively low weight percent Mg and H_2O^+.
2. Chrysotile has a relatively low Al_2O_3 content.
3. Lizardite has a large amount of Fe_2O_3 and is relatively low in FeO.
4. Both lizardite and chrysotile contain H_2O^+ in excess of the ideal formula.
5. Antigorite has the highest $FeO/FeO + Fe_2O_3 + Al_2O_3$ content and lizardite the lowest.

Other compositional variations in the serpentine mineral group are outlined in Table 1 (after Moody 1976).

The principal replacements that occur are of silicon by aluminium and of magnesium by aluminium, ferrous iron and ferric iron. Although most serpentines contain little or no nickel (0.25 wt% average), the existence of garnierite indicates that nickel may substitute for magnesium. Where serpentines are formed from peridotitic rocks, most of the iron present in the original ferromagnesian mineralogy is incorporated in magnetite or haematite and does not enter the serpentine structure. The appearance of brucite $(Mg(OH)_2)$ is important in that it defines a lower

Table 1. Compositional variations in the serpentine mineral group (after Moody 1976)

Serpentine	Al-serpentines		Fe-serpentines	Ni-serpentines	
	Intermediate compositions*	End-member compositions	End-member compositions	Intermediate compositions**	End-member compositions
Lizardite $Mg_3Si_2O_5(OH)_4$	Septechlorites $(MgAl)_3(SiAl)O_5(OH)_4$	Amesite $Mg_2Al(SiAl)O_5(OH)_4$	Greenalite $Fe_3Si_2O_5(OH)_4$ Cronstedite+ $Fe_2Fe_2SiO_5(OH)_4$	Garnierite $(MgNi)_3Si_2O_5(OH)_4$	Nepouite $Ni_3Si_2O_5(OH)_4$
Chrysotile $Mg_3Si_2O_5(OH)_4$ Antigorite++ $Mg_{2.813}Si_2O_5(OH)_{3.647}$				Garnierite $(MgNi)_3Si_2O_5(OH)_4$	Percoraite $Ni_3Si_2O_5(OH)_4$

* Substitution of Al occurs in both octahedral and tetrahedral sites with maximum substitution in amesite with 1/3 octahedral and 1/2 tetrahedral sites filled.
** Substitution chiefly in octahedral sites; small amounts of Al may or may not be present.
+ Substitution of Fe^{3+} in 1/3 octahedral and 1/2 tetrahedral sites.
++ Formula postulated from structural considerations.

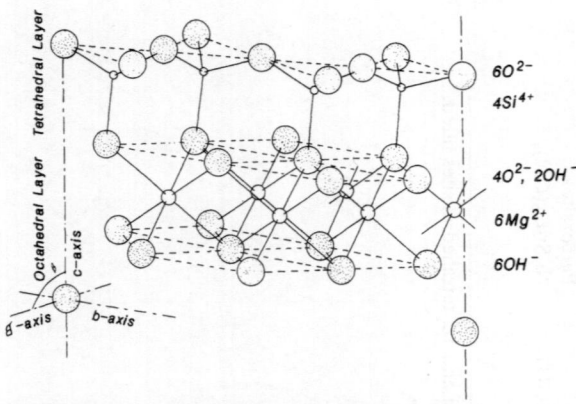

Fig. 2. Atomic structure of serpentine minerals showing SiO₄ tetrahedra linked in a sheet-form to magnesium-hydroxyl layers. Si⁴⁺ shows tetrahedral coordination, Mg²⁺ octahedral coordination.

temperature of serpentinization than does the presence of serpentine alone, c.f. equations (1) and (2):

$$2 \ Mg_2SiO_4 + 3H_2O \rightarrow$$
Forsterite
$$\rightarrow Mg_3Si_2O_5(OH)_4 + Mg(OH)_2 \qquad (1)$$
Chrysotile   Brucite
(Johannes 1968)

$$6 \ Mg_2SiO_4 + Mg_3Si_4O_{10}(OH)_2 + 9H_2O \rightarrow$$
Forsterite   Talc
$$\rightarrow 5 \ Mg_3Si_2O_5(OH)_4 \qquad (2)$$
Chrysotile
(Chernosky 1973)

Although only some of the serpentine minerals are fibrous, all of them possess a layered structure similar to clays. Serpentines, however, differ from the Kaolin group in being tri-octahedral and in the manner in which the layers are stacked (Fig. 2). One part of the serpentine layer is a pseudohexagonal series of linked SiO₄ tetrahedra all of which point the same way in the sheet. Joined to such a sheet of tetrahedra is a brucite layer in which two out of every three ions are replaced by the apical oxygen atoms of the SiO₄ tetrahedra. Many experimental results indicate that serpentine minerals with curved sheet structures exist, such as the chrysotiles with their fundamental layers curved around the x-axis to produce concentric hollow cylinders or rolls. Such

forms often occur in veins as silky fibres and are the most important source of commercial asbestos. Chrysotile fibres are generally aligned perpendicularly across such veins and may reach as much as 15 cm in length.

The process of serpentinization

To geologists and mineralogists it is important to understand serpentinization in order to determine (a) the protolith composition, (b) the pressure and temperature conditions under which serpentinization took place and therefore the environment of hydration (c) the source of fluids controlling hydration, and (d) the change in fluid composition during hydration. Physical properties such as density, seismic velocity, magnetic susceptibility and ductility of peridotites are greatly affected by serpentinization, increasing the necessity of knowing where and how the ultramafic protolith became serpentinized with respect to its emplacement in the crustal stratigraphy.

Serpentinization processes in naturally occurring ultramafic rocks have been discussed at length in the geological literature (e.g., Laurent & Hébert 1979, Coleman & Keith 1971, Komor et al. 1985). It is clear that (a) there are differences between the reaction of dunite and the reaction of pyroxenite with the fluid phase, (b) there is a definite temperature/pressure regime of hydration, (c) there is a problem of the constancy of volume and composition during serpentinization, and (d) there are different sources for serpentinizing fluids.

(a) Holstetler et al. (1966), have reported that brucite is missing from serpentinites derived from peridotites which originally contained more than 40% pyroxene. This may indicate that brucite does not form from the alteration of pyroxene-rich peridotites although Mumpton & Thompson (1975) emphasize that brucite is susceptible to weathering and rapidly alters to hydromagnetite, pyroaurite and coalingite in the surface weathering zone.

Iron in olivine redistributes during serpentinization, generally forming a separate opaque phase, e.g., magnetite, awaruite, pentlandite, ferritchromite, etc. (Ashley 1975). Wicks (1969) notes that iron may substitute for magnesium in tonicite

(6 to 72 mole% $Fe(OH)_2$), but generally magnetite is produced from the serpentinization of dunite. Magnetite production must, however, be controlled by increasing temperature at low oxygen fugacities (Moody 1976). Wicks (1969) indicates that antigorite predominates over lizardite and chrysotile in peridotites that have undergone prograde metamorphism and serpentinites consisting predominantly of antigorite have higher modal abundance of magnetite on the average than lizardite-chrysotile serpentinites (Wenner & Taylor 1971).

Chromite may be partly or completely replaced by secondary magnetite or by ferrit-chromite. The latter is commonly associated with antigorite, chlorite, tremolite, metamorphic olivine and diopside (Bliss & MacClean 1975) indicating that the serpentine has undergone a higher grade of metamorphism than is necessary for the formation of lizardite and chrysotile serpentinites, i.e., amphibolite grade as opposed to greenschist grade metamorphism.

(b) The temperatures and pressures for the production of various assemblages have been determined by reversed bracketing experiments in the presence of water. (Moody 1976, Chernosky 1973, 1975). The results are summarized in Fig. 3. The large majority of serpentinized rocks are composed of lizardite ± chrysotile. The interrelationship of these minerals depends on the starting peridotite composition since there is a close correspondence between olivine and lizardite structures which may explain why lizardite nucleates and grows easily on an olivine substrate whereas chrysotile does not.

Textural studies of naturally occurring serpentinites indicate that antigorite replaces lizardite or chrysotile if the serpentine undergoes prograde metamorphism to low amphibolite facies (Evans & Frost 1975). The higher temperature associated with antigorite formation occurs when dehydration of the mineralogy produces a lower activity of water which results in the formation of antigorite with less $Mg(OH)_2$ than found in either lizardite or chrysotile (Whittaker & Wicks 1970).

The results outlined in Fig. 3 point to an upper crustal environment for most serpentinization at a relatively low oxygen fugacity.

(c) Barnes *et al.* (1972), have classified the fluids related to serpentinization into three types on

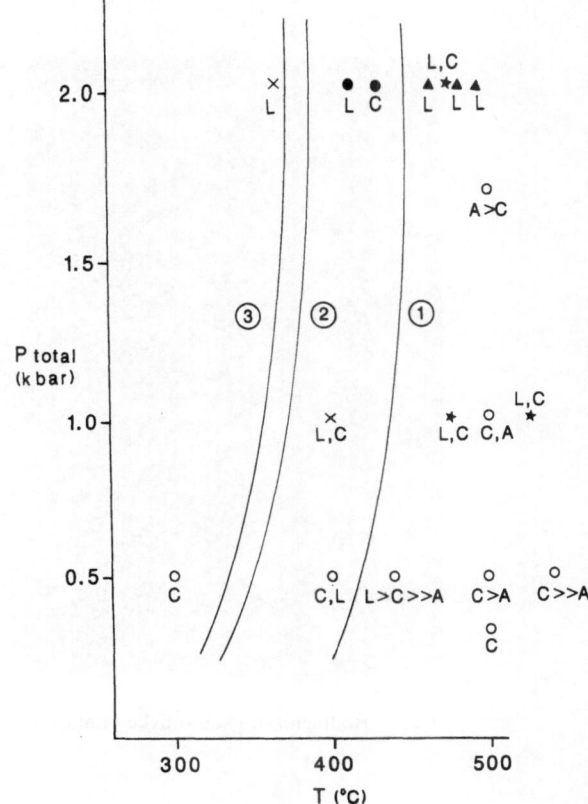

Fig. 3. Results of reversed bracketing experiments in terms of stability of polymorphs at various pressures and temperatures (L = Lizardite, C = Chrysotile, A = Antigorite).

the basis of their chemistry and isotopic compositions. The fluids must have low CO_2 contents (<5 mole percent at 500°, 1 Kb) because neither chrysotile nor lizardite are stable with respect to talc in the presence of a CO_2-bearing fluid (Johannes 1969). However, brucite is not stable in very low-CO_2 fluids and alters readily to magnesite (Johannes & Metz 1968). This is clear in rocks having undergone talc-carbonate alteration where serpentine-brucite assemblages are replaced by talc-magnesite (e.g., Naldrett 1966, Eckstrand 1975). Experimentally produced fluids are characterized by pH's in excess of 10, low Mg concentrations and high chloride contents. Such fluids have the possibility of carrying significant quantities of dissolved silica, the amount of which is controlled essentially by pH.

The chemical properties of the fluid and, naturally, its temperature and pressure, determine the elements which might be leached from the

Fig. 4. Rodingitized gabbro dyke (white) in sheared serpentinized peridotite (Lizard, Cornwall).

ultramafic rock and transported via shears and fractures. There does not, in general, appear to be a major or widespread removal of Mg or Si. The only component removed during serpentinization is Ca, which is found in the pyroxenes and plagioclases of parent rocks. Mobilization of the Ca and its localized redeposition accounts for the production of metasomatic rodingites, so characteristically associated with the altered mafic rocks found, in places, with serpentinites (Fig. 4). These rodingites are assemblages of calc-silicates including wollastonite, pectite, xonolite, hydro-grossular and prehnite. The accessibility of fluid to the ferromagnesian mineral surfaces can be increased by faulting and shearing of the ultramafic body which increases surface area and thus the rate of serpentinization: the volume of water could be an important determinant of metsomatism especially if allowed to readily flush through the ultramafic rock rather than remaining in a closed environment.

(d) Different isotopic compositions of various serpentinized ultramafic rocks (Wenner & Taylor 1973) suggest that different waters are involved in the process of serpentinization. This is especially true in the case studies of ocean floor serpentinites and their on-land equivalents of the ophiolite suite (see below). Wenner & Taylor (*op. cit.*) conclude from isotopic evidence that, if ophiolite complexes indeed represent uplifted exposures of the oceanic lithosphere, then their ultramafic portions were serpentinized essentially after their continental emplacement. Wenner & Taylor (1974), upon examination of a wide variety of serpentinites, further conclude that pervasive lizardite-chrysotile serpentinization is caused by fluids of meteoric-hydrothermal origin or by waters emanating from sedimentary formations, whereas antigoritic serpentization takes place as a result of regional metamorphism in the presence of non-meteoric waters. However, this presumes that isotopic compositions have been maintained

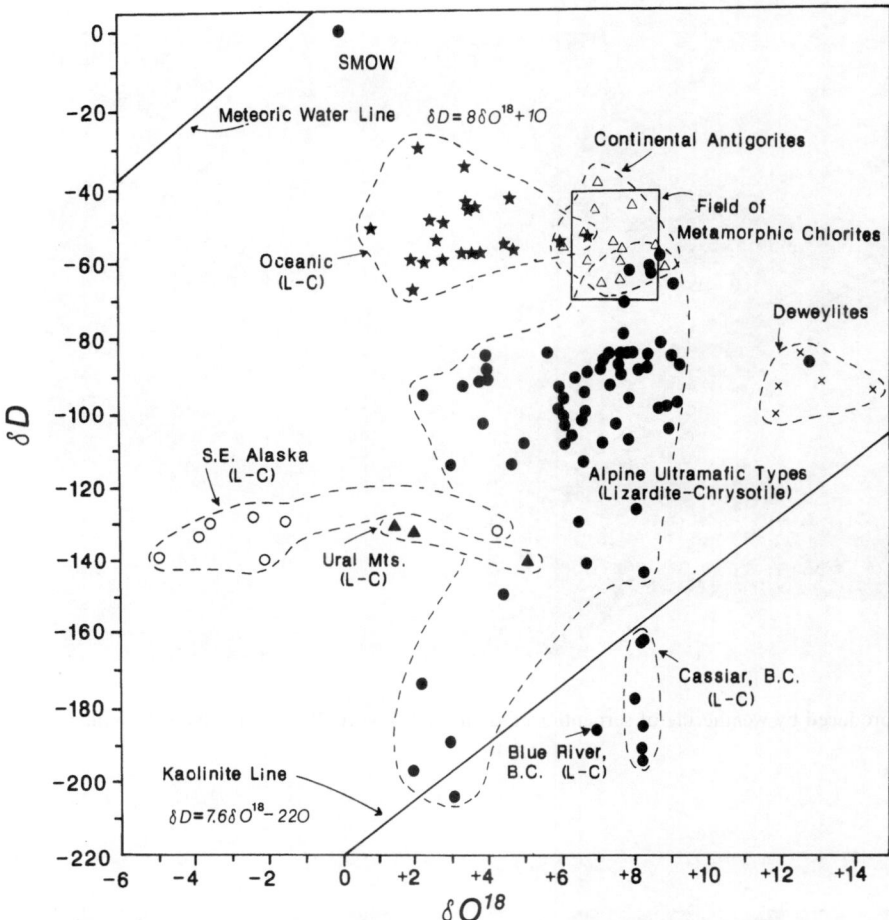

Fig. 5. δD, δO[18] isotope ratios for various serpentinized rocks. SMOW = Standard mean ocean water; L = Lizardite; C = Chrysotile.

since serpentinization. It is therefore of interest to note the wide range of δO[18] values for lizardite-chrysotile serpentines (Fig. 5) which may also indicate mixing of hydrothermal and meteoric waters in some near surface environments (e.g., Magaritz & Taylor 1974).

Textures produced during serpentinization

In the field, a number of types of serpentinite can be distinguished on the basis of their physical appearance. Three types are the most common: (a) blocky or massive serpentinite, (b) sheared serpentinite, and (c) fibrous serpentinite.

(a) Massive serpentinite retains the original geometric configuration of the olivine and pyroxene grains. In outcrop, serpentinized dunite has a smooth, yellowish-brown weathering rind up to several centimetres thick (Fig. 6). In comparison, serpentinized harzburgite or lherzolite has a rough surface produced by the more rapid weathering of serpentinized olivine compared with serpentinized pyroxene (Fig. 7). Fresh broken surfaces of these rock types are typically dark-greenish-black and have blocky subconchoidal fracture. Massive serpentinites represent most closely the original peridotite and have undergone little or no deformation during serpentinization.

The color change accompanying serpentinization can in some cases be used as a partial guide to the amount of serpentine present. Unaltered

Fig. 6. Smooth surface produced by weathering of serpentinized dunite (cf. Figure 7). Bay of Islands Complex, Newfoundland.

Fig. 7. 'Hobnail' surface produced by weathering of serpentinized harzburgite. Orthopyroxene crystals stand out above surrounding olivine crystals. Bay of Islands Complex, Newfoundland.

14

olivine and pyroxenes impact a grey-olive color to the rocks but when the rocks are more than 50% serpentinized their color is greenish-black. The remaining olivine and pyroxene may therefore produce a variegation of greyish-olive mixed with greenish-black. However, the most reliable method of determining the degree of serpentinization other than by microscopic examination is by measurement of relative specific gravity.

The fabric of serpentinized peridotite is controlled for the most part by structures that pre-date serpentinization. Where deformation in the ultramafic rock produced faults prior to serpentinization, replacement of the ferromagnesian minerals is generally complete. Where a preexisting foliation occurred in the protolith, this is often preserved in the serpentinite (Fig. 8).

(b) Sheared serpentinite is developed by progressive deformation. Fractures or faults which cross-cut blocky serpentinite often develop slip-fibre serpentinite zones at the expense of the massive serpentinite. If shearing continues then the growth of slip-fibre serpentine eventually may exceed the massive serpentine. Sheared serpentinites are often associated with brucite and magnesite (Fig. 9), and may be referred to as 'slickentite' or 'fish-scale' serpentinite. In such cases the typical yellowish-brown oxidized surface of the blocky serpentinite is replaced by smooth, shiny, bluish-green surfaces.

(c) Fibrous serpentinite develops in veins that vary from microscopic to macroscopic. Generally high veins occur in areas of undeformed blocky serpentinite and follow a joint pattern inherited from the olivine (Fig. 10). The cross-fibre form of serpentine consists of pale yellow-green fibres that have crystallized normal to the vein walls. Chrysotile is the main serpentine polymorph and when veins are several cm thick, they form the main source of commercial asbestos. If the veins are subsequently sheared, the fibres become platy and brittle and produce picrolite.

Dunites that are completely serpentinized generally assume a 'mesh-texture.' Harzburgites follow a similar process of replacement except for the orthopyroxenes, which may not be altered very much even though all the olivine is serpentinized. This differential alteration produces the hobnail surface characterized by upstanding bastite pseudomorphs.

Serpentinite protoliths: the world occurrences of ultramafic rocks

Serpentinization is associated with all ultramafic rocks to varying degrees. For the purposes of simplicity, four major world-wide descriptive classes of ultramafic complexes can be distinguished. These are: (a) stratiform; (b) concentrically zoned; (c) ophiolitic; (d) high-temperature peridotite complexes. Each can be recognized by a combination of such characteristics as tectonic setting, internal structure and overall geometric form, mineralogy and petrology, and thermal effects on country rocks.

(a) *Stratiform complexes*

Stratiform complexes, sometimes referred to as layered basic intrusions, characteristically occur in stable continental shields, and have formed throughout geologic time from earliest Precambrian to Holocene. Some of the better studied stratiform complexes include: Skaergaard Intrusion, Greenland; Bushveld Complex, South Africa; Stillwater Complex, U.S.A.; Great 'Dyke', Zimbabwe; Muskox Intrusion, Canada; and Skye and Rhum Intrusions, Scotland (for a review of these see Wager & Brown (1967)).

It was long believed that most stratiform complexes have the form of a lopolith, a large convex downwards, generally concordant intrusion with its thickness approximately 1/10 to 1/12 its width (Carmichael et al. 1974). This presumed profile resulted from observed subhorizontal, or gently convex downwards attitudes of layering. However, these layers are now interpreted to be discordant to more steeply dipping external contacts, and the general geometric form is probably more closely funnel-shaped (Fig. 11).

Most intrusions show a crude compositional zonation, with ultramafic rocks at the base, gabbros and norites in the middle, and feldspathic gabbros and ferrodiorites at the top. The larger portion of each intrusion probably crystallizes from the floor upwards by the accumulation of crystals by gravity settling – these rocks are called cumulates.

Two types of layering are present in stratiform intrusions. Visually discernible lithological layering repeated at regular intervals is called rhythmic

Fig. 8. Serpentinized deformed peridotite with stretched orthopyroxene (now bastite) delineating foliation. Lizard, Cornwall.

Fig. 9. Veins of magnesite (white) cutting sheared lizardite serpentinite. Lizard, Cornwall.

16

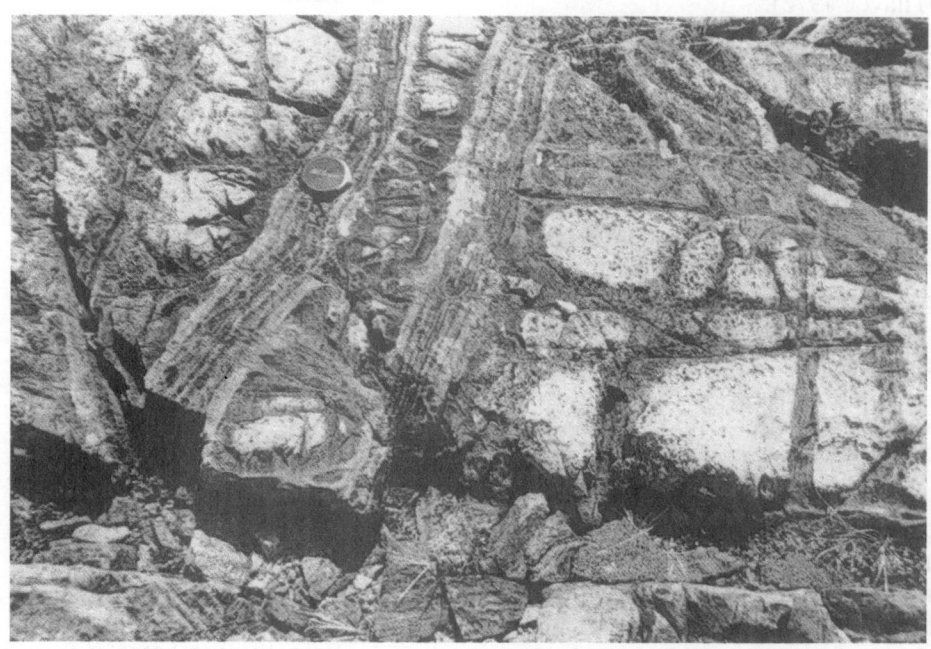

Fig. 10. Cross-fibre serpentinite cutting massive serpentinized harzburgite. Bay of Islands Complex, Newfoundland.

layering, and is controlled mainly by gravity settling, convection, and other internal processes operating within the magma chamber. Cryptic layering refers to the continuous gradual change in chemical composition of individual minerals from bottom to top. Since it is not generally possible to detect this visually, chemistry is essential in its identification. Cryptic layering is controlled by a vertical compositional gradient in successive liquid fractions. Major repetitions of grouped layers are referred to as 'cyclic units' (Jackson 1961).

Stratiform complexes comprise a diverse rock suite. Amongst the ultramafic rocks, harzburgite, orthopyroxenite and websterite are common,

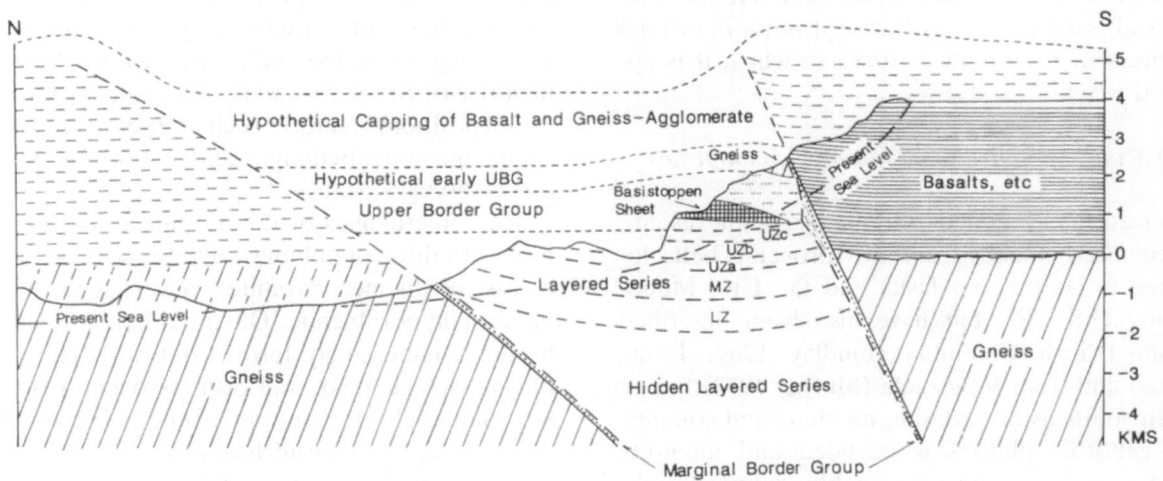

Fig. 11. Diagrammatic cross-section of a layered (stratiform complex) based upon the Skaergaard Intrusion of S. E. Greenland.
LZ = lower zone, UZ = upper zone of layered series.

while dunite, lherzolite and wehrlite are rare (Jackson & Thayer 1972).

Olivines and pyroxenes from the lowest ultramafic cumulates tend to differ significantly from those of ultramafic tectonites of ophiolites (see below). They tend to be slightly less magnesian, Fo_{85}. The Al_2O_3 content in coexisting early cumulate pyroxenes (1.67–1.70% in bronzite, 2.3–3.0% in diopside of Stillwater ultramafic rocks) overlaps some ophiolites but is notably less than that of others (Carmichael *et al.* 1974). Other differences relate to partitioning of Ca between coexisting pyroxenes.

Stratiform complexes form by the progressive fractional crystallization of a tholeiitic basaltic magma within a magma chamber. These intrusions may be divided into two categories according to the mechanism of magma supply and retention: (i) closed system – magma is emplaced in a single, rapid injection and crystallizes (e.g., Skaergaard Intrusion); (ii) open system – magma is emplaced in several pulses, with enough time between injections to allow cooling of the body and formation of cumulates. Magma may or may not leave the system (e.g., Rhum Intrusion).

Recent work has indicated that there may be problems with the cumulus theory, and it is now well established for several major intrusions that plagioclase, which has been thought as being cumulate previously, should have in fact floated in its parental magma (Bottinga & Weill 1970, Campbell *et al.* 1977). This idea prompted Irvine (1982) to redefine the term 'cumulate' so that crystal settling is a possible, but not an essential process in the origin of rocks to which it is applied.

(b) *Concentrically zoned ultramafic complexes*

Concentrically zoned ultramafic complexes are recognized mainly from two orogenic belts located in southeast Alaska and the Ural Mountains, U.S.S.R., but have also been described from British Columbia (Findlay 1969, Irvine 1968) and from Venezuela (Murray 1972).

In southeastern Alaska, an elongated complex of gabbroic plutons is intruded and intensely metamorphosed by presumably younger, sub-cylindrical ultramafic pipes (Fig. 12). Most of the larger plutons consist of a crudely concentric,

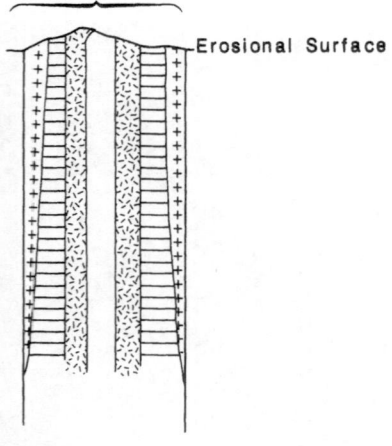

HOR.SCALE ~1–2KM

Erosional Surface

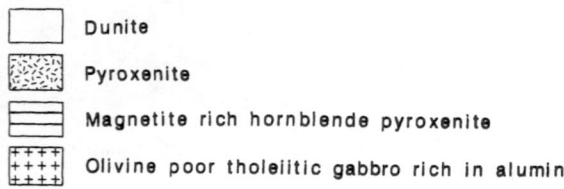

Dunite

Pyroxenite

Magnetite rich hornblende pyroxenite

Olivine poor tholeiitic gabbro rich in alumina

Fig. 12. Diagrammatic cross-section of a concentrically zoned ultramafic complex based upon Alaskan occurrences.

generally discontinuous distribution of rock types: a central core of dunite surrounded by successive shells of wehrlite, olivine pyroxenite, magnetite pyroxenite, and hornblende pyroxenite. Each ultramafic pipe is enclosed by gabbroic rocks. All shells except the hornblende pyroxenite show horizontal rhythmic layering with 'gravity settling' of olivine and pyroxene crystals. The majority of rocks are clearly cumulates, showing syn-depositional features such as gravity layering, slumping, cross-bedding, and intraformational brecciation.

The mineralogies of the ultramafic rocks contrast with those of ophiolitic ultramafic rocks, and consist of olivine, chromite, calcic augite, magnetite, and hornblende. Olivine (Fo_{93} in dunites) becomes increasingly ferrous outwards (Fo_{75} in olivine pyroxenites), and orthopyroxene is lacking completely, the only pyroxene being an aluminous diopside (Carmichael *et al.* 1974). Taylor (1967) notes that in the southeast Alaskan ultramafic pipes, no plagioclase crystallized even as an intercumulus phase throughout the ultramafic rocks, except in the late stage gabbroic pegmatites and in peripheral hornblendite. Hornblende becomes increasingly abundant with progressive fractionation, and also replaces clinopyroxene in places, but it is doubtful that it was ever a primary cumulate mineral. Abundant magnetite and minor ilmenite are typical of the pyroxenites, and minor chromite is the only oxide phase of the dunites. The mineralogical evidence shows clearly that liquidus temperatures decreased outwards through successive shells, suggesting that the pluton crystallized from the center outwards.

Intrusive contacts with the gabbroic country rocks are marked by strong thermal metamorphic aureoles. Serpentinization, where present clearly occurred after emplacement of the pluton, but most olivine remains unaltered.

These ultramafic bodies were first recognized as a distinctive association because of the rudely concentric arrangement of rock types shown by several of the larger intrusions in southeastern Alaska (Walton 1951, Ruckmick & Noble 1959). Consequently, workers referred to the complexes by such names as, 'concentrically zoned ultramafic complexes' or, more simply as 'zoned complexes' (Taylor & Noble 1960, Taylor 1967, Wyllie 1967, 1969) or 'concentric complexes' (Jackson 1971).

There are however, several objections to using any of these descriptive terms (Irvine 1974). Many zones are discontinuous and only the Blashke Islands Complex in southeast Alaska is symmetrical. Secondly, several other petrologically different ultramafic complexes have been treated as members of the same group solely because of similarities in their internal structure. Irvine (1974) therefore suggested using the name 'Alaskan-type ultramafic complexes' for want of a better name.

The petrogenesis of these zoned ultramafic complexes has been controversial ever since they were first described. Taylor (1967) proposed the successive injection of magmas of differing compositions beginning with pyroxenite magmas and ending with dunite magmas to explain the mineral paragenesis and compositions. McTaggart (1972) after re-examining the zoned ultramafic complexes of Alaska and British Columbia, concluded that the central peridotites consist of cumulates formed from basic magmas rather than

ultrabasic magmas. He suggested that they have been dropped as crude cylinders along fractures or are downward intrusions of ultrabasic crystal mushes, the zoning resulting from the subsequent metasomatism by hydrous fluids migrating through marginal fractures. He argued that contact relations between the gabbro and ultramafic rocks are often obscured by metasomatism, so that it is difficult to determine relative ages with certainty. Murray (1972) also argued against the existence of a parental ultrabasic magma, and suggested that the ultramafic complexes formed by fractional crystallization and flow differentiation under high water pressure conditions, in feeder pipes of andesitic volcanoes. He concluded that the parental magma was akin to a hydrous olivine-rich tholeiite, whose volcanic products were dominated by andesitic compositions and probably accompanied by high-alumina tholeiite. Irvine (1974), based on his detailed work on the Duke Island Complex, suggested the following order of crystallization from a critically undersaturated alkaline ultramafic magma: olivine with minor chromite, followed by a probably prolonged period of crystallization of olivine and clinopyroxene, followed finally by clinopyroxene and magnetite. The concentric zoning in the larger ultramafic pipes is attributed to the diapiric re-emplacement of rudely stratiform sequences caused by either the continuous rise of magma at depth or by tectonic compression.

(c) *Ophiolitic ultramafic rocks*

Ophiolites are believed to represent fragments of oceanic crust and mantle formed at mid-oceanic ridges, in small marginal basins, or as basement to island arcs. Throughout the world, these mafic-ultramafic rock assemblages are exposed along orogenic belts, marking areas of intense tectonic activity (Fig. 13). As such, they represent ultramafic complexes once known as Alpine Peridotites. The stratigraphy of an ophiolite consists from base to top of tectonite peridotite, cumulate peridotite, gabbro, plagiogranite, sheeted diabase dikes, pillow lavas, and pelagic sediments (Fig. 14). Such complete sequences are preserved only rarely. Good examples include the Bay of Islands Complex, Newfoundland and the Troodos ophiolite, Cyprus (*see* Malpas 1978, Malpas *et al.*

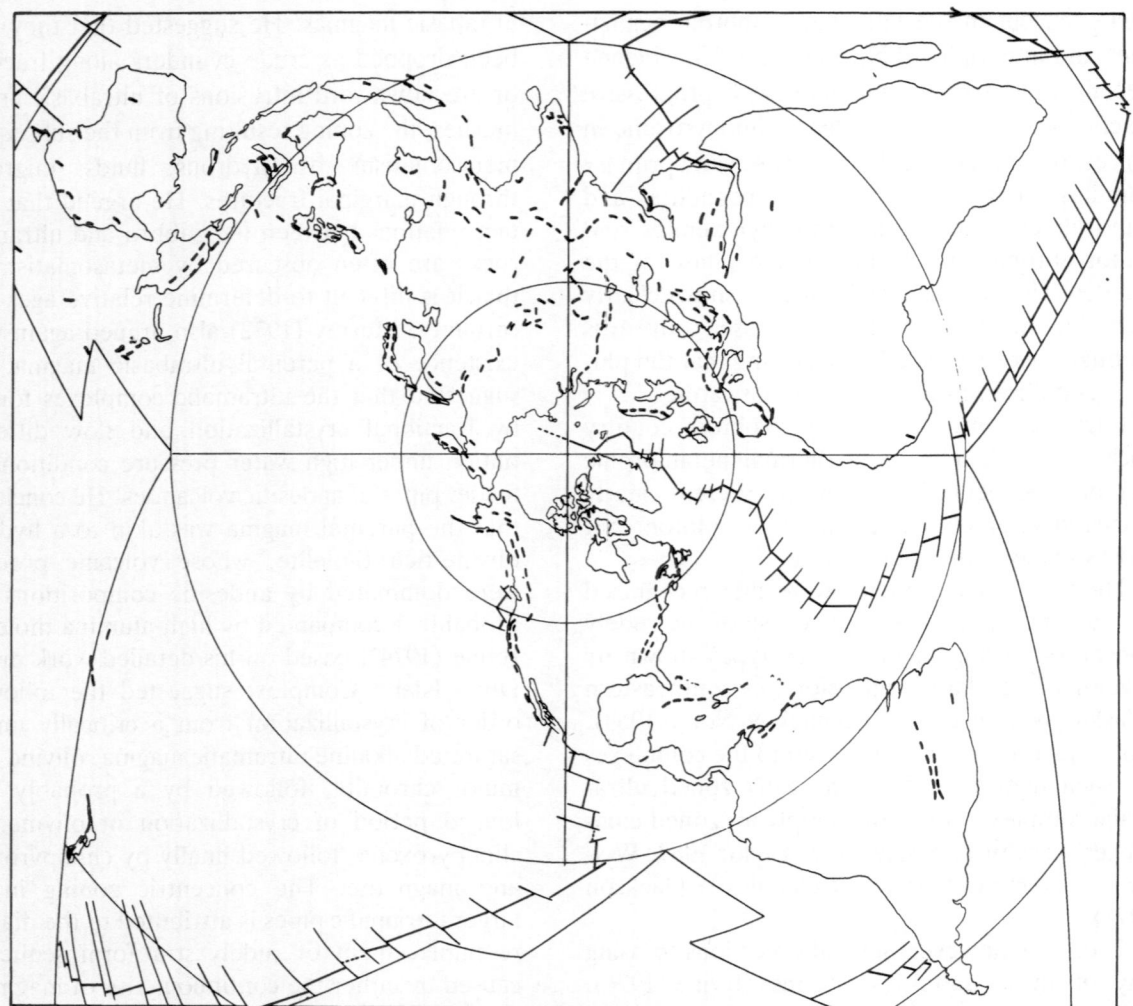

Fig. 13. North polar projection showing occurrences of ophiolitic complexes in Phanerozoic mountain belts, and the mid-ridge system (after Coleman 1977).

1990). The former displays a stratigraphic thickness of 5 km of lower Palaeozoic ultramafic rocks thought to represent the mantle that existed beneath the ancient Iapetus (proto-Atlantic) ocean. The Troodos ophiolite is somewhat younger (Mesozoic) and the ultramafic rocks are less well exposed.

In such complete ophiolite sequences, the deformed ultramafic rocks with strong gneissic fabrics lie beneath a plutonic basic sequence (Fig. 15). The ultramafic sequence is usually the thickest of the ophiolitic units and is invariably serpentinized (Fig. 16). Its base is believed to represent the surface of tectonic transport. Serpentinization takes place both in the oceanic regime where the ophiolite suite is formed and later in the continental regime into which it has been emplaced. The most common protolith of the serpentinite is harzburgite, although irregular bands and lenses of lherzolite, pyroxenite and dunite also occur. The ultramafic rocks are commonly folded with a foliation developed parallel to the axial planes of the fold traces and such deformation is generally very clear in the chromitites which are associated with dunite pods in the upper 500 m of the ultramafic assemblage. Subordinate bodies of gabbro also occur.

In the harzburgites, the well developed foliation is enhanced by the segregation of olivine and orthopyroxene into layers (Fig. 17). Grain size is variable as a result of recrystallization and textures are xenoblastic granular. Remnant oli-

20

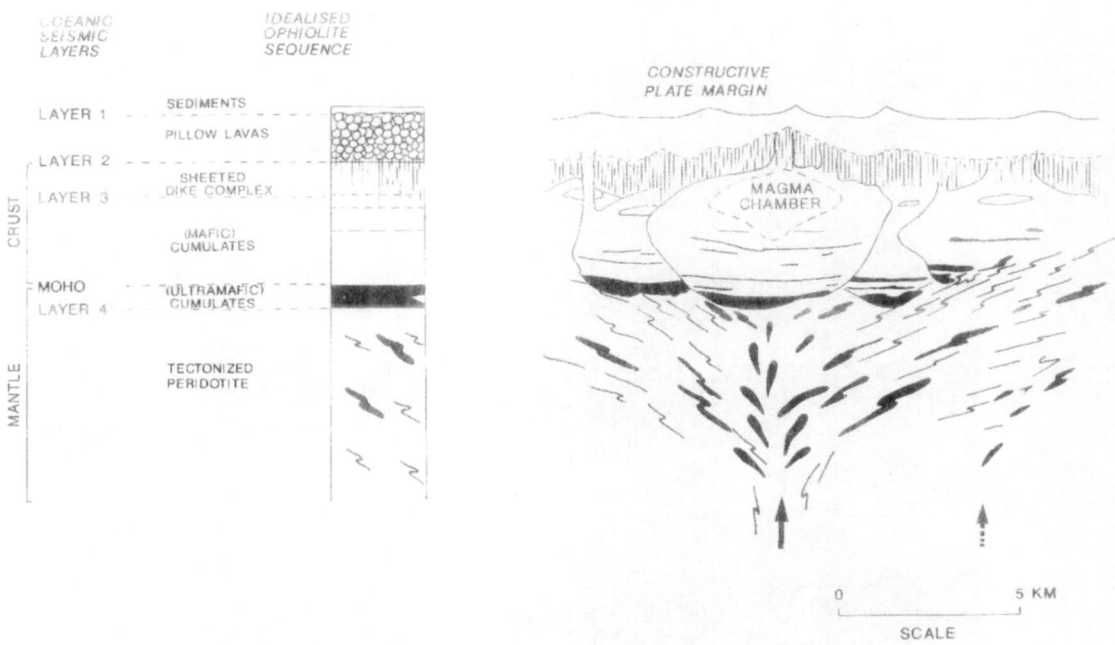

Fig. 14. Cross-section of a constructive plate margin (possibly mid-ocean ridge system) indicating nature of ophiolite generation by multiple magma-chamber intrusion. The idealized ophiolite sequence so produced may be compared with the seismic structure of the oceanic lithosphere.

Fig. 15. Deformed, banded, ultramafic rocks from the mantle sequence of the Bay of Islands ophiolite, Newfoundland. Serpentinized peridotites provide contrast in weathering with plutonic gabbros (grey) higher in the sequence. The junction between orange and grey weathering rocks is the Mohorovicic discontinuity or junction between Palaeozoic Crust and Mantle. Blow-Me-Down Mountain, western Newfoundland.

Fig. 16. View of the mantle sequence, serpentinized peridotite, forming the Table Mountain massif, Bay of Islands Complex, Newfoundland.

Fig. 17. Segregation of dunite and harzburgite layers in the mantle section of Blow-Me-Down Mountain, Bay of Islands, Newfoundland.

vine cores that have escaped serpentinization invariably show kink-banding as a result of high temperature deformation, and have compositions in the range Fo_{90-92}. Orthopyroxenes are likewise highly serpentinized, but where original mineralogy is preserved, have a composition of En_{90-91}. These orthopyroxenes often also contain clinopyroxene exsolution lamellae formed as a result of decreasing pressure. Discrete grains of clinopyroxene are present as an integral part of the lherzolite mineralogy but never form more than 5% of the harzburgites. In this case the clinopyroxene is chrome-rich and is often preserved as small grass-green crystals visible in hand specimen. Chrome spinel is a common accessory and, together with the orthopyroxene, is elongated in the foliation plane. Dunite lenses within the harzburgites contain remnant olivines with a similar composition to those in the host rock (Fo_{90-92}) but associated chrome spinel has a distinctly different chemistry and layers suggesting a cumulate origin for the layered dunite bodies.

The nature of the olivine deformation, the isoclinal folding of the tectonites and the plastic style of the deformation are taken as evidence that the recrystallization occurred at temperatures in excess of 1000°C, not far below the solidus of the ultramafic rocks. Most of the strain was accommodated by syntectonic recrystallization of olivine producing kink-bands and by the reorientation of orthopyroxene. The foliation also affects the cumulate sequence of rocks which overlie the ultramafic tectonites.

Most workers accept that the tectonized harzburgite represents depleted mantle material – a generally refractive residue from which high-magnesian basaltic melts have been extracted by partial melting (Malpas 1978). These melts are considered to have crystallized to form the overlying cumulate sequence, dikes and pillow lavas and also the lenses and pods of dunite and pyroxenite found in the tectonites. The only other possible alternative, that the ultramafic rocks represent a deformed cumulate series, can be dismissed on consideration of their thickness (sequences are over 7 km thick), and the lack of mineralogical rhythmic and cryptic variation that would be expected in any magmatic sequence. The lherzolitic relics that are present in some ophiolites can be interpreted as remains of more pristine mantle or

may represent deformed products of crystallization of the picritic melt. It can be calculated that partial melting of the order of 25% of the primary mantle material (Spinel lherzolite?) at depths in excess of 25 km would be needed to produce the harzburgite residuum and the calculated composition of picritic melt. Under these conditions, not all the clinopyroxene is necessarily removed for the parent starting material.

Overlying the ultramafic tectonites are ultramafic cumulates which are believed to be derived by fractional crystallization of mafic magma in a sub-axial magma chamber. The cumulates consist of dunite with associated chromite at the base, grading upwards into more felsic pyroxenites, and gabbroic cumulates.

The ultramafic part of the ophiolite cumulates is quite variable in thickness, and thickening and thinning by low-angle thrusting often makes estimates of thicknesses difficult or even impossible. The contact between the ultramafic cumulates and tectonites (Petrological Moho of Malpas (1978)) can be complicated by post emplacement deformation. Intercumulate plagioclase increases upwards and results in peridotites being interlayered with plagioclase-rich gabbros, troctolites, and anorthosites (Critical Zone) which marks the transition from ultramafic cumulates to mafic cumulates. Above the transition zone, peridotites are lacking and gabbroic rocks dominate.

The primary mineralogy of the ultramafic cumulates consists of clinopyroxene, orthopyroxene, olivine, and plagioclase, with significant amounts of chromite within the basal dunites. The composition of cumulate olivine overlaps that found in the tectonites, but FeO progressively increases upwards. Serpentinization is widespread and olivine may be completely replaced by serpentine and magnetite. Orthopyroxene generally forms large poikilitic grains (5–10 cm), but does not form thick monomineralic units as in stratiform complexes. Clinopyroxene is very calcic and also overlaps that found in the tectonite. Both orthopyroxene and clinopyroxene are invariably partially altered to secondary amphibole and chlorite. Plagioclase first appears as an intercumulus phase in the dunites and increases in abundance upwards. Black chromite may form layers and in some cases economic deposits near the base of the dunite.

The ultramafic cumulates have Mg #'s (0.7–0.8) lower than those of the tectonites. Most of the cumulates do not show a strong iron enrichment with progressive fractionation, which may indicate that the activity of oxygen was low and that most of the iron is contained within the silicate minerals. The wide range in composition is largely controlled by the modal proportion of calcic plagioclase. The ultramafic cumulates have a restricted range of CaO and Al_2O_3, and variable alkalies, but significantly higher amounts than the tectonites.

Serpentinization of the ultramafic rocks in ophiolites varies generally between 20% and 80% by volume. Some rocks are 100% serpentinized. The serpentinization clearly takes place in several stages. After the initial period of deformation and solid flow occurring in the upper mantle, hydration of the rocks results in the formation of amphiboles, generally of the actinolite family under the conditions of low amphibolite grade of metamorphism. Serpentinization, overlapping with the last stages of this amphibolitization, gives rise eventually to lizardite as the dominant serpentine phase, and simultaneous calcium metasomatism causes rodingitization of associated basaltic and gabbroic rocks. Aragonite is formed in fractures and cavities, probably as a result of seawater interaction with the ultramafic rocks during the last stages of oceanic serpentinization. The formation of aragonite rather than calcite may be related to the high activity of magnesium in this environment, which appears to inhibit calcite precipitation.

The hydration and serpentinization of the ultramafic rocks in the oceanic environment appears to involve both juvenile (mantle) solutions and seawater. The former are probably more important at depth while heated seawater, circulating within the oceanic crust, becomes the prevalent agent of serpentinization within a few kilometres of the ocean floor.

Subsequent stages of serpentinization, generally at lower temperatures and under the influence of both seawater initially and then circulating groundwaters, occur during the emplacement and post-emplacement periods of the ophiolites' histories.

(d) *High temperature peridotite intrusions*

According to Green (1967), 'high-temperature' peridotites are characterized by high-grade contact metamorphic aureoles, diapiric internal structure and shape, and show evidence of high pressure crystallization with later deformation. Such high-temperature peridotites have been thought of as diapiric intrusions into continental crust.

However, the presence of high temperature contact aureoles associated with high pressure crystallization and later deformation is no guarantee that these bodies are in fact intrusions and not remnants of ophiolites. For example, the Bay of Islands ophiolite exhibits these features but is clearly an ophiolite. It is believed therefore, that many bodies previously described as high-temperature peridotite intrusions, are in fact highly dismembered and poorly preserved ophiolites. These include Mt. Albert (Canada) and Lizard (UK), both described by Green (1967), and the Trinity Body (USA) (Lipman 1964), the Twin Sisters (USA) (Ragan 1963) and the Dun Mountain Body (New Zealand) (Challis 1965). However, several large peridotite bodies such as Ivrea (Italy), La Rhonda (Spain) and Benni Boucherra (Morocco) have wide contact metamorphic aureoles, and are more likely intrusions into continental crust.

The Rhonda massif located in southwest Spain, consists of approximately 95% peridotite with minor amounts of serpentinite, and pyroxene-rich mafic layers with ecologitic and gabbroic mineralogies (Obata *et al.* 1977). The massif is surrounded by a variety of contact metamorphic rocks, and petrological studies of these (Loomis 1972) indicate that the intrusion entered the crust at a temperature in excess of 1000°C. No primary igneous textures remain, and there is no evidence of cumulate processes having been important in petrogenesis. The present mineralogical assemblage is totally metamorphic, developed by syntectonic recrystallization at the time the peridotite was emplaced into the crust.

Based on their work on the Rhonda Massif, Obata *et al.* (1977) developed a petrogenetic model involving the diapiric uprise of peridotite, to explain the formation of these intrusions. As the diapir rises, it loses heat by conduction to the surroundings, so that partial melting begins inside the diapir. As melting proceeds, liquid droplets coalesce to form thin sheets, and local pockets or lenses (Dickey 1970). Since the liquids formed are gravitationally and mechanically unstable, they tend to migrate upwards. Before the liquid can escape from the diapir, it must migrate along the cooler outer zone of the diapir, where it is cooled and partially crystallizes. After the fractionated liquid escapes, the diapir is compositionally zoned, being depleted inwards to undepleted outwards. Hence, in this model, the composition of the melt generated by partial fusion of the peridotite mass is modified by fractional crystallization even before it separates from the diapir.

This model may also explain the origin of other potential high-temperature peridotite intrusions located in the western Mediterranean, which contain similar magmatic layers and broad contact metamorphic aureoles. These include Beni Bouchera (Kornprobst 1969), Totalp (Peters 1963, Peters & Niggle 1964) and Alpe Arami (Mockel 1969).

Conclusions

Serpentinization is a common process associated with the hydration of ultramafic rocks. Formation of serpentinites involves a broad range of pressure-temperature conditions and mineral assemblages provide certain clues to these conditions. The transformation of a peridotite to serpentinite involves significant changes in density, seismic velocity, rock strength and magnetic susceptibility. Since these parameters can now be routinely measured, this unique relationship between serpentinite and its protolith can be used to explain such geophysical anomalies as magnetic highs, gravity lows and variation in seismic velocity.

Ultramafic rocks occur in a wide range of geographic localities but each can be ascribed to one of four major classes. The most common occurrences are those which form part of ophiolite complexes which are invariably associated with orogenic belts. Serpentinization of these rocks reaches amounts of up to 80% (locally 100%) and results in volume changes of approximately 30%. The mineral assemblage lizardite + chrysolite + brucite + magnetite is the prevalent serpentine mineral assemblage in the ultramafic positions of many ophiolite complexes.

Even with considerable serpentinization there is no clear evidence for significant metasomatism during serpentinization. However, there may be small gains of SiO_2 and small losses of MgO, especially in dunitic and wehrlitic rocks. Some loss of Ca occurs when rodingitization takes place.

Acknowledgements

I would like to thank Terry Brace for literature research and drafting and Annie Reid for manuscript typing. This chapter greatly benefits from their help.

Appendix

Glossary

Item	Definition
ρCO_2	The partial pressure of CO_2; when two ideal gases, A and B, are contained in a Volume V at a temperature T, then Pa = na RT and Pb = nb RT. In the mixture each gas exerts a pressure that is the same as it would exert if it were present alone, and this pressure is proportional to the number of moles of the gas present. The quantities PA and PB are called the partial pressures of A and B respectively. The total pressure Pt is, Pt = PA + PB = (nA + nB)(RT).

Alkali	Any strongly basic substance, such as a hydroxide or carbonate of an alkali metal.
Antigorite	A polymorph within the serpentine group, $Mg_3Si_2O_5(OH)_4$
Aragonite	A white, yellowish, or grey orthorhombic mineral: $CaCO_3$.
Aureole	A zone surrounding an igneous intrusion in which the country rock shows the effects of contact metamorphism.
Awarnite	A mineral consisting of a natural alloy of nickel and iron.
Basalt	A dark to medium coloured, commonly extrusive (locally intrusive, as dykes), mafic igneous rock composed chiefly of calcic plagioclase and clinopyroxene in a glassy, or fine-grained groundmass; the extrusive equivalent of gabbro.
Basement	A complex of undifferentiated rocks that underlies the oldest identifiable rocks in the area.
Bastite	An olive-green, blackish-green, or brownish variety of serpentine resulting from the alteration of orthrombic pyroxene, occurring as foliated masses in igneous rocks, and characterized by a schiller (metallic or pearly lustre) on the chief cleavage face of the pyroxene.
Brucite	A hexagonal mineral, often fibrous, with the chemical formula $Mg(OH)_2$.
Chromite	A brownish to iron-black mineral of the spinel group $(Fe, Mg) (Cr, Al)_2O_4$. It occurs in octahedral crystals as a primary accessory mineral in mafic and ultramafic igneous rocks.
Chrysotile	A polymorph within the serpentine group, $Mg_3Si_2O_5(OH)_4$.
Clinopyroxene	A group name for pyroxenes crystallizing in the monoclinic system with the chemical formula $(Ca, Mg, Fe)SiO_3$.
Colour index	The classification of igneous rocks by per cent volume of dark or coloured (i.e., mafic) minerals in the rock.
Cumulus	The accumulation of crystals that precipitated from à magma without having been modified by later crystallization.
Diapirism	The process of piercing or rupturing of dome or uplifted overlying rocks by core material heated to the plastic state, either by tectonic stresses as in anticlinal folds, or by the effect of geostatic load in sedimentary strata as in salt domes or shale diapirs, or as in the case of igneous intrusions, forming diapiric structures such as plugs or batholiths.
Dunite	Peridotite in which the mafic mineral is almost entirely olivine, with accessory chromite almost always present.
Eclogitic	Said of a rock having an association of clinopyroxene and garnet with a proportion of jadeite molecule in the clinopyroxene.
En	Abbr. for the mineral enstatite $(Mg_2Si_2O_6)$ which is the Mg end-member of the orthopyroxene solid solution series, the other end-member being ferrosilite, $Fe_2Si_2O_6$. Thus the symbol En_{52} would indicate an orthopyroxene with 52 mole per cent $Mg_2Si_2O_6$, and 48 mole per cent $Fe_2Si_2O_6$.
Fo	Abbr. for the mineral Forsterite, the Mg end-member of the olivine solid solution series. It has the chemical formula Mg_2SiO_4. The other end-member of the series is fayalite, which is Fe_2SiO_4. This is a continuously variable series, in which the ratio of Mg to Fe varies from 100:0 to 0:100 and thus the symbols Fo_{xx} or Fe_{xx} indicate the molecular percentage composition. e.g. the symbol $Fo_{52}Fa_{48}$ gives full information as to the composition of a particular specimen. (May also be written $Mg_{52}Fe_{48}$).
Fractionation	i.e., fractional crystallization – separation of a cooling magma into parts by the successive crystallization of different minerals at progressively lower temperatures.
Fugacity	A thermodynamic function directly related to the chemical potential and defined in such a way that thermodynamic equations describing the behaviour of ideal gases

26

apply equally well to non-ideal gases when the fugacities of the non-ideal gases are substituted in the equations for the pressures of the ideal gases. Numerical values of fugacity are expressed in units of pressure. (F)

Gabbro	A group of dark-coloured, basic intrusive igneous rocks composed principally of plagioclase and clinopyroxene, with or without olivine and orthopyroxene.
Garnet	A group of minerals of formula $A_3B_2(SiO_4)_3$, where A = Ca, Mg, Fe^{2+}, and Mn^{2+}, and B = Al, Fe^{3+}, Mn^{3+}, and Cr.
Garnierite	A metamorphic rock consisting chiefly of an aggregate of interlocking garnet grains.
Ground mass	The interstitial material of a porphyritic igneous rock; it is relatively more fine-grained than the phenocrysts and may be glassy or both.
Harzburgite	A peridotite composed chiefly of olivine and orthopyroxene.
H_2O^+	water contained as an integral part of a crystal's structure, generally in the form of a hydroxyl group.
Hornblendite	An igneous rock composed almost entirely of hornblende, the commonest mineral of the amphibole group: $Ca_2Na(Mg, Fe^{2+})_4(Al, Fe^{3+}, Ti)(Al, Si_8O_{22})(O, OH)_2$.
Intercumulus phase	Magmatic liquid that surrounds the crystals of the cumulus, i.e., that occupies the intercumulus.
Island arc	A chain of volcanic islands rising from the deep sea floor and near to the continents.
Isocline	A fold, the limbs of which have been so compressed that they have the same dip. Isoclines occur in homogeneous rocks such as slates, and are characteristic of strong deformation.
Isopleth	A line, surface or volume on which some mathematical function has a constant value. An isopleth need not refer to a directly measurable quantity characteristic of each point in the map area.
Kink band	A type of deformation band occuring microscopically in crystals and mesoscopically in foliated rocks, in which the orientation of the lattice or of the foliation is changed or deflected by gliding or slippage and shortening along slippage planes.
Leaching	The separation, selective removal, or dissolving out of soluble constituents from a rock or orebody by the natural action of percolating water.
Lherzolite	Peridotite composed chiefly of olivine, orthopyroxene, and clinopyroxene and in which olivine is generally most abundant.
Liquidus	The locus of points in a temperature-composition diagram representing the maximum solubility (saturation) of a solid component or phase in the liquid phase. In an isoplethal study, at temperatures above the liquidus, the system is completely liquid, and at the intersection of the liquidus and isopleth, the liquid is in equilibrium with one crystalline phase
Lizardite	A polymorph within the serpentine group, $Mg_3Si_2O_5(OH)_4$.
Low sulphur sulphides	Those sulphide minerals in which the ratio of the sulphur ion to cation(s) is close to 1. e.g., covellite CuS.
Magma	Naturally occurring mobile rock material, generated within the earth and capable of intrusion and extrusion, from which igneous rocks are thought to have been derived through solidification and related processes.
Magnetite	A black, isometric, strongly magnetic, opaque mineral of the spinel group with the chemical formula $(Fe, Mg)Fe_2O_4$.
Metamorphism	The mineralogical and structural adjustment of solid rocks to physical and chemical conditions which have been imposed in response to marked changes in temperature, pressure, shearing stress, and chemical environment at depth in the earth's crust.
Metasomatism	The process of practically simultaneous capillary solution and deposition by which a new mineral of partly or wholly different chemical composition may grow in the body of an old mineral or mineral aggregate.

Meteoric water	Pertaining to water of recent atmospheric origin.
Mg#	the ratio $\dfrac{\text{MgO weight \%}}{\text{MgO} + \text{FeO weight \%}}$.
Olivine	An olive-green, greyish-green, or brown orthorhombic mineral with the chemical formula $(Mg, Fe)_2SiO_4$.
Orthopyroxene	A group name for pyroxenes crystallizing in the orthorhombic system with the chemical formula $(Mg, Fe)SiO_3$.
pH	$pH = -\log[H_3O^+]$.
Paragenesis	The sequential order of mineral formation.
Pegmatite	An exceptionally coarse-grained (most grains > 1 cm in diameter) igneous rock, with interlocking crystals, usually found as irregular dykes, lenses, or veins, especially at the margins of batholiths. Pegmatites generally have the composition of granite and represent the last and most hydrous portion of a magma to crystallize.
Peridotite	A general term for a coarse grained plutonic rock composed chiefly of olivine with or without other mafic minerals such as pyroxenes, amphiboles, and micas and containing little or no feldspars.
Phenocryst	A relatively large, conspicuous crystal in a porphyritic rock.
Picrite	A dark-coloured, generally hypabyssal rock containing abundant olivine along with pyroxene, biotite, possibly amphibole, and less than 10% plagioclase.
Pipe	A discordant pluton of tubular shape.
Plagiogranite	An igneous rock having a low potassium content, ranging in composition from quartz diorite to trondhjemite.
Pluton	An igneous intrusion.
Poikilitic	Said of the texture of an igneous rock in which small crystals of one mineral are irregularly scattered without common orientation in a large crystal of another mineral; also, said of the enclosed crystal.
Protolith	The unmetamorphosed rock from which a given metamorphic rock was formed.
Pseudomorph	A mineral whose outward crystal form is that of another mineral species; developed by alteration, substitution, incrustation, or parmorphism.
Recrystallization	The formation of new, crystalline mineral grains in a rock, essentially in the solid state, under the influence of metamorphic processes.
Rodingite	A medium- to coarse-grained, commonly calcium enriched gabbroic rock containing, as essential minerals, grossular and diallage. Altered varieties also contain prehnite or serpentine or both.
Saturation line	The line, on a variation diagram of an igneous rock series, that represents saturation with respect to silica. Rocks or magma to the right of the line are oversaturated and those to the left, undersaturated.
Solidus	On a temperature-composition diagram, the locus of points in a system at temperatures above which solid and liquid are in equilibrium and below which the system is completely solid.
Syntectonic	Said of a geological process or event occuring during any kind of tectonic activity.
Tholeiite	A group of basalts primarily composed of plagioclase, pyroxene and iron oxide minerals as phenocrysts in a glassy groundmass on intergrowth of quartz and alkali feldspar. Little or no olivine is present.
Ultramafic	Said of an igneous rock composed chiefly of mafic minerals i.e., ferromagnesian, dark coloured minerals.
Websterite	A pyroxenite composed chiefly or ortho- and clinopyroxene.
Wehrlite	A peridotite composed chiefly of olivine and clinopyroxene with common accessory opaque minerals.

28

References

Ashley, P. M. 1975. Opaque mineral assemblage formed during serpentinization in the Coolac Ultramafic Belt, New South Wales. J. Geol. Soc. Aust. 22, 91–102.

Barnes, I., J. B. Rapp & J. R. O'Neill. 1972. Metamorphic assemblages and the direction of flow of metamorphic fluids in four instances of serpentinization. Cont. Min. Pet. 35: 263–276.

Bliss, N. W. & W. H. MacLean. 1975. The paragenesis of zoned chromite from central Manitoba. Geochim. Cosmochim. Acta, 39: 973–990.

Bottinga, Y. & D. F. Weill. 1970. Densities of liquid silicate systems calculated from partial molar volumes of oxide components. Am. J. Sci., 269: 169–182.

Campbell, I. H., J. M. Dixon & P. L. Roeder. 1977. Crystal buoyancy in basaltic liquids and other experiments with a centrifuge furnace. EOS, Trans. Am. Geophys. Un. 58: 527.

Camsell, C. 1913. Geology and mineral deposits of the Tulameen District, British Columbia. Geol. Surv. Can. Mem. 26: 188p.

Carmichael, I. S. E., F. J. Turner & J. Verhoogen. 1974. Igneous Petrology. McGraw-Hill Book Company, New York. 739p.

Challis, G. A. 1965. The origin of New Zealand ultramafic intrusions. J. Pet. 6: 322–364.

Chernosky, J. V. 1973. The stability of chrysotile, $Mg_3Si_2O_5(OH)_4$ and the free energy of formation of talc, $Mg_3Si_4O_{10}(OH)_2$. Geol. Soc. Am. Abst. 1973 meeting.

Chernosky, J. V. 1975. Aggregate refractive indices and unit cell parameters of synthetic serpentine in the system MgO-SiO_2-Al_2O_3-H_2O. Am. Mineral. 60: 200–208.

Coleman, R. G. & T. E. Keith. 1971. A chemical study of serpentinization – Burno Mountain, California. J. Pet. 12: 311–328.

Dickey, J. S., Jr. 1970. Partial fusion products in alpine-type peridotites: Serrania de la Ronda and other examples. Min. Soc. Am. Sp. Pap. no. 3: 33–49.

Eckstrand, O. R. 1975. The Dumont serpentinite: a model for control of nickeliferous opaque mineral assemblages by alteration reactions in ultramafic rocks. Econ. Geol. 70: 183–201.

Evans, B. W. & B. R. Frost. 1975. Chrome spinel in progressive metamorphism – a preliminary analysis. Geochim. Cosmochim. Acta 39: 959–972.

Findlay, D. C. 1969. Origin of the Tulameen ultramafic-gabbro complex, southern British Columbia. Can. J. Earth Sci. 6: 399–425.

Green, D. H. 1967. High temperature peridotite intrusions. In: Wyllie, P. J. (ed.) Ultramafic and related rocks. Wiley and Sons Inc., New York, 212–222.

Hostetler, P. B., R. G. Coleman, F. A. Mumpton & B. W. Evans. 1966. Brucite in Alpine serpentinites. Am. Mineral. 51: 75–98.

Irvine, T. N. 1982. Observation on the origins of Skaergaard Layering. Carnegie Institution of Washington Yearbook, 82: 284–300.

Irvine, T. N. 1974. Petrology of the Duke Island ultramafic complex, southeastern Alaska. Geol. Soc. Am. Mem. 138: 240p.

Irvine, T. N. 1968. Petrologic studies of ultramafic rocks in the Aiken Lake Area, British Columbia. Geol. Surv. Can. Paper 68-1, A: 110–111.

Jackson, E. D. & T. P. Thayer. 1972. Some criteria for distinguishing between stratiform, concentric and alpine peridotite-gabbro complexes. Proc. 24th Int. Geol. Cong., Montreal, 2: 289–296.

Jackson, E. D. 1971. The origin of ultramafic rocks by cumulus processes. Fortschr. Mineralogie, 48: 128–174.

Jackson, E. D. 1961. Primary textures and mineral associations in the ultramafic zone of the Stillwater Complex, Montana. U. S. Geol. Surv. Prof. Pap. 358.

Johannes, W. 1969. An experimental investigation of the system MgO-SiO_2-H_2O-CO_2. Am. J. Sci. 267: 1083–1104.

Johannes, W. & P. Metz. 1968. Experimentelle Bestimmung von Gleichgewichlsbeziehungen in system MgO-CO_2-H_2O. Neues. Jb. Miner. Monash, 16: 15–26.

Kennedy, G. C. & M. S. Walton. 1946. Geology and associated mineral deposits of some ultramafic rock bodies in southeastern Alaska. U. S. Geol. Surv. Bull. 947-D: 65–84.

Komor, S. C., D. Elthon & J. F. Casey. 1985. Serpentinization of cumulate ultramafic rocks from the North Arm Mountain massif of the Bay of Islands ophiolite. Geochim. Cosmochim. Acta 49: 2331–2338.

Kornprobst, J. 1969. Le massif ultrabasique des Beni Bouchera (Rif. Interne. Maroc) Cont. Min. Pet. 23: 283–322.

Laurent, R. & Y. Hebert. 1979. Paragenesis of serpentine assemblages in harzburgite tectonite and dunite cumulate from the Quebec Appalachians. Can. Mineral. 17: 857–869.

Lipman, P. W. 1964. Structure and origin of an ultramafic pluton in the Klamath Mountains, California. Am. J. Sci. 262: 199–222.

Loomis, T. P. 1972. Diapiric emplacement of the Ronda high-temperature ultramafic intrusion, southern Spain. Geol. Soc. Am. Bull. 83: 2475–2496.

Malpas, J. 1978. Magma generation in the upper mantle, field evidence from ophiolite suites, and application to the generation of oceanic lithosphere. Phil. Trans. Royal Soc. Lond. A288: 527–546.

Malpas, J., E. Moores, A. Panaycotou & C. Xenophoutos, 1990. Ophiolites: Oceanic Crustal Analogues. Geol. Surv. Cyprus 733 p.

Magaritz, M. & H. P. Taylor. 1974. Oxygen and hydrogen isotope studies of serpentinization in the Troodos ophiolite complex, Cyprus. Earth Planet. Sci. Lett. 23: 8–14.

McTaggart, K. C. 1972. On the origin of ultramafic rocks: Reply. Geol. Soc. Am. Bull. 83: 3161–3162.

Mockel, J. R. 1969. The structural petrology of the garnet-peridotite of Alpe Arami (Ticino, Switzerland). Leidse Geol. Mededel. 42: 61–130.

Moody, J. B. 1976. Serpentinization: a review. Lithos 9: 125–138.

Mumpton, F. A. & C. S. Thompson. 1975. Mineralogy and origin of the Coalinga Asbestos deposit. Clays and Clay Min. 23: 131–144.

Murray, C. G. 1972. Zoned ultramafic complexes of Alaskan type: Feeder pipes of andesitic volcanoes. In: Shagam, R. E. and others, (eds). Studies in Earth and Space Sciences (Hess Volume). Geol. Soc. Am. Mem. 132: 313–335.

Naldrett, A. H. 1966. Talc-carbonate alteration of some serpentinized ultramafic rocks south of Timmins Ontario. J. Pet. 7: 489–499.

Obata, M., C. J. Suen & J. S. Dickey, Jr. 1980. The origin of mafic layers in the Ronda high-temperature peridotite intrusion, S. Spain: an evidence of partial fusion and fractional crystallization in the upper mantle. In: Allegre, C. and Aubouin, J. (eds.). Association mafiques ultramafiques dans les orogenes. Colloques Internationaux du C.N.R.S. 257–268.

Peters, Tj. & E. Niggli. 1964. Spinellfuhrende Pyroxenite (Ariegite) in den Lherzolith korpern von Lherz und Umgebung (Ariege, Pyrenean) und der Totalp (Graubunden, Schweiz), ein Vergleich. Schweiz. Min. Pet. Mitt. 44: 513–517.

Peters, Tj. 1963. Mineralogie und petrographie des Totalp Serpentins bei Davos. Schweiz. Min. Pet. Mitt. 43: 529–685.

Ragan, D. M. 1963. Emplacement of the Twin Sisters Dunite, Washington. Am. J. Sci. 261: 549–565.

Rucksmith, J. C. & J. A. Noble. 1959. Origin of the ultramafic complex at Union Bay, southeastern Alaska. Geol. Soc. Am. Bull. 70: 981–1018.

Taylor, H. P. 1967. The zoned ultramafic complexes of southeastern Alaska. In: Wyllie, P. J. (ed.) Ultramafic and Related Rocks. Wiley and Sons Inc., New York 97–121.

Taylor, H. P. & J. A. Noble. 1960. Origin of the ultramafic complexes southeastern Alaska. 21st Int. Geol. Cong. Copenhagen Rep. 13: 175–187.

Wager, R. & G. M. Brown. 1967. Layered igneous rocks. W. H. Freeman & Co., San Francisco, 588 pp.

Walton, M. S. 1951. The Blashke Islands ultrabasic complex. Ph.D. Thesis, Columbia University, New York, New York.

Wenner, D. B. & H. P. Taylor. 1974. D/H and O^{18}/O^{16} studies of serpentinization of ultramafic rocks. Geochim. Cosmochim. Acta 38: 1255–1286.

Wenner, D. B. & H. P. Taylor. 1973. Oxygen and hydrogen isotope studies of the serpentinization of ultramafic rocks in oceanic environments and continental ophiolite complexes. Am. J. Sci. 273: 207–239.

Wenner, D. B. & H. P. Taylor. 1971. Temperatures of serpentinization of ultramafic rocks based on O^{18}/O^{16} fractionation between coexisting serpentine and magnetite. Cont. Min. Pet. 32: 165–185.

Whittaker, E. J. W. & F. J. Wicks. 1970. Chemical differences among the serpentine polymorphs: a discussion. Am. Mineral. 55: 1025–1047.

Wicks, F. J. 1969. X-ray and optical studies of serpentine minerals. D.Phil. dissertation, Oxford University, England.

Wyllie, P. J. 1969. The origin of ultramafic and ultrabasic rocks. Tectonophysics 7: 437–455.

Wyllie, P. J. (ed.) 1967. Ultramafic and Related Rocks. Wiley and Sons Inc., New York, 464 p.

Author's Address:
J. Malpas
Centre for Earth Resources Research
Dept. of Earth Sciences
Memorial University of Newfoundland
St. John's
Newfoundland
Canada A1B 3X5

III

Plant life of western North American ultramafics

The red-rock forest may seem hellish to us, but it is a refuge to its flora . . . . it is the
obdurate physical adversity of things such as peridotite bedrock which often drives life
to its most surprising transformations. *David Rains Wallace, The Klamath Knot (1983).*

A. R. KRUCKEBERG

Abstract. In Western North America, ultramafics occur with decreasing abundance from California, Oregon,
Washington, to British Columbia. All the occurrences are now considered parts of ophiolite suites, and are
associated with the north-south trending cordilleras and their plate tectonics. The greatest concentrations of
ultramafics, mostly as serpentinized peridotite, are in northwestern California and southwestern Oregon.

Soils weathered from ultramafic rocks are either devoid of vegetation (barrens) or support sparse but often
distinctive floras. Cation exchange capacities range from 5.2 to 43 m · equivs $100\,g^{-1}$ dry soil; pH values are around
neutral (6.0 to 8.8); Mg/Ca quotients are invariably greater than 1.0; deficiencies of nitrogen and phosphorus are
common and can be corrected by the addition of these elements, only in the presence of adequate calcium. Tissue
analysis of serpentine plants often reveals high concentrations of magnesium and nickel.

Vegetation on ultramafic soils takes the form of distinctive variants of conifer or mixed conifer-hardwood forest,
chaparral, or grassland. Often the serpentine (S) vegetation is sharply delimited from adjacent nonserpentine (NS)
types, both by physiognomy (e.g., chaparral on serpentine, forest on nearby nonserpentine), and by species
composition. The most striking contrasts in vegetation (S *vs.* NS) are in California and Oregon. Contrasts in S–NS
vegetation are lessened in the Pacific Northwest, possibly because of increased precipitation, or the short post-
Pleistocene history of the region, or both. Proctor (Chapter VI) notes a similar lessening of contrasts in S–NS
vegetation in the United Kingdom.

Floras on ultramafic soils can be strikingly unusual. Three types of floristic elements can be found: (1) serpentine
endemics, (2) local or regional indicator species, and (3) bodenvag species, taxa widespread on S and NS habitats.
Also many NS taxa may be excluded from adjoining S soils.

The greatest concentration of species endemic to serpentine is in the Klamath–Siskiyou mountain complex of
northwestern California and southwestern Oregon, with secondary concentrations in the North Coast and South
Coast ranges and the Sierra Nevada of California. Endemics occur in all life-forms: trees and shrubs (e.g., *Cupressus
sargentii*, *Quercus durata*, *Ceanothus jepsonii*), herbaceous perennials (e.g., *Calochortus tiburonensis*, *Fritillaria
liliacea*, *Lilium bolanderi*); and annuals (e.g., *Streptanthus batrachopus*, *Layia discoidea*, *Clarkia franciscana*).
Endemics belong to genera abundantly represented in the regional flora.

Widespread species that appear as local or regional indicators of serpentine include trees like *Calocedrus decurrens*
and *Pinus jeffreyi*, shrubs (e.g., *Heteromeles arbutifolia*, *Adenostoma fasciculatum*, *Ceanothus cuneatus*) and herbs
(e.g., *Streptanthus glandulosus*, *Darlingtonia californica*, *Aspidotis densa*, *Xerophyllum tenax*).

B. A. Roberts and J. Proctor (eds), The ecology of areas with serpentinized rocks. A world view, 31–73.
© 1992 *Kluwer Academic Publishers.*

Indifferent or bodenvag species are often racially differentiated into tolerant and intolerant biotypes.

The fauna on western North American serpentines has received but scant attention, and merits closer study. Butterfly species are known to be closely tied to serpentine plants as food sources; one instance of plant mimicry of butterfly eggs is cited.

The evolution of a serpentine flora may involve a variety of speciational routes. The most probable sequence for diploid taxa could involve (1) genetic preadaptation to serpentine within a NS species; (2) racial fixation of the preadapted genotype; (3) further morphological and physiological divergence yielding an infraspecific variant; (4) attaining species status by further genetic and ecological isolation. This sequence is illustrated by *Streptanthus*, a genus of western North American crucifers, with varying degrees of fidelity and narrow endemism to California and Oregon serpentines. A more rapid mode of speciation on serpentine, saltational speciation by catastrophic selection, has been proposed.

Adaptation to ultramafic soils is likely to involve both physiological and morphological modifications. Xerophytism, nanism, glaucescence, plagiotropism and color changes (anthocyanic, chlorotic) are frequent attributes of serpentine species. A few species possess the ability to accumulate over $1000 \mu gg^{-1}$ of nickel in their foliar dry matter (hyperaccumulators).

Western North American serpentines have been exploited for minerals, timber, grazing and agriculture, with consequent effects on their floras. Mining for mercury, nickel and chromium, as well as geothermal power developments, have created the greatest disturbances to them. Only modest efforts have been made to preserve samples of serpentine vegetation. Some state and federal wilderness areas include serpentine vegetation; other serpentine areas are 'protected' either by neglect or because they are valued as watershed areas. A very few natural areas specifically for serpentine vegetation have been established in the three Pacific Coast states. None are known for British Columbia.

Introduction

Western North America from Alaska to Mexico, has been the setting for dramatic and turbulent geologic events, especially since the Cretaceous. The western edge of the continent has yielded spectacularly to plate tectonic activity – vulcanism, Cordilleran mountain-building, and consequent intrusions of ultramafics. Glacial events then added the latest of impacts on the face of the landscapes. The result has given western North America a wealth of geologic diversity (McKee 1972, Norris & Webb 1976). In no small part, the crustal deformation and mantle intrusions have been coincident with extensive displays of ultramafics – serpentinite and related ferromagnesian silicate rocks. As a consequence of its frequent surface exposures, ultramafic outcrops have had an imporant influence on the biota, especially the plant life of the region.

In western North America, the epicentre of ultramafic geology is in California, west of the Sierra Nevada crest to the Pacific. In this so-called California Floristic Province (Raven & Axelrod 1978), the serpentines recur with increasing frequency and magnitude from Santa Barbara County north to southwestern Oregon (Fig. 1). Aside from a moderate display of ultramafics in central Oregon, the next major occurrences are in western and central Washington; other ultramafic outcrops are known from British Columbia, and Alaska. This review will deal with the floristics and vegetation of the three Pacific Coast states and the province of British Columbia. There appears to be no biological information for the Alaskan ultramafics.

The ultramafics of the region occur in a variety of landforms and lithology. By far the most frequently encountered is serpentinite, in highly local to massive, extensive displays at low elevations in the Coast Ranges and foothills of the Sierra Nevada (Fig. 2). Nearly all of the California and southwestern Oregon occurrences are of this metamorphic lithology. Peridotite, and to a lesser extent, dunite, are more common in the northerly occurrences (northern California to British Columbia). Yet all locales of ultramafics show some degree of serpentinization.

There seems to be good agreement now on the origin of ultramafics in the California Floristic Province (Coleman & Irwin 1977). Plate tectonic theory accounts for these ultramafic exposures; oceanic plates, composed of mafic and ultramafic materials of the earth's mantle may ride over the continental plate, instead of being subducted under it. When this happens, the ultramafic material from the oceanic plate becomes exposed (Coleman 1977). Similar explanations are beginning to be invoked for the origin of the ultramafics north of the California Floristic Province.

The ultramafics of western North America occur in several climatic regimes, each characterized by a regional vegetation type. Throughout

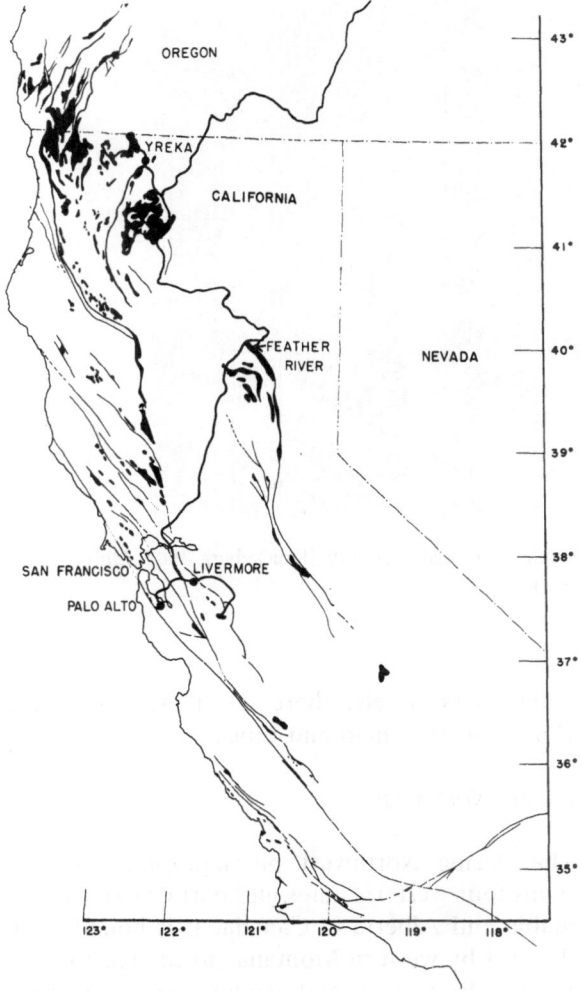

Fig. 1. Map of ultramafics in southwestern Oregon and California (Kruckeberg (1984), University of California Press).

much of California, where a Mediterranean climate prevails, most ultramafic rocks occur in regions with the chaparral, chaparral-woodland, or annual grassland types. Outcrops in northwestern California and adjacent southwestern Oregon, where a maritime climate dominates, are either in redwood forest or in mixed conifer-hardwood

forest. Ultramafics of Washington and British Columbia are mainly within the conifer forest biome.

The ultramafics of the California Floristic Province (cismontane California – west of the Sierra Nevada-Cascade axis – to southwestern Oregon) support a unique flora, rich in endemic taxa. Perhaps nowhere else on the planet but New Caledonia, does the 'serpentine syndrome' (Jenny 1980) make such a striking display. The vegetational response to ultramafics here is sharp and dramatic – with forest giving way to serpentine chaparral, and chaparral giving way to serpentine grassland or to barren 'moonscapes' nearly devoid of plant life (Fig. 2). To the north, in Central Oregon, in the mountains of Washington and British Columbia, the serpentine effect is diminished – few endemic taxa and only a quantitative shift in forest vegetation, often from productive to unproductive conifer forest (Fig. 3).

Ultramafic soils that develop in western North America usually have poor profile development, have a pH range from 6 to 7.5, a reasonably high C.E.C., and a high Mg/Ca quotient. Tissues of plants on these soils usually have high concentrations of magnesium and iron; a few taxa have been shown to be hyperaccumulators (*sensu* Brooks *et al.* 1979) of nickel. The morphological response to serpentine has not been examined, as it has in Europe, though it is well-known that some wide-ranging species are ecotypically differentiated into serpentine-tolerant and serpentine-intolerant races.

Our knowledge is still incomplete about the plant response to ultramafic soils in western North America. Much more can still be extracted from ultramafic occurrences in western North America. Besides floristics, biosystematics and ecological studies of ultramafics, especially in California, a new approach invites future investigation. Resistance, or tolerance, to ultramafic soils, while it can be cast in morphological and physiological terms for individual plants, is fundamentally a response of biota at the population and molecular genetic level. So clear is the distinction between ultramafic and non-ultramafic existence, that its study at the molecular level is bound to yield understanding of the gene-to-environment system. Why plants grow where they grow, attacked at the molecular level, is ripe for investi-

Fig. 2. An ultramafic landscape in California, "Hill 1030" near Middletown, Lake County (Kruckeberg 1984, University of California Press).

gation, and the 'serpentine syndrome' can be the focus of its rewarding study.

The substance of this chapter comes chiefly from three works (Kruckeberg 1969a,b, 1984). A number of other recent publications are also cited. Plant names follow Hitchcock & Cronquist (1973) for the Pacific Northwest, and Munz & Keck (1959) for California and southwestern Oregon.

Geology

The ultramafics of western North America are confined largely to the western Cordilleran arc: the Sierra Nevada – Cascade Range – Canadian Coast Range axis, and westward to the Pacific Ocean. This puts the bulk of ultramafic rocks in the states of California, Oregon, Washington and the province of British Columbia. Only minor

outcrops occur elsewhere, as in Montana near Hamilton, Bozeman and Libby.

Pacific Northwest

The Pacific Northwest physiographic province spans four western states and part of British Columbia and Alberta in Canada. It is bounded on the east by western Montana (to the crest of the Rocky Mountains), and extends westward across Idaho to the Pacific Coastal states of Washington and Oregon. In Canada its northern limits are central British Columbia and south-west Alberta. There are four concentrations of ultramafics in this vast territory. The southernmost one, in the Klamath-Siskiyou Mountains of southwestern Oregon is considered an extension of the ultramafic system of northwestern California (the North Coast Ranges) and will be dealt with under the California Floristic Province.

34

Fig. 3. An ultramafic landscape in the Wenatchee Mountains, Washington (Kruckeberg 1969b).

The other Oregon occurrence is in the isolated mountain ranges east of the volcanic Cascades, in Grant County. These mountains are collectively known as the Blue Mountains. The main concentration of ultramafic rocks is in the western Blue Mountains, regionally known as the Aldrich Mountains – Strawberry Range chain bordering the John Day River valley. Recent re-evaluation of ultramafics in terms of ophiolites resulting from plate tectonic activity (Coleman & Irwin 1977), suggests this Canyon Mountain Complex (Thayer 1977) is of early Permian age. For the botanist, the best and most extensive exposures are at Baldy Mountain (mostly peridotite and serpentinite) and the precipitous slopes of serpentinite bordering Canyon Creek just south of Canyon City. An earlier view (Baldwin 1959) looked upon these ultramafics as intrusions into the metamorphic complex of the Strawberry Mountains, in the form of dunite, peridotite and pyroxenite, locally altered. Baldwin (1959) considered these intrusions to be early or middle Triassic in age. The

region's rocks were mapped in detail by Thayer (1956a, b, c). But more recently Thayer (1977) conceived of the Canyon Mountain Complex as having 'originated along a spreading ridge in the ancestral Pacific Ocean, in a part isolated from terrigenous sedimentation between Early Permian and Late Triassic time . . . . and the complex is believed to have assumed its present form and structure by Lower Permian time, although it probably was raised 5000 to 6000 m by faulting during the Late Triassic and Early Jurassic.'

This westerly portion of the Blue Mountains rises up out of the arid Columbia basalt plateau country, to summits up to 2134 m. The ultramafics are mostly elevated above the shrub-steppe environment and are thus in the less xeric environment of conifer forest. The whole area east of the Oregon and Washington Cascades is called the 'dry interior' (Franklin & Dyrness 1973), a region of reduced rainfall and seasonal temperature extremes.

The Canyon Mountain peridotite and serpen-

Table 1. Chemical analyses[1] of western North American ultramafic rocks

Sample & locality	SiO_2 (%)	Al_2O_3 (%)	Fe_2O_3 (%)	MgO (%)	CaO (%)
Serpentinite, Ingalls Peak, Washington[2]	40.9	2.8	9.6	34.9	3.6
Serpentinite, Twin Sisters, Washington[2]	38.6	1.1	9.6	34.9	3.6
Ultramafic rocks, Canyon Creek Complex, Oregon[2]	42.1	1.70	2.5 (4.8)[7]	35.6	5.6
Olivine mineral, Scott Mtn. (Trinity ophiolite), California[4]	41.2	–	(11.2)[7]	48.9	–
Dunite, Red Mtn. – Del Puerto, California[5]	39.0	0.04	2.8 (5.09)[7]	46.1	0.0
Serpentinite, Josephine County, Oregon[6]	43.5	7.6	7.6	23.2	8.9

[1] Total less than 100%, since only selected values listed.
[2] Coleman (1967).
[3] Thayer (1977).
[4] Lindsley-Griffin (1977).
[5] Himmelberg & Coleman (1968).
[6] Robinson *et al.*
[7] FeO values in parentheses.

tinite occur from the foothills of the mountains to the summit of Baldy Mountain (7043 ft; 2147 m). The peridotite is dominantly harzburgite (ca. 80% olivine) and podiform chromite (Table 1).

The ultramafics of Washington state are better known geologically. All of the Washington occurrences are north of Snoqualmie Pass (Lat. 47.5° N, 121.5° W), a boundary between the volcanic Cascades to the south and the complex lithology of the Cascade Range to the north. Along an east to west traverse across the north Cascades, the first ultramafics are in the Wenatchee Mountains, a southeasterly trending massif of the main Cascade Range (Fig. 4). These montane outcrops, confined to northern Kittitas and southern Chelan counties, cluster around the spectacular Stuart Range, a serrated massif of granodiorite (the Mount Stuart Batholith). 'Old altered volcanics (greenstones), sedimentary rocks, gneisses and schists, and acid igneous granodiorite border or even interfinger with the ultramafics. The region is thus lithologically rich and complex. The terrain is rugged, with steep slopes and high ridges that culminate in ultramafic

peaks of from 450 to 640 m in altitude (Earl, Navaho, and Ingalls peaks). The clearest and most spectacular contact between ultramafic and nonferromagnesian rock types is along upper Ingalls Creek where the east boundary of peridotite at the creek abruptly gives way to the massive granodiorite of the Stuart Range (Kruckeberg 1969b). Miller considered the greenstones and associated non-ultramafic rocks as metagabbros and metadiabases, intruding a '... metabasaltic cover of massive flows, pillow lavas, and pillow breccias, with interbeds of argillite, chert and coarse breccia.'

Miller (1980) interpreted the Wenatchee Mountain ultramafics as part of an ophiolite suite, of about 400 km[2] he named the Ingalls Complex, a thrust zone of late Jurassic age which formed a '... boundary between oceanic lithosphere and metamorphic continental crust.' Miller described the ultramafics of the Ingalls Complex as a sequence, from south to north, of rocks consisting of 'milonitic lherzolites and hornblende peridotites; ... coarse-grained lherzolites.'

The ultramafic rocks here are predominantly peridotite, pyroxenite and serpentinite. In some

Fig. 4. Map of Pacific Northwest ultramafics. 1, Bralorne area, upper Bridge River; 2, Olivine and Grasshopper mountains, upper Tulameen River; 3, Eholt, above Grand Forks; above Christina Lake; 5, Sumas Mountain; 6, Twin Sisters Mountain; 7, Fidalgo and Cypress islands; 8, Double Eagle Lakes, Stillaguamish River drainage; 9, Wenatchee Mountains; 10, Fields Creek-Murderers Creek area; 11, Baldy Mountain (Kruckeberg 1969b).

places serpentinite appears only at the contacts between massive peridotite and non-ultramafic rocks. Representative mineralogical and chemical analyses are given in Table 1.

This, the largest Washington occurrence of ultramafics, covering over $300 \, km^2$ of rugged montane to alpine terrain, is a landscape of high treeless ridges, scantily forested slopes in a cool temperate montane climate (Fig. 5). The country is drier than that of the main Cascade Range, immediately to the west. This 'rain-shadow' effect, quantified by del Moral (1976), results in a low rainfall in the Wenatchee Mountain segment of the eastern flanks of the Cascades. Thus at 1700 m, an elevation typical of ultramafic sites, the annual precipitation is 600 mm while at the same elevation the western slopes receive 2600 mm (much of it as snow).

The other major ultramafic occurrences in Washington state lie west of the crest of the Cascade Range, all north of latitude 47°30′ N. Iso-

lated but large outcrops occur in the mountainous terrain of the Stillaguamish River drainages, Snohomish County (Kruckeberg 1969b, Vance 1957). These are peridotite, frequently serpentinised. The most spectacular display in this region is further north in Skagit County (Fig. 6). The Twin Sisters Mountain, southwest of volcanic Mt. Baker, is the largest exposure of dunite in North America $(880 \, km^2)$. The isolated twin peaks, flanked by the South and Middle forks of the Nooksack River, expose their dunite from the base of the range at about 300 m to the bleak, reddish-brown summits at 2123 m, the south Twin Sister. The tectonic geology of this unique outcrop was interpreted by Christiansen (1971) as of mantle origin; the dunite was emplaced cold across the Shuksan Thrust Fault in Cretaceous time. The dunite consists of the mineral olivine, and also contains small inclusions of chromite. Though Cretaceous in age, most of the Twin Sisters terrain as a habitat for colonization, must have been covered by the continental ice sheet and alpine glaciers during the Pleistocene; possibly only the summit areas were exposed as nunataks.

Between the Twin Sisters and the San Juan islands to the west, the only ultramafic outcrop of note is at Sumas Mountain, an impressive, isolated westerly outlier of the Cascade Range. Sumas Mountain consists of dunite (Huntting 1961) from near its base to the summit (792 m).

Low elevation ultramafics occur abundantly in the eastern San Juan Islands of northern Puget Sound, Washington (western Fidalgo Island and Burrows Island in Island County and on Cypress Island, Skagit County). These ultramafic exposures are currently thought to be part of an ophiolite sequence (Brown 1977). The ultramafic component of the ophiolite consists of serpentinized peridotite (Brown 1977) or alpine peridotite (Raleigh 1965), and is of Jurassic age (Brown 1977). Most are at sea level, though the Cypress Island outcrop ranges up to the summit of the Island (466 m). Minor exposures, chiefly of serpentinite occur on the coastal mainland in the Chuckanut formations of Skagit and Whatcom counties (between Bellingham and Mt. Vernon, Washington); this area is east and a little north of the San Juan Islands.

In British Columbia, ultramafics extend in-

Fig. 5. Aerial view of ultramafics (foreground) surrounded by other rock types, Wenatchee Mountains, Washington. Kruckeberg (1969b).

termittently the full length of the province, west of the Rocky Mountains Belt (Holland 1961, Monger 1977). The alpine ultramafics in British Columbia are now considered to be parts of an ophiolite assemblage, emplaced in Triassic to late Jurassic time. The rocks are mostly peridotite, with some pyroxenite and dunite, all substantially serpentinized. Major examples of ultramafics occur in the Bridge River-Shulaps area in the southern sector of the Province, Cache Creek (south central B.C.), and in the northern parts of the province, mainly in the Nahlin, Atlin and Cassiar areas (see maps in Holland 1961, Monger 1977).

I have no botanical data on the Alaskan occurrences of ultramafics. Some minor outcrops occur

in southeastern Alaska (e.g., Baranoff Island) and in the Fairbanks area, but the major occurrences are in northern and western Alaska. These latter are viewed as parts of ophiolite assemblages (Patton *et al.* 1977). They occur in three separate belts, designated as the western Brooks Range, Yukon–Koyukuk, and Rampart ophiolite belts. The ultramafic components are dunite or peridotite, each serpentinized.

California

From both geological and floristic standpoints, I consider the ultramafics of southwestern Oregon to be continuous with the ultramafics of California and in the California Floristic Province (CFP)

Fig. 6. Dunite at Olivine Bridge, Twin Sisters Mountain, North Cascades, Washington (Kruckeberg 1969b).

of Raven & Axelrod (1978). The many ultramafic bodies of this region do not have a common origin, though by 1977, the ophiolite interpretation for them was largely accepted (Coleman & Irwin 1977). 'The term ophiolite is used not for a single kind of rock but rather for a particular sequence consisting from base to top, of peridotite, gabbro, diabase dike swarm, and pillow lava. These sequences are considered to be oceanic crust stranded on land (Coleman & Irwin 1977).'

The ultramafics of the region are distributed in two major south to north-trending assemblages, the Coast Ranges and the Sierra Nevada (Kruckeberg 1984). In the Coast Ranges west of the Great Valley of California, the main ultramafic exposures begin on Figueroa Mountain (Santa Barbara County) and are intermittent to continuous throughout the Coast Ranges into the Klamath–Siskiyou mountains of northwestern California and southwestern Oregon. These Coast Range ultramafics constitute their largest representation in North America, covering more than

3885 km². The vast majority are of serpentinite rock; lesser amounts of unaltered peridotite or dunite are confined to the northwestern sector. The ultramafic bodies occur from sea level to moderate elevations (788 m on Mt. Tamalpais, Marin County, California); high altitude exposures are confined to the northwestern California-southwestern Oregon sector (e.g., Snow Mountain, 2151 m; Mt. Lassic, 1790 m; Red Mountain, 1063 m; Mt. Eddy, 2751 m; and in Oregon, Vulcan Peak, 1491 m).

Ultramafics in the Sierra Nevada of California occur in the south (Tulare County) at low elevations (below 300 m) in the western foothills of the range; northward they increase in elevation to 1933 m, as at the North Fork Feather River serpentinites.

While all the ultramafic belts of the California Floristic Province (detailed by Kruckeberg 1984) have not yet been substantiated as parts of ophiolite suites, the examples given by several authors in the Coleman & Irwin (1977) review, provide a

39

wide selection of ultramafics throughout the Coast Ranges, the Klamath-Siskiyou mountains and the Sierra Nevada. Thus it seems likely that all the ultramafics of the region can be linked to the ophiolite mode of origin. Salient features of the known ophiolite sequences include: (1) Origins of Pacific Coast ophiolites. Like other places in the world, the ophiolites of California and Oregon '. . . are exposed along belts of intense tectonism . . . and are thought to mark the sites of ancient zones of interaction between oceanic and continental crust' (Coleman & Irwin 1974). (2) Time of Emplacement. Most are of Mesozoic (Jurassic) age, except for those in the Klamath Mountains. Where at least four subparallel ophiolite belts occur, each is younger in sequence from east to west. The oldest, the Trinity ophiolite, is Paleozoic (Middle Ordovician to early Devonian, Lindsley-Griffin 1977). (3) Lithology of Ultramafics and Associated Rocks – The ultramafics of the Trinity ophiolite consist of harzburgite and dunite, both partly serpentinized; they are associated with gabbro, pyroxenite, diabase and plagiogranite. Sedimentary and volcanic rocks occur with the ultramafics. In the northern Sierra Nevada, the ultramafics of the Feather River suite are metamorphosed alpine peridotite and dunite with some gabbro (Irwin 1977). Further south in the Sierra Nevada, the Kings-Kaweah Belt is composed of serpentinized dunite and harzburgite, and associated mafic rocks, bordered by Sierra Nevada granitics (Saleeby 1977). Coast Range ophiolites, mostly upper Jurassic, are allocthonous melanges within the Franciscan formation of diverse sedimentary, metamorphic and volcanic rocks; here the ultramafics are highly serpentinized (Irwin 1977).

The extensive ultramafics of southwestern Oregon appear to be a continuation of the arcuate, north-trending rocks of the Klamath Mountains in northwestern California. Indeed, their affinity with the series of ophiolite belts to the south is confirmed (Ramp & Peterson 1979). The region's lithology is thought to be formed by subduction of an oceanic plate along a continental margin. The many major exposures of ultramafics make up much of the geology of Josephine County and parts of Jackson and Curry counties, Oregon. The exposures range from sea level (near Pistol River, Curry County) through moderate elevations in the Klamath Mountain system, to the highest summits of the Klamath–Siskiyou massifs (Chetco Peak, 1420 m; Vulcan Peak, 1419 m; Eight Dollar Mountain, 1220 m; Red Mountain, 2013 m; and Oregon Mountain, 1069 m. Most ultramafics are serpentinized, '. . . altered from peridotites varying in composition from dunite to pyroxenite. The most common variety [of serpentinite] is harzburgite (saxonite) composed mainly of olivine and orthopyroxene (Ramp & Peterson 1979).'

Soils

Ultramafic parent materials make an immediate, direct impact on the nature of the soils they foster and admirably demonstrate the Jenny (1941) model of soil-forming processes. Parent material (p) is a crucial variable in the Jenny equation, $s = f$ (climate, organisms, topography, parent material, time). Soils derived from two adjoining and contrasting lithologies, such as serpentine and granite, will differ largely as a function of the differing parent material. Most ultramafic soils in western North America occur in mountains where rocks directly underlie the soil. Such primary soils formed in residuum from ultramafic rock will reflect their bedrock source material, especially in terms of their chemical composition.

Jenny's (1980) use of the term 'serpentine syndrome' for the generalized serpentine tessera (the prism of soil plus above-ground biomass) implies a more complex relationship. His 'syndrome' refers to the interplay among the physical, chemical and biological factors that yield a serpentine soil-plant landscape.

The Mediterranean climates of southern and central California exert a particular influence on soil genesis. Northwards the cool and moist coastal temperate climates of the Klamath–Siskiyou–Coast Range country change the conditions for soil formation, especially on non-ultramafic parent materials. Altitude, slope and other topographic variables further refine the conditions for making soils, but they all show common influences of the ultramafic rock: (1) Most ultramafic soils of the region have poorly developed profiles, A–C horizons of skeletal lithosols. Current soil classification distributes ultramafic soils among the following soil orders: Entisols, Inceptisols,

Table 2. Analyses of ultramafic soils in western North America

Sample No.[2]	Locality	CEC[1]	Ca[1]	Mg[1]	Ca+Mg	Mg/Ca	pH
87	Olivine Mtn., British Columbia	20.1	6.5	13.9	20.4	2.1	6.6
3	Wenatchee Mts., Washington	18.4	4.7	11.1	15.8	2.4	7.0
9	Wenatchee Mts., Washington	27.8	4.9	16.0	20.9	3.3	6.4
63	Twin Sisters Mtn., Washington	16.1	1.5	16.5	18.0	11.0	7.1
51	Baldy Mtn., Oregon	25.0	2.3	15.2	17.5	6.6	6.6
III-1	Nickel Mtn., Oregon	34.7	1.4	24.4	25.8	17.4	–
VIII-1	Waldo Hill, Oregon	41.8	3.4	27.7	31.1	8.1	–
138	San Benito County, California	15.0	2.4	11.4	13.8	4.8	7.2
135	Lake County, California	16.0	2.8	11.8	14.6	4.2	7.0
152	Del Norte County, California	14.0	2.8	7.1	9.9	2.5	7.1
150	San Luis Obispo County, California[3]	34.0	52.5	2.7	55.2	0.1	7.5

[1] Expressed as millequivalents $100 \, g^{-1}$, 2 mm fraction, oven-dry soil.
[2] Samples 3, 9, 51, 63, 87 from Kruckeberg 1969b; samples III-1, VIII-1 from Rai, *et al.* (1970); samples 135, 138, 150, 152 from Kruckeberg (1984).
[3] A non-ultramafic soil for comparison.

Alfisols, Millisols, Vertisols and probably Aridisols (Alexander *et al.* 1985). Most upland ultramafic soils would fall into the Entisol order (Jenny 1980). (2) The parent material influences the nature of the soil colloids, either by yielding new species of colloids (chlorites, smectites) or else the parent serpentinitic minerals persist in the soil as the cationic exchange colloids (Alexander *et al.* 1985). (3) Commonly, ultramafic soils of the region possess characteristic chemical attributes (Table 2): pH values about neutral, moderate to high CEC, high concentrations of exchangeable magnesium, low concentrations of exchangeable calcium, giving a Mg/Ca quotient greater than 1.0 (averaging 3.3). In addition, low concentrations of nitrogen, phosphorus combined with moderate to high concentrations of cobalt, chromium, iron and nickel and a possible deficiency of molybdenum. (4) Any or all of the above affect the plants and animals dependent on the soil.

County soil surveys have recognized serpentine soils in California and Oregon. Ultramafic soils may be designated as a particular soil series (e.g.,

the Henneke and Dubakella Series in California) or simply called Rock Outcrop. Every soil series with ultramafic parent material has been judged infertile and considered of limited value for timber or pasturage. Their primary utility is as land for recreation (i.e. hunting), wildlife and watershed. All but a very few of the named ultramafic soil series are primary or residual soils; the alluvial soils of the Venado Series in California are one exception.

The Pacific Northwest

Ultramafic soils of central Oregon, Washington and British Columbia are nearly all formed in place, most often in mountainous or hilly terrain. In Washington and British Columbia they have developed under a rainfall of 1000–2000 mm per year; the central Oregon serpentines weather in more xeric conditions (250–750 mm per year), with hot dry summers and cold winters. These residual and colluvial soils usually have little profile development. The upper horizon is shallow and abruptly terminates at the interface with the

Table 3. Total chromium, iron and nickel in ultramafic soils ($\mu g g^{-1}$, oven-dried soil)

Source	Ni	Cr (ppm)	Fe
New Idria, California[1]	2780	1640	35000
Mt. Tamalpais, California[1]	3100	1700	52300
Nickel Mtn., Oregon[1]	2980	1730	71700
Canyon Creek area, Oregon[1]	1430	920	50200
Wenatchee Mts., Washington[1]	2020	895	39500
Santa Barbara County, California[2] ($n=6$)	1530	2080	–

[1] White (1976).
[2] Woodell *et al.* (1975).

bedrock of the C horizon. Table 2 gives some attributes of representative examples of these soils. Their pH values are around 7, most have moderate CEC, but show the typical high Mg/Ca quotients. Total concentrations of chromium, iron and nickel are high for some samples (Table 3). Non-ultramafic soils in similar terrain may have much the same physical attributes as their nearby ultramafic soils; the critical differences are in the chemistry and the consequent biological impoverishment of these residual soils (Kruckeberg 1969b). Nearly all these ultramafic soils occur in the conifer biome setting; their tree cover is invariably diminished in comparison with nearby non-ultramafic sites.

There appear to be no ultramafic soil series recorded for the Washington and British Columbia ultramafics. The soil survey for Grant County, Oregon (Dyksterhuis 1981) lists the Lemonex Series, an upland soil on the ultramafics of the Canyon Mountain Complex. The surface layer (20 cm) of this soil is a black stony clay loam; the upper 12 cm of subsoil is a very dark grayish brown gravelly clay; and the lower 35 cm is an olive gravelly clay over serpentine bedrock (68 cm). Management prescribed for this soil type is as a timber resource, for grazing and for its wildlife value.

California and Southwestern Oregon

Nearly all the ultramafic soils of this region are upland in origin, formed in place over ultramafic bedrock. Several soil series with ultramafic genesis are listed in the USDA-Soil Conservation Service County Soil Surveys. Residual (upland) series include the Henneke, Montara, Dubakella, Climara, Delpiedra, Fancher, Ipish, and Weitchpec series. Most common are the Henneke Series in central California and the Dubakella series in northwestern California and southwestern Oregon. Only four basin or alluvial series with ultramafic origins are identified; they are the Polebar, Venado, Maxwell and Conejo series.

Three major serpentine vegetation types are associated with these soils, each with its own characteristics: mixed conifer-woodland (northwestern California and southwestern Oregon), soft and hard chaparral (southern and central California), and serpentine grassland (Sierra Nevada foothills and Coast Ranges of California). Some exposures may be devoid of vegetation or only have a scanty herbaceous cover and their soils are mostly classified as lithic Argixerolls (Kruckeberg 1984).

The few studies of the properties of the ultramafic soils of the region, show a pattern. The transformation of ultramafic rock to soil can involve several things: (1) Ferromagnesium silicate minerals usually persist but may change quantitatively and mineralogically. (2) The soil usually has less magnesium than does the parent rock (Wildman *et al.* 1968a,b). The depletion of magnesium during weathering is apparently caused by dissolution and leaching in a CO_2-rich aqueous environment. (3) The near neutral pH values of the ultramafic parent material are sustained to a degree on the derived soils (Table 2).

The fate of serpentine minerals during soil genesis has been studied in soil profiles from this region. An earlier view (Proctor & Woodell 1975) was that some ultramafic soils underwent little modification of the minerals of colloidal particle size. The same phyllosilicate minerals, chrysotile and chlorite, were said to remain as the clay fraction of the soils; in other instances it was judged that new minerals of the montmorillonite type would form. The current interpretation regards the original phyllosilicate minerals of the rock to be transient, and highly susceptible to weathering (Alexander *et al.* 1985). Montmorillonite (smectite) is the common clay mineral class of ultramafic soils in California. Iron-rich montmorillonites tend to predominate in upland ultramafic soils while magnesium-rich montmorillonites prevail in basin positions adjacent to upland ultramafic soils (Senkayi 1977).

Colluvial ultramafic soils throughout the region are shallow, stony clay loams with little profile development. Koenigs *et al.* (1982) classified a number of these soils in Lake County, California as '. . . clayey serpentinitic, thermic, Lithic Rhodoxeralfs, Haploxeralfs, and Argixerolls (Soil Survey Staff 1972).' Lithic Rhodoxeralfs were associated with endemic cypress stands (*Cupressus sargentii*), while ultramafic soils of hard chaparral communities were identified as Lithic Argixerolls. An ultramafic soil, mapped as Sebastian silt loam, from the Klamath Mountains of Oregon was described by Istok & Harward (1982) as a '. . . loamy skeletal, serpentinitic, mesic Hapludalf;' this montane colluvial soil supported a cedar stand (*Chamaecyparis* sp.).

Ultramafic soils of California and southwestern Oregon are notorious for their propensity for rapid and massive erosion. All Soil Surveys call attention to this problem; severe sheet erosion can occur on undisturbed sites, and can be exaggerated where the sites are disturbed by road-building, logging, or fire.

Soil chemistry shows a consistent pattern. Low to moderate CEC, with the typical high Mg/Ca quotient (greater than 1.0). Representative soil analyses are given in Table 2. A few of the soils have been analyzed for cobalt, chromium, iron and nickel. Though the concentrations of these elements can be high, they are more variable than those of calcium and magnesium. Exchangeable cation analyses: Gordon & Lipman 1926, Kruckeberg 1951, Koenigs *et al.* 1982, McMillan 1956, Proctor & Woodell 1975, Robinson *et al.* 1935, Vlamis 1949, Walker 1954, Walker *et al.* 1955, Waring & Major 1964.

Plant physiology and morphology

The dramatic effects of ultramafic geology on plant life in many parts of the world provoke many questions. What factor or factors operate, singly or in concert, to markedly alter vegetation and to stamp the floras of ultramafics with endemism, indicator species, edaphic races and other floristic novelties? The 'serpentine syndrome' (Jenny 1980), so extreme a manifestation of the interplay between plant and substrate, has invited analysis of its components. Which of the three sets of attributes that create a serpentine habitat – physical, chemical or biological factors – is crucial in causing the serpentine effect? Better yet, ask how do these three components interact to give a serpentine syndrome its characteristic manifestation?

Scientists in North America have addressed questions about plant physiology and morphology on ultramafic soils since the turn of the century. Much of the work has focussed on the plant response to the unusual chemistry of ultramafic soils. Tissue and soil analyses, nutrient culture work and field or greenhouse trials with fertilizer amendments all have contributed to understanding the causes of the serpentine syndrome. The story in North America begins with the early work of Loew & May (1901) on the effects of high magnesium and low calcium. Wherry's (1932) studies of ash composition and the inquiry by Gordon & Lipman (1926) on the causes of infertility still left the problem unresolved. It was the landmark papers by Mason (1946a,b) on the soil specificity of certain narrow endemisms in California and elsewhere that stimulated much work in the period 1948 to 1955. Besides the vegetation studies of Whittaker (1954, 1960), and the detection of racial variation in resistance (tolerance) to serpentine (Kruckeberg, 1951, 1954), Mason's stimulus was felt by plant physiologists and soil scientists (McMillan 1956, Walker 1954, Walker *et al.* 1955, Vlamis & Jenny 1948).

Mineral nutrition studies

Plant growth on ultramafic soils can be looked at in terms of the nutrient status of the soils. Both native and crop plant species intolerant to ultramafics are thought to be either deficient in essential nutrients or susceptible to toxic levels of magnesium and heavy metals. Walker's studies (1948 1954, Walker *et al.* 1955) on the nutrition of serpentine and non-serpentine species provided the first experimental support of the importance of nutrition.

Walker (1954) discussed five hypotheses that

might explain the resistance or lack of it of plants to ultramafic soils: (1) Deficiency of major nutrients, nitrogen and phosphorus especially, contributed to poor growth. But poor growth still occurred on ultramafic soils which had been fertilized with these elements. Good growth could only be obtained by providing a low magnesium/calcium quotient. (2) Gordon & Lipman (1926) gave the high alkaline reaction of ultramafic soils as the cause of infertility. Because not all ultramafic soils are alkaline, and because some non-ultramafic but alkaline soils of California are quite fertile, Walker rejected this hypothesis. (3) Walker's (1948) discovery of the first molybdenum deficient soil for North America, was based on greenhouse tests of California ultramafic soils, using tomato plants for the bioassay. Walker felt that there was as yet no clear evidence that molybdenum was deficient in the field, either for crop plants on ultramafic alluvial soils derived from serpentine or for native species resistant to ultramafic soils. (4) Robinson *et al.* (1935) had suggested that nickel and chromium toxicities were possible causes of infertility. Walker (1954) found that under conditions of optimal macronutrient supply, intolerant species grew well on experimentally modified ultramafic soils that still retained nickel and chromium. Walker concluded that although he could not induce nickel and chromium toxicity in the greenhouse, the possibility of this toxicity could not be ruled out.

Thus, of the possible causes of impaired growth of non-serpentine species as well as adaptation of serpentine taxa, Walker's (1954) fifth hypothesis singled out low calcium of the soils as the critical factor. Plants native to California ultramafic soils (e.g., two species of *Streptanthus* (Cruciferae) and *Helianthus bolanderi* Gray subsp. *exilis* (Compositae)), seemed adapted to the low soil calcium by a greater absorptive capacity for the ion (Fig. 7). The calcium connection has been confirmed by others (Vlamis & Jenny 1948, Walker *et al.* 1955, Kruckeberg 1954, and Main 1974).

Yet there is growing evidence that calcium deficiency is not a universal determinant of infertility on ultramafic soils. Turitzin (1981) showed that nitrogen and phosphorus limited growth of herbs on the ultramafic soils of Jasper ridge near Stanford, California. He could detect little or no effect on growth from additions of calcium, potassium or sulphur. Proctor (1970) has discussed the role of calcium and has concluded that since it is now known to be required in micro-quantities by plants and that acidic soils have no unusual vegetation, calcium deficiency is unlikely to explain the peculiar ultramafic flora. Proctor *et al.* (1981) have analyzed soil solutions and found very low calcium concentrations in some of the ultramafic examples and admitted that true calcium deficiency is a possibility. However, the toxicity of magnesium and nickel, which is severe at low calcium concentrations, is always a more likely explanation of acute effects on plant growth. Hence the roles of high magnesium levels, low nitrogen and phosphorus, molybdenum deficiency, heavy metal toxicities cannot be ruled out. Plant existence on serpentine should be framed in Jenny's (1980) context of the 'serpentine syndrome.' The survival on serpentine soil then is viewed as a summed response to several factors, physical and chemical. And of course, tolerance to ultramafic or any other habitat is a product of evolutionary accommodation to all the relevant environmental factors. This holistic view – that the total is greater than the sum of its parts – applies to ultramafic biota in two ways. One is the influence of several causal factors operating at any given site, and the other sees different specific environmental factors exerting a dominating influence at particular ultramafic locales. Both facets serve as cautionary reminders that no single factor should be claimed as the decisive determinant for survival.

The high values of exchangeable magnesium in most serpentine soils, of western North America make this cation suspect as another factor in creating the serpentine syndrome. How do plants cope with the exceptionally high concentrations of magnesium? Is magnesium ever toxic? There are no papers on magnesium toxicity for serpentines of western North America, although Proctor (1970) has evidence for its toxicity in a Scottish ultramafic soil. He (personal communication) has shown acute magnesium toxicity in *Avena sativa* grown in soil from the ultramafic barren area on Clear Creek, San Benito County, California. A requirement for high concentrations of magnesium has been suggested in two studies. Madhok & Walker (1969) found that the serpentine endemic,

Fig. 7. Growth of two sunflower species on ultramafic soil modified to varying levels of calcium saturation (from left to right). Field soil at left. Upper: Common sunflower (*Helianthus annuus*); lower: *H. bolanderi* (serpentine species) (Walker *et al.* 1955).

Helianthus bolanderi ssp. *exilis*, could tolerate high concentrations of external magnesium, even though internal values were not high. Apparently this endemic sunflower requires high external magnesium to achieve sufficient internal concentrations. A similar behavior for magnesium was found by Main (1970); the grass, *Poa curtifolia*, endemic to the ultramafic soils of the Ingalls Complex, Washington, appeared to have a high requirement for magnesium. Clearly the nature

of the role of magnesium in the well-being of serpentine plants merits further study.

Cobalt, chromium, iron and nickel can occur in high concentrations in the ultramafic soils of western North America (Table 3), as they do elsewhere in the world. However there has been no decisive evidence that shows any of these elements affects plant growth in western North America. Nickel might be implicated in the serpentine syndrome, especially in ecotypic toler-

Table 4. Co, Cr, Fe and Ni concentrations in above-ground tissues of plants from ultramafic soils in western North America (μg g^{-1} oven dry matter, mean values, ranges in parentheses)

Source	Ni	Cr	Co	Fe
Four woody spp., California (Koenigs *et al.* 1982)	10.1 (4.5–14.0)	0.66 (∅.38–0.89)	–	–
Eleven woody spp., British Columbia (Brink *et al.* 1972)	14.8 (9.2–25.2)	1.02 (0.4–3.0)	0.0	21.1 8.4–45.1
Nine herbaceous spp., British Columbia (Brink *et al.* 1972)	19.2	0.88	0.43	31.25
Eight spp., California (Funkhouser 1980)	0.04 (0.0–0.58)	– (0.0–1.7)	–	–
Thirty-four *Streptanthus-Caulanthus* taxa with < 1000 ppm (Reeves *et al.* 1981)				
15 taxa	39.4(5–95)	–	–	–
19 taxa	n.d.*			
Streptanthus polygaloides (Reeves *et al.* 1981)	8910	–	–	–
Fifty-four herbs, <1000 ppm Reeves *et al.* 1983	48.4 (1–257)	–	–	–
Thlaspi montanum >1000 ppm, *n*=3 Reeves *et al.* 1983	8220 (5530–11200)	–	–	–
Linanthus androsaceus *n*=5; Woodell *et al.* (1975)	29.4 (8.3–100)	6.7 (2.7–17.9)	–	–
Five California shrub spp. (*n*=112) Wallace *et al.* (1982)	9.4 (1.0–21.6)	0.8 (0.2–5.6)	–	–

* n.d.=not detectable; – (dash)=no data.

ance, in narrow endemism, as well as a cause of serpentine avoidance by non-resistant species. Values for nickel and chromium (Table 3) are typically higher on ultramafic than on non-ultramafic soils. Mercury might also be locally toxic, as it is often associated with ultramafic outcrops in California. Ore dumps and mine tailings around quicksilver mines in ultramafic belts are known to contain high levels of volatile mercury (Kazantzis 1980). For most plants, soil pH values may be too high to render the heavy metals available for uptake. Yet hyperaccumulator species seem to be an exception to this.

It has been shown that plants native to ultramafic soils vary greatly in their tissue nickel concentrations (Table 4). The New Zealand team of biogeochemists, Brooks, Reeves and associates, have analysed tissue (largely herbarium specimens) from a wide array of sites around the world. Some serpentine species appear to take up excessive amounts (>1000 μg g^{-1} = a hyperaccumulator species) of nickel, while others, often of the same genus, do not. Reeves *et al.* (1981) analysed several species of the western North American cruciferous genus, *Streptanthus*, some of which are narrowly endemic to ultramafic soils in California. Only one species, *S. polygaloides*, was a hyperaccumulator, with 8910 ppm of nickel. A wider survey of western North American taxa, both on and off ultramafic soils (Reeves *et al.* 1983) showed few hyperaccumulators. Out of fifty taxa of serpentinophytes, only two (two varieties of *Thlaspi montanum*, Cruciferae) were nickel hyperaccumulators. A comparable result was obtained by Samiullah *et al.* (1985) for tissue samples from plants on Washington state ultramafics. Only one out of forty-two taxa sampled was a hyperaccumulator: *Arenaria rubella* from the dunite of the Twin Sisters Mountain accumulated 1360 ppm.

The species-specific response to nickel is most puzzling. Why do some taxa accumulate nickel while others seemingly exclude it? Why are hyperaccumulators so prevalent only in certain genera like *Thlaspi* or *Alyssum*, or families like the Cruciferae or Caryophyllaceae?

Molecular genetic and physiological answers to these questions are needed in order to understand the evolutionary, ecological and systematic aspects of heavy metal response, both tolerance and accumulation. A promising molecular genetic approach is being taken with copper tolerance (Lolkema *et al.* 1984 and Rauser *et al.* 1980). These and other studies find that a metal-chelating protein, a metallothionein, serves to render copper inocuous to plant metabolism. However, little is known of the biochemical mechanisms by which the plants resist or tolerate nickel toxicity.

Crops and timber: fertility problems

Ultramafic soils are infertile for crop plants and in western North America some attempts have been made to improve fertility for agriculture. Martin *et al.* (1953) were able to alter the Mg/Ca quotients of a California alluvial ultramafic soil by adding gypsum. Only with 2880 kg of gypsum/ hectare could a short-term improvement be gained; only the top 15 cm had sufficient calcium for satisfactory growth of cultivated barley. Pot tests with *Trifolium subterraneum* by Jones *et al.* (1976, 1977) gave similar results. Added calcium supplied as $CaSO_4$ improved yield; phosphorus and sulfur, in the presence of added calcium, also increased yields. They did pot tests on over twenty different samples of lithic Argixerolls of the Henneke series, ultramafic soils of primary origin in California.

Elsewhere in western North America, the other references to the agronomic potential of ultramafic soils are in the various county Soil Surveys, prepared by the Soil Conservation Service. Resid-

ual ultramafic soils are fit only for limited livestock grazing; the Soil Conservation Service deems these soils to be of value only for watershed and wildlife. Timber is harvested from some, but in the main, '. . . serpentine areas within forest zones are classed as non-commercial and not included in the timber base of the local management units (Rai *et al.* 1970).' Regeneration is poor after cutting, in stands of Jeffrey pine (Rai *et al.* 1970). Attempts to improve upland ultramafic soils for grazing by brush removal and fertilizer amendments have been unsuccessful. Agronomists of the Hopland Field Station (a branch of University of California, at Davis) have conducted field trials where native chaparral is removed, the soils fertilized and then planted with forage species. Improved forage is only short-term; nutrient-enriched grasses either succumb selectively to wildlife predation or do not maintain good growth in subsequent seasons (M. B. Jones, personal communication).

A few studies with ultramafic soils in the Pacific Northwest have given similar results. Calcium and magnesium nutrition of Wenatchee Mountain ultramafic soils has been investigated. In a study by Walker *et al.* (1955), growth responses were reported for two non-serpentine species (*Helianthus annuus* and *Fagopyron esculentum*) and for the serpentine sunflower (*Helianthus bolanderi* ssp. *exilis*) grown on ultramafic soils with experimentally altered calcium concentrations. The two non-serpentine species were much depressed in growth at low calcium concentrations; the serpentine sunflower showed good growth, albeit with leaf symptoms of calcium deficiency at even the lowest calcium concentrations (2.8% of the base saturation). Using nutrient culture on sand, Main (1974) compared the growth of cloned serpentine and non-serpentine races of *Agropyron spicatum* at varying concentrations of calcium and magnesium. On low calcium (high magnesium) substrates, yields of the serpentine race were higher than those of the non-serpentine accession; further, Main found significantly higher concentrations of calcium in the tissue of the serpentine population when grown on low calcium. According to Proctor (personal communication), these experiments are all consistent with the view that it is magnesium that is toxic, more so at low concentrations of calcium.

Intraspecific variability on ultramafic soils

The work of Main (1974), gave the first evidence that a native grass species (*Agropyron spicatum*) can show differences in resistance to ultramafic soils. It is appropriate at this point, then, to review the genecological studies of serpentinicolous species, the edaphic counterparts of the classical work of Clausen, Keck & Hiesey (1940, 1948) on climatic races in California. 'Wide-ranging species (bodenvag = indifferent or ubiquist taxa) that live in a variety of environments may be expected to have either a broad phenotypic/genotypic tolerance, or to have evolved local and differential genotypic accommodation to each variant habitat (Kruckeberg 1984).' Races resistant to ultramafic soils, within dicotyledonous herbaceous 'bodenvag' species of California were reported by Kruckeberg (1951, 1954, 1984). Such races were found in *Gilia capitata*, *Salvia columbariae*, *Achillea lanulosa*, and others (Fig. 8). The one grass tested (*Sitanion jubatum*, Kruckeberg 1951) showed no differences in resistance for the several populations tested.

Similar studies (Kruckeberg 1967) demonstrated variation in resistance for several herbaceous Pacific Northwestern species (e.g., *Achillea millefolium*, *Antennaria racemosa*, *Fragaria virginiana* and *Prunella vulgaris*). Ecotypic differentiation in woody species is less clearly demonstrable. While Kruckeberg (1967) found differences in resistance to ultramafic soils in *Pinus contorta* from a range of provenance, no such differentiation into ultramafic and non-ultramafic races could be shown for *P. sabiniana* (Griffin 1965). Yet another species of pine, *P. ponderosa*, gives hints of racial variation. Jenkinson (1974) compared growth of yellow pine seedlings from a variety of ultramafic and granitic soils in California. Those plants collected from ultramafic sites appeared to grow better on ultramafic test soils than those from granitic sites. Jenkinson could not determine if the differences in resistance were simply due to inherent variability in all populations or whether the resistant progeny were the result of introgression with the serpentinicolous pine, *P. jeffreyi*.

If seed plants regularly show racial differentiation, then one might expect the same of microorganisms. Pegtel (1980) found that the bacteria (*Rhizobium lupini*) in the nodules of ultramafic-resistant (*Lupinus* spp. of the Wenatchee Mts, Washington) are resistant to high nickel concentrations. Pegtel suggested that resistance of *R. lupini* have been selected by ultramafic soils. That some microorganisms are not resistant to ultramafic soils was shown by the results of Tadros (1957) who found that damping-off fungi which cause decay in populations of the California annual, *Emmenanthe penduliflora*, are not present in the ultramafic soils where the *Emmenanthe* thrives. Some plants may not form nodules on ultramafic soils. The absence of nodules on *Ceanothus cuneatus* growing in ultramafic soils of southwestern Oregon (White 1967) suggests lack of resistance by the microorganism.

An ultramafic flora is made up of more than endemics. Many plant species that range widely over a variety of habitats, may also be found on ultramafic soils (Kruckeberg 1984). The few examples (cited above) that have been tested nearly always show an inherited resistance to the substrate while their non-ultramafic counterparts do not. It seems likely that most bodenvag (indifferent or ubiquist) species would be genotypically differentiated into resistant and non-resistant strains. Could this racial differentiation be a prelude to the initial stage in edaphically induced speciation? It seems plausible and may be inferred from examples from the California flora. The genera, *Ceanothus*, *Clarkia*, *Gilia*, *Phacelia* and *Streptanthus* have serpentine endemic taxa that have close relatives on non-ultramafic soils. Acquisition of resistance may be the first stage in the sequence of events leading to speciation. Of course, it is equally likely that ecotypic differentiation of wide-ranging taxa need not proceed beyond the stage of racial differentiation.

Resistance to ultramafic soils appears to assure long-term survival to not only endemic taxa but to racial variants of wide-ranging, edaphically versatile species. The resistance seems to be a manifestation of a genetically controlled capacity for growth and reproduction in a nutritionally adverse environment. Yield, vigour, growth rates, and continued reproductive ability are all components of inherited resistance. Enough endemic taxa and resistant races of bodenvag species now have been tested from ultramafic habitats of western North America to provide some general expectations. Annuals in such genera as

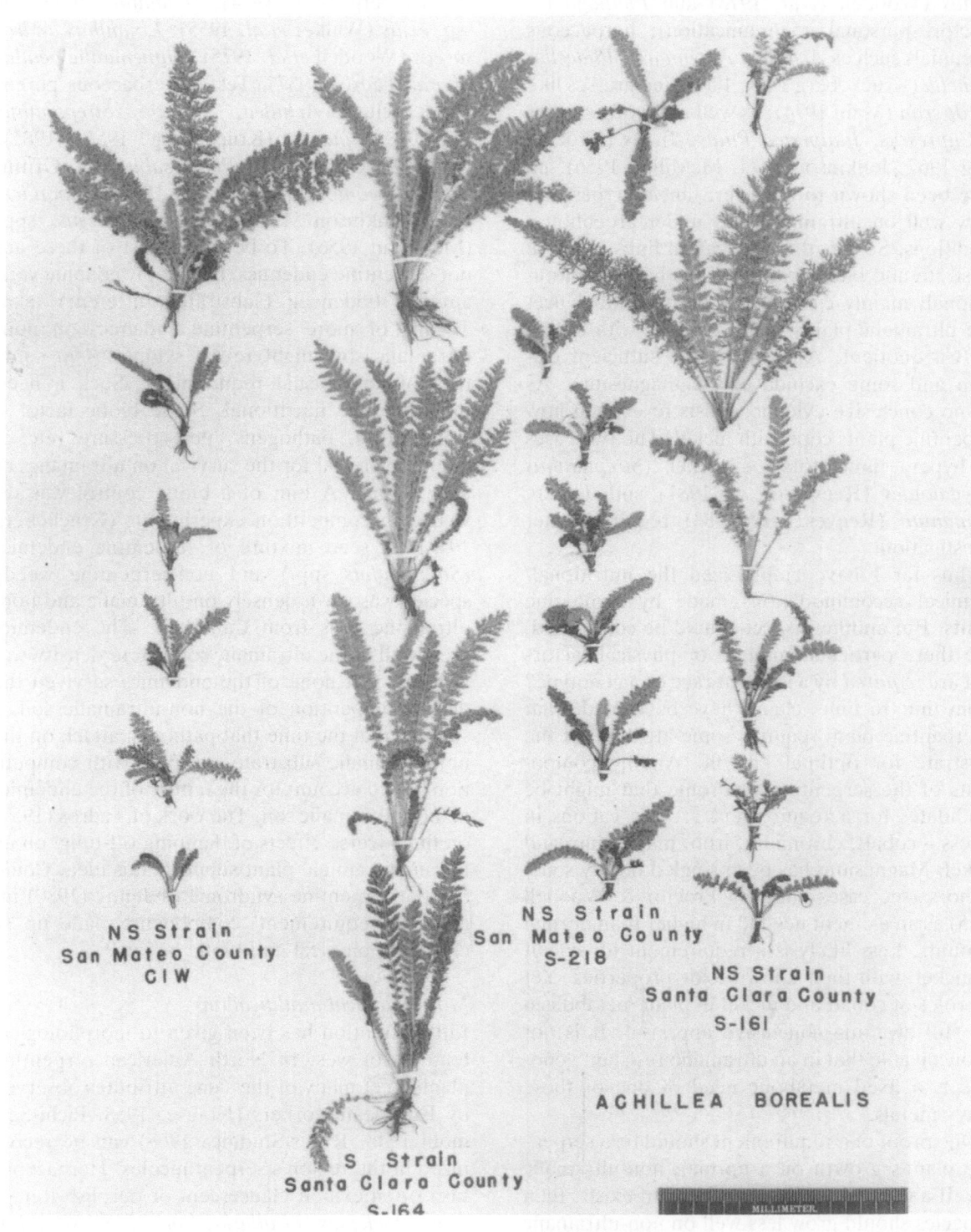

NS Strain
San Mateo County
CIW

NS Strain
San Mateo County
S-218

NS Strain
Santa Clara County
S-161

ACHILLEA BOREALIS

S Strain
Santa Clara County
S-164

MILLIMETER

Fig. 8. Racial differences in resistance to ultramafic soil, *Achillea millefolium.* Note single tolerant (preadapted) plants among non-serpentine progeny (Kruckeberg 1951).

Gilia, *Salvia*, (Kruckeberg 1951, 1954), *Strep-tanthus*, *Helianthus* (Walker *et al.* 1955); *Lin-anthus* (Woodell *et al.* 1975) and *Plantago* (J. Proctor, personal communication); herbaceous perennials such as *Achillea*, *Antennaria*, *Phacelia*, *Prunella* (Kruckeberg 1954, 1967) and grasses like *Agropyron* (Main 1974); as well as woody species in *Cupressus*, *Juniperus*, *Pinus*, *Taxus* (Krucke-berg 1967, Jenkinson 1974, McMillan 1956), all have been shown to have resistant biotypes that grow well on ultramafic soils under greenhouse conditions. So far, the only causal links between substrate and biological tolerance have been nu-tritional, mainly calcium/magnesium imbalance. The ultramafic plants thrive on soils with a high Mg/Ca quotient. All can extract sufficient cal-cium and some exclude excess magnesium. As yet no conclusive evidence exists to explain how serpentine plants cope with nickel. The two cases of hyperaccumulators of nickel (*Streptanthus polygaloides* (Reeves *et al.* 1981) and *Thlaspi montanum*, (Reeves, *et al.* 1983), require further investigation.

Thus far I have emphasized the nutritional/chemical accommodations made by serpentine plants. But another aspect should be considered. Are there particular mineral or physical factors that are *required* by a tolerant race or a genotype? From time to time, claims have been made that a serpentine plant requires some attribute of the substrate for optimal growth. Among compo-nents of the serpentine syndrome that might be candidates for a requirement are the cations in excess – cobalt, chromium, iron, magnesium and nickel. Magnesium has been singled out by some authors (see cases cited by Proctor & Woodell 1975) as an element needed in higher than normal amounts. Less likely is a requirement for cobalt or nickel, with their known toxic properties. Yet the roles of cobalt and nickel in plant metabolism are still awaiting conclusive appraisal. It is not inconceivable that in an ultramafic-resistant geno-type is a fixed metabolic need of one of these heavy metals.

One proof of a requirement should be a serpen-tine plant's growth on a normal, non-ultramafic soil. If a need for such a requirement exists, then a species should grow less well on non-ultramafic soils. Resistant plants have been tested on non-ultramafic soils repeatedly and have always grown better in them. Some sampled biotypes include annuals – species of *Gilia*, *Streptanthus*, *Salvia* (Kruckeberg 1951, 1954); *Helianthus bolanderi* ssp. *exilis* (Walker *et al.* 1955); *Linanthus andro-saceus* (Woodell *et al.* 1975) *Emmenanthe pendu-liflora* (Tadros 1957). Tested herbaceous peren-nials include *Achillea*, *Sitanion*, *Streptanthus tortuosus optatus* (Kruckeberg 1951, 1985); woody species include *Pinus sabiniana* (Griffin 1965), *P. contorta* (Kruckeberg 1967). *P. ponder-osa* (Jenkinson 1974), and *Cupressus* spp. (McMillan 1956). To be sure, most of these are not serpentine endemics, but simply edaphic vari-ants of bodenvag (substrate-indifferent) taxa. Testing of more serpentine endemics on non-ultramafic soil might reveal symptoms in some taxa of an unusual requirement. Such a need might not be nutritional. Some biotic factor – competition, pathogens, pest pressure, etc. – might be crucial for the survival on ultramafics of an endemic. A hint of a biotic control was re-vealed in competition experiments (Kruckeberg 1954). A seed mixture of serpentine endemics (*Streptanthus* spp.) and non-serpentine weedy species was sown densely on ultramafic and non-ultramafic soils from California. The endemics grew well in the ultramafic soil where density was reduced; but none of the endemics survived the dense competition of the non-ultramafic soil. I suggested at the time that pathogen attack on the non-ultramafic substrate, coupled with competi-tion, could account for the failure of the endemics on non-ultramafic soil. The work of Tadros (1957) on the adverse effects of damping-off fungi on an ultramafic annual plant supports this idea. Could not the serpentine syndrome of Jenny (1980) in-clude a 'requirement' component, made up of chemical, physical *and* biotic factors?

Morphological modifications
Little attention has been given to morphological features of western North American serpentine plants. Yet many of the same attributes observed by European workers (Krause, 1958, Pichi-Ser-molli 1948, Ritter-Studnička 1968) can be recog-nized in the region's serpentinicoles. Herbaceous taxa often exhibit glaucescent or purplish foliage (e.g., *Asclepias*, *Fritillaria*, *Lomatium*, *Mimulus*, *Streptanthus*, in Kruckeberg 1984). Dwarfism (nanism) and plagiotropism are also characteristic

of serpentine floras of the region. Woody species of serpentine chaparral formations are often sclerophyllous shrubs, with small leathery leaves. While these features are not unique to serpentine floras in the west, they do reach more extreme manifestations on ultramafic soils. Whittaker (1960) observed that several woody species in the Siskiyou Mountain flora were shrubs on ultramafic soils but trees on non-ultramafic substrates. Some of these shrubby variants are recognized taxonomically (e.g., *Lithocarpus densiflora* var. *echinoides*, *Quercus chrysolepis* var. *vaccinifolia* and *Q. garryana* var. *breweri*, while a shrub form of *Umbellularia californica* has not been named).

The physiognomy of serpentine vegetation, especially the woody components, can be used in mapping landscapes by photogrammetry, and more recently by remote sensing. The Remote Sensing Laboratory of the University of Washington (Department of Geology) under the leadership of Dr. John Adams, has been mapping vegetation of the Klamath-Siskiyou mountains, using satellite photos. Here, where contrasts between serpentine and non-serpentine landscapes are sharp, remote sensing can give an accurate assessment of vegetation. Part of this study has been to discover those morphological properties of the woody components of vegetation that give the plant cover its unique reflectance properties (S. Cooke, personal communication). Reflectance of visible and infrared light for serpentine versus non-serpentine habitats differs with respect to species composition and plant architecture (both for individuals and for the contrasting plant communities).

Vegetation

Landscapes with ultramafic rocks are unmistakeable, especially in their more extreme forms. They display barren stony ground with the glistening mottled rock exposed everywhere, with widely spaced, low-growing herbs as the only signs of plant life; and where woody vegetation is present it occurs as sparse open stands of shrubs and often with stunted conifers.

This review will summarize what is known about the vegetation ecology of serpentine for the Pacific Northwest (including central Oregon) and then for California and southwestern Oregon.

Serpentine vegetation of the Pacific Northwest Region

The several occurrences of ultramafic rocks in British Columbia, Washington state and central Oregon are intimately associated with the coniferous forest biome. From the crest of the Cascade Range westward, conifer forest on adjoining non-ultramafic soils produce impressive stands of Douglas fir (*Pseudosuga menziesii*), species of true fir (*Abies* spp.), hemlock (*Tsuga* spp.), and cedar species (*Chamaecyparis nootkatensis* and *Thuja plicata*. East of the crest, Douglas fir and yellow pine (*Pinus ponderosa*) dominate the non-ultramafic perimeter. This conifer setting for the ultramafics of the region is described by Franklin & Dyrness (1973). The conifer influence extends onto the ultramafics in a variety of ways: all the ultramafic soils of the region are stocked with conifers, but to varying degrees. Further, some species associated with coniferous forests occupy ultramafic soils; others are excluded from them. On the more extreme of ultramafics, a stunted, sparse remnant of the nearby forest appears, with much the same conifers. In other instances the open forest on ultramafic soils can have a unique mix of conifer species, some the same as those nearby, and others not in the adjacent forests (Kruckeberg 1969b and Franklin & Dyrness (1973)).

Del Moral (1972, 1974, 1982) used principal components analysis on field data to determine the relative importance of environmental gradients for the distribution of species. He looked at vegetation on both ultramafic and non-ultramafic sites in the Wenatchee Mountains, in much the same areas sampled by Kruckeberg (1969b). The interplay between substrate and moisture was found to operate differently on the vegetation of ultramafic serpentine versus non-ultramafic (sandstone) sites. Soil moisture was more important in the control of species distribution on ultramafic soils than on sandstone soils, where light was important. Ultramafic soils excluded non-resistant species, but those species that extended onto the ultramafic soils responded to the moisture gradient. Mesic species tended to be restric-

Fig. 9. Sparse, open stand of conifers on ultramafics, dense conifer forest on non-serpentine just beyond. Wenatchee Mountains, Washington (Kruckeberg 1969b).

ted to non-ultramafic soils, while xeric species usually occupied ultramafic soils.

In the Wenatchee Mountains, where treeless barrens occur on ultramafic soils surrounded by mixed conifer forest on non-ultramafic soils, the sere landscapes take on an alpine aspect (Figs 9, 10). Such vegetation is restricted to herbaceous perennials that scantily clad the bare ultramafic exposures; the widely spaced and sparsely popul-ated herb cover consists of 80% hemicrypto-phytes, 14% chamaephytes and 6% geophytes (Kruckeberg 1969b). Less severe mid-elevation ultramafic rocks can harbour exceptional mixes of conifer species. Subalpine species (*Pinus al-bicaulis*, *Abies lasiocarpa* and *Juniperus com-munis*) extend downward, and at the same time lower montane conifers (e.g., *Pinus ponderosa*, *P. contorta* var. *latifolia*, *P. monticola* and others) range upward on the ultramafics, beyond their usual altitudinal occurrences (Kruckeberg 1969b). Besides the stark physiognomy, species composition can also serve as a measure of the effect of ultramafic substrates in the Wenatchee Mountains. Both Kruckeberg (1969b) and del Moral (1974) noted that several species on non-ultramafic soils did not occur on nearby ultramafic soils. Certain families and genera, well repre-sented on non-ultramafic soils are missing or poorly represented on ultramafics (e.g., Scrophu-lariaceae, Leguminosae, Rosaceae, Liliaceae; *Potentilla*, *Lupinus*, *Trifolium*). Avoidance of ultramafic soils by species of the surrounding nor-mal soils has often been overlooked as one com-ponent of the unusual flora and vegetation.

Though few in number, endemic species can form a major component of serpentine vegetation in the Wenatchee Mountains. The serpentine en-demic fern, *Polystichum lemmonii* may form ex-tensive clumps in mesic rock outcrop areas, and three angiosperm herbs, *Poa curtifolia*, *Loma-tium cuspidatum* and *Chaenactis thompsonii*) may be the only plants on the most barren of outcrops.

Fig. 10. Sparse, open stand of conifers on dunite, Twin Sisters Mountain, Washington (Kruckeberg 1969b).

Other important components of the vegetation cover are the several woody and herbaceous species that are predominantly serpentinicolous in the Wenatchee Mountains but that grow on other substrates elsewhere in the Pacific Northwest. These can be regarded as indicator species for serpentine; examples are *Aspidotis densa, Salix brachycarpa, Ledum glandulosum, Eriogonum pyrolifolium, Arenaria obtusiloba, Ivesia tweedyi* and *Anemone drummondii.*

Conventional wisdom has it that azonal habitats, like those of ultramafic soils, are more stressful than nearby habitats on zonal (normal) soils. Therefore plants on ultramafic soils might be expected to show important differences in growth rates. Higgins (1984) has tested differences in productivity and competitive ability between plants of contrasting substrates in the Wenatchee Mountains of Washington, habitats described by Kruckeberg, 1969b and by del Moral, 1972. Higgins measured differences in productivity for *Abies lasiocarpa* by comparing water loss and

photosynthetic rates on the populations on ultramafic and zonal soils. He could find no consistent differences in water loss between the populations, but detected genetic differences in photosynthesis.

Higgins (1984) tested the relationship between competitive ability and stress tolerance by comparing serpentine and non-serpentine races of *Achillea millefolium* var. *lanulosa.* Attributes measured included rates of early growth, leaf production, total biomass, germination rate and seedling survival; competition was tested using deWit replacement plots. He hypothesized that plants from the zonal sites would be better competitors than those from serpentine. The zonal plants showed better competitive ability only at high densities and on fertilized substrates. He concluded that '. . . no functional relationship can be described between predicted traits of good competitors and their competitive ability.'

The conifer setting for serpentines of the Pacific Northwest suggests that a mycorrhizal deficiency may play a part in the reduction in the conifer presence on ultramafic soils. If the fungal component of the mycorrhizal association is intolerant of the serpentine substrate, depressed growth of conifers could result. A survey of the macrofungi in the contrasting edaphic habitats of the Wenatchee Mountains – the same sites studied by Kruckeberg (1969a) and del Moral (1982) – was made by Maas and Stuntz (1969). They found no significant differences in mycofloras of the contrasting substrates. Species diversity on ultramafic soils was like that of a typical forest community. Fewer fungal species were found on the ultramafic soils, with fifty-three species compared with 146 species on the non-ultramafic soils. There was no apparent avoidance of ultramafics by mycorrhizal fungi; in fact, 'more species tended to be mycorrhizal, lignicolous, or parasitic on serpentine than on non-serpentine soil.' The absence of a clear-cut sorting of the mycoflora by substrate may have a simple explanation. These fungi are inhabitants of the organic layers of the soil and thus may not be 'sensing' that portion of the soil derived from parent rock. Fungi that inhabit mineral soil (e.g., certain ascomycetes, phycomycetes and fungi imperfecti) are more likely to show degrees of resistance to ultramafic soils. There do not appear to be any studies on such soil fungi (but see Tadros 1957).

Serpentine vegetation in other parts of the Pacific Northwest has received no study beyond Kruckeberg (1969b). It exhibits some of the features encountered on the Wenatchee serpentines. None of the other areas have endemic species of flowering plants. Three ferns, *Polystichum lemmonii*, *Aspidotis densa* and *Adiantum pedatum* ssp. *calderi* appear commonly as indicators, as do a few other flowering plants.

Serpentine vegetation in California and southwestern Oregon

All serpentine areas in California and southwestern Oregon are encompassed by the California Floristic Province (CFP) of Raven and Axelrod (1978). It includes all of cismontane California to the Rogue River drainage in Oregon; only the California deserts and the country east of the Sierra Nevada crest are excluded. The ultramafics within this floristically rich area occur mainly in the Coast Ranges from Santa Barbara County north to the Klamath–Siskiyou mountains in the north, plus a substantial south-to-north sequence of exposures in the western Sierra Nevada. This large area (latitude 34°–43° N), with its topographic and geological diversity, includes several vegetation types (Barbour & Major 1977). Grassland, chaparral, oak woodland, mixed evergreen forest and coniferous forest – each vegetation type on normal soils is challenged by ultramafic soils throughout the California Floristic Province. Except for the Klamath–Siskiyou mountains, the ultramafic outcrops occur in a Mediterranean climate where winter rains yield to progressively more xeric conditions for the balance of a growing season. Thus the xeric character adds an additional stress. The resulting plant life on ultramafic soils is sharply delimited from adjacent vegetation.

Ultramafic outcrops often occur as islands or elongate peninsulas in a sea of non-ultramafic lithology (Kruckeberg 1984). Some outcrops – barrens – are devoid of plant cover; or, the barren may support a meagre herb flora, with widely spaced individuals. Still other outcrops may support serpentine versions of grassland, chaparral or evergreen forest, each with its characteristic physiognomy and mix of endemic, indicator or indifferent species. Contact zones between serpentine and non-serpentine are usually sharp, the creations of abrupt shifts in density and species composition.

Usually the serpentine – nonserpentine contrasts take the form of sudden shifts in life form and physiognomy. The most frequent shifts are from blue oak (*Quercus douglasii*), digger pine (*Pinus sabiniana*) to hard chaparral or a sparse grass – forb cover, or from chamise (*Adenostoma fasciculatum*) chaparral to a *Ceanothus – Arctostaphylos* scrub. Whittaker (1975) comments on this shift in life-form spectra as a worldwide phenomenon: 'In Quebec, Canada, the shift is from taiga to tundra; in Oregon, U.S.A., it is an abrupt transition from Douglas fir to open pine woodland; in California, it is from oak woodland to chaparral; in Cuba and New Caledonia, the shift can be dramatic, from tropical forest to savannah-scrub; and in New Zealand it is from southern beech forest to tussock grassland.'

Two serpentine plant communities – serpentine

Fig. 11a. Serpentine chaparral. Mt. Tamalpais, California (Kruckeberg 1984, University of California Press).

grassland and serpentine chaparral, are recognized by Barbour & Major (1977). 'Serpentine chaparral is an open, low type associated with serpentine soils from San Luis Obispo Co. northward through the Coast Ranges and foothills of the northern Sierra Nevada. The shrubs are characterized by apparent 'xeromorphism' (peinomorpism, serpentinomorphism) and dwarfed stature resulting from reduced productivity and growth (Whittaker 1954). The dominant shrubs are *Adenostoma fasciculatum* and *Heteromeles arbutifolia*, but noteworthy are several localized endemic shrub species, *Arctostaphylos viscida* and *Ceanothus jepsonii*.

"*Cupressus sargentii*, *Garrya congdonii*, and *Quercus durata* are unmistakeable 'indicator species' because of their typical restriction to, and numerical dominance on, serpentine soils (Kruckeberg 1954). Serpentine chaparral may be associated with foothill woodland (*Pinus sabiniana*) or montane coniferous forest (*P. jeffreyi*, *P. ponderosa*, *P. attenuata*, *Pseudotsuga menziesii*),

as an understory (Whittaker 1954). The thousands of hectares of serpentine chaparral in the North Coast Ranges are easily distinguished from the oak-grasslands on hills of non-serpentine origin (Walker 1954). Wells (1962) has described serpentine chaparral in San Luis Obispo Co., as has Vogl (1973) for the Santa Ana Mts., Orange Co." Figs 11 and 12 shows views of serpentine chaparral in the Coast Ranges.

Serpentines of the Coast Ranges and Sierra Nevada also support grassland, usually with their own unique floras. Howell (1970) describes the grasslands on serpentine from limited samples near sea level just north of San Francisco. These grasslands harbour some of the most restricted endemics of California: *Streptanthus niger*, *Castilleja neglecta*, and *Calochortus tiburonensis*. McNaughton (1968) compared a serpentine grassland with one on sandstone, just south of San Francisco. He found that the non-serpentine grassland had a greater productivity and higher annual yield of biomass, but that the serpentine

Fig. 11b. Serpentine chaparral. Western Tehama County, California (Kruckeberg 1984, University of California Press).

grassland showed greater species diversity. Productivity and species diversity of the sandstone grassland were correlated with the habitat gradient (cool, moist to warm, dry); no such relationship was found for the serpentine grassland.

Although chaparral vegetation is the commonest life-form to develop on serpentine, there has been little study of the serpentine variant of the chaparral in California. Gray (1979) examined vegetation sequences in the Coast Ranges and the Sierra Nevada. Breaks in the sequence at Snow Mountain in the North Coast Ranges were demarcated by lithology: chaparral woodland on lower elevation ultramafics abruptly changed upward to montane coniferous forest on non-ultramafic substrates; changes in both species composition and life-form coincided with the shift from serpentine to non-serpentine parent materials.

Areas in the South Coast Ranges have a great diversity of habitats generated by topography and substrate. Wells (1962) examined the patterns of woody vegetation in the diverse habitats of the San Luis Obispo region in relation to substrate and to the effects of fire. Ultramafic soil, one of the major substrates in the study area, elicits more of a floristic than a physiognomic response in the chaparral vegetation. Wells concluded that fire and other disturbances had so markedly altered the vegetation as to obscure the interplay of substrate and vegetation.

A curious contradiction exists in the literature on serpentine vegetation. On the one hand, it is often described as more xeric than neighboring vegetation on zonal sites; yet other authors (Wells 1962, Hardham 1962, Kruckeberg 1969b) have observed that ultramafic soils often support a mix of xeric and mesic, even riparian species. Certainly, in western North America, 'islands' of moist substrate are frequent; seeps, springs and permanent streams can occur with greater frequency on ultramafic substrates. Just how drought-stressed the ultramafic soil of a barren area may be is not known. Is the xeric 'look' of landscapes on ultramafics more the manifestation

56

Fig. 12. Nearly pure stand of *Cupressus sargentii* on serpentinite, Sonoma County, California (Kruckeberg 1984, University of California Press).

of physiological 'drought' due to nutritional imbalances than to actual moisture stresses?

Apart from Tadros' (1957) study of the effects of damping-off fungi on higher plants, I know of only one other study of the microbial flora. Lipman (1926) discovered that all three samples of serpentine soils from the Mount Diablo region (central California) contain very small populations of bacteria; further, no nitrifying bacteria could be isolated.

Forest vegetation on ultramafic soils of the CFP

That forests exist on ultramafics may seem to be a paradox, given the expected barrenness of many ultramafic outcrops. Yet forests on these substrates have received more attention than has the grassland and chaparral. The contrasts between regional forest on normal soils and those on ultramafics are dramatic. There are increases in the openness of the sparsely stocked stands, consequent reductions in biomass, and changes in the species composition of the forests. It was on ser-

pentine forest vegetation in the CFP that Whittaker (1954, 1960) tested his gradient analysis approach to vegetation. Whittaker's study area in the Siskiyou Mountains of southwestern Oregon is one of the richest botanical areas in western North America (Whittaker 1961, Wallace 1983). The ancient Klamath–Siskiyou mountain systems are a mosaic of topographic and lithological diversity, with a long history of continuous exposure. Coniferous forest dominates the region; Whittaker's intensive study of forest on diorite and on ultramafic rocks quantifies the dramatic visual differences. The ultramafic vegetation at low elevations tracks faithfully a moisture gradient from a mesic mixed conifer forest (Lawson cypress, Douglas fir, western white pine), '. . . through very distinctive forest-shrub stands with several conifers and two-phase undergrowth of sclerophyll shrubs and grass, to *Pinus jeffreyi* woodlands' (Whittaker 1960). Increase in light penetration within the more sparse, open tree canopy permits a much greater development of undergrowth (shrub and herb strata) than on the dior-

57

ite. 'Floristically, the undergrowth on serpentine in the Siskiyous is richer in species than that on diorite; in relation to physiognomy, the serpentine undergrowth is of greater coverage and more massive' (Whittaker 1954). Whittaker emphasized the evergreenness of the serpentine vegetation; deciduous trees and shrubs are virtually absent (Whittaker 1954).

Ultramafic soils create striking contrasts in the Pacific Coast redwood belt of northwestern California and were among the more effective environmental determinants of vegetation in the gradient study of Waring & Major (1964). Ultramafic soils of the Dubakella Soil Series support stands of incense cedar – Jeffrey pine in close proximity to mixed redwood – conifer forests on normal soils. Each of the four environmental gradients studied (moisture, nutrients, light and temperature) gave index values (see below) that clearly separated the incense cedar – Jeffrey pine stands from the redwood stands.

Calcium was the critical nutrient that gave a great difference in nutrient index values. However, Waring & Major emphasized the interactive nature of the four operational gradients in the above table. The resultant vegetation response is a net effect of each factor acting together.

Waring has used the environmental gradient approach to reveal the partitioning of forest types in the Klamath–Siskiyou mountains (Waring 1969, Waring *et al.* 1975). Among the several forest types he identified on the basis of moisture stress and temperature effectiveness, the Jeffrey Pine Type stands apart from all the rest by its fidelity to ultramafic soils. Douglas fir, incense cedar and the Siskiyou endemic, *Picea breweriana*, are often associated with the Jeffrey pine, but are not restricted to serpentine. Brewer's spruce is resistant to ultramafic soils despite its unusual chemistry, by succeeding under reduced competition. Waring suggested that the endemic spruce may have survived in post-Pleistocene times by occupying these soils.

Another conifer, Lawson cypress (*Chamae-*

cyparis lawsoniana) may be common on ultramafic soils for similar reasons (Zobel & Hawk 1980). Lawson cypress usually is confined to sites of low moisture stress (saplings with minus 11 bars at predawn) on both ultramafic and non-ultramafic sites. Many Siskiyou Mountain ultramafics have seeps and springy places here and there; the combination of wet ultramafic substrates and the absence of competition from other conifers seems to provide Lawson cypress a hospitable habitat. It is not known if these ultramafic-resistant populations are genetically predisposed for life on ultramafics.

The genus *Cupressus* in California is well known for its fidelity to unusual substrates (McMillan 1956). McMillan found that *Cupressus* species endemic to either sterile acid sands or to ultramafics were successful as seedlings on ultramafic soils. The common factor for those species of *Cupressus* in a genus that live on both acid and ultramafic soils might be the low calcium status of each substrate. MacMillan's results suggest that the species respond individualistically to different substrates, an indication that tolerance ranges vary even among species of the same species-complex.

Since Whittaker (1954, 1960), the only major study of serpentine vegetation in the Klamath–Siskiyou region was the intensive soil – nutrient – plant diversity analyses by White (1971). He examined thirty-six transects from non-ultramafic to ultramafic habitats in Josephine County, Oregon, by first recording species differences, then by evaluating soil and tissue nutrient changes across each transect. Floristic composition showed marked and consistent differences on either side of a contact zone. White discovered that calcium concentrations in the soil were more variable than those for magnesium, chromium and nickel. Ultramafic soils had both the lowest and the highest calcium concentrations. White used correlations of interactions among species composition, substrate and plant nutrient concentrations to reach the conclusion that species pre-

Index Values for

Vegetation type	Moisture	Nutrient	Light	Temperature
Incense cedar – Jeffrey pine	9.0	23.1	45.3	3.3
Redwood – Douglas fir	47.8	57.7	4.2	4.5

ferences were determined, not by calcium differences, but by concentrations of magnesium, nickel and chromium, in that order of importance.

R. B. Walker (personal communication) advises caution in regard to using soil calcium values. Cation availability is dependent on percent saturation of the base exchange complex by calcium and magnesium; as well, the availability of both nutrient ions and nickel is affected by pH. The degree to which the 'serpentine syndrome' manifests itself depends in part on the actual availability to plants of the mineral elements in soil and rock. Serpentine soils *do* vary widely in their ionic status, especially in calcium and magnesium.

Serpentine floras

The kinds of plants that can occur on ultramafics are inevitably influenced by the composition of the adjacent non-ultramafic flora. The surrounding flora has its effect on serpentine plant life in several ways. The climate of a region imposes tolerance limits on plants of ultramafic and non-ultramafic substrates alike. But within the regional climatic framework, the selective forces of the challenging substrate have resulted in stocking ultramafics with unique assemblages of species. In most instances, plants of ultramafics are related to taxa on the normal soils. The uniqueness of serpentine floras derives from particular mixes of four types of floristic elements (Kruckeberg 1984). These are: (1) endemics, narrowly restricted to serpentine; (2) local indicator species showing high fidelity to but not restricted to serpentine; (3) wide-ranging species that occur in a variety of habitats underlain by both normal (zonal) and ultramafic soils; (4) species that are excluded from ultramafics. I will review these four modes of floristic expression, first for the Pacific Northwest Region and then for the CFP.

Serpentine floras of the Pacific Northwest

These have been described in Kruckeberg (1964, 1969a,b). Only five species, all herbaceous perennials, are confined to ultramafic soils in the Pacific Northwest (Figs 13, 14). One, *Luina serpentina*, is from Grant County, Oregon; the others (*Poa curtifolia*, *Lomatium cuspidatum*, *Chaenactis thompsonii* and *Polystichum lemmonii*, are from the Wenatchee Mountains of central Washington. The fern, *P. lemmonii*, is a strict serpentine plant throughout the region from British Columbia (Cody & Britton 1984) to northern California; in fact it is the only serpentine endemic to occur in both floristic provinces. A novel serpentine form of *Adiantum pedatum* L., now recognized as *A. pedatum* var. *calderi* (Cody 1983) closely parallels in the western sector of its range, the distribution of *P. lemmonii*.

Approaching endemic status in the Wenatchee Mountains are *Salix brachycarpa*, *Claytonia megarhiza* var. *nivalis*, *Douglasia nivalis* var. *dentata*, *Cryptantha thompsonii* and *Chaenactis ramosa*. These species are mostly on ultramafic soils but can be occasionally found on nearby unstable talus of non-ultramafic origins. *Salix brachycarpa* has a remarkable distribution; it is discontinuous from eastern Canada to the Pacific Coast, occupying edaphically unusual habitats throughout its range (Hitchcock *et al.* 1964, v. 2). This low shrubby willow is only on ultramafic soils in central Washington. These species show a fidelity to ultramafics, intermediate between the endemic type and the local indicator type.

Peterson (1971) defined 'local indicator' as a class of '. . . widely distributed species which grow and predominate over an ore deposit.' The category is well represented in our area; high fidelity to ultramafic soils may involve extension of normal range on to these soils, beyond the main, non-ultramafic portion of the distribution; or, it may be manifested as a local increase in abundance on ultramafics of a species that is infrequent elsewhere; or, it may take the form of local fidelity of a species that is restricted to another edaphically unusual habitat elsewhere. Examples are best known from the Wenatchee Mountain ultramafic soils (Kruckeberg 1969b): The fern, *Aspidotis densa*, and several flowering plants (*Ledum glandulosum*, *Eriogonum pyrolifolium*, *Polygonum newberryi*, *Arenaria obtusiloba*, *Anemone drummondii*, *Thlaspi fendleri*, *Ivesia tweedyi*, and *Castilleja elmeri*) all exhibit this pattern.

Bodenvag (also called ubiquist or indifferent)

Fig. 13. Lomatium cuspidatum: endemic of Wenatchee Mountains ultramafics.

species are common on the regions' ultramafic soils. Some are known to be racially adapted to ultramafics and exhibit resistant and non-resistant genetic races (Kruckeberg 1967). Certain dominant species of the regional flora are indifferent to the substrate. Conifers such as Douglas fir, yellow pine, western white pine, lodgepole pine, common juniper, western yew, are frequent colonizers of less barren ultramafic outcrops. Shrubs like *Arctostaphylos nevadensis*, *Sorbus sitchensis* and *Vaccinium* spp. cohabit with the tree species on ultramafic outcrops. Bodenvag herbs may form a dominant component of some more favorable ultramafic sites; examples from Kruckeberg (1969b) and Main (1974) that are known to have ecotypic variants include *Agropyron spicatum*, *Achillea millefolium*, *Fragaria virginiana*, *Antennaria racemosa*, *Senecio pauperculus* and *Eriophyllum lanatum*. Other bodenvag species, not yet known to be racially adapted to ultramafic soils include *Silene parryi*, *S. menziesii*, *Thlaspi fendleri*, *Lupinus laxiflorus*, *Lomatium brandegei*,

Pyrola picta, *Arnica* spp., *Gilia aggregata*, *Hieracium albiflorum* and *Senecio integerrimus*. Bodenvag species are less likely to occur on the most barren of exposures where only the narrow endemics or the local indicator species are likely to be found.

What does *not* grow on serpentine is a part of the total floristic manifestation. The avoidance of ultramafics by a number of species potentially recruitable from adjacent non-ultramafic sites, changes the composition of the remaining serpentine flora. All life-forms – trees, shrubs, grasses and forbs – have species that avoid ultramafics (Kruckeberg 1969b). Avoidance is not absolute. Some taxa may rarely occur on ultramafic, yet others are never found there. Clear-cut 'avoiders' in the Wenatchee Mountains include the shrubs *Ceanothus velutinus*, *Acer glabrum* and *Pachystima myrsinites*; the ferns, *Cheilanthes gracillima* and *Cryptogramma crispa*; and herbs like *Sitanion jubatum*, *Penstemon* spp., *Balsamorhiza saggitata*, and two *Luinas*, *L. hypoleuca* and *L. nar-*

Fig. 14. Douglasia nivalis: endemic of Wenatchee Mountains ultramafics.

dosmia. The crustose lichen, *Rhizocarpon geographicum*, is so strict an avoider of ultramafic rock that it can be used to identify non-ultramafic erratics or other intrusive rocks in amongst ultramafics.

Serpentine Flora of the California Floristic Province (CFP)

The rich diversity of the flora of the CFP is well known (Stebbins & Major 1965, Raven & Axelrod 1978). 48% of the flora of the region is endemic (2133 endemic species out of a total of 4427 species, Raven & Axelrod 1978). Given the wide latitudinal and areal extent (1% of the CFP land area) of ultramafic exposures, as well as the many contacts with other lithologies, the significant ultramafic contribution to the overall floristic richness is to be expected. I calculated (Kruckeberg 1984) that 9 percent of the total endemism (189 taxa out of 2133 endemic species) is restricted to ultramafics. These ultramafics are enriched by species other than endemics; local or regional indicators and serpentine-tolerant ubiquists further embellish the plant life of the Province. A detailed account of serpentine floristics is given in Kruckeberg (1984).

Serpentine endemics

Although California botanists had been busy describing species endemic to ultramafic substrates since the late 19th century – often ignoring the substrate! – it was Herbert Mason (1946a,b) who first fully appreciated this narrow edaphic endemism. New serpentine endemic taxa, especially from northwestern California, are still being described.

The serpentine endemics of the CFP take the form of trees, shrubs, herbaceous perennials and annuals (Figs 15, 16). Only two tree species (*Cupressus sargentii* and *C. macnabiana*) are largely confined to ultramafics. The small number of shrubby endemics include *Quercus durata*, *Cean-*

61

Fig. 15. Berberis pumila (Del Norte County): endemic of California ultramafics.

othus jepsonii, *C. pumilus* and *Garrya congdonii*. The bulk of the endemics are annual and perennial herbs. Table 5 (modified from Kruckeberg 1984) summarizes the floristic and distributional features of the endemics. Some serpentine endemics can be very local, restricted to a small area or to even a single outcrop (e.g., *Calochortus tiburonensis*, *Streptanthus batrachopus*, *S. niger*. Others remain faithful to ultramafic soils over more extensive areas (e.g., *Streptanthus polygaloides*, *S. breweri*, *Ceanothus jepsonii*, *Quercus durata*). While serpentine endemics can be found in a variety of vascular plant families, there is a tendency for them to be well represented in certain genera or families. The Liliaceae (*Calochortus*, *Allium*, *Fritillaria*, *Lilium*) are well represented on ultramafics, while there are very few endemics in the Gramineae, Cyperaceae and Juncaceae, families richly diversified in California. Certain genera of dicotyledons (e.g., *Streptanthus*, *Lomatium*, *Phacelia*, *Linum* and *Navarretia*) show much endemism on ultramafics. Sixteen out of eighteen taxa (species, varieties and subspecies) of the subgenus Euclisia (*Streptanthus*) are serpentine endemics (Kruckeberg & Morrison 1983); eight taxa in subgenus Hesperolinon (*Linum*) are serpentine endemics (Sharsmith 1961). The significant taxonomic bias among serpentine endemics is recounted in Kruckeberg (1984).

Throughout the CFP, proximity to the Pacific coast, elevation and latitude all confer particular character on the serpentine floras (Table 5). Inland, from southern California to the interior (inner) North Coast Ranges, the lower rainfall exaggerates the xeric nature of serpentine plants. In more mesic coastal areas, plant cover on ultramafics increases and life-form diversifies. The area richest in endemics is in northwestern California and adjacent Oregon, where there are over sixty endemic taxa in the combined North Coast Ranges and the Klamath – Siskiyou mountain regions. Through the South Coast Ranges from Santa Barbara County to west-central California the many, often discontinuous outcrops support a large number of endemics, forty species. The ultramafics of the Sierra Nevada, though extensive, support the fewest number of narrow endemics, thirteen species.

Indicator species on ultramafic soils

Many plants, though not narrow endemics, show high substrate fidelity. One such group of species

Fig. 16. Eriogonum alpinum (Mt. Eddy): endemic of California ultramafics.

Table 5. Taxa endemic to the ultramafics within the California Floristic Province[1]

Region	Number of taxa	Examples
Sierra Nevada	13	*Allium sanbornii, Streptanthus polygaloides*
Sierra Nevada and Coast Ranges	13	*Garrya congdonii, Helianthus exilis*
Coast Ranges (inclusive)	192	*Cupressus sargentii, Allium falcifolium*
South Coast Ranges	40	*Streptanthus insignis, Calochortus obispoensis*
Bay Area	19	*Streptanthus niger, Clarkia franciscana*
Napa, Sonoma, Lake counties	27	*Streptanthus brachiatus, Senecio clevelandii*
North Coast Ranges	14	*Allium hoffmanii, Asclepias solanoana*
Klamath–Siskiyou area	19	*Phacelia dalesiana, Streptanthus barbatus*
NW California–SW Oregon	30	*Fritillaria glauca, Lilium bolanderi*
NW California–Pacific NW	2	*Polystichum lemmonii*

[1] 215 endemic taxa, including ferns, conifers and flowering plants; species and infraspecific taxa (from Kruckeberg 1984).

is clearly faithful to ultramafics in one part of their range, yet grow on other substrates elsewhere. A good example is *Pinus jeffreyi*, which is exclusively on ultramafics in the Klamath–Siskiyou country but is found mostly on granitic substrates in the eastern Sierra Nevada (Griffin & Crichfield 1972). Regional fidelity to ultramafics is also shown by *Calocedrus decurrens*, *Pinus attenuata*, *Quercus vaccinifolia*, and several herbaceous species (e.g., *Aspidotis densa*), *Darlingtonia californica* and *Xerophyllum tenax*; these species are common over a range of soil types, but in northwestern California and Oregon are usually restricted to ultramafic soils. Another type of indicator seems not to show the regional preference of the first group. Rather, the species are widespread throughout the California Floristic Province; they occur only on ultramafics when the surrounding vegetation is inhospitable. Thus, chaparral species like *Ceanothus cuneatus*, *Adenostoma fasciculatum* and *Arctostaphylos* species can be dominant on islands of ultramafic outcrop surrounded by mixed conifer – hardwood forests where the chaparral species are absent. *Pinus sabiniana* can also conform to the same pattern: locally absent from surrounding non-serpentine vegetation but present on ultramafic soils.

Indifferent or bodenvag species

Plant life on ultramafics can be enriched with those species that can be found on both sides of an ultramafic contact. On less barren outcrops some elements of the serpentine flora are clearly members of the regional flora, that elsewhere occupy a variety of substrates. Woody taxa of this capability include several conifers (Douglas fir, Lawson cypress, *Pinus lambertiana* and *P. monticola*); woody dicotyledonous plants such as *Arbutus menziesii*, *Rhododendron occidentale*, *Quercus sadleriana*, *Cercocarpus betuloides*, *Garrya fremontii*, *Calycanthus occidentalis*, and many herbaceous taxa (e.g., annuals like *Gilia capitata*, *Linanthus* spp., *Phacelia* spp., *Salvia columbariae*, *Cordylanthus* spp., *Mimulus* spp., *Collinsia* spp., etc. Many herbaceous perennials fit this same relationship (e.g., *Achillea millefolium*, *Eriogonum umbellatum*, *Thlaspi montanum*, *Scutellaria* spp., *Lupinus* spp., *Trifolium* spp., *Polygala* spp., etc.).

Mentioned earlier was Whittaker's (1960) observation of the bodenvag behavior of certain woody species; often taxonomically recognized, they are the serpentine counterparts of species occurring as trees on normal soils. While the bodenvag capability of many taxa does not (and need not) have taxonomic recognition, there is good reason to expect that genotypic accommodation has occurred within the taxonomic species. Ultramafic resistant races of bodenvag species have been discussed earlier; it need only be remarked that ecotypic differentiation for ultramafic resistance may well be the common means of achieving habitat plasticity for a species. Less often does a bodenvag species show no racial differentiation (e.g., *Pinus sabiniana*, Griffin 1965).

Exclusion – the avoidance of serpentine

Spores, seed, and other propagules must reach ultramafics intermittently from surrounding floras, yet the ultramafic soils retain their distinctive species compositions – endemics and indicators. Migrations across a contact are largely unsuccessful. Only a selected number of indifferent (bodenvag) taxa (just discussed) reveal the influence of normal floras nearby. Hence, a major feature of ultramafic habitats is their lack of species found just beyond the contact. The terms 'avoidance' and 'exclusion' (Kruckeberg 1984) embrace this notion of missing taxa.

In the CFP wherever forest, mixed woodland, chaparral or grassland encircle ultramafic outcrops, we see the evidence of exclusion. A majority of the species on the non-ultramafic substrates do not appear on the ultramafic soils. The transect studies of White (1971) in southwestern Oregon portray well this exclusion phenomenon; species composition changes dramatically along each of the 36 transects, from non-ultramafic and contact zone to ultramafic. In California, when chaparral occurs on either side of a contact, replacement of one species by another in the same woody genus may occur (Kruckeberg 1984). *Quercus dumosa* and *Q. wizlizenii* are replaced by *Q. durata* on serpentine; similarly *Ceanothus purpureus* is replaced by *C. jepsonii*, *Arctostaphylos canescens* by *A. montana*, and *Garrya fremontii* by *G. congdonii*. Other chaparral species may not be represented at all by their ultramafic counterparts.

The replacement phenomenon may not involve simply the substitution of a non-resistant species by a resistant one. Where the habitat (e.g., substrate) is intermediate in character, a population intermediate in resistance and in morphological features may develop by hybridization. This outcome is nicely illustrated by the case of the serpentinicolous *Quercus durata* and its congener of non-ultramafic soils, *Q. dumosa*. On an exposed ridge in central California with a nutrient-deficient *non-ultramafic* soil, Forde & Faris (1962) described a shrub oak population clearly introgressant between the two species.

As in the earlier description of excluded species on Pacific Northwest ultramafics, the exclusion on California – Oregon ultramafics can have a taxonomic bias. Certain genera (e.g., *Ranunculus*, *Delphinium*, *Phlox*, *Cryptantha*, *Astragalus*, *Penstemon*, *Lupinus* and *Trifolium*) or families (e.g., Ranunculaceae, Rosaceae, Leguminosae, Primulaceae, Scrophulariaceae) show a decided avoidance of ultramafic soils.

The avoidance of ultramafic soils must be the result of lack of resistance to these soils. But such an ecophysiological explanation begs for an evolutionary explanation, too. If we assume that resistance to ultramafics evolved from non-resistance by a change in genetic make-up, then avoiders of ultramafics must not have evolved the requisite resistance. Genetic preadaptedness to a new environmental challenge must be the usual beginning for an established resistant population; this should hold for resistance to ultramafics just as it does for resistance to DDT by houseflies. Since the raw material for such preadaptation (mutations, etc.) arises at random, it is not surprising that in some plants, the requisite preadaptations have not come about.

Weed species on ultramafics

The CFP has been inundated by a host of weeds, mostly of Mediterranean origin (Raven and Axelrod 1978). Some now occur on ultramafic outcrops, especially in disturbed areas (e.g., roadcuts, quarries, etc.). Annual grasses like *Bromus mollis*, *B. rubens*, *Festuca* (*Vulpia*) *megalura* and *Avena* spp. are frequent in such habitats (Proctor

& Woodell 1975); dicot herbs like *Erodium cicutarium*, *Hypochaeris radicata*, *Rumex acetosella*, *Cardamine oligosperma*, *Centaurea solstialis*, *Daucus carota*, *Spergularia rubra*, and others, also cohabit with native serpentine species under disturbance. It is likely that some introduced weeds are now ecotypically resistant to ultramafic soils as was shown for *Prunella vulgaris* and *Rumex acetosella* on Pacific Northwest ultramafics (Kruckeberg 1967). Remarkably few weeds occupy undisturbed ultramafic outcrops even when ample bare ground is available.

Fauna on ultramafics

The most dramatic biotic response to ultramafic outcrops is by plants. But the universal dependence of animals on primary producers leads us to expect that animals are influenced by ultramafic outcrops and their vegetation. Yet next to nothing is known about this dependence – interdependence of plants and animals on ultramafics. Proctor & Woodell (1975) end their review of the scanty literature by saying: 'Why the animals of serpentine have been so little studied we do not know, but they offer a fascinating field for further work.'

What information there is largely pertains to animals found on California ultramafics (Kruckeberg 1984); nothing has been published on the effects of ultramafics on animals for the Pacific Northwest.

Ehrlich and collaborators at Stanford, California, have focused on the ecology and population dynamics of the butterfly, *Euphydryas editha*. It is chiefly restricted to ultramafics in the Bay Area of central California, where it appears to be localized by its host plant, *Plantago erecta*, common on ultramafics. It could not be demonstrated that the butterfly is tied to the ultramafic soil through the chemistry – or availability – of the host plant. Rather, the race of *E. editha* appears to be suited to the low humidity and high temperatures of the ultramafic habitat, as well as to the absence of pathogens (Johnson *et al.* 1968).

The most remarkable association of fauna with serpentine flora is the dependence of certain pierid butterflies on serpentine plants in the cruciferous genus *Streptanthus* (Shapiro 1981). Certain species preferentially feed on *Streptanthus*. This close relationship has evoked an unusual response on the part of the host plant. Callosities on the margin of *Streptanthus* leaves look like insect eggs; these egg-mimicking structures serve to deter egg-laying on the host plant, thus reducing damage by herbivory.

There are two instances of vertebrate ecologies influenced by ultramafics. The pocket gopher (*Thomomys bottae*) uses the corms of *Brodiaea* spp. (Liliaceae) as its chief food; the authors (Proctor & Whitten 1971) report a higher than normal intake of magnesium, but do not comment on the possible metabolic consequences. Disturbance by pocket gophers of these same serpentine grasslands has been found to aid in perpetuating the floristic composition of this community type (Hobbs & Mooney 1985). Another vertebrate, the California salamander (*Ensatina eschscholtzii*), though widespread in California, shows local intergradation between subspecies on ultramafics (R. Stebbins 1949).

As Proctor & Woodell (1975) remarked, the ways in which animals may interact with serpentine and/or its flora have barely been explored. Herbivory of plants that accumulate magnesium and nickel must lead to some intriguing biochemical specialization in the herbivores. Other questions arise: Are there endemics among the fauna of ultramafic outcrops? Do edaphic races of wide-ranging animal species develop? The field for investigation of plant – animal interactions here invites further study.

Evolutionary considerations

The remarkable endemism of the CFP has stimulated speculation on the modes of origin of the endemic diversity of the region. I will review here the several models or hypotheses proposed to account for the evolution of serpentine endemics in California; the analysis is based on my recent critique (Kruckeberg 1984).

At the outset, we need to recognize three related problems. (1) An evolutionary consideration of serpentine flora consists of two facets: first, the evolution of genetic resistance proceeds initially at the microevolutionary level, in contrast with the genesis of distinct endemic species at the macroevolutionary level of speciation. (2) There is little likelihood that a single mode of origin for serpentine endemics can be supported (Stebbins 1980). (3) Do the unique attributes of serpentine plants come into being via unique evolutionary processes? Or, can the several known modes of speciation proposed for other biota serve to account for the evolution of serpentine taxa. We address these questions first by a historical retrospective, followed by a synthesis of current ideas, in the form of models, for the origins of serpentine species.

History of ideas on serpentine endemism

The papers by Stebbins (1942) and by Mason (1946a,b) set the stage for subsequent discussions on the origin of narrow endemics. Stebbins (1942) advanced the notion of the fixation of a reduced gene pool by biotype depletion as the cause of genetic delimitation and fixation of endemics. Much later, Stebbins (1980) modified the biotype depletion hypothesis to allow for those recently described instances where rare taxa did not show a reduced genetic variation. Yet, in some form, the biotype depletion hypothesis remains intact. Mason (1946a) advanced the idea of narrowed spans of genetically fixed tolerance as a crucial basis of narrow edaphic specialization. These landmark papers of the 1940s helped formulate possible solutions to the problem of evolutionary accommodation to serpentines (Kruckeberg 1951, 1954).

The problems of endemism in the California flora were subjected to the first full-scale analysis by Stebbins & Major (1965). They invoked the notions of paleo- and neo-endemics, substantiated by levels of polyploidy and geological age of endemic areas. No single mode of origin could account for serpentine endemics, a point re-emphasized by Stebbins (1980): 'Like every other problem of evolution, that of the nature and occurrence of rare species is not a simple one that can be solved by applying indiscriminately one or a few general principles.'

The latest major contribution to the evolutionary origins of California endemics by Raven and Axelrod (1978), devotes a chapter to edaphic endemism, emphasizing ultramafic areas. They categorize three groups of serpentine endemics, based on degree of restriction, life-form and age. Their third group, herbaceous species wholly restricted to ultramafics, contains some species whose origins may have come about by saltational speciation (Raven 1964). Drastic reduction of population size may coincide with major cytogenetic reorganization of the gene pool of the remnant population. Should such a marginal population persist on an ultramafic soil or other edaphically unique site, it may become in time a recognized edaphic endemic. Saltational species resulting from catastrophic selection, has been demonstrated only for species of *Clarkia*; Raven (1964) thinks it a suitable model for other herbaceous endemics (e.g., in *Streptanthus*, *Phacelia* or *Linum*).

Possible modes of origin

It is reasonable to expect that the first evolutionary step in adaptation to ultramafic soils is the acquisition of resistance to some facet of the serpentine syndrome. As discussed earlier, the genesis of racial resistance is well known for bodenvag species on and off ultramafics. Indeed such ecotypic differentiation is now well known for a similar edaphic stress: resistance to copper, lead and zinc (Antonovics *et al.* 1971). This resistance story may provide a useful model for the initial stages of origins of serpentine endemics. Further, the subsequent isolation of the edaphic races may follow the route elucidated for case-histories in the copper-lead-zinc resistance literature (Antonovics *et al.* 1971). We can construct a model of one mode of origin of serpentine endemics from this analogue. The model draws on components of similar models devised by White (1978), Antonovics *et al.* (1971), and Bradshaw (1976). *Stage 0.* Some preadaptation for serpentine tolerance exists in certain non-serpentine populations. *Stage 1.* Disruptive selection effectively separates a species into ultramafic-soil resistant and non-resistant gene pools. Jeffrey pine (*Pinus jeffreyi*) in the CFP illustrates this stage. Furnier & Adams (1986) have shown that one Sierra Nevada population on ultramafics is more similar in allozyme allele frequency to the ultramafic populations of the Klamath Mountains than it is to nearby non-ultramafic populations of the Sierra Nevada. *Stage 2.* Further genetic divergence for function and structure occurs within the resistant part of the effectively discontinuous population. *Stage 3.* Isolation between the serpentine and non-serpentine segments of the species becomes genetically fixed; the two populations are unable to exchange genes. *Stage 4.* Further reinforcement of genetic-ecologic isolation and consequent further divergence of the serpentine population may occur, generated by the initial genetic discontinuity. Such a model invoking disruptive selection at an edaphic boundary leads logically – and perhaps realistically – to consider the ultimate speciational event as sympatric (White 1978, Dickinson & Antonovics 1973).

Restriction to ultramafic soils may not be the final, persisting outcome of the speciational events. Habitats intermediate between ultramafic and non-ultramafic can be hospitable to introgressants arising from hybridization between the serpentine and non-serpentine species (Forde & Faris 1962).

Some possible modes of origin for ultramafic tolerance and for endemic speciation may have been acted out in certain California genera. *Streptanthus* (subgenus Euclisia) and *Linum* (subgenus Hesperolinon) appear to have taxa that exemplify various stages in the evolution of serpentine endemism (Kruckeberg 1984).

Preadaptation for tolerance to low calcium might have arisen as a response to calcium stress quite apart from ultramafic soils (Walker, personal communication). Low calcium in acid soils for example (McMillan 1956) could 'prepare' a population for serpentine resistance. Could the rather high incidence of ericaceous shrubs on ultramafic soils be a legacy of this possible sequence?

Adaptations to serpentine

Adjustment to life on ultramafics is sure to have been the result of evolutionary processes, all the way from the development of strikingly distinct endemics to the genesis of more subtle physiological tolerance in serpentine races. But it becomes more hazardous both to ascribe particular mechanisms to such evolution and to single out particular traits as adaptive. Such caution is urged on evolutionists by Gould & Lewontin (1978) who take a dim view of those who write 'adaptationist programmes' for particular traits. They urge a whole organism view for the interpretation of adaptation, plus a recognition of the constraining nature of an organism's history and its endowed architectural 'Bauplan.' What does this leave us to say about adaptation to life on ultramafic soil?

Perhaps it is safest to cast the problem of adaptation – and selection – for existence on ultramafic soils as hypotheses or questions. (1) Is overall tolerance to serpentine genetic, and thereby subject to selection? (2) Is the end-product of selection for resistance an adaptation to ultramafic soil? (3) Are the morphological attributes of serpentine plants (e.g., dwarfism, glaucescence, stenophylly, etc.) adaptive responses? (4) Are the various alternative responses to ultramafics (e.g., hyperaccumulation *vs.* exclusion of nickel) the result of lineage constraints in the Gould-Lewontin sense described above? These are grand questions, that could be posed for almost any structure – function specialization – and therefore answerable mostly at the philosophical level. The facile answer would be that any whole organism as well as all the members of a population represent an adaptive accommodation to ultramafic soil. What we cannot yet say is whether particular parts of an adapted organism can be *the* critical adaptations. Differences in allele frequency and heterozygosity between ultramafic and non-ultramafic populations of Jeffrey pine have been interpreted as adaptive responses to ultramafic soils (Furnier & Adams 1986).

Exploitation of ultramafic habitats and the effects on the biota

Ultramafic landscapes in western North America, like most other terrains, have felt the hand of man either by intention or by accident. As a resource, the rock, the soil and its mantle of vegetation have suffered a variety of disturbances (Kruckeberg 1984). Mining, forestry and agriculture, each has taken its toll on these biologically unusual areas.

The search for and extraction of minerals has probably been the most pervasive type of disturbance. In California, mercury mines are closely associated with ultramafic outcrops, even though the mercury ore is not a constituent of ultramafic rocks. The historic mines of San Benito and Santa Clara counties, altered adjoining areas of ultramafic rocks in a variety of ways: the felling of the sparse timber on ultramafics for use in the mines, road building, etc. Extraction of chromium, nickel, magnesite and asbestos from other ultramafic deposits continues throughout the region, especially in northern California and southwestern Oregon. Disturbance of vegetation attends all of these activities. The most recent threat to large areas of serpentine vegetation in California is a unique mining exploitation, the utilization of geothermal power in the ultramafic belt of the Mayacamas Mountains.

Timber taken from ultramafic habitats, despite the lower yields and slow regeneration, has modified the vegetation particularly in the Klamath–Siskiyou Mountains. Although forest managers usually recognize that most ultramafic soils support timber which is of marginal value, old-growth stands have been selectively cut and the companion vegetation disturbed.

Attempts to grow crops on alluvial soils of ultramafic origin have not been successful (Martin *et al.* 1953). Even with substantial additions of fertilizer, the improvement of yields of legumes and pasture grasses is usually not sustained. It is for grazing of livestock that most ultramafic habitats have been used. Serpentine chaparral and grassland on Henneke and Dubakella series soils afford a low level of forage. Attempts to improve grazing by eradicating serpentine chaparral and seeding with grasses have met with only limited success (Jones *et al.* 1976).

Even recreational use of these areas can be destructive. The urge to drive all-terrain vehicles over ultramafic landscapes at New Idria has jeopardized the existence of *Madia discoidea*, a narrow serpentine endemic. Even the danger to the 'motorists' of inhaling asbestos in the dust they cause has not diminished this pastime.

The least destructive uses of these lands are as wildlife and watershed resources. U.S.D.A. Soil Conservation Service Soil Surveys consider these the most appropriate uses of ultramafic soils.

Conservation and management of ultramafic habitats

All of the uses described in the previous section, except as wildlife and watershed resources, can substantially affect the plant life on ultramafics. Stands of the unusual vegetation types – serpentine chaparral, serpentine grassland or woodlands – can be modified or destroyed. At the same time, the disturbance may threaten the persistence of a rare species on serpentine (Kruckeberg 1984). The geothermal power development, mentioned above, poses a threat to at least one narrow endemic in the area, *Streptanthus brachiatus*. This case in fact has become a *cause celebre*; the California Rare and Endangered Species Program requires that rare taxa be protected against extirpation. Since the taxonomic status of *S. brachiatus* is in doubt, the Bureau of Land Management funded a study of the plant; here, the need to develop a resource became a unique motive for supporting taxonomic research.

Many areas with ultramafic soils in the west are passively spared destruction, by 'benign neglect'. They are remote, often barren, and of limited economic value. Since they often support unusual vegetation and may harbour rare taxa, there should be an active effort to preserve important samples. To date only a very few preserves have been set aside expressly for their serpentine biota: two in California (Frenzel Creek Federal Research Natural Area in Colusa County, and The Nature Conservancy's Ring Mountain Preserve in Marin County), two in Washington state (Olivine Bridge Natural Area in Skagit County by the Department of Natural Resources and the Federal Research Natural Area at Eldorado Creek in Kittitas County); none as yet has been set aside in Oregon. However, the Rough and Ready Creek Wayside and the Kalmiopsis Wilderness in Josephine County, Oregon, set aside for their unique species, are mostly on ultramafics. Efforts to preserve parts of the remarkable flora of Eight Dollar Mountain in Josephine County, Oregon, have been partially successful, especially in the face of claims on the land for mining.

A substantial number of serpentine endemic taxa appear on the federal and state rare and endangered species lists for California (Smith *et al.* 1980) and Oregon (Siddall *et al.* 1979). Since some of these rarities will be given statutory protection, the rare species and their habitats will be preserved, thus assuring the preservation of samples of serpentine plant communities.

Land managers who have jurisdiction over ultramafic habitats can anticipate certain consequences of management practices. If passive management is emphasized for wildlife and watershed, the ultramafic areas should maintain themselves with only a minimum of effort. Fire and erosion are the only two major threats to serpentine vegetation, when not otherwise disturbed for resource extraction. Should a serpentine habitat be set aside as a natural area preserve, either for its vegetation or for particular rare taxa, management need not involve much more than a 'hands off' policy. The vegetation of ultramafic soils, unlike some seral communities on normal soils, appear to be steady-state ecosystems (Whittaker 1960, Proctor & Woodell 1975). The preserves should need no special treatment to retain their present climax status.

Future research

The profound contrasts between environments with ultramafic and non-ultramafic soils offer many opportunities for further work in comparative biology. Indeed, it is surprising that the biological consequences of the sharp differences have not been used for fundamental comparative studies in the areas of ecophysiology, population genetics and molecular biology. Here are just some of the problems that such studies can be expected to elucidate. (1) Gene flow across a sharp edaphic boundary; (2) sympatric divergence and the processes of reproductive isolation leading to species formation; (3) the genetic basis of serpentine resistance – oligogenic or polygenic? (4) the cellular and molecular basis of tolerance to low nutrients, and high magnesium or to nickel concentrations; (5) plant – animal interactions in ultramafic habitats (e.g., is the insect fauna conditioned in some way by ultramafic-resistant plants?). Of course, any of these prospects for future research could be put to the test on ultramafics of other parts of the world. And indeed they should be. But given the proximity of these remarkable habitats to centers of botanical research, western North American ultramafics should be coaxed into yielding still more secrets. It goes without saying that these ultramafics are national heritages, meriting protection both for their unique biota and as preserves of research value for the future.

Acknowledgements

Some of the author's research cited here has been funded by the National Science Foundation (G-1323, GB-4579, GB-11710) and the University of Washington Graduate Research Fund. Special thanks go to Professor Richard B. Walker who has both contributed substantially to this review by his major studies on the mineral nutrition of ultramafic soils and plants, and by his wise counsel during the preparation of the manuscript. I am inevitably led to honor Professor Hans Jenny, eminent soil scientist, whose wisdom I have drawn on repeatedly in these pages and over the years.

References

Alexander, E. B., W. E. Wildman & W. C. Lynn. 1985. Ultramafic (Serpentinitic) mineralogy class. In: Kittrick, J. A., et al. (eds.), Mineral Classification of Soils, Soil Sci. Soc. Amer., Madison, WI.

Antonovics, J., A. D. Bradshaw & R. G. Turner. 1971. Heavy metal tolerance in plants. In: Advances in Ecological Research 7: 1–85.

Baldwin, E. M. 1959. Geology of Oregon. University of Oregon Cooperative Bookstore, Eugene, Oregon.

Barbour, M. G. and J. Major. 1977. Terrestrial Vegetation of California. John Wiley and Sons, New York.

Bradshaw, A. D. 1976. Pollution and evolution. In: Mansfield, T. A. (ed.), Effects of Air Pollutants on Plants, pp. 135–159. Cambridge Univ. Press.

Brink, V. C., K. Fletcher & S. Parmar. 1972 (?). Trace element levels in plants growing on ultramafic (ultrabasic) rocks in British Columbia. First Report (mimeographed). Univ. of British Columbia, Vancouver.

Brooks, R. R., R. S. Morrison, R. D. Reeves, T. R. Dudley & Y. Akman. 1979. Hyperaccumulation of nickel by Alyssum Linnaeus (Cruciferae). Proc. Roy. Soc. Lond. Sec. B, 203: 387–403.

Brown, E. H. 1977. Ophiolite on Fidalgo Island, Washington. In: Coleman, R. G. and W. P. Irwin (eds.), North American Ophiolites, pp. 67–73. Bulletin 95, Oregon Department of Geology and Mineral Industries, Portland, Oregon.

Christiansen, N. I. 1971. Fabric, seismic anisotropy, and tectonic history of the Twin Sisters dunite, Washington. Geol. Sci. Amer. Bull. 82: 1681–1994.

Clausen, J., D. D. Keck and W. M. Heisey. 1940. Experimental studies on the nature of species. I. Effect of varied environments on western North American plants. Carnegie Inst. publ. no. 520, Washington.

Clausen, J., D. D. Keck & W. M. Heisey. 1948. Experimental studies on the nature of species. III. Environmental responses of climatic races of Achillea. Carnegie Inst. publ. no. 581, Washington.

Cody, W. J. 1983. Adiantum pedatum ssp. calderi, a new subspecies in north-eastern North America. Rhodora 85: 93–96.

Cody, W. J. and D. M. Britton. 1984. Polystichum lemmonii, a rock shield-fern new to British Columbia and Canada. Canadian Field-Naturalist 98: 375.

Coleman, R. G. 1977. Ophiolites. Ancient Oceanic Lithosphere. Springer-Verlag, Berlin.

Coleman, R. G. & W. P. Irwin. 1974. Ophiolites and ancient continental margins. In: Burk, C. A. & C. L. Drake (eds.), The Geology of Continental Margins, pp. 921–931, Springer-Verlag, New York.

Coleman, R. G. & W. P. Irwin (eds.). 1977. North American Ophiolites. Bulletin 95. Oregon Department of Geology and Mineral Industries, Portland, Oregon.

Dickinson, H. and J. Antonovics. 1973. Theoretical consideration of sympatric divergence. Amer. Nat. 107: 256–274.

Dyksterhuis, E. L. 1981. Soil Survey Grant County, Oregon. Central Part. U.S.D.A. Soil Conservation Service, Washington, D.C.

Forde, M. B. & D. G. Faris. 1962. Effect of introgression on the serpentine endemism of Quercus durata. Evolution 16: 338–347.

Franklin, J. F. & C. T. Dyrness. 1973. Natural vegetation of Oregon and Washington. U.S.D.A. Forest Service General Technical Report PNW-8 U.S. Government Printing Office, Washington, D.C.

Funkhouser, A. 1980. [Unpublished data of student project – College of the Pacific, Stockton, California.]

Furnier, G. R. and W. T. Adams. 1986. Geographic patterns of allozyme variation in Jeffrey pine. Amer. J. Bot. 73: 1009–1015.

Gordon, A. & C. B. Lipman. 1926. Why are serpentine and other magnesian soils infertile? Soil Sci. 22: 291–302.

Gould, S. J. and R. Lewontin. 1978. The spandrels of San Marco and the Panglossian paradigm: A critique of the adaptationist programme. Proc. R. Soc. London 205: 581–598.

Gray, J. G. 1979. The vegetation of two California mountain slopes. Madrono 25: 177–185.

Griffin, J. R. 1965. Digger pine seedling response to serpentinite and non-serpentinite soil. Ecology 46: 801–807.

Griffin, J. R. & W. B. Critchfield. 1972. Distribution of forest trees in California. U.S.D.A. For. Serv. Res. Paper PSW-82.

Hardham, C. B. 1962. The Santa Lucia Cupressus sargentii groves and their associated northern hydrophilous and endemic species. Madrono 16: 173–178.

Higgins, S. S. 1984. A comparison of plant responses to stress and competition on serpentine and zonal soil. PhD Thesis, Wash. State Univ., Pullman, Washington.

Higgins, S. S. and R. N. Mack. 1987. Comparative responses of Achillea millefolium ecotypes to competition and soil type. Oecologia 73: 591–597.

Himmelberg, G. R. & R. G. Coleman. 1968. Chemistry of primary minerals and rocks from the Red Mountain-Del Puerto Ultramafic mass, California. U.S. Geol. Survey Prof. Paper 600-C: C18–C26.

Hitchcock, C. L. & A. Cronquist. 1973. Flora of the Pacific Northwest. University of Washington Press, Seattle.

Hitchcock, C. L., A. Cronquist, M. Ownbey & J. W. Thompson. 1964. Vascular Plants of the Pacific Northwest. vol. 2 (Salicaceae to Saxifragaceae). Univ. of Wash. Press, Seattle.

Hobbs, R. J. & H. A. Mooney. 1985. Community and population dynamics of serpentine grassland annuals in relation to gopher disturbance. Oecologia 67: 342–351.

Holland, S. S. 1961. Jade in British Columbia. Annual Rept. Mines and Petroleum Resources, British Columbia, Victoria.

Howell, J. T. 1970. Marin Flora, 2nd Ed. Univ. of Calif. Press, Berkeley.

Huntting, M. T. 1961. Geological map of the State of Washington. Wash. State Div. of Mines and Geol., Olympia.

Irwin, W. P. 1977. Ophiolitic terranes of California, Oregon, and Nevada. In: Coleman, R. G. and W. P. Irwin (eds.), North American Ophiolites, pp. 75–92. Bulletin 95. Oregon Depart. of Geol. and Mineral Industries, Portland, Oregon.

Istok, J. D. & M. E. Harward. 1982. Influence of soil moisture on smectite formation in soils derived from serpentinite. Soil Sci. Soc. of Amer. Journ. 46: 1106–1108.

Jenkinson, J. L. 1974. Ponderosa pine progenies: differential response to ultramafic and granitic soils. U.S.D.A. For. Serv. Res. Paper PSW-101.

Jenny, H. 1941. Factors of Soil Formation. McGraw-Hill, New York.

Jenny, H. 1980. The Soil Resource: Origin and Behavior. Springer-Verlag, New York.

Johnson, M. P., A. D. Keith & P. R. Ehrlich. 1968. The population biology of the butterfly, Euphydryas editha. VII. Has E. editha evolved a serpentine race? Evolution 22: 422–423.

Jones, M. B., C. E. Vaughn & R. S. Harris. 1976. Critical Ca levels and Ca/Mg ratios in Trifolium subterraneum L. grown on serpentine soil. Agronomy Journ. 68: 756–759.

Jones, M. B., W. A. Williams & J. E. Ruckman. 1977. Fertilization of Trifolium subterraneum L. growing on serpentine soils. Soil Sci. Soc. Amer. Journ. 41: 87–89.

Kazantzis, G. 1980. Mercury. In: Waldron, H. A. (ed.), Metals in the Environment, pp. 221–261. Academic Press, New York.

Koenigs, R. L., W. A. Williams & M. B. Jones. 1982a. Factors affecting vegetation on a serpentine soil. I. Principal components analysis of vegetation data. Hilgardia 50: 1–14.

Koenigs, R. L., W. A. Williams, M. B. Jones & A. Wallace. 1982b. Factors affecting vegetation on serpentine soil. II. Chemical composition of foliage and soil. Hilgardia 50: 15–26.

Krause, W. 1958. Andere Bodenspezialisten. Handb. Pflanzenphysiol. 4: 755–806.

Kruckeberg, A. R. 1951. Intraspecific variability in the response of certain native plant species to serpentine soil. Amer. Journ. Bot. 38: 408–419.

Kruckeberg, A. R. 1954. The ecology of serpentine soils: A symposium. III. Plant species in relation to serpentine soils. Ecology 35: 267–274.

Kruckeberg, A. R. 1964. Ferns associated with ultramafic rocks in the Pacific Northwest. Amer. Fern Journ. 54: 113–126.

Kruckeberg, A. R. 1967. Ecotypic response to ultramafic soils by some plant species of northwestern United States. Brittonia 19: 133–151.

Kruckeberg, A. R. 1969a. Soil diversity and the distribution of plants, with examples from western North America. Madrono 20: 129–154.

Kruckeberg, A. R. 1969b. Plant life on serpentinite and other ferromagnesian rocks in northwestern North America. Syesis 2: 15–114.

Kruckeberg, A. R. 1979. Plants that grow on serpentine – A hard life. Davidsonia 10: 21–29.

Kruckeberg, A. R. 1984. California serpentines: Flora, vegetation, geology, soils and management problems. Univ. of Calif. Publs. in Botany 78: 1–180.

Kruckeberg, A. R. and J. L. Morrison. 1983. New Streptanthus taxa (Cruciferae) from California. Madrono 30: 230–244.

Lindsley-Griffin, N. 1977. The Trinity ophiolite, Klamath Mountains, California. In: Coleman, R. G. & W. P. Irwin (eds.), pp. 107–120, North American Ophiolites. Bulletin 95. Oregon Depart. of Geol. and Mineral Industries, Portland, Oregon.

Lipman, C. B. 1926. The bacterial flora of serpentine soils. Journ. of Bacteriology 12: 315–318.

Loew, O. & D. W. May. 1901. The relation of lime and magnesia to plant growth. I. Liming of soils from a physiological standpoint. II. Experimental study of the relaton of lime and magnesia to plant growth. U.S. Dept. of Agr. Plant Ind. Bull. 1, 53 pp.

Lolkema, P. C., M. H. Donker, A. J. Schouten and W. H. O. Ernst. 1984. The possible role of metallothioneins in copper tolerance of Silene cucubalus. Planta 162: 174–179.

Maas, J. L. and Stuntz, D. E. 1969. Mycoecology on serpentine soil. Mycologia 61: 1106–1116.

Madhok, O. P. and R. B. Walker. 1969. Magnesium nutrition of two species of sunflower. Plant Physiol. 44: 1016–1022.

Main, J. L. 1974. Differential responses to magnesium and calcium by native populations of Agropyron spicatum. Amer. Journ. of Botany 61: 931–937.

Main, J. L. 1981. Magnesium and calcium nutrition of a serpentine endemic grass. Amer. Midland Naturalist 105: 196–199.

Martin, W. E., J. Vlamis and N. W. Stice. 1953. Field correction of calcium deficiency on a serpentine soil. Agronomy Journ. 45: 204–208.

Mason, H. L. 1946a. The edaphic factor in narrow endemism. I. The nature of environmental influences. Madrono 8: 209–226.

Mason, H. L. 1946b. The edaphic factor in narrow endemism.

II. The geographic occurrence of plants of highly restricted patterns of distribution. Madrono 8: 241–257.

McKee, B. 1972. Cascadia: The Geological Evolution of the Pacific Northwest. McGraw-Hill, New York.

McMillan, C. 1956. The edaphic restriction of *Cupressus* and *Pinus* in the Coast Ranges of central California. Ecol. Monogr. 26: 177–212.

McNaughton, S. J. 1968. Structure and function in California grasslands. Ecology 49: 962–972.

Miller, R. B. 1980. Structure, petrology and emplacement of the ophiolitic Ingalls Complex, North central Cascades, Washington. PhD Thesis, Univ. of Wash., Seattle.

Monger, J. W. H. 1977. Ophiolitic assemblages in the Canadian Cordillera. In: Coleman, R. G. & W. P. Irwin (eds.), pp. 59–65, North American Ophiolites. Oregon Dept. of Geol. and Mineral Industries, Portland, Oregon.

Moral, R. del. 1972. Diversity patterns in forest vegetation of the Wenatchee Mountains, Washington. Bull. Torrey Bot. Club 99: 57–64.

Moral, R. del. 1974. Species patterns in the upper North Fork Teanaway River drainage, Wenatchee Mountains, Washington. Syesis 7: 13–30.

Moral, R. del. 1976. Wilderness in ecological research: an example from the Alpine Lakes. Pp. 173–194, in Terrestrial and Aquatic Ecological Studies of the Northwest, ed. R. A. Soltero, Eastern Washington State College Press, Cheney, Wash.

Moral, R. del. 1982. Control of vegetation on contrasting substrates: herb patterns on serpentine and sandstone. Amer. Journ. Botany 69: 227–238.

Moral, R. del, A. F. Watson & R. S. Fleming. 1976. Vegetation structure in the Alpine Lakes region of Washington State: Classification of vegetation on granitic rocks. Syesis 9: 291–316.

Munz, P. A. & D. D. Keck. 1959. A California Flora. Univ. of Calif. Press, Berkeley and Los Angeles.

Norris, R. M. & R. W. Webb. 1976. Geology of California. John Wiley & Sons, New York.

Novak, F. A. 1928. Quelques remarques relatives au problem de la vegetation sur les terrain serpentiniques. Preslia 6: 42–71

Patton, W. W., Jr., I. L. Trailleur, W. P. Brosge & M. A. Lanphere. 1977. Preliminary report on the ophiolites of northern and western Alaska. In: Coleman, R. G. and W. P. Irwin, pp. 51–57, North American Ophiolites. Bull. 95, Dept. of Geol. and Mineral Industries, Portland, Oregon.

Pegtel, D. M. 1980. Evidence for ecotypic differentiation in *Lupinus*-associated *Rhizobium*. Acta Bot. Neerl. 29: 429–441.

Peterson, P. J. 1971. Unusual accumulations of elements by plants and animals. Science Progress, Oxford 59: 505–526.

Pichi-Sermolli, R. 1948. Flora e vegetazione delle serpentine e delle altre ofioliti dell'alta valle del Trevere (Toscana). Webbia 6: 1–380.

Proctor, J. 1970. Magnesium as a toxic element. Nature 227: 742–743.

Proctor, J., W. R. Johnston, D. A. Cottam & A. B. Wilson. 1981. Field-capacity water extracts from serpentine soils. Nature 294: 245–246.

Proctor, J. & S. R. J. Woodell. 1975. The ecology of serpentine soils. Adv. Ecol. Res. 9: 255–265.

Proctor, J. & Whitten, K. 1971. A population of the valley pocket gopher on a serpentine soil. Amer. Midl. Nat. 85: 517–521.

Rai, D., G. H. Simonson & C. T. Youngberg. 1970. Serpentine-derived soils in watershed and forest management. Report to the U.S. Dept. of Interior, Bureau of Land Management. Dept. of Soils, Oregon State Univ., Corvallis, Oregon (Mimeographed).

Raleigh, C. B. 1965. Structure and petrology of an alpine peridotite on Cypress Island, Washington, U.S.A. Beitr. zur Mineralogie und Petrographie 11: 719–741.

Ramp, L. and N. V. Peterson. 1979. Geology and Mineral Resources of Josephine County, Oregon. Bull. 100, Oregon Dept. of Geol. and Mineral Industries, Portland, Oregon.

Rauser, W. E. & N. R. Curvetto. 1980. Metallothionein occurs in roots of *Agrostis* tolerant to excess copper. Nature 287: 563–564.

Raven, P. H. 1964. Catastrophic selection and edaphic endemism. Evolution 18: 336–338.

Raven, P. H. & D. I. Axelrod. 1978. Origin and relationships of the California flora. Univ. of Calif. Publs. in Botany 72: 1–134.

Reeves, R. D., R. R. Brooks & R. M. Macfarlane. 1981. Nickel uptake by Californian *Streptanthus* and *Caulanthus* with particular reference to the hyperaccumulator *S. polygaloides* Gray (Brassicaceae). Amer. Journ. Bot. 68: 708–712.

Reeves, R. D., R. M. Macfarlane and R. R. Brooks. 1983. Accumulation of nickel and zinc by western North American genera containing serpentine-tolerant species. Amer. Journ. Bot. 70: 1297–1303.

Ritter-Studnička, H. 1968. Die Serpentinomorphosen der Flora Bosniens. Botanische Jahrb. 88: 443–465.

Robinson, W. O., G. Edgington and H. G. Byers. 1935. Chemical studies of infertile soils derived from rocks high in magnesium and generally high in chromium and nickel. U.S. Dept. of Agric. Tech. Bull. 471, pp. 1–28.

Saleeby, J. 1977. Fracture zone tectonics, continental margin fragmentation, and emplacement of the Kings-Kaweah ophiolite belt, southwest Sierra Nevada, California. In: Coleman, R. G. and W. P. Irwin (eds.), North American Ophiolites, pp. 141–159, Bull. 95, Dept. of Geol. and Mineral Industries, Portland, Oregon.

Samiullah, Y., A. R. Kruckeberg and P. J. Peterson. 1985. Hyperaccumulation of nickel by *Arenaria rubella* (Wahlenb.) J. R. Smith (Caryophyllaceae) from Washington, U.S.A. Unpbl. manuscript.

Senkayi, A. L. 1977. Clay mineralogy of poorly drained soils developed from serpentinite rocks. PhD Thesis, Univ. of Calif., Davis.

Shapiro, A. M. 1981. Egg-mimics of *Streptanthus* (Cruciferae) deter oviposition by *Pieris sisymbrii* (Lepidoptera: Pieridae). Oecologia 48: 142–143.

Sharsmith, H. 1961. The genus *Hesperolinon* (Linaceae). Univ. of Calif. Publs. in Botany 32: 235–314.

Siddall, J. L., K. L. Chambers & D. H. Wagner. 1979. Rare, threatened and endangered vascular plants in Oregon – an interim report. Oregon Natural Area Preserves Advisory Committee, Salem, Oregon.

Smith, J. P., Jr., R. J. Cole & J. O. Sawyer, Jr., in collaboration with W. R. Powell. 1980. Inventory of rare and endangered vascular plants of California. Special Publ. No. 1 (2nd ed.) Calif. Native Plant Soc., Berkeley.

Soil Survey Staff. 1975. Soil Taxonomy. U.S.D.A. Agric. Handbook 436.

Stansell, V. 1980. *Darlingtonia californica*. Geographical distribution, habitat and threats. Endangered Species Office, U.S. Fish and Wildlife Serv., Portland, Oregon.

Stebbins, G. L., Jr. 1942. The genetic approach to problems of rare and endemic species. Madrono 6: 241–272.

Stebbins, G. L. 1980. Rarity of plant species: A synthetic viewpoint. Rhodora 82: 77–86.

Stebbins, G. L., Jr. and J. Major. 1965. Endemism and speciation in the California flora. Ecol. Monogr. 35: 1–35.

Stebbins, R. 1949. Speciation in salamanders of the plethodontid genus *Ensatina*. Univ. of Calif. Publs. in Zool. 48: 377–526.

Tadros, T. M. 1957. Evidence of the presence of an edaphobiotic factor in the problem of serpentine tolerance. Ecology 38: 14–23.

Thayer, T. P. 1956a. Preliminary Geologic Map of the Aldrich Mountain Quadrangle, Oregon. Mineral Investigations. Field Studies Map MF 49. U.S. Geol. Surv., Washington, D.C.

Thayer, T. P. 1956b. Preliminary Geologic Map of the Mt. Vernon Quadrangle, Oregon. Mineral Investigations. Field Studies Map MF 50. U.S. Geol. Surv., Washington, D.C.

Thayer, T. P. 1956c. Preliminary Geologic Map of the John Day Quadrangle, Oregon. Mineral Investigations. Field Studies Map FM 51. U.S. Geol. Surv., Washington, D.C.

Thayer, T. P. 1977. The Canyon Mountain Complex, Oregon, and some problems of ophiolites. In: Coleman, R. G. and W. P. Irwin (eds.), North American Ophiolites, pp. 93–105. Bulletin 95, Oregon Dept. of Geol. and Mineral Industries, Portland, Oregon.

Turitzin, S. N. 1981. Nutrient limitations to plant growth in a California serpentine grassland. Amer. Midland Naturalist 107: 95–99.

Vance, J. A. 1957. The geology of the Sauk River Area in the northern Cascades of Washington. PhD Thesis, Univ. of Wash., Seattle.

Vlamis, J. 1949. Growth of lettuce and barley as influenced by degree of Ca saturation of soil. Soil Sci. 67: 453–466.

Vlamis, J. & H. Jenny. 1948. Calcium deficiency in serpentine soils as revealed by absorbent technique. Science 107: 549–551.

Vogl, R. J. 1973. Ecology of knobcone pine in the Santa Ana Mountains, California. Ecol. Monogr. 43: 125–143.

Walker, R. B. 1948. Molybdenum deficiency in serpentine barren soils. Science 108: 473–475.

Walker, R. B. 1954. Factors affecting plant growth on serpentine soils. In: Whittaker, R. H. (ed.), The ecology of serpentine soils: A symposium. Ecology 35: 258–266.

Walker, R. B., H. M. Walker, and P. R. Ashworth. 1955.

Calcium-magnesium nutrition with special reference to serpentine soils. Plant Physiol. 30: 214–221.

Wallace, A., M. B. Jones & G. V. Alexander. 1982. Mineral composition of native woody plants growing on a serpentine soil in California. Soil Sci. 134: 42–44.

Wallace, D. R. 1983. The Klamath Knot: Explorations of Myth and Evolution. Sierra Club Books, San Francisco.

Waring, R. H. 1969. Forest plants of the eastern Siskiyous: Their environmental and vegetational distribution. Northwest Science 43: 1–17.

Waring, R. H., W. H. Emmingham, and S. W. Running. 1975. Environmental limits of an endemic spruce, *Picea breweriana*. Canadian Journ. of Bot. 53: 1599–1613.

Waring, R. H. and J. Major. 1964. Some vegetation of the California coastal redwood region in relation to gradients of moisture, nutrients, light and temperature. Ecol. Monogr. 34: 167–215.

Wells, P. V. 1962. Vegetation in relation to geological substratum and fire in the San Luis Obispo quadrangle, California. Ecol. Monogr. 32: 79–103.

Wherry, E. T. 1932. Ecological studies of serpentine-barren plants. I. Ash Composition. Proc. Pennsylvania Acad. Sci. 6: 32–38.

White, C. D. 1967. Absence of nodule formation in *Ceanothus cuneatus* in serpentine soils. Nature 215: 875.

White, C. D. 1971. Vegetation – soil chemistry correlations in serpentine ecosystems. PhD Thesis. University of Oregon, Eugene.

White, M. J. D. 1978. Modes of speciation. Wh. H. Freeman and Co., San Francisco.

Whittaker, R. H. 1954. IV. The vegetational response to serpentine soils. In: Whittaker, R. H. (ed.). The ecology of serpentine soils: A symposium. Ecology 35: 275–288.

Whittaker, R. H. 1960. Vegetation of the Siskiyou Mountains, Oregon and California. Ecol. Monogr. 30: 279–338.

Whittaker, R. H. 1961. Vegetation history of the Pacific Coast states and the "central" significance of the Klamath region. Madrono 16: 5–23.

Whittaker, R. H. 1975. Communities and ecosystems. 2nd Ed. MacMillan Publ. Co., Inc., New York.

Wildman, W. E., M. L. Jackson & L. D. Whittig. 1968a. Serpentinite rock dissolution as a function of carbon dioxide pressure in aqueous solution. Amer. Mineralogist 53: 1252–1263.

Wildman, W. E., M. L. Jackson & L. D. Whittig 1968b. Iron-rich montmorillonite formation in soils derived from serpentinite. Soil Sci. Soc. of Amer. Proc. 32: 787–794.

Woodell, S. R. J., H. A. Mooney & H. Lewis. 1975. The adaptation to serpentine soils in California of the annual species *Linanthus androsaceus* (Polemoniaceae). Bull. Torrey Bot. Club 102: 232–238.

Zobel, D. G. and G. M. Hawk. 1980. The environment of *Chamaecyparis lawsoniana*. Amer. Midl. Nat. 103: 280–297.

Author's Address:
A. R. Kruckeberg
Botany KB-15
University of Washington
Seattle WA 98195
U.S.A.

IV

Ecology of serpentinized areas,

Newfoundland, Canada

B. A. ROBERTS

Abstract. The area covered by the serpentinized Ophiolite rocks approaches about 3% of the Island of Newfoundland (total area 106,000 Km^2) and forms the most striking physiographic features in Atlantic Canada. The geology and tectonic zones in Newfoundland can be traced along the full length of the Appalachians in the USA and to the Northeast to the British Caledonides.

The major soils of the serpentinized areas of Newfoundland are Orthic Regosols, Gleyed Regosols, and Rego Gleysols with a peaty phase in the fen areas. Two major soil series are prominent. The Blomidon series consists of moderately well to imperfectly drained Gleyed Regosols, cryoturbic phase, which occur mainly on the bottom flats. The Round Hill series comprises well to moderately well drained Orthic Regosols, cryoturbic phase, that are common on moderate to fairly steep slopes. Together they cover an area of more than 7700 ha in western Newfoundland as well as extensive areas in the northern and central portions of the province. Soil textures are loam to sandy loam. The soils are derived from base-rich very stony, sandy loam colluvium and till derived mainly from serpentinized rocks. These deposits vary in thickness and fall into the depth categories of shallow and deep and shallow and bare with horizon development restricted by frost churning. Patterned ground in the form of stone polygons and stripes is common. The cryoturbation factor combined with high Mg concentrations (12–16%, total), (6–8 – 16.64 meq/100 g, available) and low essential macronutrients create adverse conditions for plant growth. In addition there are present possibly toxic quantities of Ni (0.308 – 0.389% total) (0.008–0.031 meq/100 g available). The Mg/Ca quotients are (50–120 total) and (2–20 available) and are among the highest reported for soils of serpentinized areas.

The vegetation (made up of between 120–190 basicolous and acidicolous species) is sparse and covers less than 10% of the area. Six major plant communities occur but the vegetation contains a mixture of only two prominent subassociations primarily related to soil moisture. The serpentinized alpine plateau is believed to have not been glaciated during the last glaciation and some of the plant species may be relics of an older flora.

Arenaria humifusa and *A. marcescens* have been identified as hyperaccumulators (greater than 1000 µg g^{-1} dry wt) of Ni as well as *Senecio pauperculus* and *Solidago hispida*. The three other members of the Caryophyllaceae are strong Ni accumulators with *Cerastium beeringianum*, *Lychnis alpina*, and *silene acaulis* having 572, 913 and 898 µg g^{-1} Ni dry wt in above ground parts. Five other plant species have greater than 500 µg g^{-1} dry wt Ni. In sampling from the same transects in two successive years these strong accumulators had concentrations that stayed similar or in most cases became slightly less. In the hyperaccumulators, both species of *Arenaria* had similar concentrations of Ni in successive years.

B. A. Roberts and J. Proctor (eds), The ecology of areas with serpentinized rocks. A world view, 75–113.
© 1992 *Kluwer Academic Publishers*.

The ability of the members of the Caryophyllaceae to tolerate and possibly concentrate large quantities of these toxic elements and also tolerate low macronutrients may be a major reason for their success.

The pine family showed the lowest accumulation of Co, Cr, and Ni and this in combination with having one of the highest uptake rates of macronutrients is a major reason for its success on serpentine soils. The other successful families are related to tolerance of lower macronutrient conditions and a tolerance of heavy metals.

Comparison is made with soils of serpentinized areas of Portugal, Poland, Sweden, Great Britain, the USA and other countries with similarities and differences noted.

Introduction

The serpentinized areas of Newfoundland, particularly the western part, are perhaps the most striking physiographic features in Atlantic Canada. Geologists, for example, have hailed them as 'the eighth wonder of the world' Neale (1974), Rogerson (1983), because of their spectacular geology and physiography. These areas stand out primarily as reddish-brown coloured rock deserts against a surrounding green backdrop of the Boreal forest zone.

The sparse but botanically interesting floras of the Newfoundland serpentinized areas have received attention by botanists and ecologists since the early visits to Newfoundland by Fernald (Fernald 1911). His initial impressions of the serpentinized table land from Woody Point reflects the interest botanists have long had: 'From Woody Point one looks back to the Canadian forest on the slope of Lookout Mountain and across to the alternately wooded and bare slopes on the opposite side of the Arm. Farther up the Arm, but standing some miles to the south, is The Tableland, a flat-topped and great, seemingly naked block of serpentine (2336 feet high), weathered pinkish – to yellowish-ochre and looking like a gigantic pale brick wall, only streaked down its side with lines of white snowfields. Lesser rounded knobs lie between it and the "bottom" we should say the head) of the Arm and, towering above them, like a small Matterhorn, rises the sharp pinnacle of the Peak of Teneriffe. That was our view upon landing and we were impatient to get everywhere at once.'

This chapter reviews the location and extent of serpentinized areas in Newfoundland, their climate, geology, soils and soil development. A review of their interesting flora and the nutrition and growth of plants on serpentinized areas are also added.

Location and extent of the study areas

The locations and extent of serpentinized areas of Newfoundland are outlined in Fig. 1 (Modified from Williams & Malpas 1972, Williams 1979, Dunning 1981). The area covered by the serpentinized ophiolitic rocks approaches about 3% of the island of Newfoundland (total area 106,000 km^2). The serpentinized areas fall within about five different physiographic divisions (Roberts 1983) as shown in Fig. 2, but are very prominent in only two:
1. the serpentinized hills of Hare Bay #1; and
2. the Bay of Islands serpentinized Range #3.

Most of the detailed study by the author has been confined to the Bay of Islands serpentinized Range, however, observations and review will cover most serpentinized areas and the main fea-

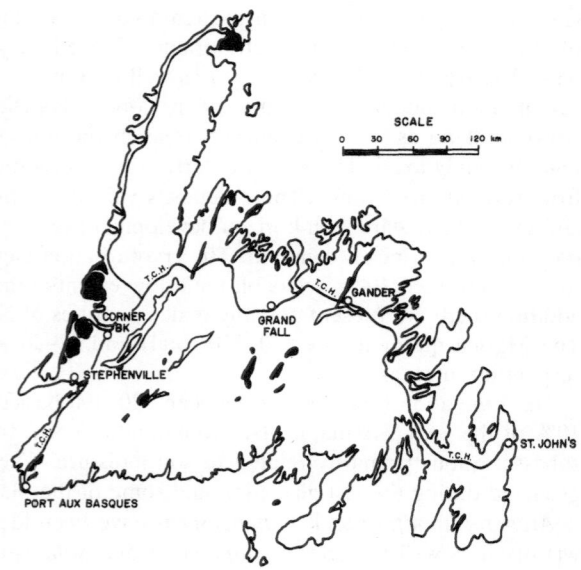

Fig. 1. Location and extent of serpentinized areas in Newfoundland, modified after Williams (1979) and Dunning (1981).

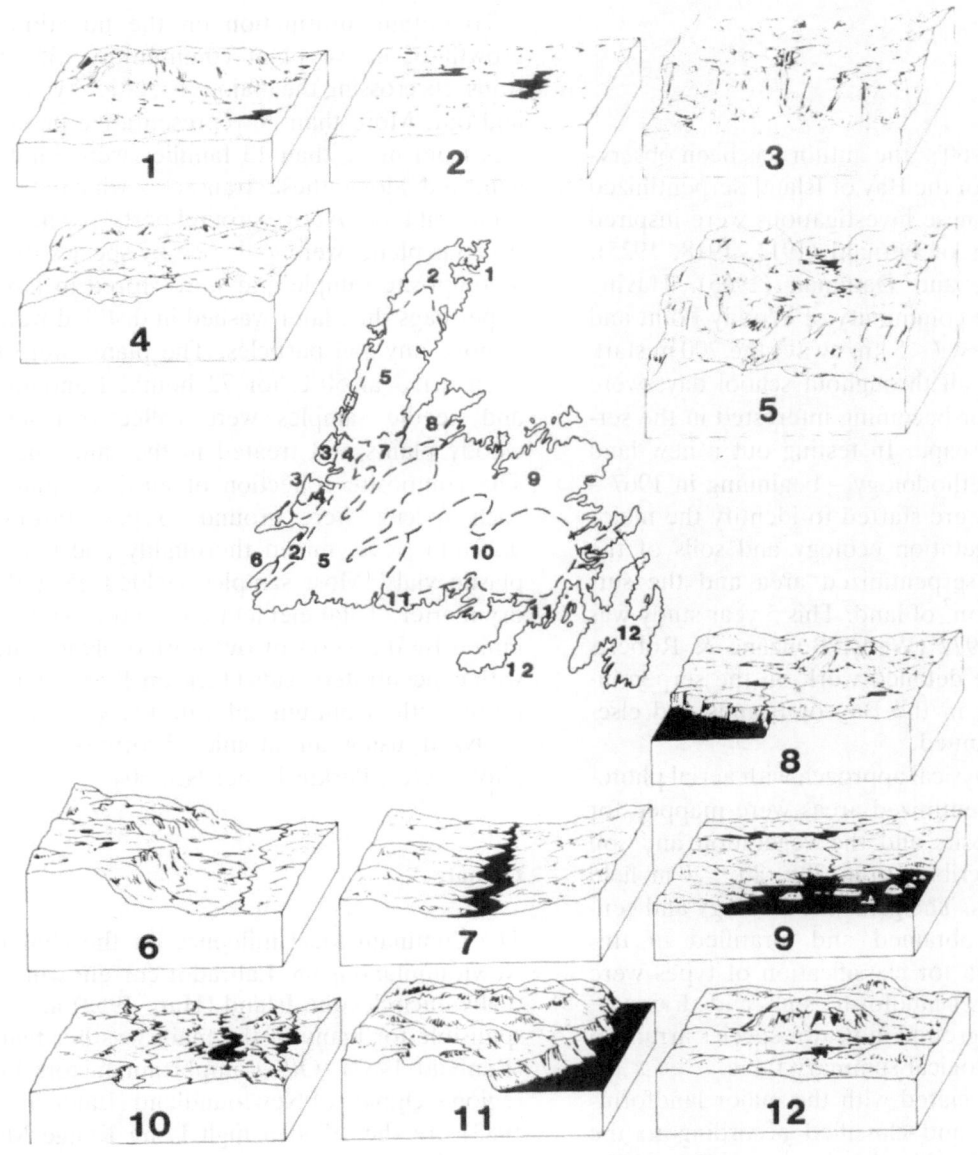

1. Serpentinized Hills of Hare Bay
2. West Coast Lowland
3. Bay of Islands Serpentinized Range
4. West Coast Calcareous Uplands
5. Long Range Mountains
6. Anguille Mountains

7. Grand Lake – White Bay Basin
8. Burlington Peninsula
9. North East Trough
10. Central Plateau
11. South Coast Highlands
12. Eastern Upland

Fig. 2. Physiographic divisions of Newfoundland (Roberts 1983).

tures of geology, soils and flora will be described for each area.

Methods

Since the mid 1960's, the author has been observing the ecology of the Bay of Island Serpentinized areas. Initially these investigations were inspired by early reports of Fernald (1911, 1918, 1925), Cooper (1936), and Damman (1965). Having grown up in the community of Woody Point and being able to view (<5 km away) the 700 m stark table mountain all throughout school days were other reasons for becoming interested in the serpentinized landscape. In testing out a new land classification methodology – beginning in 1967 – formal studies were started to identify the major land types, vegetation ecology and soils of the Bay of Island serpentinized area and the surrounding 2500 km^2 of land. This 5 year study was completed in 1972 (Wells, Bouzane & Roberts 1972) and more detailed work on the serpentinized areas both in the Bay of Islands and elsewhere has continued.

Using a biophysical approach with aerial photographs the serpentinized areas were mapped for superficial deposits and the vegetation and soil types were described from 205, 20 × 20 m field plots in all areas. The gross morphology and generalities were obtained and stratified in this manner. Criteria for classification of types were differences in (1) landform, soils and drainage; (2) presence of ground indicator species arranged in a phytosociological summary table.

The soils associated with the major landforms were described and classified according to the Canadian Soil classification System (1974, 1978). The sequence of horizons, chemical and physical properties were assessed on several hundred soil samples according to methods outlined by Roberts (1980). Water extracts were obtained on selected soil samples by following methods outlined by Proctor *et al.* (1981).

In all plots vegetation relevés for phytosociological analysis were conducted according to the Zurich–Montpellier (Z–M) method (Mueller–Dombois & Ellenberg 1974) using a modified Braun–Blanquet scale. More than 100 relevés of vegetation in the serpentinized areas have been

taken and many of the important findings are in Appendix 1.

To obtain information on the nutrition and growth of native plant communities, 35 meter transects crossing the major vegetation types were laid out. More than 30 representative plant species from more than 13 families were randomly collected along these transects, which had the same soil type. Above-ground parts of whole herbaceous plants were collected by species to form a composite sample and were stored in labelled paper bags then later washed in distilled water to remove any soil particles. The plants were then oven dried at 50°C for 72 hours. Random leaf and needle samples were collected from the woody plants and treated in the same manner. The composite collection of air-dried plants of each species were ground to pass through a 0.25 mm sieve, mixed thoroughly and stored in plastic vials. Most samples yielded 25–100 g of dry matter. Total element concentration was obtained by digestion of oven-dried plant material with concentrated acids (Jackson 1958). After dilution with strontium chloride the extracts were analyzed using an atomic absorption spectrophotometer, Perkin Elmer No. 304.

Climate

The dominant local influence on the climate of Newfoundland is the Labrador current which virtually encircles the Island (Hare 1952) and is responsible for many of the main weather features (Banfield 1983). Other important factors in the regional climate of Newfoundland (Banfield 1981, 1983) are the >700 m high Long Range Mountains of western and northern Newfoundland and to the west and south-west the Gulf of St. Lawrence and Cabot Strait. These are extensive enough to significantly influence the heat and moisture properties of air flowing east of the Canadian Mainland. The large number of bays and inlets are capable of generating localized weather variations in the eastern areas (Banfield 1983).

The climate of the Newfoundland serpentinized areas is varied because they cover a wide geographic area. It is also complicated by the elevation aspect: serpentinized deposits range from near sea level to more than 700 meters. The cli-

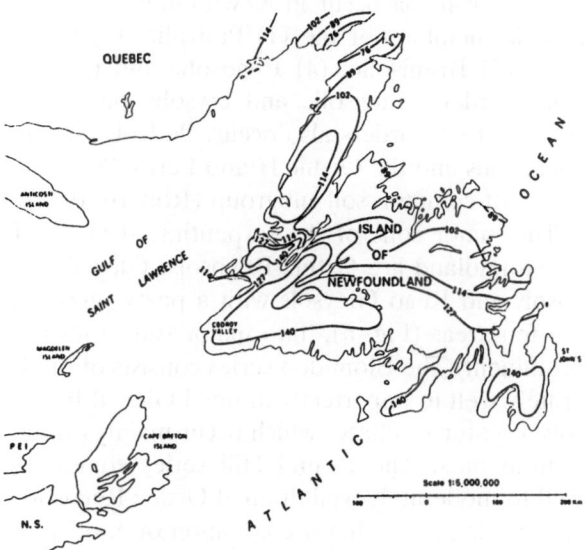

Fig. 3. Mean annual precipitation (89 to 140 cm) for New-foundland serpentinized areas (Roberts 1983.)

matic maps show that for the major serpentinized areas, the mean annual precipitation (ranges from 89 to 140 cm) (Fig. 3) the average freeze-free period (11 to 140 days) (Fig. 4) and the growing

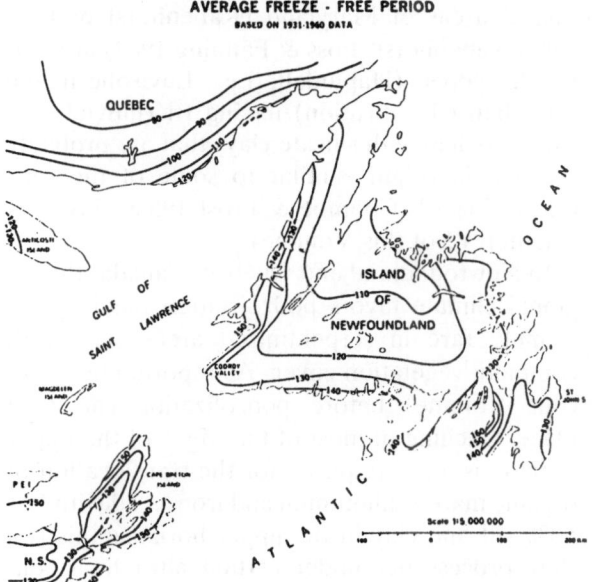

Fig. 4. The average freeze-free period (110–140 days, Roberts 1983.)

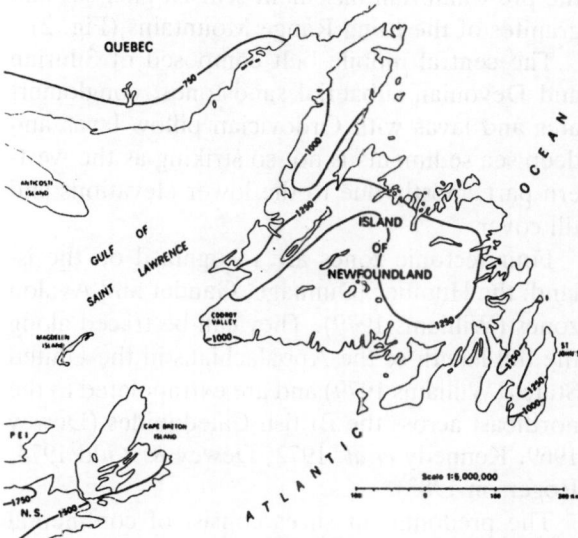

Fig. 5. The growing degree days above 5°C (1000–1250, Roberts 1983.)

degree days above 5°C (1000 to 1250) (Fig. 5). The precipitation and potential evapotranspiration curves for Stephenville, Corner Brook and Deer Lake in western Newfoundland are given in Wells, Bouzane & Roberts (1972) and for the central portion of the Island in Damman (1964).

Geology

The island of Newfoundland lies at the northeasternmost limit of the Appalachian mountain system in North America. Rogerson (1983) reports that it has in fact been a major field centre from which tectonic zones have been identified and traced throughout the Appalachian Orogenesis (Williams 1979). The importance of the island as a focus of geological interest stems from Wilson's (1966) observation that the island probably represents the consequences of Paleozoic continental drift and collision. The major features identified were the 'Proto-Atlantic' in central Newfoundland, separated by two continental remnants, western Newfoundland and the Avalon (Rogerson 1983). Earlier, it was recognized (Williams & Malpas 1972) that the western platform is the Klippen or thrust sheets of Cambrian and Ordovician serpentinized rocks. Other features in the crest are the Cambrian and Ordovician shallow water sediments of limestone and sandstone and the pre-Cambrian basement schists, gneisses and granites of the Long Range Mountains (Fig. 2).

The central mobile belt composed of Silurian and Devonian subaerial sandstones, conglomerates and lavas with Ordovician pillow lavas and deep sea sediments is not so striking as the western part, mostly due to the lower elevations and till cover.

Four tectonic zones are recognized on the island: the Humber, Dunnage, Gander and Avalon zones (Williams 1979). They can be traced along the full length of the Appalachians in the United States (Williams 1979) and are extrapolated to the northeast across the British Caledonides (Dewey 1969, Kennedy *et al.* 1972, Dewey & Kidd 1974, Rogerson 1983).

The predominant slices consist of continental sediments, shale melange and, uppermost of all, ophiolotic suites of mantle periodotites and ocean crust volcanics from the Western Continental Platform or Humber Zone. The sequences were assembled in the east at a time when the continental margin was isostatically depressed. The ophiolites were rafted over the continental margin on top of pre-assembled sedimentary slices (Rogerson 1983) and the westward transportation of the piles or emplacement of the Klippen, took place by gravity-sliding as the eastern margin of the continent was uplifted.

The White Hills peridotite is composed of an interlayeral sequence of spinel lherzolite and hartzburgite and subordinate dunite occurs as layers parallel to spinel lherzolite and hartzburgite and dykes and lenses that lie oblique to the tectonite fabric (Talkington & Malpas 1981). The processes of serpentinization and related geology in Newfoundland and elsewhere is reviewed in this volume (Malpas Chapter 2).

Soils

Classification and Pedogenesis

The soils are described and classified according to the Canadian System of Soil Classification (1978), and are brought to the subgroup (level 3) and series (level 4) where possible.

Five of the nine soil orders (level 1) which occur in Canada occur in Newfoundland. These include members of the (1) Podzolic; (2) Gleysolic; (3) Brunisolic; (4) Regosolic and (5) Organic Orders. Luvisolic and crysolic phases of some of these orders also occur. Podzols are the zonal soils and the Orthic Humo-Ferric Podzol is the most common soil subgroup (Roberts 1983).

The major soils of the serpentinized areas of Newfoundland are Orthic Regosols, Gleyed Regosols and Rego Gleysols with a peaty phase in the fen areas (Fig. 6). Two major soil series are prominent. The Blomidon series consists of moderately well to imperfectly drained Gleyed Regosols, cryoturbic phase, which occur mainly on the bottom flats. The Round Hill series comprises well to moderately well drained Orthic Regosols, cryoturbic phase, that are common on moderate to fairly steep slopes. Together they cover an area of more than 7,700 ha in western Newfoundland as well as extensive areas in the northern and central portions of the province (Roberts 1980, 1986b).

The major soils of Newfoundland serpentinized areas are virtually identical to those described by Sirois (Chapter V of this volume) where cryoturbation, low plant cover and changes in drainage as a result of topographic position give rise to ultramafic Regosols.

The soils of the Southern Appalachian serpentinized areas of Maryland (Rabenhorst & Foss 1981, Rabenhorst, Foss & Fanning 1982) are better developed (Hapludalfs i.e., Luvisolic in the Canadian Classification) having Bt (illuvial horizon enriched with silicate clay) that are probably Aeolian in origin, similar to some of the soils on the Lizard (Coombe & Frost 1956b, Proctor, Chapter VI of this Volume).

In Newfoundland and Quebec, Canada, the regional climate favors podzolization but it is extremely rare in serpentinized areas even with complete vegetation cover. As reported by Sirois (this Volume) before podzolization can take place, leaching of most of the Mg^{2+} of the upper horizon is a pre-requisite for the translocation of organic matter, aluminum and iron. And with 50–60% silt and clay in the upper horizons this is a slow process but under certain altitudinal conditions with different degrees of erosion, rainfall and land use (Ragg & Ball 1964) conditions for

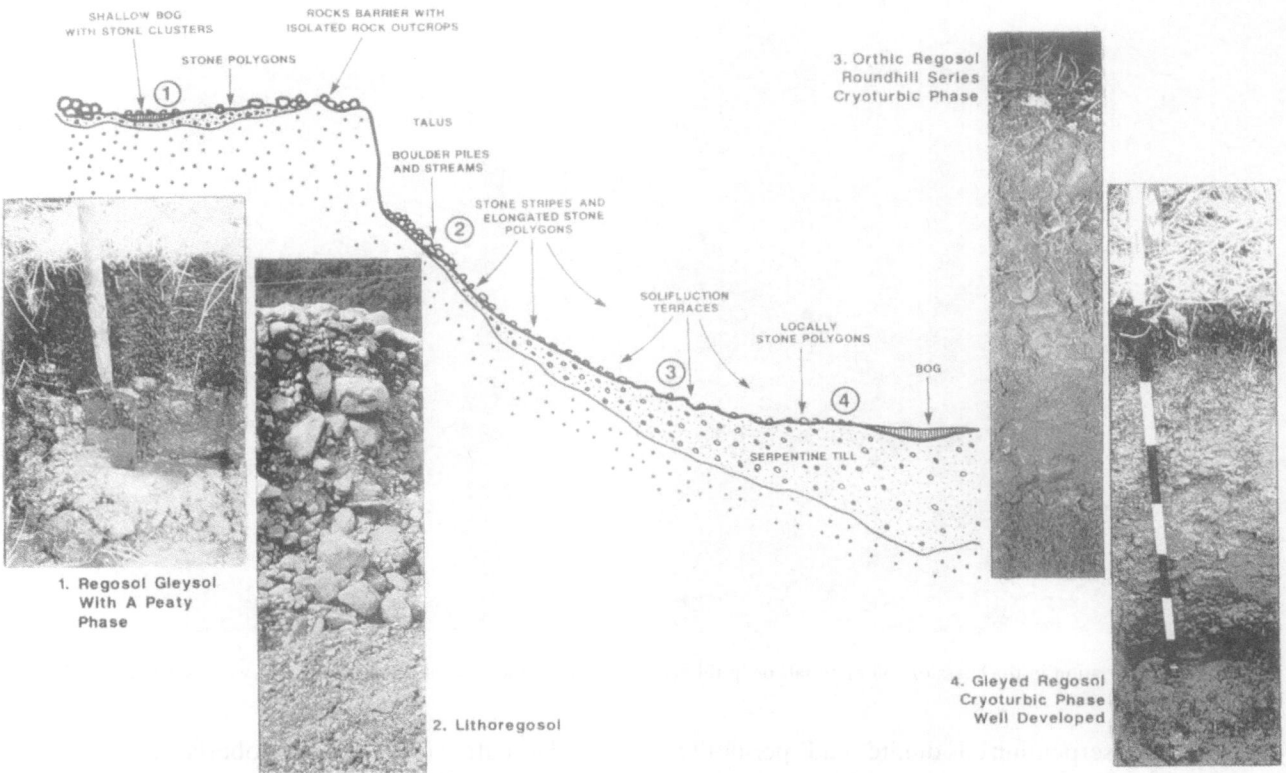

Fig. 6. The major toposequence of soils on west Newfoundland serpentinized areas (Modified from Damman 1967, Roberts 1980, 1986.)

the establishment of atypical Podzols are enhanced.

The soils of the serpentinized areas from Newfoundland, therefore, are similar to those described by Sirois for Quebec, Canada (Chapter V) and Great Britain (Coombe & Frost 1956b), Spence 1957, Ragg & Ball 1964, Wilson & Berrow 1978, all of which are reviewed by Proctor, Chapter VI). There are however, also many nutritional, moisture and drainage differences in addition to toxic metal content, and direct country comparison, as pointed out by Proctor & Woodell (1975), is therefore difficult. There are also some similarities to the soils of serpentinized areas of Sweden (Rune 1953), Finland (Vuokko 1978), Germany (Rhodenkirchen & Roberts, in prep.), Spain (Ojea & Lopex 1980, van den Born *et al.* 1985, and DeSequeira 1969), Portugal (this Volume) and Japan (Suzuki, Mizuno & Kimura, 1971, Mizuno & Nosaka, this Volume). The serpentinitic Cambisols of Switzer-

land (Juchler & Stichter 1986) are better developed and are influenced by a more complete vegetation cover. Soils from Central America (Halstead 1968) showed similar Mg/Ca values to Newfoundland soils but the latter have lower N and CEC values.

Physical and chemical properties

Soil textures (Roberts 1980) are loam to sandy loam for most serpentinized sites in Newfoundland with 40–60% sand content. Clay content varied 11–25% and was certainly influenced by frost churning. Clay minerals are mainly kaolinite under the better drained mineral soil areas where vegetation is discontinuous and frost churning is common, and montmorillonite under the completely vegetated sedge fen areas (Dearden 1979).

The mineral soils are derived from base-rich very stony, sandy loam colluvium and till derived

Fig. 7. Variation in thickness of soil material, deep till and colluvium at the base of Table Mountain, west Newfoundland.

from the serpentinized dunite and peridotite. There is great variation in thickness and both depth categories of shallow and deep and shallow and bare (Roberts 1980) are reported (Figs 7, 8).

As in Quebec ultramafics, cryoturbation tends to homogenize the mineral soil by vertical mixing. There is plenty of free moisture, especially in the fall and spring when the ground is often bare of snow cover and the imperfectly drained Regosols are at full field capacity. Cryoturbation also keeps a relatively high Mg/Ca quotient (fresh material from below) in addition to providing an unstable substrate and is a main factor in slowing down the colonization of serpentinized areas (Roberts 1980, Roberts 1982, Sirois Chapter V, Birse 1982).

Pattern ground in the form of stone polygons and stripes (Wells, Bouzane & Roberts 1972, Roberts 1980) is common at elevations of less than 150 m (Fig. 9) in the western Newfoundland serpentinized areas. Cryoturbation is also quite evident in true alpine areas over 700 m with no major vegetation change.

A typical soils transect from western Newfoundland (Fig. 6) was analyzed for chemical properties and presented in Table 1. It contains similar data presented by Roberts (1980). The total and available NPK values are generally low but are similar to many parent materials of forested sites in western Newfoundland (Roberts & Khalil 1980). The saturation of Mg^{++} ions for the mineral soils is greater than 60% and often 90% or over.

The high Mg content (12–16% total) and 7.18 to 16.6 meq/100 g available is very typical of values over three different soil subgroups. The variation within soil subgroups (over 50 analysis) shows Mg differences of less than 5% in total and less than 3% available. Total Ca values for the same soils transect (Table 1), except for peat covered soils, range from 0.12–0.26% and 0.12 to 1.76 meq/100 g and vary up to 13% within the same soil type and up to 25% variation in available Ca within the same soil type. The Mg/Ca quotients are commonly 50–100 for total values and 3–25 for availables.

In the Regosol Great Group (profiles 3 & 4, Table 1) the average pH is 6.8 ± 0.06. Total Ni averaged 3030 ± 150 $\mu g\,g^{-1}$ dry wt. and available Ni 3.4 ± 0.3 $\mu g\,g^{-1}$ dry wt. which is about 0.1% of the total. The Mg/Ca quotient is 11.9 ± 1.5 and the organic matter content <4.5%. The CEC

Fig. 8. Large areas of shallow and bare boulder colluvium with serpentine weathered from Dunite boulders.

values range from 11–19.5 meq/100 g and averaged 15.6 meq/100 g (Table 2).

The relationship between pH, total and available Ni, the Mg/Ca quotient and the organic matter content (%) for a typical transect (#34) is given in Table 3. In Rego-Gleysols with >50% organic matter and pH below 6.5, the availability of Ni was 2–3.6% of the total amount. The Mg/Ca quotient was 0.4–9.6 but generally less than 1 except for one value at the low pH of 5. In Orthic Regosols where pH values exceeded 6.6 and organic matter was between 5–20%, Ni availability was less than 1% even though there was 3–4 times as much Ni in the total values. The Mg/Ca quotient increased to 26–68 which was lower than values shown in Table 1.

The relationship between total exchangeable bases, cation exchange capacity and saturation with Mg is compared to the Mg/Ca quotients for total, available and water extract analysis for two of the major soil series and is shown in Table 4. The total values have the highest (50–100) Mg/Ca quotients; Mg/Ca quotients (2–30) for availables are less but the lowest Mg/Ca (less than 10) were

recorded for the water extracts. Proctor *et al.* (1981) found that Mg/Ca quotients were generally higher in water extracts, whereas our results for the two main soil series show a reduced Mg/Ca quotient except for one soil horizon in the rh series. In that Cy2 horizon (Table 4) the saturation with Mg^{++} ions was only 62%, organic matter was less than 1% and the pH was 7.1.

In all profiles only trace quantities (0.2–0.5 $\mu g\,g^{-1}$ dry wt) of Ni were available in the water extract analysis suggesting that Ni may play a lesser role in terms of toxicity in Newfoundland serpentine soils.

Calcium Seep – (travertine)

Of great importance for plant distribution (discussed in Vegetation Section on Serpentinized areas) are the few surface deposits produced by the precipitation of calcite. The travertine occurs sporadically on the lower flats and on the solifluction terraces where due to frost action and soil creep, calcium deposits work their way to the surface (Figure 10).

(a)

(b)

Fig. 9. (a, b) Sorted polygons with sparse vegetation cover. Behind sheltered boulders and in crevice vegetation cover is more complete.

Table 1. Total element concentration (% oven dry basis), Ca/Mg, Mg/Ca, (T.E.B., C.E.C.) meq/100 gm pH, organic matter %, selected metals, μg g⁻¹

Horizon	Depth (cm)	N	P	K	Ca	Mg	Na	T.E.B.	Ca/Mg	Mg/Ca	pH	OM%	C.E.C.	Al	Co	Cr	Cu	Ni	Zn
1.*																			
Om[1]	35-0	2.0070	0.0515	0.0932	0.0965	1.282	0.0213		0.07	13.3	5.0	56.5		7803	43	309	23	1069	45
Cg1	0-15	0.1740	0.0034	0.0187	0.1483	7.183	0.0123		0.02		6.1	4.5		11,862	129	500	-	2705	81
Om[2]	35-0	0.0566	0.0004	0.3938	1.1377	10.904	0.0687	13.0448	0.17	9.6			107.2	-	Tr.	Tr.	Tr.	38.6	2.2
Cg1	0-15	0.0008	0.0001	0.0379	0.2315	8.473	0.0461	8.8891	0.03	36.6			14.1	-	Tr.	0.6	Tr.	67.2	0.2
3.*																			
Cy1[1]	0-10	0.1330	0.0170	0.0425	0.1440	12.638	0.0195		0.01	87.8	6.9	2.2		7930	180	1270	30	3610	70
Cy2	10-25	0.1470	0.0180	0.0580	0.1500	12.205	0.0190		0.01	96.8	7.0	3.6		8630	160	1090	40	3560	80
Cgy1	25-40	0.1400	0.0200	0.0370	0.1280	12.385	0.0190		0.01	96.7	7.0	3.1		9360	170	1260	30	3600	80
Cgy2	40-50	0.2000	0.0200	0.0850	0.1390	13.163	0.0170		0.01	94.7	6.5	4.4		8590	170	1050	30	3890	90
Cgc	50-60*	0.0220	0.0080	0.0180	0.1380	15.990	0.0160		0.01	115.9	7.3	0.6		7040	140	860	20	3090	50
Cy1[2]	0-10	0.0003	0.0002	0.0340	0.8950	13.654	0.0790	14.6620	0.07	10.7	7.8		14.9		Tr.	Tr.	0.6	3.9	1.2
Cy2	10-25	0.0003	0.0003	0.0290	0.8190	16.638	0.0760	17.5620	0.05	11.3	7.9		17.3		Tr.	Tr.	0.8	4.4	1.8
Cgy1	25-40	0.0003	0.0003	0.0360	0.8790	16.282	0.0850	17.2820	0.05	18.5			17.6		-	Tr.	-	3.9	-
Cgy2	40-50	0.0006	0.0002	0.0230	0.7160	15.971	0.0630	16.7730	0.04	22.3			18.5		-	Tr.	-	1.9	-
Cgc	50-60	0.0004	0.0006	0.0220	0.5410	9.800	0.0680	10.4040	0.06	18.1			11.0		-	Tr.	-	1.9	-
4.*																			
LF[1]	2-0	0.2070	0.0111	0.0127	0.1419	12.320	0.0098		0.011	86.7	6.4	3.5		-	271	1447	-	2278	108
Aeh	0-13	0.0985	0.0104	0.0027	0.2112	13.966	0.0087		0.015	66.1	6.6	1.8		-	240	1611	-	2565	88
Cy1	13-30	0.1255	0.0127	0.0068	0.2626	12.926	0.0128		0.020	49.2	6.7	2.4		-	344	1206	-	2459	104
Cy2	30-45	0.0615	0.0077	0.0034	0.2108	13.276	0.0097		0.016	62.9	6.6	1.5		-	304	1505	-	2924	96
Cy3	45-60	0.0625	0.0086	0.0028	0.1951	13.449	0.0085		0.015	68.9	6.8	1.2		-	274	1079	-	2876	90
Cy4	60-75	0.0605	0.0087	0.0036	0.1977	13.658	0.0095		0.015	69.1	6.8	1.0		-	279	965	-	2750	89
Cg	75*	0.0595	0.0085	0.0030	0.2005	13.818	0.0094		0.014	68.9	6.7	0.9		-	267	1094	-	2758	85
LF[2]	2-0	0.0005	0.0005	0.1507	2.8980	13.952	0.3364	17.3374	0.20	4.8			19.5		Tr.	Tr.	-	4.4	-
Aeh	0-13	0.0002	0.0004	0.0513	1.7645	11.028	0.3241	13.2373	0.16	6.3			13.7		Tr.	Tr.	-	2.4	-
Cy1	13-30	0.0002	0.0003	0.0495	1.4314	13.282	0.5885	15.3520	0.11	9.3			16.7		Tr.	Tr.	-	3.6	-
Cy2	30-45	0.0001	0.0003	0.0439	1.4418	11.164	0.3948	13.0344	0.13	7.7			15.7		Tr.	Tr.	-	4.0	-
Cy3	45-60	0.0001	0.0003	0.0465	0.9486	10.520	0.3262	11.8417	0.09	11.1			14.3		Tr.	Tr.	-	3.2	-
Cy4	60-75	0.0001	0.0003	0.0425	0.9497	10.426	0.2893	11.7079	0.09	11.0			14.7		Tr.	Tr.	-	4.4	-
Cg	75*	0.0001	0.0003	0.0380	0.9221	11.198	0.2927	12.4514	0.08	12.1			13.4		Tr.	Tr.	-	2.4	-

Total nitrogen by standard Kjeldahl distillation method (Jackson 1958). Total phosphorus, digestion of 80-mesh soil with concentrated acids (Jackson 1958). All total elements were read from this extraction using an atomic absorption spectrophotometer Perkin–Elmer No. 304. Organic matter content by the Walkley-Black method (Jackson 1958). Available nitrogen by a steam distillation method (Bremner and Keeney 1965). Available phosphorus by the chlorostannous – reduced, molybdophosphoric blue color method with dilute sulfuric acid (Jackson 1958). Available elements were determined using the above atomic absorption unit, Perkin–Elmer No. 304 after extraction with neutral ammonium acetate. Exchange cations by the centrifuge method using neutral ammonium acetate. Cation-exchange capacity by the sodium saturation method (Black 1965). [1]Total nutrients (%). [2]Available nutrients meq/100 g.
*Numbers correspond to the locations in Fig. 6.

Table 2. Total and available Ni $\mu g\,g^{-1}$, pH, organic matter percent and CEC in typical soils of serpentinized areas in western Newfoundland

Horizon	Depth (cm)	pH	Total Ni	Available Ni	Available %	OM%	CEC meq/100 g
Mineral Soil Profile							
FH	3–0	6.6	731	7.3	1.00	66.1	103.0
Cg1	0–7	6.8	2699	10.2	0.38	9.1	42.6
Cg2	7–28	6.8	3389	13.1	0.38	5.9	37.7
Cg3	28–58	6.9	3451	9.3	0.27	7.6	35.7
Cg4	58–90	6.9	3292	8.2	0.25	10.2	38.5
Fen Profile (Deep)							
Oh1	10–0	6.4	1242	25.5	2.05	60.7	64.8
Oh2	25–10	6.5	673	17.2	2.56	68.6	62.1
Oh3	25–50	6.3	639	16.4	2.57	66.4	63.3
Oh4	95 + − 75	6.4	589	11.9	2.02	66.8	58.4

Table 3. The pH, total and available Ni ($\mu g\,g^{-1}$), Mg/Ca quotient and organic matter % of surface soils along transect #34

pH	Total Ni	Available Ni	% Average of Total Ni	Mg/Ca	OM%
5.0	1069	38.6	3.6	9.6	56.5
6.1	2705	67.2	2.5	36.6	4.5
6.3	639	16.4	2.6	0.4	66.4
6.4	589	11.9	2.0	0.9	66.8
6.4	1242	25.5	2.0	0.5	60.7
6.5	673	17.2	2.6	0.8	68.6
6.6	2158	16.5	0.8	26.9	21.0
6.8	2699	10.2	0.4	33.0	9.1
6.8	3389	13.1	0.4	46.0	5.9
6.9	3451	9.3	0.3	68.4	7.6
6.9	3292	8.2	0.3	57.9	10.2

Table 4. The total exchangeable bases (TEB), cation exchange capacity (CEC) meq/100 g; base saturation and saturation with Mg^{++} ions (%) and the Mg/Ca quotient for total, available and water extract analysis for the two main serpentine soil series (Roberts 1980) of western Newfoundland

Horizon	Depth (cm)	TEB	CEC	Base Saturation	Saturation with Mg^{++} ions	Mg/Ca Total	Mg/Ca Available	Mg/Ca Water Extract
Blomidon Series (bm)								
Cy1	0–6	14.662	14.86	99	92	97.5	15.3	9.1
Cy2	6–30	17.562	17.31	100	96	81.4	20.3	8.0
Cgy1	30–40	17.282	17.55	98	93	96.7	18.5	9.9
Cgy2	40–50	16.773	18.47	91	87	94.6	22.3	9.1
Cgc1	50–60	10.404	11.01	94	89	115.8	18.1	7.9
Cgc2	60+	10.104	10.78	94	90	118.2	33.8	7.9
Round Hill Series (rh)								
Cy1	0–12	7.883	10.24	76	67	80.4	7.1	6.7
Cy2	12–22	9.409	10.95	85	62	60.4	2.8	7.5
Cy3	22–30	10.687	12.22	87	80	52.6	11.6	5.1
Cc	30+	9.161	10.58	87	79	105.9	11.9	5.9

Fig. 10. Calcium seep material (travertine) at base of soliflucted stone, west Newfoundland.

Flora and vegetation

Many researchers agree that the floras of serpentinized areas are unique, botanically interesting and often contrast strongly with more luxuriant vegetation types of adjacent areas. The cause of the unusual plant associations of serpentinized areas was the subject of considerable dispute in the first half of this century and was still largely unresolved (Lyon *et al.* 1968) until the review of Proctor & Woodell (1975). In more recent years, however, there is a decreasing tendency to try and find single factors which are the cause of differences in serpentinized areas. There are a number of generalities that are common in the literature on temperate areas which sum up some of the characteristics of these areas.

a) The serpentine flora is relatively poor in species diversity and abundance with many species occurring disjunctively (Rune 1953).

b) On serpentine, several species are represented by particular races (ecotypes) differing morphologically from the type races of the species (Rune 1953, Walker 1954, Kruckeberg 1954).

c) The serpentine flora contains basicolous as well as acidicolous plants which often grow together (Rune 1953, Coombe & Frost 1956a).

d) The serpentine flora has a relatively xerophytic character, (Rune 1953, Whittaker 1954, Carter, Proctor & Slingsby 1987).

e) The serpentine is often dominated by a certain family or genera e.g. *Caryophyllaceae* in E. North America (Roberts 1980, 1981) and northern Europe (Rune 1953, Coombe & Frost 1956a, Proctor and Woodell 1971). Whittaker (1954) noted that grasses are more easily able to occupy serpentine environments than are many other herbs and suggests that some grass species may do so without ecotypic differentiation.

f) The physiognomy of serpentine vegetation is characterized by stunting and impoverishment of growth. Serpentine sites tend to be open with low plant cover (Walker 1954, Kruckeberg 1954, Whittaker 1954, Roberts 1980, 1986a, Carter, Proctor & Slingsby 1987).

g) The existence of rare arctic-alpine species confined to the serpentine mainly because of competition (Rune 1953, Fernald 1911, Whittaker 1954, Damman 1965, Roberts 1981, Zika & Dann 1985)

87

General description of the vegetation of Newfoundland serpentinized areas

This section describes the serpentine vegetation and flora of Newfoundland and attempts to compare it with serpentine floras from other parts of the world.

On the ultramafic areas of Newfoundland, vegetation is sparse and cover is less than ten percent. Tamarack, *Larix laricina*, attains a height of three to six metres in a few areas. Scrub tamarack and black spruce, *Picea mariana* (<1 m in ht.), growing densely, occurs in large discontinuous patches and provides shelter for such lower shrubs as *Juniperus communis, J. horizontalis, Potentilla fruticosa, P. tridentata, Betula pumila, Andromeda glaucophylla, Ledum groenlandicum* and *Salix arctica;* the more exposed areas where patterned ground is common are dominated by herbs of the Caryophyllaceae namely *Arenaria humifusa, A. marcescens, Cerastium alpinum, C. beeringianum* and *Silene acaulis*. Other herbs such as *Armeria labradorica, Senecio pauperculus*, the grasses *Deschampsia caespitosa* and *D. flexuosa* are also scattered throughout the areas as well as the lichens *Cladonia rangiferina, C. alpestris, C. uncialis* and *Cetraria islandica*. The moss *Rhacomitrium lanuginosum* is common to exposed and non-exposed areas (Roberts 1979, 1980a, b, 1981).

Pollett (1972) describes the peatland communities over serpentinized areas as peaty mounds, admixed with barren serpentine soils. Here the plant association *Junco-Selaginelletum selaginoides* is prominent. Although similar to the rich fens (on limestone) of northern Newfoundland it is differentiated by the species group *Juncus filiformis, Deschampsia flexuosa* and *Arenaria marcescens. Armeria labradorica* and *Lychnis alpina* are also typical of these fens (Fig. 11a, b). In addition these wet areas have an almost continuous vegetation cover of plants well known from bogs and fens outside the serpentine area, such as *Sarracenia purpurea, Myrica gale, Sanguisorba canadensis, Senecio pauperculus, Potentilla fruticosa, Andromeda glaucophylla, Pinguicula vulgaris, Juniperus horizontalis* and *Scirpus cespitosus*.

Dearden (1975, 1979) describes six major communities on serpentine near Table Mountain, western Newfoundland.

Type A: Arctic-alpine communities found in the most exposed location

This subdivision represents the most xeric and exposed sites found on the serpentine, the vegetation being characterised by its extremely sparse and stunted nature. The most inhospitable colonised sites are dry rock crevices, often caused by the weathering out of the softer serpentine from the dunite host rock. Such sites may be colonised by *Arenaria humifusa, Androsace septentrionalis*, and other hardy arctic-alpine species. The fern *Adiantum pedatum* var. *aleuticum* is a common occupant of more moist and sheltered crevices (Fig. 12a).

The location of the arctic-alpine communities is closely related to the local topography and the associated features of patterned ground development. The initial coloniser is typically *Rhacomitrium lanuginosum*, followed by such species as *Rhododendron lapponicum, Cerastium terrae-novae, C. beeringianum, Sagina nodosa, Arenaria humifusa, A. marcescens*, and various *Salix* species. The central fine areas of the polygons are colonised by chamaephytes, such as *Silene acaulis, Armeria labradorica* var. *submutica, Diapensia laponica*, and *Saxifraga oppositifolia* (Fig. 12b).

The third component of the Type A communities are those areas which experience slightly less environmental stress, such as the banks of the solifluction terraces. The bank vegetation is dominated by prostrate tree species such as *Betula pumila* and *Salix arctica*, their extensive root systems helping to stabilise the soil, along with various shrub species including *Vaccinium uliginosum* var. *alpinum, Potentilla fruticosa* and *Juniperus communis*. The herbaceous arctic-alpine species previously mentioned are common along with *Conioselinum pumilum* and *Solidago hispida* (Fig. 12c, 12d).

Type B: Arctic-alpine communities in less exposed locations

These communities have a more diverse and closed vegetation cover than the pioneer communities, and yet still retain sufficient openness so that arctic-alpine species, such as *Lychnis alpina* var. *americana* and *Saxifraga aizoides*, are not entirely eliminated.

(*a*)

(*b*)

Fig. 11. (a) Shallow fen peats form on flat areas at the base of steep slopes or on the relatively flat alpine plateaus, west Newfoundland. (b) Typical pool pattern with *Junus filiformis* in fens of serpentinized areas, west Newfoundland.

Fig. 12(a).

Fig. 12(b).

Fig. 12(*c*).

Fig. 12(*d*).

Fig. 12. (a) *Adiantum pedatum* in a moist sheltered crevice. (b) Part of frost polygon colonised with *Silene acaulis*. (c, d) *Potentilla fruticosa* and *Juniperus communis* also provide shelter for the common alpine herbs and help stabilize the frost churning soils with their extensive roots.

Fig. 13. Patches of *Rhacomitrium* moss, with some arctic-alpine species where patterned ground has a more complete vegetative cover.

The distinct patterned ground features of the more exposed areas are here almost totally vegetated, with large areas colonised by *Rhacomitrium lanuginosum*. The wetter habitats, particularly of the coarser polygon boundaries, are colonised by *Scirpus cespitosus*, *Carex echinata* and *Juncus trifidus*, *R. lanuginosum* being replaced by *Pleurozium schreberi*, *Hylocomium splendens* and *Dicranum* spp. in wetter areas. In small sheltered depressions on the plateau the community has additional shrub species, in particular *Andromeda glaucophylla*, *Myrica gale*, *Juniperus horizontalis* (Fig. 13).

Type C: Ecotone communities dominated by shrubs and sedges

This type forms the ecotone between Type B and the peatland and sedge meadow type communities and is found at bottom slope locations with better soil conditions and free moisture. Common species are *Juniperus communis*, *Ledum groenlandicum*, *Myrica gale*, *Betula pumila* with *Scirpus cespitosus*, *S. hudsonianus*, *Carex scirpo-*

ides, *Juncus balticus* and *Thalictrum alpinum* (Fig. 14).

Type D: Peatland and Sedge Meadow Communities

As described by Pollett (1972), with the addition of a description of several large fen hummocks over a meter above the water table, the base of the hummocks is composed of various *Sphagnum* species (*S. papillosum*, *S. nemoreum*, *S. fuscum*, *S. palustre*), topped with deep clumps of black, undecomposed *Rhacomitrium lanuginosum*. The fens (above 700 m) are dominated by various *Carex* species, including *Carex buxbaumii*, *C. echinata*, *C. exilis*, *C. limosa*, *C. paupercula*, *C. scirpoides*, in association with grasses such as *Festuca rubra*, *Eriophorum angustifolium*, *E. chamissonis*, and the rushes *Juncus balticus*, *J. filiformis* and *J. trifidus*. The more exposed fens have abundant members of *Betula michauxii*, *Vaccinium uliginosum* var. *alpinum*, *V. oxycoccus* and *Empetrum nigrum*, all of which are not common in lower altitude fens (Fig. 11).

92

Fig. 14. Ecotone between type B and peatland communities.

Type E: Fellfield Communities

These are areas of isolated coarse mineral substrates on relatively gentle slopes (<5°) commonly found at low altitude. Floristically they display great uniformity, having a large number of species that occur in the same relative positions in each clump. Thus, the foremost parts of the clump usually contain prostrate *Alnus crispa*, followed by other dwarf tree species, such as *Larix laricina* and *Betula pumila*, which precede the large 'tail' of *Juniperus communis* stretching back downslope (Fig. 15).

Type F: Scarp-foot communities dominated by trees and shrubs occurring in sheltered localities

Although some dominant species of the Type F community, notably *Larix laricina*, *Juniperus communis* and *J. horizontalis*, occur in other habitats, many species (e.g. *Acer rubrum*, *Betula papyrifera*, *Nemopanthus mucronata*, *Abies balsamea* and *Pinus strobus*) are restricted to these locations. Similarly, very common shrubs in this habitat (*Gaylussacia dumosa*, *Lonicera villosa*, *Vaccinium vitis*, *Ideae*, *V. angustifolium*, *V. mac-*

rocarpon, *Rhododendron canadense*, *Kalmia polifolia*, *Gaultheria hispidula* and *Epigaea repens*) are found very rarely elsewhere. Herbs are dominantly those of the understorey, well adapted to low light supplies, such as *Circaea alpina*, *Cornus canadensis*, *Maianthemum canadense* and *Trientalis borealis* (Fig. 16).

This habitat is not always underlain by sepertinized rocks and till from shale and gabbro are mixed under this type. The drainage and seepage of this area is then frequently from non-serpentinized rocks.

Using Z-M techniques, Meades (1983) described two sub-associations in the arctic-alpine heath formation:

LYCHNETUM TYPICUM – Lt. The TYPICUM subassociations are differentiated from the ADIANTETO-sum by the lack of the differential species of the latter subassociation. This heath type occurs on the rapidly drained boulder talus. It is similar to the 'Arctic alpine communities of less exposed locations' described by Dearden (1979). *Lychnis alpina* var. *americana*, *Arenaria humifusa*, *A. rubella*, *A. marcescens*, *Solidago hispida* var. *lanata*, *Juniperus horizontalis* and *Rhacomitrium lanu-*

93

Fig. 15(*a*).

ginosum occur in isolated clumps that collectively cover less than 10–15 percent of the felsenmeer. The soils are ultramafic Regosols with little or no profile development (Roberts 1980). Large sorted boulder polygons are common throughout the level terraces.

LYCHNETUM ADIANTETOSUM – La. The ADIAN-TETOSUM is differentiated from the TYPICUM sub-association by the occurrence of the following species groups:

– Sanguisorba canadensis
– Sarracenia purpurea
– Scirpus cespitosus
– Adiantum pedatum
– Drosera rotundifolia
– Anaphalis margaritacea

This habitat represents the late snowbed communities and drainage channels that retain significant quantities of moisture throughout the season. Quite often snowbeds last all year but rarely two or more years without complete melt has ever been observed by the author in the last 25 years (Fig. 17a, b). *Scirpus hudsonianus*, *Sanguisorba canadensis*, *Adiantum pedatum* var. *aleuticum* are dominant in these flushes. Dearden (1979) referred to these communities as 'ecotone com-munities dominated by shrubs and sedges.' Krummholz dominated by *Larix laricina* and *Picea mariana* (1–2 m) is frequently associated with the ADIANTETOSUM snowbed communities. The krummholz sometimes have clumps of dwarf birches, *Betula pumila* and *Betula borealis*, along WITH DWARF SHRUBS (I.E., *Myrica gale*, *Ledum groenlandicum*, *Nemopanthus mucronata*) in smaller clumps. The herb layer is formed of *Sanguisorba canadensis*, *Geocaulon lividum*, *Goodyera repens* and *Mitella nuda* and *Osmunda regalis*. *Epigaea repens*, a southern coastal plain element, reaches it furthest northern distribution in this habitat.

In the best developed flushes there is a continuous veneer of sedge peat (5–10 cm) over the ultrabasic serpentine boulders. In the krummholtz communities this peat sufficiently masks the underlying substrate to permit the establishment of the more acidophilous species already listed (Fig. 17).

Dyras integrifolia (Fig. 19) and *Cypripedium calceolus* are often found on the calcium seep deposits (Fig. 10) but also spread to the adjacent serpentine soils if soil depth and moisture are available.

Fig. 15(b).

Fig. 16(a).

Fig. 15. (a) Dwarf trees *Larix laricina* and *Betula pumila* form the most extensive patches of vegetation on gentle slopes. (b) The maximum development of *Larix laricina* (5–10 m) on a sheltered serpentinized slope near northwest Gander River, central Newfoundland.

Floristic and ecological comparison of the Newfoundland ultramafics with other sites

The serpentine flora of western Newfoundland is fairly rich in species but generally low in frequency of occurrence and biomass. The number of species making up the structure of the communites are given in Table 5 and a complete list in Appendix 1. The number of species per vegetation group on serpentinized areas in Western Newfoundland (Table 5) approaches 190 species but some areas within the perimeters are only partially serpentine in origin.

In the Monocotyledoneae, herbs account for about 12% and 100% of the sedges and grasses.

In the Dicotyledoneae herbs account for more than 85% and 90% of the shrubs; 10% of the shrubs are from the Gymnospermae. The adjacent alpine flora (Robertson & Roberts 1982) contains some 250 vascular plants representing more than 45 plant families.

Although the Newfoundland serpentinized flora is characterized by only a few serpentinicolous species, they are botanically interesting. Bouchard, Hay & Rouleau (1978) and Meades (1983) reported on the obvious ties between the Scandinavian and Eastern North American serpentinized floras having several species in common including *Arenaria humifusa* and *A. rubella*, *Lychnis alpina* and *Adiantum pedatum* var. *aleuticum*. The large patches of the moss *Rhacomitrium lanuginosum* are a common feature of some Eastern North American and Scandinavian (Rune 1953, Vuokko 1978) serpentinized areas, as are the grass *Deschampsia cespitosa* and the

Fig. 16(b).

herb *Silene acaulis*. *Lychnis alpina* is one of the most common species throughout most of Boreal to Alpine serpentinized areas. *Arenaria marcescens* and *A. humifusa* also form a link with the Quebec, Scandinavian and Newfoundland serpentinized floras. Some of the other species which occur on limestone as well as on serpentinized areas are listed by Damman (1965). They include: *Arenaria humifusa*, *A. rubella*, *Armeria labradorica*, *Castilleja septentrionalis*, *Cerastium beeringianum*, *Rhododendron lapponicum*, *Salix cordifolia*, *Solidago multiradiata*.

The serpentine vegetation of west Newfoundland resembles that of Quebec and Great Britain in terms of individual species present. Many of the plants occurring on Newfoundland serpentinized areas are found on the Vermont, USA, ultramafics where they are considered rare because of limited habitat (Zika & Dann 1985).

Appendix 1 lists constant plant species that occur on Newfoundland's serpentine areas. It can be seen that more than 13 plant species and 20 genera are common to Scottish (Spence 1970, Proctor & Woodell 1971, Birse 1982) and Newfoundland serpentines. As regards England six species and 17 genera are common to both (Coombe & Frost 1956a). It should be noted that long species lists are not common in literature sources and hence a detailed comparison cannot always be made. It should also be noted that Scotland and Newfoundland are in the same general climatic belt and this factor is important in looking at species composition.

Betula spp. and *Salix* spp. are found on serpentines of Czechoslovakia. *Silene inflata* is common on Polish serpentines (Sarosiek 1964).

Kruckeberg (1954, 1969, Chapter III of this Volume) states that of all the genera the most characteristic of serpentine over all of central California is *Streptanthus* and gives evidence to show that this genus exhibits genetic differences on serpentine sites, which is reflected in the growth characteristics. The non-serpentine races of this genus grew only one third the size when both serpentine and non-serpentine races were grown in the same serpentine soil. *Streptanthus polygaloides* is known as a hyperaccumulator of Ni (Reeves, Brooks & MacFarlane 1981) as are three varieties of *Thlaspi montanum* (Reeves, MacFarlane & Brooks 1983) among 57 taxa of serpentine-tolerant species.

Most authors including Proctor & Woodell (1971), Kruckeberg (1969) and Rune (1953) state that some plants occur disjunctively on serpentine. Proctor & Woodell list some eight species that are disjunct in Scottish serpentines. The most notable is *Lychnis alpina* which is also disjunct in Newfoundland (Fig. 20).

Most serpentine floras contain basicolous as well as acidicolous species. Damman (1965) reports that many acidicolous species occur on the basic serpentine soils of Newfoundland because of lack of competition. Proctor & Woodell (1971) suggest that the ability of both these groups of plants to produce healthy root systems in soil of pH 5–6 together with the lack of competition and the physical factors present are the reasons why these two groups of plants are found on serpentine areas, and I tend to agree.

Xeromorphism as Proctor & Woodell (1971)

Fig. 16(c).

Fig. 16. (a) Contrast in well vegetated shale deposits left of fault within the serpentinized area. (b) Serpentinized ultrabasic rocks right of fault. (c) The faulted contact between the shale bedrock (left) and serpentinized bedrock (right) is the visible location for the very rare mineral Xonotlite.

suggest is probably related more to the low tissue production associated with low nutrient levels than to extensive drought. However Carter, Proctor & Slingsby (1987) point out that the physical nature of the skeletal soils and the absence of groundwater are likely to result in frequent drought for plants of this more extreme habitat. On Newfoundland serpentines rainfall is common several times a week throughout the growing season. Roberts (1980) has shown that the dominant serpentine soils in Newfoundland have significant quantities of available moisture; however, the fellfield plant communities may occasionally suffer from drought stress in mid summer.

Whittaker (1954) shows that serpentine vegetation of the Siskiyou mountains (S.W. Oregon) is not impoverished if adjacent vegetation on diorite is compared. The approximate number of species on both rock types is about 100, the same as Newfoundland; however cover is much greater, leading Whittaker to speculate that some serpentine floras are a seral stage still evolving to a climax type. This statement is worthy of consideration. When observing the serpentine vegetation of west Newfoundland I have noted that the vegetation cover is increasing on the west Newfoundland serpentines over the past twenty years and is much greater than it was in Fernald's time, in 1911. Colonization is extremely slow due to the many adverse soil conditions described earlier. A sample of *Larix laricina* incremental cores also showed higher growth rates in the past 25 years.

In Pennsylvania serpentinized barrens during periods when moisture and nutrient supply are low, growth of serpentine-restricted plants (Hart 1980) is favoured over weed species but within a particular range of moisture and nutrient availability both types can co-exist. Moisture and nutrition are also very important in the determination of vegetation types in Western North America (Kruckeberg, this Volume), especially in California (Turitzin 1981) and Oregon.

The members of the Caryophyllaceae are conspicuous rather than dominant in serpentine

97

(a)

(b)

Fig. 17. Melting snow provides moisture throughout most of the growing season for the *Lychnetum adiantetosum Subassociation*: (a) Large cirque, western Newfoundland, showing snow conditions in mid June. (b) Large crevices and narrow steep sided valleys fill completely (10–50 m) with windblown snow in winter and may persist all year.

Fig. 18. In the best developed flushes there is a continous veneer of sedge peat (5–20 cm) over the serpentinized boulders and colluvium.

Fig. 19. Dryas integrifolia on serpentinized bedrock having spread from adjacent calcium seep material (Fig. 10).

99

Table 5. Number of species per plant group on serpentinized areas in western Newfoundland

Trees	3–6	(wB, bS, wS, wP, and bF)
Shrubs	41–52	
Herbs	49–86	
Sedges	15–20	(largely Carex)
Grasses	6–13	
Rushes	4–6	(Juncus)
	3–7	
Vascular cryptograms	121–190	
*Mosses/Liverworts	83	
Lichens	8–10	

Table based on collections of Roberts and Dearden.
N.B. Some are as only partially serpentine in origin, however, and it depends greatly where lists or specimens are collected. Table 5 reports the typical range in number of plants.
*Following Belland & Brassard (1988).

Fig. 20. Lychnis alpina on serpentinized rocks, central New-foundland.

areas. Cover values are low but frequency of occurrence in northern Europe (Proctor & Woodell 1971) and in Newfoundland show this, while shrubs and grasses have higher biomass.

The ability of the Caryophyllaceae family and other plant species to tolerate serpentine soils will be discussed in the next section in conjunction with their nutrient uptake.

Origin of the flora of Newfoundland serpentinized areas

In addition to identifying and describing the Newfoundland flora, Fernald is well known for his outspoken views on the origin of the flora and how this was related to past glaciations. His early thoughts included a coastal plain element in the Newfoundland flora that had migrated northwards along the coast. In his later investigations he proposed the existence of nunataks (a hill or mountain sticking out above the ice) as the reason for the occurrence of the cordilleran element in the flora. This idea coincided with A. P. Coleman's studies of geomorphology and initially received major approval (Roberts 1987).

From a geomorphological point of view mapping by Grant (1969) drew attention to smooth, rounded, felsenmeer-covered summits rising above the roche moutonnées surface of the Long Range Mountains at the northern end of Northern Peninsula. These were regarded as nunataks projecting through the local ice cap during the latest glacial expansion ca. 11,000 years B.P. By 1977 Grant (1977) expanded the scope by treating the remainder of the Northern Peninsula where several dozen more summits of felsenmeer (delimited by ice-marginal features) were depicted as having remained above 'a late Pleistocene readvance of piedmot glaciers'. These extended from the Long Range ice cap that was deployed between coastal highlands onto the lowlands as contiguous expanded-foot glaciers. The serpentinized area of western and northern Newfoundland therefore can be regarded as nunataks. The nunatak theory appears to explain the presence of certain bryophytes (Belland 1987a, b) in the Long Range Mountains, but still most of the distributions of the higher plants can be satisfactorily explained by correlations with present climatic and edaphic conditions. Other peculiar distributions can be explained by assuming that the species were widespread during the post-glacial climatic optimum, as reported by Damman (1965), and the former Islands of land now below sealevel off Sable Island and Newfoundland can help explain the distribution of the red pine (Roberts 1985).

Concentrations of macro-nutrients (macro-elements) in native plants growing on ultramafic soils from western Newfoundland

In almost all early studies of serpentinized floras, attention was given to the actual chemistry of the soils on which the plants grow rather than to the nutrient uptake or element concentrations that are present in the foliage of these plants. Gerloff, Moore & Curtis (1966) stated that the selectivity in the absorption and accumulation of mineral elements, although recognized for many years, continues to be one of the most interesting but least understood aspects of plant nutrition. In more recent years, however, with the use of sophisticated atomic absorption and gas chromatography equipment, research on the uptake of elements and the selectivity shown by various plant species to particular elements has increased dramatically.

Although reports on the ability of plants on serpentinized areas to exclude or accumulate elements was common (Roberts 1979, 1980, 1981) Baker (1981) articulated the relevant definition into 3 categories of plants: (1) *Accumulators*, where metals from low or high soil levels are concentrated in above-ground plant parts; (2) '*Indicators*', where uptake and transport of metals to the shoot are regulated so that internal concentration reflects external levels; and (3) *Excluders*, where metal concentrations in the shoot are maintained constant and low over a wide range of soil concentrations, up to a critical soil value above which the mechanism breaks down and unrestricted transport results. Brooks (1983) reviewed much of the literature to this date on what factors effect elemental accumulation by plants and their respective mechanisms. Much of this work was related to trace elements and will be discussed in the next section.

The macro-elements in thirty-one species re-

presenting 14 families growing on serpentinized soils from western Newfoundland are given in Table 6. Nitrogen values are given in percent dry weight, the remaining elements in $\mu g\, g^{-1}$ dry weight. The Mg/Ca quotient is calculated for each species and is also shown in Table 6. The average soil conditions along a transect are shown in Table 7.

The Caryophyllaceae on average show the lowest nitrogen and phosphorus contents with the exception of a few species. *Cladonia* and *Rhacomitrium* which are both cryptogams, lacking vascular tissue and largely depending upon rain and splash water for their nutrition, are also low: *Anaphalis margaritacea*, *Scirpus cespitosus* and *Senecio pauperculus*, generally adapted to low levels of N and P elsewhere off serpentine, are also low.

Anaphalis is not common on all serpentine soils but is found along hiking trails and roadsides scattered throughout the area. Characteristically this species in Newfoundland exists along roadways where it is probably adapted to receiving low nitrogen amounts. The *Scirpus* sample was from a peat deposit and sampled nearly at the end of its growing season. These two factors probably account for the low N and P concentration. *Sarracenia*, also growing on peat, had a higher N and P concentration because of its insectivorous nature. Slowly growing species characteristic of infertile soils usually exhibit a low absorption rate per plant. In comparison with species from high nutrition environment, species from infertile soils absorb considerably less nutrients under high-nutrient conditions but similar quantities and in some cases even more nutrients at extremely low availability, (Chapin 1980). This may explain the high uptake of N and P in *Aster*. *Cladonia alpestris* growing on an adjacent shale deposit showed an almost identical amount of both nitrogen (0.55%) and phosphorus (840 $\mu g\, g^{-1}$). *Cladonia* growing in a forest community in Minnesota (Henry 1973) average 0.7 and 0.8% nitrogen, and 1500 and 1100 $\mu g\, g^{-1}$ dry wt. phosphorus.

It should be noted that the two plant families on Newfoundland serpentine soils with only one species each, thrift, *Armeria labradorica* (Plumbaginaceae) and harebell, *Campanula rotundifolia* (Campanulaceae) have significant amounts of nitrogen and phosphorus. The quantities of N and P in harebell (Henry 1973) from Minnesota

Table 6. Element concentration of native plants growing on serpentine soils from western Newfoundland ($\mu g\,g^{-1}$ dry wt except for nitrogen values in percent)

Family	Scientific name	N%	P	K	Ca	Mg	Na	Mn	Mg/Ca
Caryophyllaeae	*Arenaria humifusa*	0.240	105	2310	1617	148500	91	765	91.8
	Arenaria marcesens	0.360	245	2453	1298	121500	180	816	93.6
	Lychnis alpina	0.960	383	7623	4683	43000	163	368	9.2
	Cerastium beeringianum	0.800	798	4730	11088	26500	95	171	2.4
	Silene acaulis	–	249	1137	8878	22023	239	559	2.5
Compositae	*Aster nemoralis*	1.680	1598	17794	4402	11000	290	138	2.5
	Solidago hispida	0.980	713	10164	4998	46500	215	540	9.3
	Senecio pauperculus	0.630	490	9429	3570	82000	359	918	23.0
	Anaphalis margaritaceae	0.560	433	3020	8400	6030	480	71	0.7
Plumbaginaceae	*Armeria labradorica*	0.900	608	3036	2948	26250	137	268	8.9
Campanulaceae	*Campanula rotundifolia*	1.620	1103	10353	3927	16250	165	164	4.1
Polypodiaceae	*Adiantum pedatum*	1.410	403	9912	4578	31750	180	202	6.9
Pinaceae	*Larix laricina* I	1.950	863	2630	2350	2550	410	391	1.1
	Larix laricina II	1.910	805	2820	2470	3090	290	274	1.3
	Larix laricina III	1.670	850	2750	2400	3160	400	213	1.3
	Picea mariana	0.990	753	2480	3070	2031	460	126	0.7
	Juniperus communis	0.990	753	2450	5190	2010	600	61	0.4
	Juniperus horizontalis	1.000	930	6840	5260	3200	720	72	0.6
Corylaceae	*Betula michauxii*	1.250	655	1840	6630	4520	500	81	0.7
	Betula pumila	1.240	605	960	8020	4420	460	114	0.6
	Alnus crispa	1.440	255	420	5230	4480	184	66	0.9
	Alnus rugosa	1.760	675	1570	6900	2910	360	159	0.4
Sarraceniacea	*Sarracenia purpurea*	1.320	905	1790	2280	4540	300	123	2.0
Cyperaceae	*Scirpus cespitosus*	0.450	95	1200	2040	1760	250	221	0.9
Ericaceae	*Andromeda glaucophylla*	0.990	493	3010	5820	13620	119	253	2.3
	Ledum groenlandicum	1.150	683	6340	6340	3220	410	78	0.5
	Rhododendron lapponicum	1.380	820	3616	5085	2400	614	138	0.5
Rosaceae	*Potentilla fruticosa*	1.060	568	1190	4700	7290	450	85	1.6
	Potentilla tridentata	1.020	690	4350	5590	7980	360	230	1-4
	Sanguisorba canadensis	0.900	505	1670	9580	4530	420	60	0.5
Gramineae	*Deschampsia caespitosa*	–	–	6204	887	7310	846	134	8.2
Grimmiaceae	*Rhacomitrium lanuginosum* I	0.480	170	920	1740	26300	188	454	15.1
	Rhacomitrium lanuginosum II	0.430	160	830	1390	26390	158	417	19.0
Cladoniaceae	*Cladina stellaris*	0.550	843	1400	1150	3020	127	48	2.6

were 1.94 and 1.08% for N and 2300 and 1900 $\mu g\,g^{-1}$ P respectively. Both these P values are higher than in the individuals on serpentine (Table 6).

The Corylaceae (known to be nitrogen fixers) also have highest amount of nitrogen in its foliage. Both species of *Alnus* have the ability to fix atmospheric nitrogen and the role of this family in improving site quality is well known (Tarrant & Trappe 1971). The N and P concentration of the pine family was nearly the same as for individuals on adjacent shale deposits (Roberts, unpublished), with the exception of *Larix* which showed almost twice the amount of nitrogen in its foliage on the serpentinized areas and may partially explain why this species is the only tree species to

attain a height of more than a meter on these areas.

Soil pH values from the transect sites range from 5.0 to 7.5 (Table 7) with mean value of about 7.0 and this is the optimum range in which N and P total concentrations are in available form thus favouring maximum uptake of these two elements.

Much of the research on nutrient uptake by plants from serpentine soil has been done by growing agricultural crops on serpentine soil brought into a greenhouse. Proctor & Woodell (1975) cover this aspect. These authors also summarize what work has been done on the uptake of nutrients by native plants, with most of all early work confined to the uptake of calcium and

Table 7. Element concentration (percent) of serpentine soils from western Newfoundland, Transect No. 1

Horizon	Depth (cm)	N	P	K	Ca	Mg	Na	Mn	Mg/Ca	pH
1:										
Total Nutrients										
Cy1	0–10	0.1330	0.0170	0.0425	0.1440	12.6380	0.0195	0.1170	87.7	7.2
Cy2	10–25	0.1400	0.0200	0.0370	0.1280	12.3850	0.0190	0.1430	96.8	7.0
Available Nutrients										
Cy1	0–10	0.0003	0.0002	0.0012	0.0171	0.1841	0.0018	–	10.8	
Cy2	10–25	0.0003	0.0003	0.0014	0.0176	0.1980	0.0020	–	11.3	
2:										
Total Nutrients										
Om	0–10	2.0070	0.0515	0.0932	0.0965	1.2828	0.0213	0.0199	13.3	5.0
Available Nutrients										
	0–10	0.0566	0.0004	0.0154	0.0228	0.1320	0.0140	–	5.8	

Soils:

Soil No. 1 is a moderately well to imperfectly drained Gleyed Regosol, cryoturbic phase, from the Blomidon series occurring mainly on the bottom flats (Roberts 1980).

Soil No. 2 is a Typical Mesisol (Classification CSSC 1978). The soils were sampled near Winterhouse Brook, lat. 49°28′ N, long. 57°58′ W, 3.5 km SW of Woody Point, Bonne Bay, and at the foot of the Blow-Me-Down Range, lat. 49°04′ N, long. 58°15′ W, 16 km west of Corner Brook, Newfoundland.

magnesium. The Mg/Ca quotient is a quick index to judge the toxicity of a serpentinized site, as the major factor of toxicity (Proctor & Woodell 1975) often relates to the amount of calcium (Proctor 1970, 1971a, 1971b). Table 8 summarizes the Mg/Ca relationships on the average for parent rock total, soils, total and available, and the Mg/Ca quotients for the Caryophyllaceae family (which represent on the average the highest quotients for plants). Proctor & Woodell (1975), citing other sources, suggest that for healthy plant growth, especially crops, a quotient of vegetation and soil of not more than unity is necessary. Various native serpentine species (Proctor & Woodell 1975) showed no yield depression at 6% Ca satur-

ation. However <20% average saturation of Ca produced depressed yield in crop plants.

Mg saturation in Newfoundland serpentinized soils (Roberts 1980) is on the average 90–96% but can be as low as 62–80%. The upper values indicate less than 6% calcium by saturation, hence correspond to the values (Table 8) in the Caryophyllaceae family. The Mg/Ca quotient of 1 or less in the Corylaceae and Pinacea may indicate that the biomass of these two families is greater because they are able to absorb Ca and reduce magnesium. Also the ability of the Carophyllaceae to exist on such low Ca and tolerate and concentrate large quantities of Mg is perhaps the reason for their success in many temperate

Table 8. Mg/Ca quotients of rocks, soils and the Caryopyllaceae family for western Newfoundland

Component	Ca	Mg	Mg/Ca
Serpentine rock	1.12%	39.38%	35.2
Serpentine regosols total	0.164%	13.716%	83.6
Serpentine regosols available	0.895 meq/100 g	14.662 meq/100 g	16.4
Caryophyllaceae family			
Arenaria hamifusa	1,617 µg g^{-1}	148,500	91.8
Arenaria marcensens	1,298 µg g^{-1}	121,500	93.6
Lychnis alphina	4,683 µg g^{-1}	43,000	9.1
Cerastium berringianum	11,088 µg g^{-1}	26,500	2.4
Silene acaulis	8,878 µg g^{-1}	22,023	2.5

areas. These conclusions are similar to the ones reached by Lyon *et al.* (1968) and Lyon *et al.* (1971) (Lee this volume). In studying the element concentrations of plants growing on New Zealand serpentine soils it was also found that some species were able to exclude the high level of magnesium while others appeared able to preferentially accumulate calcium. The same species are not found on Newfoundland serpentines because of climate differences but the mechanism and strategy are probably much the same.

The Mg/Ca quotients for black spruce, larch and speckled alder growing on non-serpentinized soils from Western Newfoundland are 0.35, 0.26 and 0.38 respectively. The latter is practically the same for both serpentine and non-serpentine sites. However the two former have ratios of more than twice the serpentine ratios (see Table 6).

Studies on the uptake of Ca and Mg in Scotland (Proctor & Woodell 1975) show the Ca/Mg quotient of *Juniperus communis* to be 2.48, 2.1 and 1.86 which is identical to our values. Likewise, *Silene* spp, Ca/Mg Scotland 0.42, 0.45, 0.38, 0.44 are similar to Newfoundland values of 0.40. This section has attempted to explain some of the factors associated with plant nutrition which is much more complex on serpentine soils than on any other soil types. Further discussion will follow as the trace element contents of these same plant species will be discussed in the next section.

Concentrations of some trace elements in native plants growing on serpentinized soils from western Newfoundland

One of the most interesting aspects in the study of plant nutrition is the trace element balance which is required for healthy plant growth. Much of the work relating to this subject has been confined to agricultural crops for obvious reasons, but quite early on Robinson, Edgington & Byers (1935) introduced the limiting factors of high Cr and Ni in addition to low macronutrients and high Mg as causes for growth problems on soils from serpentinized areas.

The trace element content, and in particular Co, Cr, Ni, of serpentine soils were documented world-wide (Proctor & Woodell 1975), and the

possible toxic effects of these three elements seem to be a factor in the unsuitability. There are a number of reasons (Proctor & Woodell 1975) why heavy metal toxicity is so variable as an ecological factor in serpentine soils. It includes differences in geological origin, composition and weathering, pH, and great variability interspecific and intraspecific response of plants.

Much of the data on the uptake of heavy metal elements and other trace elements summarized by Proctor & Woodell (1975), is related to the growth of agricultural crops in serpentine soils (Halstead, Finn & MacLean 1969). The early report of Minguzzi & Vergnano (1948), Vergnano Gambi, this Volume) of high Ni accumulation in the native serpentine plant *Alyssum bertolonii*, has led to a large number of studies on heavy metal accumulation in plants from serpentinized areas, e.g., Crooke & Inkson (1955), Crooke (1956) and Paribok & Alexeyeva-Popova (1966). In the 1960s, Cole (1973, this Volume) identified more Ni accumulating plants as did Wither & Brooks (1977). Brooks *et al.* (1977) used dried herbarium specimens to identify Ni accumulating plants and termed those of 1000 μg g^{-1} dry wt (0.1%) Ni as hyperaccumulators. By 1983, Brooks (1983) had identified more than 40 hyperaccumulators of Ni and many other strong accumulators (100–1000 μg g^{-1} dry wt) of Ni. The forms of Ni in various hyperaccumulators and potential methods for selective accumulating (Still & Williams 1980) were also investigated and work has continued to the present.

The trace elements (Fe, Zn, Cu, Ni, Cr, Co, Al) of 31 representative plant species from 14 families are given in Table 9 (values are μg g^{-1} dry wt). Table 10 gives the average of the total and available trace elements which are found in the soil from which the transect was taken. The average pH value for 19 mineral soil samples along the transect was 7.2.

From the results in Table 9 it can be seen that the Caryophyllaceae family has the highest average concentration of Fe, Ni, Cr, Co and Cu. *Arenaria humifusa* and *A. marcesens* were identified by Roberts (l980b, 1981) as hyper-accumulators as well as *Solidago hispida* and *Senecio pauperculus* (Table 9). The three other members of the Caryophyllaceae are strong Ni accumulators with *Cerastium beeringianum*, *Lychnis al-*

104

Table 9. Micronutrient (trace element) concentrations and the Mg/Ca ratio of native plants growing on serpentine soils from western Newfoundland (μg g^{-1}/dry wt)

Family	Scientific name	Fe	Zn	Cu	Ni	Cr	Co	Al	Mg/Ca
Caryophyllaeae	*Arenaria humifusa*	43911	76	45	2330	530	113	3010	91.8
	Arenaria marcesens	44370	46	21	2365	416	123	5060	93.6
	Lychnis alpina	13107	56	41	913	174	56	1500	9.7
	Cerastium beeringianum	7293	41	13	572	138	28	2420	2.4
	Silene acaulis	15360	41	12	898	180	60	1277	2.5
Compositae	*Aster nemoralis*	4641	55	24	396	94	19	138	2.5
	Solidago hispida	20859	60	22	1023	362	68	6820	9.3
	Senecio pauperculus	37995	63	24	1903	406	120	4920	23.0
	Anaphalis margaritaceae	1530	46	8	40	10	0	100	0.7
Plumbaginaceae	*Armeria labradorica*	13413	40	17	726	168	38	3790	8.9
Campanulaceae	*Campanula rotundifolia*	1742	45	17	561	102	24	1460	4.1
Polypodiaceae	*Adiantum pedatum*	8925	33	15	616	130	30	710	6.9
Pinaceae	*Larix laricina* I	740	20	6	29	0	0	120	1.1
	Larix laricina II	690	21	6	32	0	0	100	1.3
	Larix laricina III	820	23	7	37	0	0	190	1.3
	Picea mariana	570	34	6	35	0	0	50	0.7
	Juniperus communis	620	20	8	32	0	0	50	0.4
	Juniperus horizontalis	1080	19	5	62	10	0	90	0.6
Corylaceae	*Betula michauxii*	960	88	6	48	0	0	20	0.7
	Betula pumila	690	58	6	63	0	0	40	0.6
	Alnus crispa	940	50	9	77	10	0	50	0.9
	Alnus rugosa	1140	32	10	22	10	0	50	0.4
Sarraceniaceae	*Sarracenia purpurea*	2520	33	16	92	32	0	330	2.0
Cyperaceae	*Scirpus cespitosus*	640	17	5	31	0	0	20	0.0
Ericaceae	*Andromeda glaucophylla*	9150	36	9	500	126	20	1050	2.3
	Ledum groenlandicum	1740	23	8	54	28	0	220	0.5
	Rhododendron lapponicum	290	21	12	20	7	6	138	0.5
Rosaceae	*Potentilla fruticosa*	2670	30	7	98	21	0	420	1.6
	Potentilla tridentata	5220	35	10	206	60	0	1020	1.4
	Sanguisorba	790	20	10	44	11	0	40	0.5
Gramineae	*Deschampsia caespitosa*	1896	16	8	95	37	8	222	8.2
Grimmiaceae	*Rhacomitrium lanuginosum* I	16200	40	10	685	203	40	2040	15.1
	Rhacomitrium lanuginosum II	18620	42	10	719	246	44	1900	19.0
Cladoniaceae	*Cladina alpestris*	2910	33	8	73	24	0	630	2.6

pina, and *Silene acaulis* having 572, 913, and 898 μg g^{-1} Ni dry wt in above ground parts (Table 9). Five other plant species have greater than 500 μg g^{-1} dry wt Ni (Table 9). In samples from the same transects in two successive years the strong accumulators either had very similar concentrations or, in most cases, slightly lower. In the hyperaccumulators both species of *Arenaria* had also similar levels of Ni as shown in Table 9.

Shewry & Peterson (1976) reported that Ni in shoots of *Silene acaulis* was 5 times that for *S. maritima* and that *Cerastium nigrescens* also accumulated high levels of Ni. Proctor & Johnston (1977) found that *Lynchis alpina* (possible varieties) in Britain had some uptake of Ni, but only about 1/10 of Newfoundland values. In studies of the uptake of heavy metals by Scandinavian varieties of *Lychnis alpina* and *Silene dioica* Brooks & Crooks (1980) reported that there was no difference in the tolerance and uptake of four heavy metals including Ni for both serpentine and non-serpentine varieties of *Lychnis* and *L. alpina* var. *serpentinicola* (Rune 1953) may not be really a variety. Both *Lychnis* varieties tolerated and accumulated over 7,000 μg g^{-1} dry wt of Ni in their tissue making it a strong hyperaccumulator in terms of tolerance.

The ability of the members of the Caryophyllaceae to tolerate and possibly concentrate large quantities of these toxic elements and also tolerate low macronutrients (Roberts 1981) is a major reason for their success. As expected, Ni concen-

Table 10. Micronutrient (trace element) concentrations ($\mu g\,g^{-1}$/dry wt) of serpentine soils from western Newfoundland, Transect No. 1

Horizon	Depth (cm)	Fe	Zn	Cu	Ni	Cr	Co	Al	Mg/Ca	pH
1:										
Total Nutrients										
Cy1	0–10	96,450	70	30	3610	1270	180	7930	87.7	7.2
Cy2	10–25	91,540	80	40	3560	1090	160	8630	96.8	7.0
Avaliable Nutrients										
Cy1	0–10	2.6	1.2	0.6	3.9	tr.	tr.	7.8	10.8	
Cy2	10–25	2.8	1.8	0.8	4.4	tr.	tr.	7.9	11.3	
2:										
Total Nutrients										
Om	0–10	27,610	45	23	1069	309	43	7803	13.3	5.0
Available Nutrients										
	0–10	16.40	2.2	–	38.6	tr.	tr.	–	5.8	

Soils:
Soil No. 1 is a moderately well to imperfectly drained Gleyed Regosol, cryoturbic phase, from the Blomidon series occurring mainly on the bottom flats (Roberts 1980).
Soil No. 2 is a Typic Mesisol (Classification CSSC 1978). The solis were sampled near Winterhouse Brook, lat. 49°28′ N. long. 57°58′ W, 3.5 km SW of Woody Point, Bonne Bay, and at the foot of the Blow-Me-Down Range, lat. 49°04′ N, long. 58°15′ W, 16 km west of Corner Brook, Newfoundland.

trations in the Newfoundland soil (Table 10) were the highest for the three so labeled toxic elements, basically because of its more soluble properties (discussed in Roberts 1980).

Rhacomitrium moss, which is so common on several of the world's serpentinized areas, is a strong accumulator of Ni in Newfoundland and the concentrations in the edge of clumps usually exceed the centre of the patch (Roberts 1981). This suggests that splash water is an important contributor of metals since mosses lack vascular tissue. In British serpentine sites Johnston & Proctor (1977) showed *Rhacomitrium* a strong Ni accumulator (472 $\mu g\,g^{-1}$) but these values are less than Newfoundland values of 685 and 719 $\mu g\,g^{-1}$ (Table 9).

In Newfoundland only two species of pine are indigenous (Roberts 1985). *Pinus strobus* was observed a few times on serpentinized areas in western Newfoundland but the pine family is well represented with larch, spruce and juniper species.

Collectively the Pinaceae showed the lowest accumulation of all trace and heavy elements reported in Table 9, and this in combination with having one of the highest tissue concentrations of macronutrients (Table 6) is a major reason for their success on serpentine soils. The other successful families are related to tolerance of lower macronutrient conditions and a tolerance of heavy metals (Table 9).

With the exception of the Pinaceae and Corylaceae, large amounts of Fe (Table 9) are found in all the native plants of this serpentine site. The Caryophyllaceae must have the ability to concentrate these large quantities of Fe as due to soil pH (7.2) only small quantities are available to plants of this serpentinized habitat.

Fe, Ni, Cr, Co values in leaf tissue of *Alnus rugosa* growing on an adjacent non-serpentine site showed 470 $\mu g\,g^{-1}$ of Fe but now Ni, Cr or Co concentrations. Individuals growing on serpentine (Table 9) showed 1,440 $\mu g\,g^{-1}$ of Fe, 22 $\mu g\,g^{-1}$ of Ni, 10 $\mu g\,g^{-1}$ of Cr and no Co. *Larix laricina* growing on an adjacent non-serpentine site had concentrations of 500 $\mu g\,g^{-1}$ Fe, 15 $\mu g\,g^{-1}$ Ni, no detectable Cr or Co concentrations. This is approximately two thirds the concentration shown in this species on serpentine (Table 9). *Picea mariana* values on non-serpentine were almost identical to the *Larix laricina* values of the non-serpentine site, while *Cladonia alpestris* from the non-serpentine site showed no detectable Ni, Cr or Co. The same species on serpentine had 73 μg g^{-1} Ni, 24 μg g^{-1} Cr and no detectable Co.

Proctor & Woodell (1975) report Ni concentrations in some native plant species. *Betula* spp. showed 895 μg g^{-1} dry wt. which was more than 10 times as much shown for our values (Table 9) but in the range of the other dominant families with the exception of the pine family. Average Ni concentrations for the pine family (Proctor & Woodell 1975) were 9.2–15 μg g^{-1} dry wt. which is about one-half the concentration shown for this family on our serpentinized areas (Table 9). Zinc and Cu values follow the above trends. Aluminum tissue concentration is highest when high Ni concentrations exist.

Acknowledgements

I thank Dr. J. A. Munro, Regional Director General and A. B. Case, R. S. van Nostrand, Program Directors of Forestry Canada, Newfoundland & Labrador Region for encouragement and laboratory support. My father introduced me to the 'Tablelands' serpentinized site at a very early age and accompanied me there numerous times. Dr R. E. Wells and Dr D. Bazjak provided leadership on the CLI land classification surveys and on early introduction to the Bay of Islands ultramafic sites. I thank E. R. Dawe, E. M. Pike, J.-G. Zakrevsky, D. Trenholm and M. Maher of the Soils Laboratory for completing many of the chemical analysis over a long period of time. The author has benefited greatly from discussions with many Forestry Canada and university colleagues, especially Drs W. J. Meades, E. D. Wells, T. L. Chow and Drs A. W. H. Damman, I. Brookes, R. Belland and L. K. Thompson.

I would also like to thank E. Woodrow, Dr A. U. Mallik and Dr J. Proctor for comments on the earlier manuscript. I am grateful to Ms M. MacDonald and Ms M. Gillingham for typing services. R. Ficken copied many of my slides to produce the black and white prints and Ms B. A. Janes provided excellent technical assistance over the long period of study. The author is most grateful for the support of the publisher and technical editor over the past four years.

Appendix I

A list of the plants checklisted for the Newfoundland serpentinized areas. Abundance cover as follows: r = rare, + = sparse, 1 = <5%, 2 = <25%, 3 = 25–50%, 4 = 50–75%, 5 = 75–100%.

Species	Abundance
TREE LAYER	
Abies balsamea (L.) Milll.	+
Betula papyrifera Marsh.	+
Larix laricina (Du Roi) K. Koch.	1–2
Picea glauca (Moench) Voss	+
P. mariana (Mill.) BSP.	+
Pinus strobus L.	1
SHRUBS	
Acer rubrum L.	+
Alnus crispa (Ait.) Pursh	1
A. rugosa (Du Roi) Spreng.	r
Andromeda glaucophylla Link	1
Arctostaphylos alpina (L.) Spreng.	+
A. rubra (Rehd. & Wils.) Fern.	+
A. Uva-ursi (L.) Spreng.	1
Betula Michauxii Spach	+
B. pumila L. var. *renifolia* Fern.	1
Chamaedaphne calyculata (L.) Moench var. *angustifolia* (Ait.) Rehd.	1
Diapensia lapponica L	r
Dryas integrifolia Vahl	r
Empetrum nigrum L.	+
Epigaea repens L.	+
Gaultheria hispidula (L.) Bigel.	+
G. procumbens L.	r
Gaylussacia baccata (Wang.) K. Koch.	+
G. dumosa (Andr.) T. & G.	+
Juniperus communis L.	1–2
J. communis L. var. *depressa* Pursh	+
J. communis L. var. *saxatilis*	

Pallas	+	var. *submutica* (Blake)	
J. horizontalis Moench	+	H. F. Lewis	+ − 2
Kalmia angustifolia L.	+	*Artemisia borealis* Pall	+
K. polifolia Wang.	+	*A. canadensis* Michx.	+
Ledum groenlandicum Oeder	+	*Aster nemoralis* Ait.	1
Loisleuria procumbens (L.) Desv.	r − +	*A. novi-belgii* L.	+
Lonicera villosa (Michx.) R. & S.	r	*Astragalus eucosmus* Robins	+
Myrica Gale L.	1	*Campanula rotundifolia* L.	+ − 1
Nemopanthus mucronata (L.)		*Castilleja septentrionalis* Lindl.	+
Trel.	+	*Cerastium arvense* L.	+ − 1
Potentilla fruticosa L.	1	*C. beeringianum* C. & S.	+ − 1
P. tridentata Ait.	1	*C. terrae-novae* Fern. & Wieg.	+
Pyrus decora (Sarg.) Hyland	+	*Circaea alpina* L.	r
P. floribunda Lindl.	+	*Cirsium vulgare* (Savi) Tenore	r
Rhamnus alnifolia L'Her.	+	*Clintonia borealis* (Ait.) Raf.	r
Rhododendron canadense (L.)		*Conioselinum chinense* (L.) BSP.	+
Torr.	+	*C. pumilum* Rose	+
R. lapponicum (L.) Wahlenb.	1	*Comandra Richardsiana* Fern.	r
Rosa nitida Willd.	+	*Coptis groenlandica* (Oeder)	
Salix arctica Pallas var.		Fern.	r
kophophylla (Schneid.)		*Cornus canadensis* L.	+
Polunin	1	*Cypripedium Calceolus* L.	r
S. calcicola Fern. & Wieg.	+	*Drosera intermedia* Hayne	+
S. cordifolia Pursh	+	*D. rotundifolia* L.	+
S. glauca L.	+	*Epilobium latifolium* L.	r
S. glaucophylloides Fern.	+	*Erigeron hyssopifolius* Michx.	+
S. reticulata L.	+	*Euphrasia Randii* Robins. forma	
S. Uva-ursi Pursh	1	*iodantha* (Fern. & Wieg.)	
S. vesitita Pursh	+	Fern.	r
Shepherdia canadensis (L.) Nutt.	+	*Fragaria virginiana* Duchesne	+
Vaccinium angustifolium Ait.	1	*Geocaulon lividum* (Richards.)	
V. macrocarpon Ait.	+	Fern.	+
V. uliginosum L. var. *alpinum*		*Gnaphalium sylvatium* L.	+
Bigel.	1	*Habenaria dilatata* (Pursh) Hook.	+
V. Vitis-Idaea L.	1	*H. hyperborea* (L.) R. Br.	+
Viburnum cassinoides L.	+	*H. psycodes* (L.) Spreng.	+
		H. viridis (L.) R. Br.	+
HERBS		*Iris versicolor* L.	+ − 1
Achillea borealis Bong.	+	*Kobresia simpliciuscula*	
A. Millefolium L.	+	(Wahlenb.) Mackenz.	+
Anaphalis margaritacea (L.) C.		*Leontodon autumnalis* L.	+
B. Clarke	+	*Ligusticum scothicum* L.	+
Androsace septentrionalis L.	+	*Linnaea borealis* L.	+
Anemone parviflora Michx.	+	*Linum catharticum* L.	+
Antennaria eucosma Fern. &		*Lychnis alpina* L. var. *americana*	
Wieg.	+	Fern.	+ − 1
Arenaria humifusa Wahlenb.	+ − 1	*Maianthemum canadense* Desf.	+
A. marcescens Fern.	+ − 1	*Mitella nuda* L.	+
A. rubella (Wahlenb.) Sm.	r	*Nuphar variegatum* Engelm.	+
Armeria labradorica Wallr.		*Oxytropis terrae-novae* Fern.	+

Pinguicula vulgaris L.	+
Prenanthes trifoliolata (Cass.) Fern.	+
Primula laurentiana Fern.	+
P. mistassinica Michx.	+
Pyrola asarifolia Michx.	+
P. rotundifolia L.	+
P. secunda L.	r
Rhinanthus borealis (Sterneck) Chabert	+
Rubus acaulis Michx.	+
R. Chamaemorus L.	+ – 1
R. pubescens Raf.	+
Sagina nodosa (L.) Frenzl	+
Sanguisorba canadensis L.	+ – 1
Sarracenia purpurea L.	+ – 1
Saxifraga aizoides L.	+
S. Aizoon Jacq.	+
S. oppositifolia L.	r
Senecio pauciflorus Pursh	+
S. pauperculus Michx.	+ – 1
Silene acaulis L. var. *exscapa* (All.) DC.	+ – 1
Smilacina stellata (L.) Desf.	+
Solidago hispida Muhl. var. *lanata* (Hook.) Fern.	+ – 1
S. multiradiata Ait.	+
Stellaria longifolia Muhl.	+
Tanacetum huronense Nutt.	+
Thalictrum alpinun L.	+ – 1
T. polygamum Muhl. var. *hebecarpum* Fern.	+ – 1
T. pubescens Pursh	+
Tofieldia glutinosa (Michx.) Pers.	+
T. pusilla (Michx.) Pers.	+
Trientalis borealis Raf.	+
Triglochin maritima L.	+
T. palustris L.	+
Viola cucullata Ait.	+

GRASSES

Calamagrostis inexpansa Gray var. *robusta* (Vasey) Stebbins	+ – 1
C. Pickeringii Gray	+ – 1
Danthonia intermedia Vasey	+
Deschampsia atropurpurea (Wahlenb.) Scheele	+
D. cespitosa (L.) Beauv.	+ – 1
D. cespitosa (L.) Beauv. var. *littoralis* (Reut.) Richter	+
D. cespitosa (L.) Beauv. var.	

glauca (Hartm.) Lindm.	+
D. flexuosa (L.) Trin.	+ – 1
Festuca rubra L.	+
F. scabrella Torr.	+
Poa. glauca Vahl.	+
P. palustris L.	+
Schizachne purpurascens (Torr.) Swallen	+

SEDGES

Carex aquatilis Wahlenb.	+
C. Buxbaumii Wahlenb.	+ – 1
C. capillaris L. var. *major* Blytt	+
C. disperma Dew.	+
C. echinata Murr.	+
C. exilis Dew.	+
C. flava L.	+
C. glacialis Mackenz.	+
C. lasiocarpa Ehrh.	+
C. limosa L.	+
C. livida (Wahlenb.) Willd.	+
C. paupercula Michx.	+
C. rupestris Bellardi	+
C. scripoidea Michx. var. *scirpiformis* (Mackenz.) O'Neill & Duman	+ – 1
C. vaginata Tausch	+
Eriophorum Chamissonis C. A. Mey.	+ – 1
E. viridi-carinatum (Engelm.) Fern.	+ – 1
Scirpus cespitosus L.	1 – 2
S. hudsonianus (Michx.) Fern.	+ – 1

RUSHES

Juncus alpinus Vill.	+
J. articulatus L.	+
J. balticus Willd.	+
J. castaneus Sm.	+
J. filiformis L.	+
J. trifidus L.	+

VASCULAR CRYPTOGRAMS

Adiantum pedatum L. var. *aleuticum* Rupr.	+
Equisetum arvense L.	+
E. scirpoides Michx.	+
Lycopodium annotinum L.	+
L. Selago L.	+
Osmunda regalis L.	+ – 1
Selaginella selaginoides (L.) Link	+

References

Baker, A. J. M. 1981. Accumulators and excluders – strategies in the response of plants to heavy metals. Journal of Plant Nutrition 3: 643–654.

Banfield, C. E. 1983. Climate Monographiae Biologicae 48: 37–106. In Biogeography and ecology of the Island of Newfoundland. Edited by G. R. South, Dr. W. Junk Publishers, The Hague.

Belland, R. J. 1987a. The moss flora of the Gulf of St. Lawrence Region: Ecology and phytogeography. Journ. Hattori Bot. Lab. 62: 205–267.

Belland, R. J. 1987b. The disjunct moss element of the Gulf of St. Lawrence Region: Glacial and postglacial dispersal and migrational histories. Journ. Hattori Bot. Lab. 63: 1–76.

Belland, R. J. & G. R. Brassard. 1988. The bryophytes of Gros Morne National Park, NFLD, Canada, ecology and phytogeography. Lindbergia 14: 97–118.

Black, C. A. (Ed.) 1965. Methods of soil analysis, Part 2. Chemical and microbiological properties. Agron. Ser. No. 9, Am. Soc. Agron. Inc., Madison, Wis.

Birse, E. L. 1982. Plant communities on serpentine in Scotland. Vegetatio 49: 141–162.

Bouchard, A., S. Hay & E. Rouleau. 1978. The vascular flora of St. Barbe South District, Newfoundland: An interpretation based on biophysiographic areas. Rhodora 80: 228–308.

Brooks, I. A. 1982. Ice marks in Newfoundland: A history of ideas. Geographie Physique et quaternaire 36: 139–163.

Brooks, R. R. 1983. Biological methods of prospecting for minerals. John Wiley & Sons, Toronto. 322 pp.

Brooks, R. R. 1987. Serpentine and its vegetation: A multidisciplinary approach. Dioscorides Piess, Portland, Oregon. 454 pp.

Brooks, R. R. & H. M. Crooks. 1980. Studies on uptake of heavy metals by the Scandinavian "Kisplanten" Lychnis alpina and Silene dioica. Plant and Soil 54: 491–496.

Canada Soil Survey Committee 1974. The system of soil classification for Canada. Can. Dept. of Agriculture, Ottawa, Ontario, Pub. No. 1455, 122–124.

Canadian Soil Survey Committee 1978. The Canadian system of soil classification. Research Branch, Canada Department of Agriculture. Ottawa, Publication 1646. 164 pp.

Carter, S. P., J. Proctor & D. R. Slingsby. 1987. Soil and vegetation of the keen of Hamar Serpentine, Shetland. Journal of Ecology 75: 21–42.

Chapin, F. S. III 1980. The mineral nutrition of wild plants. Ann. Rev. Ecol. Syst. 11: 233–260.

Cole, M. M. 1973. Geobotanical and biogeochemical investigations in the sclerophyllus woodland and scrub associations of the Eastern Goldfields area of Western Australia with particular reference to the role of Hybanthus floribundus (Lindl.) F. Muell. as nickel indicator and accumulator plant. J. appl. Ecol. 10: 269–330.

Cooper, J. R. 1936. Geology of the southern half of the Bay of Islands Igneous Complex. Nfld. Dept. of Natural Resources, Geological Section, Bulletin No. 4. 62 pp.

Coombe, D. E. & L. C. Frost. 1956a. The Heaths of the Cornish Serpentine. Journal of Ecology, 44: 226–256.

Coombe, D. E. & L. C. Frost. 1956b. The nature and origin of the soils over the Cornish serpentine. J. Ecol. 44: 605–15.

Crooke, W. M. and R. H. E. Inkson. 1955. The relationship between nickel toxicity and major nutrient supply. Plant and Soil 6: 1–15.

Crooke, W. M. 1956. Effect of soil reaction on uptake of nickel from a Serpentine Soil. Soil Science, 81: 269–276.

Damman, A. W. H. 1964. Some forest types of central Newfoundland and their relationship to environmental factors. Forest Science Monograph No. 8, 62 pp.

Damman, A. W. H. 1965. The distribution patterns of northern and southern elements in the flora of Newfoundland. Rhodora, 67: 363–392.

Damman, A. W. H. 1967. The forest vegetation of western Newfoundland and site degradation associated with vegetation change. Ph.D. Thesis, University of Mich., U.S.A. 319 pp.

Dearden, P. 1975. The biogeography of Table Mountain, Bonne Bay, Newfoundland: An investigation of plant community composition and distribution on a serpentine bedrock. M.Sc. Thesis, Memorial University, St. John's, Newfoundland.

Dearden, P. 1979. Some factors influencing the composition and location of plant communities on a serpentine bedrock in western Newfoundland. J. of Biogeography. 6: 93–104.

DeSequeira, E. M. 1969. Toxicity and movement of heavy metals in serpentinic soils (North-Eastern Portugal). Agronomia Lusitana, 30: 115–154.

Dewey, J. F. 1969. The evolution of the Caledonian/Appalachian orogen. Nature 222: 124–128.

Dewey, J. F. & W. S. F. Kidd. 1974. Continental collisions in the Appalachian/Caledonian orogenic belt: variations related to complete and incomplete suturing. Geology 2: 543–546.

Dunning, G. R. 1981. The Annieopsquotch Ophiolite belt, southwest Newfoundland. In Current Research, Part B, Geological Survey of Canada, Paper 81-1B. pp. 11–15.

Ernst, W. 1972. Ecophysiological studies on heavy metal plants in South Central Africa. Kirkia 8, 125–145.

Fernald, M. L. 1911. A botanical expedition to Newfoundland and southern Labrador. Rhodora 13: 109–162.

Fernald, M. L. 1918. The contrast in the floras of eastern and western Newfoundland. Amer. J. Bot. 5: 236–247.

Fernald, M. L. 1925. Persistence of plants in unglaciated regions of Boreal America. Mem. Amer. Acad. Sci. N.S. 15: 237–242.

Fernald, M. L. 1933. Recent discoveries in the Newfoundland flora. Rhodora 35: 80–107.

Fernald, M. L. 1950. Gray's manual of botany. 8th edition. American Book Company, New York. 1632 p.

Gerloff, G. C., D. G. Moore & J. T. Curtis. 1966. Selective absorption of mineral elements by native plants of Wisconsin. Plant and Soil 25: 393–405.

Gough, L. P., T. Hansford Shacklette & A. A. Case. 1979. Element concentrations toxic to plants, animals and man. U.S. Dept. of the Interior, Geological Survey Bulletin 1466. 80 p.

Grant, D. R. 1969. Late Pleistocene readvance of piedmont glaciers in western Newfoundland. Marit. Sed. 5: 126–128.

Grant, D. R. 1977. Glacial style and ice limits, the Quaternary stratigraphic record, and changes of land and ocean level in the Atlantic Provinces, Canada. Géogr. phys. Quat. 31: 247–260.

Halstead, R. L. 1968. Effects of different amendments on yield and composition of oats grown on a soil derived from serpentine material. Can. J. Soil Sci. Vol. 48, 301–305.

Halstead, R. L., B. J. Finn & A. J. Maclean. 1969. Extractability of nickel added to soils and its concentration in plants. Can. J. Soil Sci. 49: 335–342.

Hare, F. K. 1952. The climate of the island of Newfoundland. A geographical analysis. Canada Dept. Mines Techn. Surv., Geogr. Bull. 2: 36–89.

Hart, R. 1980. The Co-existence of Weeds and Restricted Native Plants on Serpentine Barrens in Southeastern Pennsylvania. Ecology 61(3): 688–401.

Henry, D. G. 1973. Foliar nutrient concentrations of some Minnesota forest species. Minnesota Forestry Research Notes. No. 241. 4 pp.

Hobbs, R. J. & H. A. Mooney. 1985. Community and population dynamics of serpentine grassland annuals in relation to gopher disturbance. Oecologia 67: 342–351.

Jackson, M. L. 1958. Soil chemical analysis. Prentice-Hall Inc., Englewood Cliffs, New Jersey.

Johnston, W. R. & John Proctor. 1977. Metal concentrations in plants and soils from two British serpentine sites. Plant and Soil 46: 275–278.

Juchler, S. and H. Sticher. 1986. Soils on serpentinite near Davos/Switzerland (Station Ch-5, Delenwakl). Mitteilgn. Dtsch. Bodenkundl. Gesellschaft., 48: 91–105. XIII. Congress of the International Society of Soil Science. Hamburg, Germany. August 13–20, 1986.

Kennedy, M. J., E. R. W. Neale & W. E. A. Phillips. 1972. Similarities in the early structural development of the northwestern margin of the Newfoundland Appalachians and Irish Caledonides. 24th International Geol. Congress, Montreal, Section 3, 516–531.

Koenigs, R. L., W. A. Williams & M. B. Jones. 1982. Factors affecting vegetation on a serpentine soil. I. Principal components analysis of vegetation data. Hilgardia 50: 1–14.

Koenigs, R. L., W. A. Williams, M. B. Jones & A. Wallace. 1982. Factors affecting vegetation on serpentine soil. II. Chemical composition of foliage and soil. Hilgardia 50: 15–26.

Kruckeberg, A. R. 1954. The ecology of serpentine soils. III. Plant species in relation to serpentine soils. Ecology 35: 267–274.

Kruckeberg, A. R. 1969. Soil diversity and the distribution of plants with examples from western North America. Madrono, A West American, Journal of Botany 20: 129–154.

Lyon, G. L., P. J. Peterson, R. R. Brooks & G. W. Butler. 1976. Calcium, magnesium and trace elements in New Zealand serpentine flora. Journal of Ecology. 59: 421–429.

Lyon, G. L., R. R. Brooks, P. J. Peterson & G. W. Butler. 1968. Trace elements in a New Zealand serpentine flora. Plant and Soil 29: 225–240.

Meades, W. J. 1983. Heathlands. Monographiae Biologicae 48: 267–318. In Biogeography and ecology of the island of Newfoundland, edited by G. R. South. Dr. W. Junk, Publishers, The Hague.

Minguzzi, C. & O. Vergnano. 1948. Il contenuto di nichel nelle ceneri di Alyssum bertolonii Desv. Memorie Soc. tosc. Sci. nat. A55: 49–77.

Muëller-Dombois, D. & E. Ellenberg. 1974. Aims and methods of vegetation ecology. John Wiley and Sons, New York. 547 pp.

Neale, E. R. W. 1974. The eighth wonder of the world. Geos. Summer 1974: 2–4.

Ojea, F. G. & M. I. Lopez Lopez. 1980. Soils of the Spanish humid zone, morphology and general characteristics. Anales de Edafologia Y Agrobiologia 39: 403–415.

Pollett, F. C. 1972. Classification of peatlands in Newfoundland. Environment Canada, Can. For. Ser., Nfld. Forest Research Centre. Proceedings of the 4th International Peat Congress I–IV, Helsinki, 1972. pp. 101–110.

Proctor, J. 1970. Magnesium as a toxic element. Nature 227: 742–743.

Proctor, J. 1971. The plant ecology of serpentine. III. The influence of a high magnesium/calcium ratio and high nickel and chromium levels in some British and Swedish serpentine soils. Journal of Ecology 59: 827–842.

Proctor, J. & S. R. J. Woodell. 1975. The ecology of serpentine soils. In Advances in Ecological Research 9: 256–347.

Proctor, J. & I. D. McGowan. 1976. Influence of magnesium on nickel toxicity. Nature, London. 260: 134.

Proctor, J. & W. R. Johnston. 1977. Lychnis alpina L. in Britain. Watsonia 11: 199–204.

Proctor, J., W. R. Johnston, D. A. Cottam & A. B. Wilson. 1981. Field-capacity water extracts from serpentine soils. Nature 294: 245–246.

Paribok, T. A. & N. V. Alexeyeva-Popova. 1966. Content of some chemical elements of wild plants of the polar Urals as related to the problem of the serpentine. Botanicheskii Zhurnal 51: 339–353.

Rabenhorst, M. C., J. E. Foss & D. S. Fanning. 1982. Genesis of Maryland soils formed from serpentinite. Soil Sci. Soc. Am. J. 46: 607–616.

Rabenhorst, M. C. & J. E. Foss. 1981. Soil and geologic mapping over mafic and ultramafic parent materials in Maryland. Soil Sci. Soc. Am. J. 45: 1156–1160.

Ragg, J. M. & D. F. Ball. 1964. Soils of the ultra-basic rocks of the island of Rhum. Journal of Soil Science 15: 124–133.

Reeves, R. D., R. R. Brooks & R. M. Macfarlane. 1981. Nickel uptake by Californian streptanthus and Caulanthus with particular reference to the hyperaccumulator, S. polygaloides Gray (Brassicaceae) Amer. J. Bot. 68: 708–712.

Reeves, R. D., R. M. Macfarlane & R. R. Brooks. 1983. Accumulation of nickel and zinc by western North American genera containing serpentine tolerant species. Amer. J. Bot. 70: 1297–1303.

Roberts, B. A. 1979. Concentrations of macro-elements (macro-nutrients) in native plants growing on serpentine soils from western Newfoundland. For presentation at the C.B.A. Meetings, Ottawa, June 1979. Abstract published. The Canadian Botanical Association Bulletin. Vol. 12, No. 3, p. 36.

Roberts, B. A. 1980a. Some chemical and physical properties of serpentine soils from western Newfoundland. Can. J. Soil Sci. 60: 231–240.

Roberts, B. A. 1980b. Concentrations of micro-nutrients

(trace elements) in native plants growing on serpentine soils from western Newfoundland. Paper presented at the Botany 80 symposium at the University of British Columbia, Vancouver, 12–16 July 1980. Abstract published. Botanical Society of America, Miscellaneous Publication 158, p. 95.

Roberts, B. A. & M. A. K. Khalil. 1980. Detailed soil and site assessment of hybrid poplar planting areas in Newfoundland. Environment Canada, Bi-monthly Research Notes 36: 16–18.

Roberts, B. A. 1981. Ecology of serpentinized areas, west Newfoundland, Canada. Paper presented at the XIII International Botanical Congress, Sydney, Australia, August 1981. Abstract published, p. 328.

Roberts, B.A. 1982. Some characteristics of the soils of the serpentinized areas of western Newfoundland. Paper presented at the annual meeting of the Canadian Society of Soil Science, AIC, held at the University of British Columbia, Vancouver, B.C., July 11–15, 1982. Abstract published. p. 28, titles and abstracts for the annual meeting.

Roberts, B. A. 1983. Soils, Monographiae Biologicae 48: 107–161. In Biogeography and ecology of the island of Newfoundland, edited by G. R. South, Dr. W. Junk Publishers, The Hague.

Roberts, B. A. 1985. Distribution and extent of *Pinus resinosa* Ait. in Newfoundland. Rhodora 87: 341–356.

Roberts, B. A. 1986. Stress in serpentinized areas. Invited paper presented at the stressed ecosystem, Canadian Botanical Association, 22nd Annual Meeting, Sudbury, June 1986. Abstract published #2, p. 21.

Roberts, B. A. 1986. Soils of serpentinized areas, their ecology and distribution. Paper presented at the 13th Congress of the International Society of Soil Science, 13–20 August, Hamburg, Germany, Extended summary published Commission V, p. 1253–1254.

Roberts, B. A. 1987. Merritt Lyndon Fernald (1873–1950); his investigations and discoveries in the Newfoundland flora. pp. 148–157. In Early Science in Newfoundland and Labrador, edited by D. H. Steele, Avalon Chapter of Sigma xi, St. John's, Nfld. ISBN No. 0-9693100-0-5.

Robertson, A. & B. A. Roberts. 1982. Checklist of the Alpine flora of Western Brook Pond and Deer Pond areas, Gros Morne Nat. Park. Rhodora. 84: 101–115.

Robinson, W. O., G. Edgington & H. G. Byers. 1935. Chemical studies of infertile soils derived from rocks high in magnesium and generally high in chromium and nickel. U.S.D.A., Technical Bulletin No. 471. 29 pp.

Rodenkirchen, H. & B. A. Roberts. In prep. Some characteristics of plants and soils from the serpentinized area of Haidberg/Zell, Northeast Bavaria, Germany.

Rogerson, R. J. 1983. Geological evolution, Monographiae biologicae 48: 5–35. In Biogeography and ecology of the island of Newfoundland, edited by G. R. South, Dr. W. Junk Publishers, The Hague.

Rune, O. 1953. Plant life on serpentine and related rocks in the north of Sweden. Acta Phytogeographica Suecica, 139 pp.

Sarosiek, J. 1964. Ecological analysis of some plants growing on serpentine soils in Lower Silesia. Monographic Botanicae 18: 97–105.

Shewry, P. R. & P. J. Peterson. 1976. Distribution of chromium and nickel in plants and soil from serpentine and other sites. Journal of Ecology 64: 195–212.

Spence, D. H. N. 1957. Studies on the vegetation of Shetland. I. The serpentine debris vegetation in Unst. Journal of Ecology 45: 917–945.

Spence, D. H. N. 1959. Studies on the vegetation of Shetland. II. Reasons for the restriction of the exclusive pioneers to serpentine debris. Journal of Ecology 47: 641–649.

Spence, D. H. N. 1970. Scotish serpentine vegetation. Oikos 21: 22–31.

Suzuki, S., N. Mizuno & K. Kimura. 1971. Distribution of heavy metals in serpentine soils. Soil Science and Plant Nutrition 17: 194–198.

Still, E. R. & R. J. P. Williams. 1980. Potential methods for selective accumulation of nickel (II) ions by plants. Journal of Inorganic Biochemistry 13: 35–40.

Talkington, R. W. & J. Malpas. 1980. Spinel phases of the White Hills peridotite, St. Anthony complex, Newfoundland: Part I. Occurrence and Chemistry, pp. 607–619. Proceedings of the International Ophiolite Symposium, Cyprus 1979. Edited by A. Panayiotou, Cyprus Geology Dept. 781 p.

Tarrant, P. F. & J. M. Trappe. 1971. The role of Alnus in improving the forest environment. Plant and Soil, Special Volume 1971. pp. 335–348.

Turitzin, S. N. 1981. Nutrient limitations to plant growth in a California serpentine grassland. American Midland Nat. 107: 95–99.

Van den Born, G. J., A. K. Bregt, H. Kok & J. Zijlstra. 1981. Weathering and soil formation on mafic and ultramafic rocks in N. Galicia, Spain. Research Project J050–816 Report. Second Edition. Dept. of Soil Science and Geology, Agricultural Univ., Wagengen, The Netherlands. pp. 1–193.

Vuokko, S. 1978. The vegetation of ultramafic areas in North Finland. Luonnon Tutkija 82: 131–134.

Walker, R. B. 1954. The ecology of serpentine soils. II. Factors affecting plant growth on serpentine soils. Ecology 35: 259–266.

Wallace, A. 1982. Additive, protective and synergistics effects on plants with excess trace elements. Soil Science 133(5): 319–323.

Wallace, A., M. B. Jones & G. V. Alexander. 1982. Mineral composition of native woody plants growing on a serpentine soil in California. Soil Science 134(1): 42–44.

Wells, R. E., J. P. Bouzane & B. A. Roberts. 1972. Reconnaissance land classification of the Corner Brook area, Newfoundland. Newfoundland Forest Research Centre, St. John's, Newfoundland. Information Report N-X-83. 123 pp.

Whittaker, R. H. 1954. The ecology of serpentine soils. IV. The vegetational response to serpentine soils. Ecology 35: 275–288.

Williams, Harold. 1979. Appalachian orogen in Canada. Canadian Journal of Earth Sciences. Vol. 16: 792–807.

Williams, H. and J. Malpas. 1972. Sheeted dikes and brecciated dike rocks within transported igneous complexes, Bay of Islands, Western Newfoundland. Can. J. Earth Sci. 9: 1216–1229.

Wilson, J. T. 1966. Did the Atlantic close and then re-open? Nature 211: 676–681.

Wilson, M. J. & M. L. Berrow. 1978. The mineralogy and heavy metal content of some serpentinite soils in Northeast Scotland. Chemie der Erde B37: 181–205.

Wither, E. D. & R. R. Brooks. 1977. Hyper-accumulation of nickel by some plants of southeast Asia. Journal of Geochemical Exploration 8: 579–583.

Zika, P. F. & K. T. Dann. 1985. Rare plants on ultramafic soils in Vermont. Rhodora 87: 293–304.

Author's Address
B. A. Roberts
Forestry Canada
P.O. Box 6028
St. John's
Newfoundland
Canada A1C 5X8

V

A phyto-ecological investigation of the Mount

Albert serpentine plateau

L. SIROIS & M. M. GRANDTNER

Abstract. Principal coordinate analysis of soil data collected from Mount Albert serpentine outcrops suggests that the highest Mg/Ca quotients are found in the well to moderately well drained soils. Twelve taxa are found to be more or less restricted to serpentine-rich soils. Of these, 6 are exclusive (*Adiantum pedatum* var. *aleuticum*, *Polystichum mohrioides*, *Cheilanthes siliquosa*, *Arenaria marcescens*, *Armeria maritima* var. *labradorica*, *Lychnis alpina*) and 6 are preferentials (*Arenaria humifusa*, *Arenaria sajanensis*, *Artemisia campestris* var. *borealis*, *Festuca altaica*, *Rhodendron lapponicum*, *Salix arctica*). Most of these taxa are found to occur more frequently in communities dominated by *Rhacomitrium lanuginosum*, which are widely distributed in the alpine tundra. Such communities are found principally on moderately well drained soils having a high Mg/Ca quotients and which are subject to intensive frost heaving. Less well drained soils are colonized by *Vaccinium uliginosum–Betula glandulosa* and by *Scirpus caespitosus* communities, which themselves contain fewer serpentinicolous plants. On serpentine, the boundary between alpine tundra and krummholz is at a lower altitude than on other substrates of the region. The ecological position of the serpentinicolous plants and the altitudinal distribution of the vegetation show analogies with other boreal serpentine outcrops.

Introduction

Mount Albert (Fig. 1a) contains one of the most important serpentine formations in Eastern Canada. Its distinct floristic and ecological characteristics have been briefly described by several authors including Fernald (1907, 1925), Dansereau & Raymond (1948), Raymond (1950), Scoggan (1950), Rune (1953, 1954), Moisan (1956) and Marie-Victorin (1964). Sirois (1984a, b) has recently completed a more intensive ecological study on which the present text is based and for which a vegetation map has been made (Sirois 1984a).

The Mount Albert plateau is almost entirely composed of serpentinized ultramafic and is bordered to the north by a narrow ridge of amphibolite. This geological contact coincides with the boundary between alpine tundra vegetation and stunted conifers (krummholz). This fact, observed by Scoggan (1950), Raymond *et al.* (1954) and Gray & Brown (1979) suggests that the plateau's vegetation is strongly influenced by edaphic factors.

The goal of this chapter is to analyse the principal characteristics of the soils, the flora, and the vegetation of the plateau to define their specificity and to assess the influence of the ultramafic soil on the vegetation.

Study area

Situated in Gaspé Provincial Park, Québec (Fig. 1b), Mount Albert is part of the Chic-Chocs mountains, an upheaval of the Appalachian plateau whose highest summits reach about 1000 m. The study area is a 160 ha plateau shaped like a horseshoe. The vegetation has a physiognomy characteristic of the alpine belt. The altitude of the study area is between 1150 and 900 m. The latter height corresponds to the lower limit of the sub-alpine vegetation on non-ultramafic soils of the adjacent Mount Jacques-Cartier (Boudreau 1981).

Climate

The alpine vegetation belt (which has a representative meteorological station at Mount Logan) has a cold and humid climate (Fig. 2). This is largely due to the combined effects of altitude and the proximity of the Gulf of St. Lawrence. In the alpine tundra, the mean temperature ranges from $-3°$ to $-5°C$ (Gray & Brown 1979, and Fig. 2). The annual precipitation is about 1660 mm, of which one third falls as snow, and is the highest in Southern Québec. Fog covers the Chic-Chocs 200 days per year. In the best conditions the growing season extends from the beginning of June to mid-September, although from mid-August frosts commonly occur (Gagnon 1970).

Geology and geomorphology

Almost all of the study area is formed of serpentinized peridotite of the harzburgite (90%) or dunite (10%) type. The ultramafic material rose during the lower Ordovician period (Wanless *et al.* 1973). A mafic rock, amphibolite, encircles the ultramafic formation, except in the south (Fig. 3). The amphibolite belt forms a ridge in the north, the remainder being composed of slightly convex or concave planes which control drainage and snow cover. The study area is covered with till and the deposits are derived from the ultramafic rock.

The fine fraction of these deposits is dominated by silts and clays. Some rockfields are found with a thick layer of boulders ($\pm 0.5\,m^3$). A small pocket of till of 3 ha and several blocks of amphibolite transported onto the ultramafic soils attest to glacial activity of unknown age and origin. However, recent works indicate, that at the maximum of the last glaciation (20,000 B.P.) the highest summits in the region were covered by ice (Payette *et al.* 1981, Payette & Boudreau 1984).

The alternate freezing and thawing, associated with the presence of permafrost (Gray & Brown 1979), or the impermeability of the parent material (Sirois & Grandtner 1982) or both, are responsible for the plateau's widespread cryoturbation. In mineral soils, this expresses itself principally by the formation of sorted polygons, stripes and mud boils. In soils covered with thick organic layers, earth core hummocks are common (Lebrun & Thériault 1979).

116

Fig. 1(a).

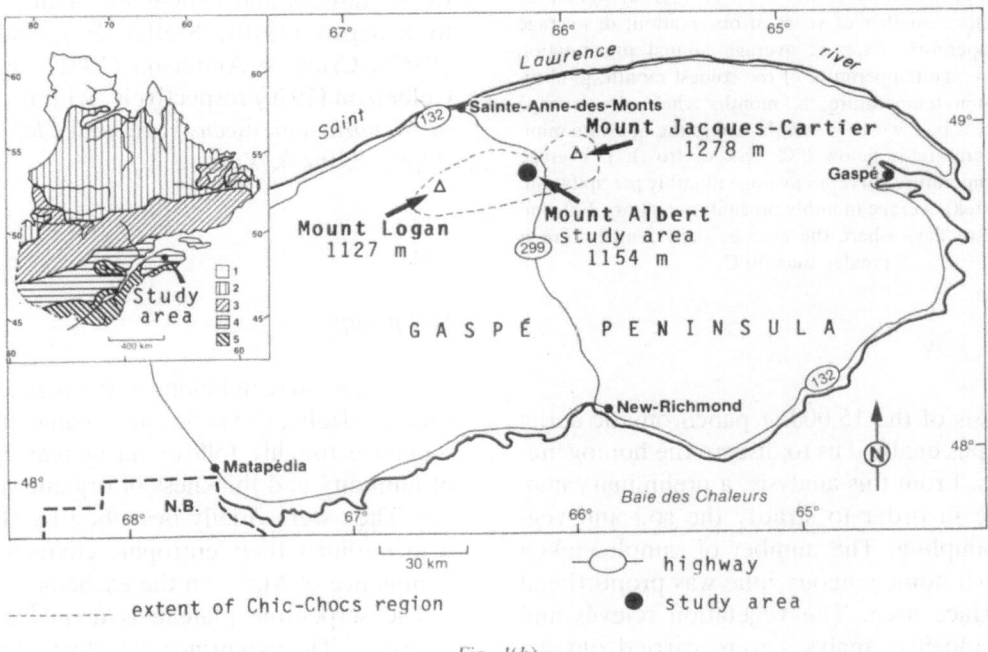

Fig. 1(b).

Fig. 1. Location of study area in relation to the Gaspé peninsula and to the natural vegetation zones of Québec–Labrador.

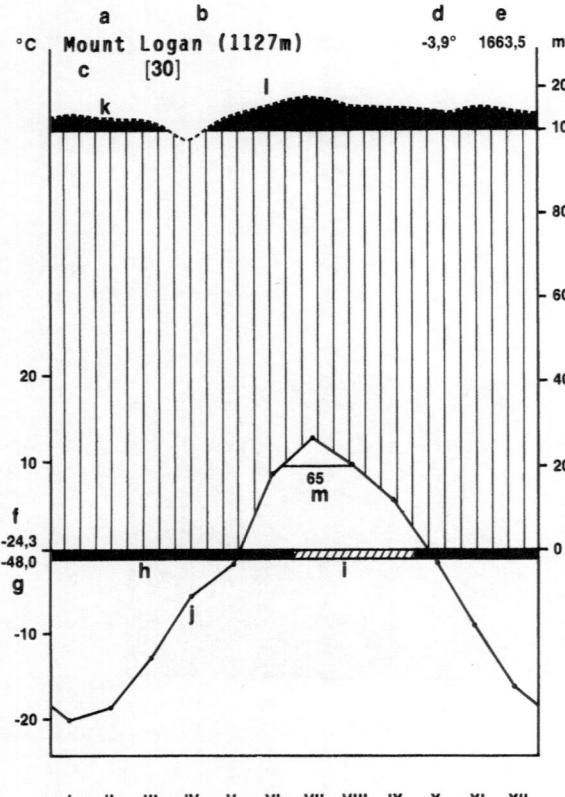

°C Mount Logan (1127m)   -3,9° 1663,5   mm

Fig. 2. Climatogram of the Mount Logan Station, alpine level of the Chic-Chocs (from the data of Gagnon (1970) following the method proposed by Walter *et al.* (1975)). a: station; b: altitude (m); c: number of years of observation; d: average annual temperature (°C); e: average annual precipitation (mm); f: average temperature of the coldest month; g: absolute minimum temperature; h: months where the average temperature is below 0°C; i: months where the absolute minimum temperature is below 0°C (risk of frost); j: average monthly temperature curve; k: average monthly precipitation; l: (shaded area) average monthly precipitation above 100 mm; m: number of days where the average daily temperature is greater than 10°C.

Methods

An analysis of the 15,000: 1 panchromatic aerial photographs enabled us to discern the homogeneous zones. From this analysis, a preliminary map was made in order to stratify the soil and vegetation sampling. The number of samples taken within each homogeneous zone was proportional to its surface area. The vegetation relevés and phytosociological analyses were carried out according to the Zürich–Montpellier (ZM) method (Mueller-Dumbois & Ellenberg 1974) employing the abundance-dominance coefficients proposed by Braun-Blanquet (1932). Most sample areas were 5×5 m according to the minimal area curves obtained on the plateau. For economy of space, the original vegetation tables (Sirois 1984b) were condensed into a synthesis table, retaining only the most important (frequency >60%) species. The description of soil profiles, using Canadian standards (CCP 1972, 1978), was made on eighty-two of the 160 relevés of vegetation studied (Sirois 1984b). A sample of each horizon was taken for physico-chemical analysis. After sieving, the fraction less than 2 mm was tested for the following properties: granulometry (densimetrical method of Bouyoucos, 1936), pH ($CaCl_2$ 0.01 M, 1:2.5 P/V), Fe and Al percentage (Na pyrophosphate extraction 0.1 M), and Ca and Mg (ammonium acetate extraction according to McKeague 1978).

Part of the edaphic data was subjected to principal coordinate analysis with the use of Gower's coefficient (Legendre & Legendre 1979). This coefficient was chosen as it allows the simultaneous use of quantitative, semi-quantitative and qualitative edaphic descriptors (Frontier 1983) which is advantageous for eco-pedological research (Bergeron *et al.* 1982). Vascular plants, liverworts, mosses and lichens are named according to Scoggan (1978), Stotler & Crandall-Stotler (1977), Crum & Anderson (1981), and Hale & Culberson (1970) respectively, with the exception of *Arenaria marcescens* Fern. and *Cladina stellaris* (Opiz.) Pouz & Vezda.

Soils

Soil groups

Soils of the plateau belong to the regosolic, brunisolic, podzolic, gleysolic and organic orders. This sequence roughly follows an increasing gradient of humidity and thickness of organic surface layers. They were briefly described by Butt (1979) who outlined their eutrophic character and the dominance of Mg^{2+} on the exchange complex.

The serpentine plateau is mainly covered by Regosols. The two principal factors which impede the differentiation of the horizons are, as in the ultramafic Regosols of Newfoundland (Roberts

118

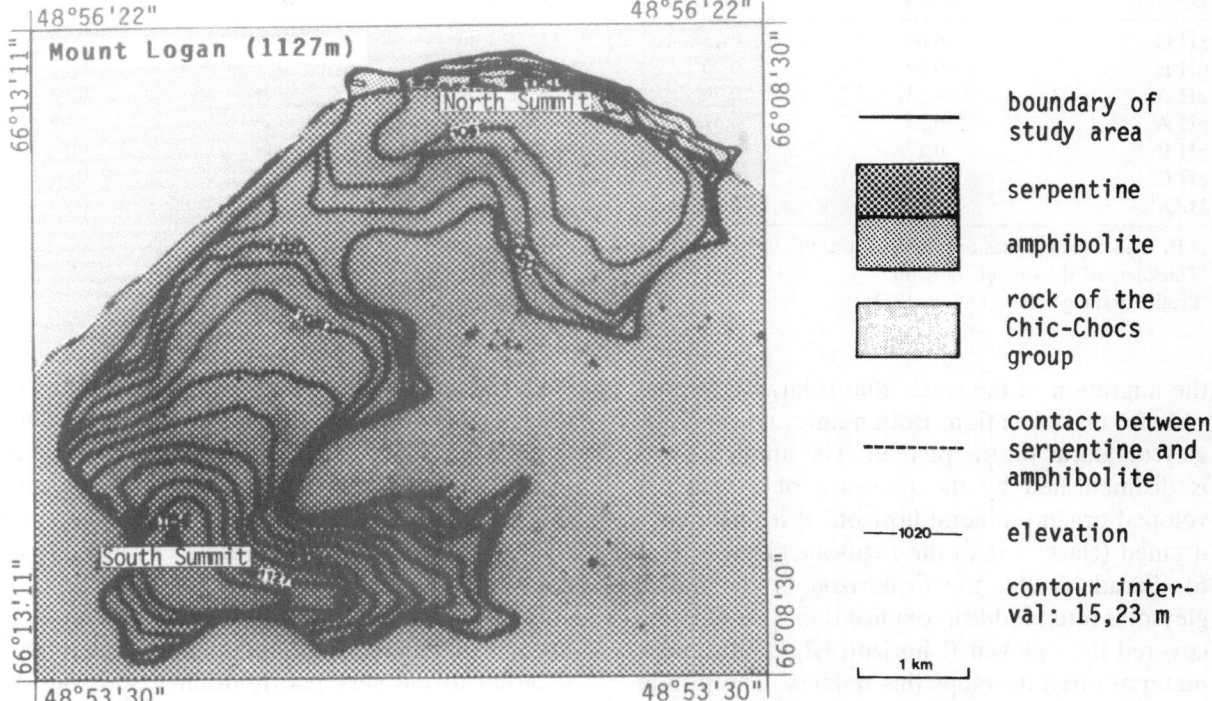

Fig. 3. Geological map of the Mount Albert area.

1980), cryoturbation and low plant cover. Cryo-turbation tends to homogenize the soil by vertical mixing of soil particles. The regosols of the better drained sites (drainage classes 2 and 3) belong to the Orthic subgroup, whereas the Humic Orthic Regosols are associated with imperfect drainage (class 4) and/or a higher plant cover. They are also subject to frost heaving.

Whereas amphibolite parent material gives rise to dystric Brunisols, the ultramafic parent material gives rise to eutric Brunisols. These are generally moderately well drained and fall into the orthic subgroup. In the imperfectly drained sites (class 4), Gleyed Eutric Brunisols are seen and sometimes subject to cryoturbation.

Even though the regional climatic conditions favour podzolization, this process does not occur frequently on ultramafic soils, even with a coniferous cover. The ultramafic rock limits podzolization of the deposits arising from it (Pédro & Bitar 1966). The leaching of most of the Mg^{2+} of the upper mineral horizon is a prerequisite for the translocation of organic matter, aluminum and iron toward the illuvial horizon (Buol *et al.*

1980, Rieger 1983). The high initial magnesium content in the parent material thus slows the process of podzolization. Furthermore, the high proportion of fine minerals (silt and clay $=61\% \pm 10$, $n = 8$) in the ultramafic soils slows down the leaching process. In effect, there is a general absence of an eluvial (Ae) horizon. However, some ultramafic podzols were found and belonged to humo-ferric orthic podzols. These podzols are characterized by a relatively low organic carbon content in the illuvial horizon (Bf). The environments where the ultramafic podzols are found are in krummholz formations or snow-beds dominated by *Ericaceae*. Cryoturbation is not seen in these soils.

The concave areas, about one quarter of the study area, had gleysols. These soils are generally covered with continuous organic horizons, and are less subject to cryoturbation than the brunisols and regosols. The surface accumulation of a substantial layer of organic matter tends to reduce the incidence of cryoturbation (Rieger 1983). This organic horizon constitutes a thermal insulator that slows frost penetration and, as a result,

119

Table 1. Spearman's correlation coefficient between the edaphic descriptors and the first two axes of the P.C.A.

Descriptor	Axis I	Axis II	Descriptor	Axis I	Axis II
pH O	0.08	0.13	Mg/Ca O	−0.38	0.33
pH H	0.09	0.90	Mg/Ca H	0.15	0.53
pH Ah	0.13	0.79	Mg/Ca Ah	0.28	0.60
pH A	−0.33	0.52	Mg/Ca A	−0.29	−0.09
pH B	0.17	0.69	Mg/Ca B	−0.07	0.57
pH C	−0.19	0.77	Mg/Ca C	0.17	0.54
M.O.[1]	0.62	−0.06	DRN[2]	−.20.88	−0.23

N.B. Underlined values are significant at the 0.001 level.
[1]Thickness of the organic horizon.
[2]Profile drainage class (C.C.P. 1972).

the migration of the water that is largely responsible for cryoturbation. Both humic and regosolic gleysols occur on the plateau. The humic gleysol is distinguished by the presence of a well developed organo-mineral horizon. It is also better drained (class 5) than the regosolic gleysol (class 6). Paradoxically, the C horizon of the humic gleysol is often reddish (oxidized) in spite of being covered by a gleyed B horizon (Bg). Ultramafic material often develops this redness. The iron in the C horizon can be kept in the ferric (oxidized) form, in spite of the water saturation, if the soil solution is highly basic (Buol et al. 1980). The regosolic gleysol is characterized by the presence of a peaty horizon, about 25 cm thick, arising largely from the accumulation of Sphagnum and sedge debris.

The peaty deposits of the plateau are not very thick. Thus, they are mainly terric humisols in the humisol great group. They occupy a small area with the regosolic gleysols, from which they are differentiated only by thicker peat layers.

Soil chemical and physical properties

We studied pH, drainage, and absolute and relative Ca and Mg concentrations. These soil factors for each horizon, as well as the thickness of the organic horizon and the mineralogy of the parent rock, were used for the principal coordinate analysis (P.C.A.). This analysis was conducted in order to identify which of the descriptors best explained the variability of the soil profiles, and to detect the existence of edaphic gradients. The results are presented in Table 1. The first two principal axes explain, respectively, 31% and 17% of the total variability of the soil profiles.

The calculation of Spearman's correlation coefficient between the descriptors and the values of the coordinates of the first two principal axes (Table 1) reveals that the first axis shows a declining gradient of drainage while the second axis shows increasing gradient of the pH value and of the Mg/Ca quotient.

Figure 4 illustrates the transition from moderate and well drained soils on the left of the scatter diagram to the very poorly drained soils on the right. The dispersal of the pedogenetic orders (Fig. 5) quite accurately reflects this trend. Here, under better drainage conditions, the soils are generally in a regosolic stage. Under meso-hygrophilic conditions (classes 3 and 4), the drainage variable plays a less important role in determining the differentiation of the soils which belong predominantly to the brunisols and the podzols. Further to the right of the diagram, poor drainage (class 5) corresponds to the development of humic gleysols, endowed with a distinct organomineral horizon. The worst drainage conditions (class 6) favour the development of rego gleysols with thick organic horizons, and of organosols. The latter are grouped together on the extreme right of the diagram, with the exception of 2 profiles situated in the lower left which correspond to folisols developed on amphibolite.

The lithologic origin of the parent material considerably influences the dispersion of profiles in the diagram (Fig. 6). Almost all of the soils derived from pure serpentine material (marked P in Fig. 6), and even those which contain a mixture of the two types of material, are found in the upper part all across the diagram. Soils generated from amphibolite are grouped mainly in the bottom toward the left. According to the scatter dia-

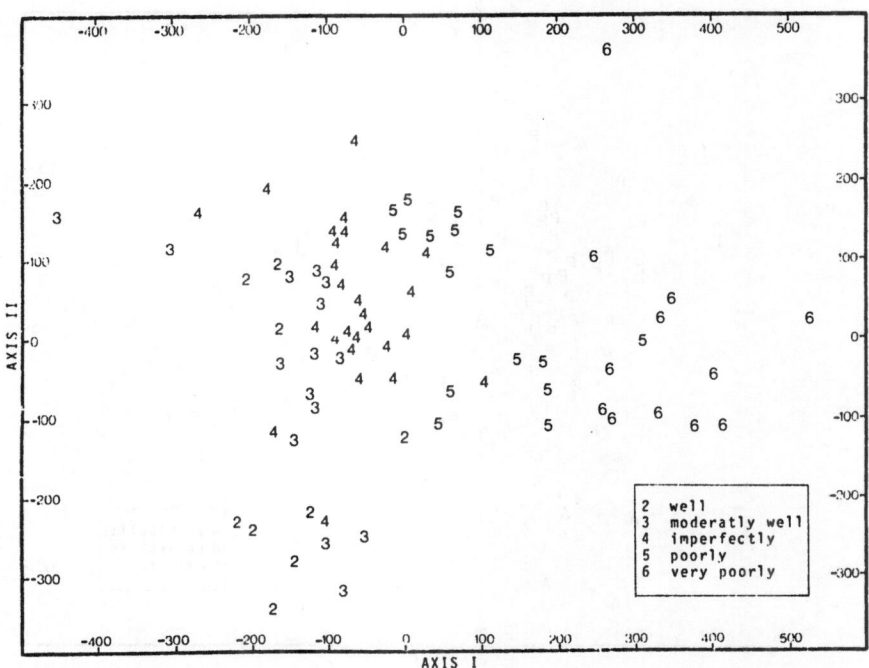

Fig. 4. Soil profile scatter diagram according to the first 2 axes of the P.C.A.; drainage class (CCP 1972).

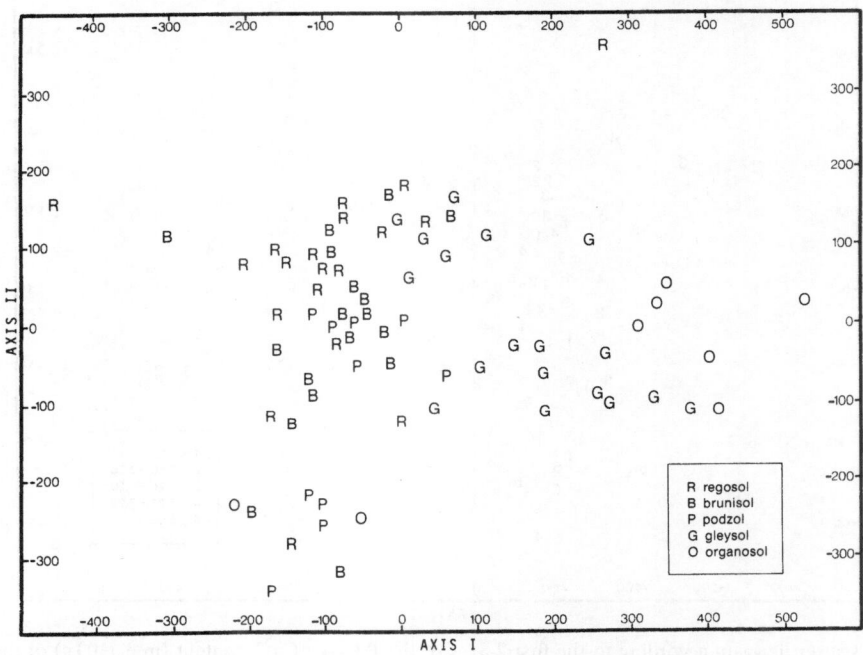

Fig. 5. Soil profile scatter diagram according to the first 2 axes of the P.C.A.; soil order.

gram of Ca concentration in the surface soil (Fig. 7), the profiles of serpentine soils (Fig. 6) are distributed across the different classes of calcium concentration. It is known that ultramafic soils are not uniformly poor in this element (Proctor & Woodell 1975, Butt 1979) and this is supported by our data. These results also indicate that profiles of ultramafic soils with the highest calcium concentration in the surface horizon are distributed independently of drainage and pedogen-

121

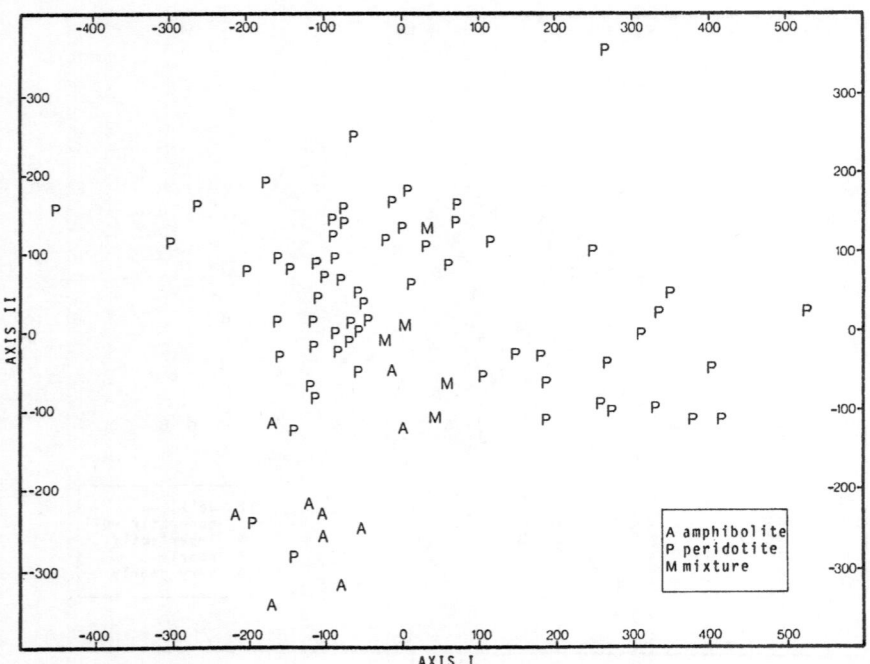

Fig. 6. Soil profile scatter diagram according to the first 2 axes of the P.C.A.; lithologic origin of the bed rock.

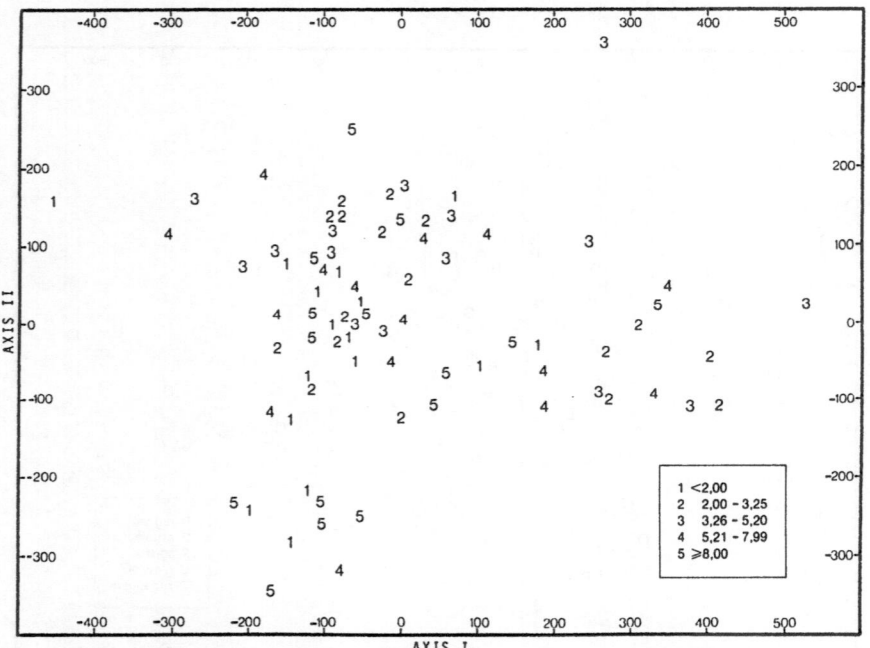

Fig. 7. Soil profile scatter diagram according to the first 2 axes of the P.C.A.; Ca^{2+} content (m.e./100 g) of the surface horizon.

esis. There is a correlation between the highest calcium concentration and amphibolite parent material except for two sites which are subject to deep snow cover and to severe leaching.

In other respects, when the distribution of pro-

files according to their surface Mg^{2+} concentration is examined (Fig. 8), there is little apparent grouping together of profiles amongst the ultra-mafic soils. Nevertheless, imperfectly drained profiles are concentrated to the left of the dia-

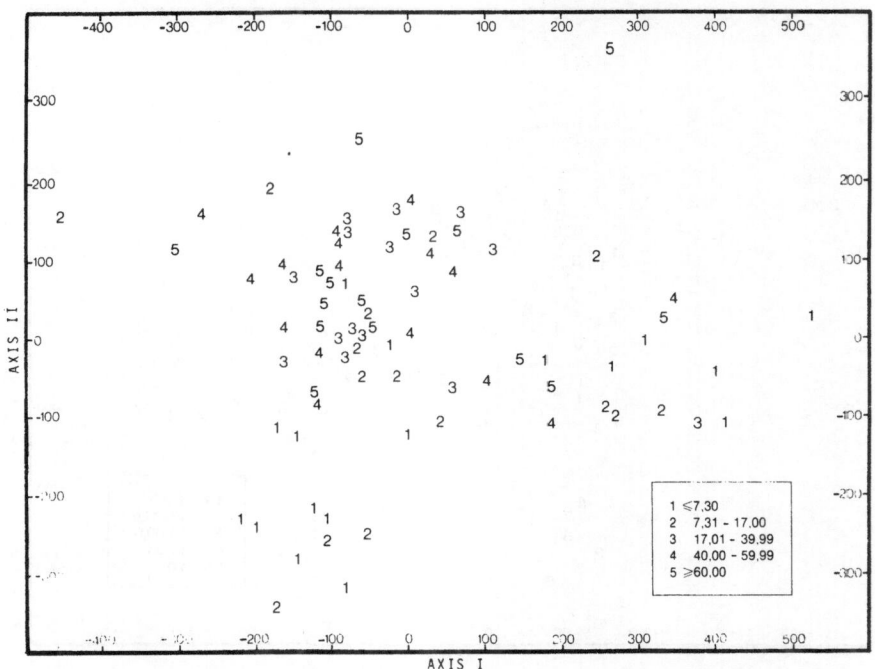

Fig. 8. Soil profile scatter diagram according to the first 2 axes of the P.C.A.; Mg^{2++} content (m.e./100 g) of the surface horizon.

gram. This suggests a tendency of the better drained soils (Fig. 4) to be richer in Mg in the upper horizon. Furthermore, it is noted that amphibolite soils are found among the classes having the lowest Mg concentration.

Finally, the comparison of Figs. 4, 5, 6 and 9 shows that in ultramafic soils, the lowest Mg/Ca quotients are found principally in the poorly drained soils such as organosols and gleysols whereas the quotient is higher among the better drained soils. It is also noteworthy that the lowest Mg/Ca quotients are always associated with soils derived from amphibolite.

If, as has been suggested by Loew & May (1901), Proctor (1971), and Proctor & Woodell (1975), a high Mg/Ca quotient is adverse to plant growth, these results suggest that, the phytotoxicity of the surface horizon diminishes along a gradient of decreasing drainage. Moreover, with respect to Mg/Ca quotient it appears that amphibolite-derived soils are more favourable for plants.

Flora

Serpentine soils exert, according to the region, a more or less pronounced selective pressure on the floristic composition of the vegetation (Proctor & Woodell 1975). Fernald (1907) was the first to underline the peculiar character of the flora of Mount Albert. The effect of the serpentinized ultramafic rock is demonstrated by the exclusive presence of certain species on Mount Albert which are absent from the other summits in the Chic-Chocs region, and by the absence on Mount Albert of species common elsewhere in the region.

The classification proposed here follows that of Rune (1953). From observations of the relative abundance of species on and off the ultramafic soil of the Mount Albert plateau, the degrees of serpentine affinity of vascular plants were defined. The taxa can be categorized as follows:
A) *Serpentinicoles*: taxa that are more frequent or abundant on ultramafic soils than on other

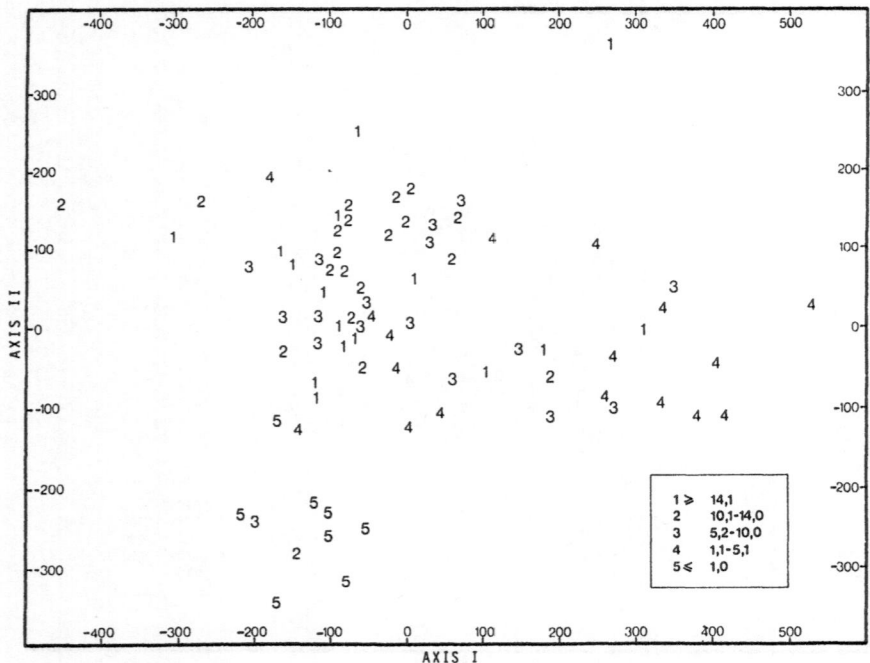

Fig. 9. Soil profile scatter diagram according to the first 2 axes of the P.C.A.: Mg/Ca quotient of the surface horizon.

substrates in the region. Thus defined, the concept of a serpentinicole has a regional validity and cannot be applied generally to the whole of the geographic distribution of these plants. In the context of this study, the designated region is the Chic-Chocs mountains (Fig. 1).

This group can be subdivided into:

  i) *exclusives*: which are never found off ultramafic soils in the region;
  ii) *preferentials*: which are found more often on ultramafic soils in the region.

B) *Indifferents*: taxa that are as frequent on ultramafic soils as on other substrata in the region.

C) *Serpentinifuges*: taxa never found on ultramafic soils in spite of their frequency on contiguous other substrata.

Table 2 lists the serpentinicoles species of Mount Albert which account for about 10% of the flora.

Lychnis alpina and *Armeria maritima* var. *labradorica* are exclusive serpentinicoles on Mount Albert which is at the southern limit of their distribution in North America (Rousseau 1974). According to Payette & Lepage (1977), *A. maritima* var. *labradorica* is more frequent as a maritime species in the Hudson Bay region. Because they

are subject to sea spray which has a high Mg/Ca quotient, coastal soils are in some ways similar to ultramafic soils (Butt 1979). The coastal soils generally have high concentrations of magnesium which often is the preponderant cation. *Armeria maritima* var. *labradorica* could be preadapted to ultramafic soils because of its maritime affinity. The other exclusive serpentinicoles (Table 2) are restricted to ultramafic soils or, more rarely, dolomitic formations (Scoggan 1950, Legault & Blais 1968, Rousseau 1974, Wagner & Rouleau 1984) throughout Eastern North America. In Western North America as well, *Adiantum pedatum* var. *aleuticum*, *Cheilanthes siliquosa*, and *Polysticum mohrioides* show a tendency to grow in serpentinic environments (Rune 1953).

The classification of the preferential taxa represents a first approximation and, as such, is subject to revision. *Rhododendron lapponicum*, is present on other summits in the region (Scoggan 1950, Gervais 1964, 1982, Boudreau 1981). However, Rune (1953) pointed out that it is most common on Mount Albert. Moreover, Payette & Lepage (1977) underlined its affinity for dolomitic substates in the Hudson Bay region and, according to Rousseau (1974), this could be true for all of the southern part of its distribution. *Festuca*

Table 2. Serpentinicoles of Mount Albert

Exclusives	Preferentials
Lychnis alpina	*Rhododendron lapponicum*
Armeria maritima var. *labradorica*	*Festuca altaica*
Arenaria marcescens	*Artemisia campestris* ssp. *borealis*
Adiantum pedatum var. *aleuticum*	*Arenaria sajanensis*
Polystichum mohrioides	*Arenaria humifusa*
Cheilanthes siliquosa	*Salix arctica*

altaica is another species which has occasionally been noted on non-ultramafic substrata in the region (Gervais 1964, 1982). Rousseau (1974) outlined this plant's affinity for dolomitic or serpentized substrates in Eastern North America. *Arenaria humifusa*, *A. sajanensis* and *Artemisia campestris* ssp. *borealis* have been collected occasionally on other summits in the region (Gervais 1964, 1982). Nevertheless, several authors, including Scoggan (1950) and Rune (1953, 1954), have emphasized their affinity for the calcareous or ultramafic soils. *Salix arctica* is very rarely encountered on the other Chic-Chocs summits, whereas it occurs widely on the ultramafic regosols of Mount Albert. We have therefore included it as a preferential serpentinicole.

The serpentinifuges are found on the amphibolite ridge. These include *Arctostaphylos alpina*, *Carex capillaris* spp. *capillaris*, *Diapensia lapponica*, *Salix uva-ursi* and *Vaccinium vitis-idaea* var. *minus*.

Sirois (1984b) showed that the serpentinicoles of Mount Albert belong to the phytogeographic group of Asian and Cordilleran species or, to the arctic-alpine group in the wide sense. In being more subject to cryoturbation than most other materials (Birse 1980), ultramafic soils resemble soils of higher latitudes or altitudes. The success of the arctic-alpines could be linked to low competition maintained by cryoturbation.

Despite the fact that most of the vascular plants of the plateau belong to circumboreal or American boreal groups, none of these showed a serpentinicolous affinity.

The result of this phytogeographic analysis supports the hypothesis of Rune (1954) that the presence, on Mount Albert, of rare taxa or disjunct populations is principally attributable to the ultramafic substrate.

Vegetation

The Mount Albert plateau is part of the fir forest great domain (Grandtner 1966), coinciding with the Acadian section of the boreal forest (Rowe 1972). The vegetation of the ultramafic part of the plateau is composed of tundra or krummholz. The tundra was subdivided into tundra on mineral soil and tundra on peaty soil. The krummholz cover is only a small area of the plateau.

Tundra on mineral soil

The four types of plant communities described below are arranged along a gradient of decreasing soil drainage. Along this gradient, plant cover goes from discontinuous to continuous due to a decrease of cryoturbation.

Rhacomitrium lanuginosum association
This association (Table 3) is dominated by *Rhacomitrium lanuginosum*, the only bryophyte considered to be a preferential serpentinicole in the Chic-Chocs. This moss plays a crucial role on the ultramafic plateau. At the beginning of plant succession (Dearden 1979), it occupies the moderately well drained sites and initiates pedogenesis. *Rhacomitrium lanuginosum* is tolerant of a wide range of exposure, duration of snow cover and the level of exchangeable bases (Tallis 1964). It is more likely that a regular supply of water strongly influences its distribution (Kallio & Heinonen 1973). On Mount Albert, it seems to benefit from snow free locations which are highly exposed to wind and actively cryoturbic. Thus, this species holds an important place in all open vegetation on the ultramafic plateau.

The *Rhacomitrium* association is otherwise characterized by the constant presence of shrubs

Table 3. Synthetic phytosociological table of the vegetation of the Mount Albert plateau

Phytosociological units Number of relevés		(1) 21	(2) 13	(3) 16	(4) 10	(5) 9	(6) 7	(7) 6	(8) 6	(9) 6
Shruby layers										
Rhododendron lapponicum	LS[a]	5.1[b]	5.1	3.1	3.2	1.+	1.+	.	.	.
Salix arctica	LS	4.+	4.1	2.1	3.+	2.+	1.+	1.+	.	.
Betula glandulosa	LS	5.2	5.2	5.2	5.2	5.2	5.2	5.2	4.1	4.1
Vaccinium uliginosum	LS	5.1	5.2	5.3	5.3	5.2	5.2	5.2	5.2	5.+
Ledum groenlandicum	LS	3.1	5.+	5.2	5.1	4.2	5.1	5.+	5.1	4.1
Andromeda glaucophylla	LS	1.+	1.+	3.1	5.2	5.2	3.1	5.2	.	.
Salix brachycarpa	LS	5.+	5.1	5.1	4.1	3.1	4.1	.	.	.
Kalmia polifolia	LS	.	1.+	1.+	1.1	.	2.+	4.+	4.+	5.+
Empetum nigrum	LS	1.+	1.+	1.+	.	.	3.1	2.+	4.1	5.+
Loiseleuria procumbens	LS	1.1	3.1	1.1	1.2	.	3.1	.	.	.
Juniperus communis	LS	1.+	3.1	1.2	1.1	.	3.1	.	.	.
Oxycoccus quadripetalus	LS	.	.	.	1.+	.	.	4.+	.	.
Abies balsamea	LT	.	.	.	.	.	.	.	.	1.1
Abies balsamea	TS	.	.	.	.	.	1.+	.	.	2.2
Abies balsamea	MS	.	.	.	.	.	.	.	2.1	4.4
Abies balsamea	LS	.	.	.	.	.	.	.	1.+	4.3
Picea glauca	LT	.	.	.	.	.	.	.	.	1.1
Picea glauca	TS	.	.	.	.	.	.	.	.	2.1
Picea glauca	MS	.	.	.	.	.	.	.	.	2.1
Picea mariana	MS	.	.	.	.	.	.	.	4.5	3.2
Picea mariana	LS	.	.	.	.	.	.	2.2	2.5	2.3
Amelanchier bartramiana	MS	.	.	.	.	.	.	.	.	4.1
Amelanchier bartramiana	LS	.	.	.	.	.	.	.	2.1	3.2
Vaccinium angustifolium	LS	.	.	.	.	.	.	.	1.2	5.1
Phyllodoce caerulea	LS	.	.	.	.	.	.	.	4.1	3.+
Herbaceous layer										
Arenaria sajanensis		1.+	1.+	.	.	.	.	.	.	.
Artemisia campestris ssp. *borealis*		4.+	4.+	1.+	1.+	.	.	.	.	.
Arenaria humifusa		3.+	3.+	2.+	2.+	2.+	.	.	.	.
Arenaria marcescens		5.+	5.+	3.+	3.+	1.+	1.+	.	.	.
Armeria maritima var. *labradorica*		5.+	4.+	4.+	3.+	1.+	3.+	.	.	.
Lychnis alpina		1.+	4.+	4.+	2.+	1.+	3.+	.	.	.
Festuca altaica		.	1.+	2.+	1.+	.	3.1	.	.	.
Scirpus caespitosus		5.1	5.1	5.2	5.2	5.4	5.3	5.4	.	.
Juncus trifidus		5.+	5.1	4.1	3.+	.	3.1	.	.	.
Carex scirpoidea		5.+	5.+	5.1	3.+	.	3.+	.	.	.
Campanula rotundifolia		5.+	5.+	3.+	3.+	.	4.+	1.+	.	.
Silene acaulis		5.+	5.+	2.+	1.+	.	1.+	.	.	.
Deschampsia caespitosa		4.1	4.1	5.1	4.2	3.1	5.+	2.+	.	.
Solidago multiradiata		2.+	4.+	3.+	3.+	.	5.+	.	.	.
Carex bigelowii		2.1	1.1	4.2	3.2	3.1	4.1	4.2	.	.
Cerastium arvense		3.+	3.+	4.+	2.+	2.+	2.+	.	.	.
Selaginella selaginoides		.	4.+	2.+	1.+	.	5.+	1.+	.	.
Deschampsia flexuosa		1.+	1.+	1.+	1.+	.	2.+	1.1	5.+	4.1
Drosera rotundifolia		.	.	.	.	.	.	4.1	.	.
Smilacina trifolia		.	.	.	.	.	.	4.2	.	.
Melampyrum lineare		.	.	.	1.+	.	.	.	.	4.+
Maianthemum canadense		.	.	.	.	.	.	.	.	4.+
Solidago macrophylla ssp. *thyrsoidea*		.	.	.	.	.	1.+	.	.	4.+
Streptopus roseus		.	.	.	.	.	.	.	.	3.+
Schizachne purpurascens		.	.	1.+	1.+	.	.	.	.	3.+
Cornus canadensis		.	.	.	.	.	.	.	1.+	4.+
Lycopodium annotinum		.	.	.	.	.	.	.	2.1	5.+

Table 3. Continued

Phytosociological units	(1)	(2)	(3)	(4)	(5)	(6)	(7)	(8)	(9)
Number of relevés	21	13	16	10	9	7	6	6	6
Linnaea borealis	.	.	.	.	.	.	.	2.+	5.+
Trientalis borealis	.	.	1.+	.	.	.	.	3.+	5.+
Coptis trifolia ssp. *groenlandica*	.	1.+	.	.	.	2.+	.	4.+	5.+
Clintonia borealis	.	.	.	.	.	.	.	3.+	5.1
Rubus pubescens	.	.	1.1	.	.	.	2.+	1.+	3.+
Gaultheria hispidula	.	.	.	.	.	.	.	5.+	4.1
Moss layer									
Rhacomitrium lanuginosum	5,3	5.2	4.1	4.2	3.1	3.2	.	.	.
Drepanocladus uncinatus	1.+	1.+	3.1	4.2	4.2	3.+	2.3	.	.
Campylium stellatum	1.2	2.+	3.1	4.2	5.3	3.1	2.1	.	.
Dicranum scoparium	1.+	1.+	2.+	1.1	1.+	4.1	2.1	.	.
Ptilidium ciliare	2.+	1.+	2.1	1.+	2.+	3.+	2.+	.	.
Pleurozium schreberi	.	1.+	2.1	1.2	1.+	3.1	2.3	4.4	5.3
Sphagnum nemoreum	.	1.+	1.1	.	.	.	1.4	.	.
Sphagnum russowii	.	.	.	.	.	.	1.4	.	.
Sphagnum fuscum	.	.	1.1	.	.	.	2.3	.	.
Sphagnum angustifolium	.	.	.	.	.	.	1.2	.	.
Sphagnum centrale	.	.	.	.	.	.	1.2	.	.
Lophozia lycopodioides	.	1.+	.	.	.	.	.	2.1	3.2
Dicranum fuscescens	.	1.+	.	.	1.+	1.+	1.+	4.2	3.+
Hylocomium splendens	.	.	1.+	.	.	.	1.+	3.+	4.1
Lophozia hatcheri	.	2.+	1.+	1.+	.	2.+	.	1.+	4.1
Lophozia ventricosa	.	.	.	.	.	.	.	4.+	1.+
Lichen layer									
Cetraria laevigata	5.1	5.2	5.2	4.2	2.2	5.2	2.2	3.+	2.+
Cladina mitis	5.+	5.1	5.+	3.+	2.+	2.1	1.+	2.+	1.+
Cladonia uncialis	5.+	5.+	2.+	2.+	1.+	3.+	1.+	2.1	3.+
Cetraria delisei	3.+	3.1	2.+	3.2	2.2	.	1.+	.	.
Ochrolechia frigida	5.1	4.+	1.+	3.1	.	1.+	1.+	.	.
Thamnolia subuliformis	5.+	4.2	2.+	2.+	2.+	.	.	.	.
Cladina rangiferina	3.1	3.1	4.+	2.1	.	3.1	2.1	4.2	5.+
Cetraria cucullata	4.+	3.+	3.+	2.+	.	.	.	.	.
Cladonia gracilis	4.1	5.+	2.+	2.1	1.+	1.+	.	2.+	1.+
Cladonia coccifera	3.+	4.+	3.+	2.+	1.+	2.+	.	2.1	2.+
Cetraria nivalis	5.+	4.+	1.+	1.+	.	.	.	2.+	1.+
Stereocaulon alpinum	2.+	2.+	1.+	1.+	.	1.+	.	.	.
Cladonia amaurocraea	2.+	2.+	1.+	2.+	1.+	.	.	.	.
Stereocaulon paschale	1.+	2.+	1.+	1.+	.	3.+	.	.	.
Cladonia pyxidata	2.+	2.+	1.+	1.+	.	1.+	.	.	.
Cladonia crispata	.	1.+	.	.	.	1.+	.	2.1	2.+
Peltigera aphtosa	.	.	.	.	.	.	.	1.+	2.+

(1) *Rhacomitrium lanuginosum* association.
(2) *Rhacomitrium lanuginosum* variant.
(3) *Ledum groenlandicum* sub-association.
(4) *Andromeda glaucophylla* variant.
(5) *Scirpus caespitosus-Campylium stellatum* association.
(6) *Selaginella selaginoides* variant.
(7) *Scirpus caespitosus-Sphagnum* spp. variant.
(8) Krummholz dominated by *Picea mariana*.
(9) Krummholz dominated by *Abies balsamea*.
[a]LS low shrub (0–0.3 m), MS medium shrub (0.3–2.5 m), TS tall shrub (2.5–5 m). LT low tree (5–10 m).
[b]The first number shows the presence class (Braun–Blanquet, 1932) and the second is the mean cover value.

such as *Betula glandulosa, Vaccinium uliginosum, Rhododendron lapponicum, Salix arctica* and *S. brachycarpa* which are more important as cryoturbation becomes less pronounced. The herbaceous stratum contains most of the serpentinicolous species, notably *Arenaria humifusa, Arenaria marcescens, Arenaria sajanensis, Armeria maritima* var. *labradorica, Artemisia campestris* ssp. *borealis, Lychnis alpina, Rhododendron lapponicum* and *Salix arctica*. Several of these species establish themselves on a bare substrata. This stratum also contains several indifferent species such as *Campanula rotundifolia, Carex scirpoidea, Juncus trifidus, Scirpus caespitosus*, and *Silene acaulis*.

Finally, in the lichen stratum, only *Cetraria laevigata* and *Ochrolechia frigida* are abundant although several others are frequent. The former apparently succeeds *Rhacomitrium* as the shrub strata develops.

As defined here, the association is confined to serpentine and is not found elsewhere in the Chic-Chocs. In the boreal hemisphere, its equivalent can be found on some exposed ultramafic outcrops of Newfoundland (Dearden 1975, 1979) and Sweden (Rune 1953). Communities of *Rhacomitrium lanuginosum* with similar physiognomy, but of different floristic composition are reported for other substrates in arctic regions (Billings & Mooney 1968, Hartman 1980).

Vaccinium uliginosum–Betula glandulosa association

The two shrubs which dominate this association have a wide ecological range. Thus, it has been subdivided into several units that characterize different habitats. Three of these units are hereafter described.

The variant with *Rhacomitrium* (Table 3) is distinguished from the previous association more by structure (Table 4) and by the relative importance of constituent species (Table 3), than by the presence of differential taxa. As indicated in Table 4, the cover of shrubby and herbaceous species increases as the *Rhacomitrium* carpet becomes less continuous.

There is a greater cover of lichen (mainly *Cetraria laevigata*) in the variant. Serpenticoles con-

Table 4. Average cover by strata in the *Rhacomitrium* association (1) and the *Rhacomitrium* variant of the *Vaccinium uliginosum-Betula glandulosa* association (2)

Strata/Vegetation Unit	Cover (%)	
	(1)	(2)
Shrubby	17	47
Herbaceous	9	17
Mossy	51	22
Lichen	15	33

tinue to be a characteristic element in this variant. The success of several of these could be connected to the cryoturbation factor that maintains patches of bare soil. This variant resembles the type B vegetation, 'arctic-alpine communities in less exposed locations' reported by Dearden (1975, 1979) for Newfoundland.

The *Ledum groenlandicum* sub-association (Table 3) is characteristic of the imperfect to poorly drained parts of the study area, on a flat or slightly convex slope, moderately exposed to wind.

The herbaceous layer is marked by the constancy and abundance of sedges and grasses like *Carex scirpoidea, Deschampsia caespitosa* and *Scirpus caespitosus*, which are adapted to wet soil conditions. The serpentinicoles *Arenaria* spp., *Artemisia campestris* ssp. *borealis, Rhododendron lapponicum*, and *Salix arctica* are present but infrequent. Conversely, the increased incidence of *Lychnis alpina* suggests that this serpentinicole is relatively hygrophilous. The serpentinicoles exploit primarily the small hummocks and the patches of bare soil in this habitat.

Within the *Ledum groenlandicum* sub-association there is a variation which is characterized by the hygrophilous shrub *Andromeda glaucophylla* (Table 3) and by a more closed vegetation with fewer serpentinicoles. This variant corresponds to a slightly wetter environment and it forms with the *Ledum* subassociation, a mosaic which is difficult to map.

Tundra on peaty soil

Fine textured deposits on the concave slopes of the plateau, frequently saturated with water, present conditions favourable to the development of regosolic gleysols and organosols. Aside from some phytosociological units of minor spatial importance, these soils support vegetation essentially dominated by *Scirpus caespitosus* and by a well developed moss layer.

Scirpus caespitosus–Campylium stellatum association

In the wettest sites, this association (Table 3) is dominated by *Scirpus caespitosus* that propagates vegetatively using a thick network of rhizomes. It is accompanied by shrubs such as *Ledum groenlandicum* and *Andromeda glaucophylla*, all growing in a moss carpet of *Campylium stellatum* and *Drepanocladus uncinatus*. Couillard (1978) reported this association in a boreal peat bog on limestone bedrock. This bog has *Drepanocladus revolvens* rather than *D. uncinatus*, a species considered serpentinicole by Rune (1953).

The *Selaginella selaginoides* variant (Table 3), corresponds to a stonier, slightly better drained environment. It has a higher species diversity. *Selaginella selaginoides*, the differential of this variant, is generally accompanied by *Solidago multiradiata* and *Deschampsia caespitosa*. *Festuca altaica* is frequent here and seems to be the only serpentinicole species whose optimum distribution is in a clearly wet environment. By comparison with the *Scirpus* association, the lichen stratum is more important than the moss stratum attesting to the more rapid drainage found in the variant.

Scirpus caespitosus–Sphagnum spp. association

In this association (Table 3) dominated by *Scirpus caespitosus*, the moss stratum is formed of one or another of the following species: *Sphagnum nemoreum*, *S. centrale*, *S. angustifolium*, *S. fuscum* and *S. russowii*. Although the *Sphagnum* presence is indicative of wetter conditions, their respective ecological preferences are difficult to establish (Gauthier 1980). This justifies the identification of this phytosociological unit on the basis of the presence and abundance of the *Sphagnum* genus. Some peatland species which characterize this association are: *Oxycoccus quadripetalus*, *Drosera rotundifolia*, and *Smilacina trifolia*. Communities dominated by *Scirpus caespitosus* and various *Sphagnum* species have also been described for peat bogs of boreal environments, on basic (Couillard 1978) as well as acidic (Gauthier 1980) substrata.

Krummholz

The krummholz corresponds to eroded or prostrate shrubby coniferous formations. The diverse growth forms adopted by conifers in subalpine and alpine environments are subjected to the erosive and desiccating action of winter winds (Payette 1974). The height that they reach is a good indicator of the severity of the local climate and the thickness of the snow cover at the site. Furthermore, the dominance of one or another conifer species is, of itself, ecologically significant and it is on this criterion which we based our classification. In most krummholz, the feather moss carpet, principally *Pleurozium schreberi*, is well developed. The plants that make up the krummholz are of boreal affinity. An exception to this is the serpentinicole *Adiantum pedatum* var. *aleuticum* which is present in certain stony sites especially crevices.

Krummholz dominated by Picea mariana

The krummholz formations dominated by *Picea mariana* (Table 3) are those found at the highest altitudes. They are geneally situated in wetter environments than the krummholz dominated by *Abies balsamea* and are at the interface of the lower and upper alpine belts (*sensu* Boudreau 1981). These harsher ecological conditions are also reflected by a less diverse herbaceous stratum. *Melampyrum lineare*, *Maianthemum canadense*, *Solidago macrophylla* var. *thyrsoidea*, *Streptopus roseus* and *Schizachne purpurascens*, generally present in *Abies balsamea* dominated krummholz are excluded here.

Krummholz dominated by Abies balsamea

Fir dominates amphibolite based krummholz formations (Table 3). It also dominates certain ultramafic soils, but only where the topography confers good protection in the form of winter snow. On the amphibolite ridge, the fir is regularly accompanied by *Picea glauca* up to 1050 m. On the ultramafic plateau *Picea glauca* is very rare above 900 m. Moreover, on other summits in the region,

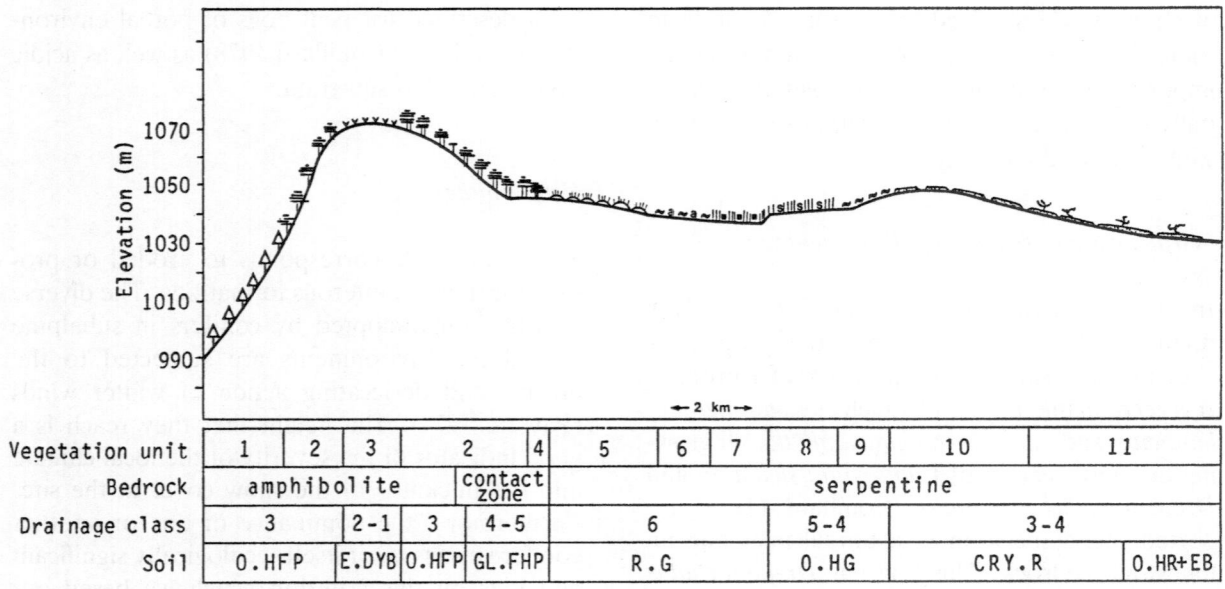

Vegetation unit	1	2	3	2	4	5	6	7	8	9	10	11
Bedrock	amphibolite			contact zone		serpentine						
Drainage class	3		2-1	3	4-5	6			5-4		3-4	
Soil	O.HFP		E.DYB	O.HFP	GL.FHP	R.G			O.HG		CRY.R	O.HR+EB

Fig. 10. Toposequence of a section of the Mount Albert plateau. Vertical scale exageration: 5 X. 1. Subalpine forest (not studied); 2. *Abies balsamea* dominated krummholz; 3. Alpine vegetation on amphibolite; 4. *Picea mariana* dominated krummholz; 5. *Scirpus caespitosus-Sphagnum* spp. association; 6. *Andromeda glaucophylla variant*; 7. *Scirpus caespitosus-Campylium stellatum* association; 8. *Selaginella selaginoides* variant; 9. *Ledum groenlandicum* sub-association; 10. *Rhacomitrium lanuginosum* association; 11. *Rhacomitrium lanuginosum* variant; CRY.R cryic regosol; EB eutric brunisol (sometime cryic); E. DYB eluviated dystric brunisol; GL.FHP gleyed ferro-humic podzol; O.HFP orthic humo-ferric podzol; O.HG orthic humic gleysol; O.HR orthic humic regosol; R.G. rego gleysol.

the *P. glauca* often grows at altitudes above 1100 m (Boudreau 1981). This suggests the existence of an interaction between the properties of ultramafic soils and climatic effects, in determining the altitudinal limit of *Picea glauca* on Mount Albert. The presence of this species is evident on well developed krummholz that takes the form of tall shrubs. For lower krummholz and on ultramafic soils *Picea mariana* is co-dominant.

Discussion

In the climatic context of the Mount Albert plateau, the influence of soil factors on the vegetation is seen at two levels. Firstly, at the physiognomic level, on mesic ultramafic soils, krummholz formations are rarely found above 1000 m. This is lower than the altitude attained by the krummholz of the adjacent amphibolite ridge (1050 m, Fig. 10). This is even lower than on other summits in the region, such as the McGerrigle Mountains. The soils here, princi-

pally derived from granite rock, support krummholz formations above 1200 m (Boudreau 1981). Thus, on the ultramafic soils of Mount Albert, the limit between krummholz and tundra is at a lower altitude than on other substrates in the region. These observations confirm that ultramafic soils often support vegetation of a physiognomy found at higher altitude on other substrata (Proctor & Woodell 1975). Secondly, the influence of soil factors on the floristic composition of the vegetation varies in intensity in different environments. The examination of Table 3 and the toposequence (Fig. 10) shows that most plants considered to be serpentinicoles are more frequent in vegetation units occurring on well to moderately well drained regosols and brunisols, often in a cryoturbated phase. On the other hand, communities of wetter soils are, for the most part, composed on species of Boreal affinity, common on other substrates in the region. The principal coordinate analysis of local soil conditions shows a negative correlation between Mg/Ca quotient and increse in soil wetness.

In sum, a drainage going from good to moderate, the occurrence of cryoturbation, and high Mg/Ca quotient in the soil are some of the ecological factors which account for the success of most of Mount Albert's serpentinicoles. This situation is similar to that of Sweden (Rune 1953) and of Scotland (Spence 1957, Birse, 1982) where it is reported that the plants characteristic of serpentine belong primarily to pioneer communities of cryoturbated skeletal soils. This is the case in Newfoundland where Dearden (1975, 1979) described, in ecologically similar situations, communities composed of *Rhacomitrium lanuginosum* and other arctic alpine-species, several of which, in our opinion, can be considered serpentinicoles. The cryoturbation and the Mg/Ca quotient can be considered interdependent ecological factors. As suggested by Rune (1954), cryoturbation may lead to a more or less pronounced regeneration of the surface horizon by bringing up less leached materials with a higher Mg/Ca quotient from the deeper horizons. This characteristic can slow down the colonization of the ultramafic soils.

The results of this study allow the conclusion that the ultramafic soils of Mount Albert accentuate the response of the vegetation to climatic conditions, such that alpine tundra occurs at a lower altitude than is usual. In this tundra, the ecological position of the majority of serpentinicole plants is analogous to the situations reported for several other serpentines in the northern hemisphere like those of Newfoundland, Scotland, Finland and Sweden. Thus, the example of Mount Albert helps clarify ecological analogies existing on other boreal serpentines.

Acknowledgments

This work has been done with the help of the administration of Gaspé Provincial Park. François Lutzoni and Renaud Mimeault contributed to the fieldwork and Valerie Huff translated the text from the French version. The suggestions of an anonymous reviewer greatly improved the manuscript. Financial support was granted by the Québec Environment Department and from Laval University to M.M.G. and by an NSERC scholarship to L.S.

References

Bergeron, Y., C. Camiré, A. Bouchard & P. Gangloff 1982. Analyse et classification des sols pour une étude écologique intégrée d'un secteur de l'Abitibi, Québec. Géogr. phys. Quat., XXXVI (3): 291–305.

Billings, W. D. & H. A. Mooney 1968. The ecology of arctic and alpine plants. Biol. Rev. 43: 481–529.

Birse, E. L. 1980. Suggested amendments to the world soil classification to accommodate Scottish mountain and aeolian soils. J. Soil Sci. 31: 117–124.

Birse, E. L. 1982. Plant communities on serpentine in Scotland. Vegetatio 49: 141–162.

Boudreau, F. 1981. Écologie des étages alpin et subalpin du mont Jacques-Cartier, Parc de la Gaspésie, Québec. M.Sc. thesis, Université Laval, Québec, 185 p.

Bouyoucos, G. L. 1936. Direction for making mechanical analysis of soils by hydrometer method. Soil Sci. 42: 225–229.

Braun-Blanquet, J. 1932. Plant sociology (English translation of Pflanzensoziologie by G. D. Fuller and H. S. Conard). McGraw Hill, New York, 439 p.

Buol, S. W., F. D. Hole, & R. J. McCracken. 1980. Soil genesis and classification. Iowa State University Press, Ames, 404 p.

Butt, G., 1979. Plant response to certain base-rich soils in eastern Canada. M.Sc. thesis, McGill University, Montréal, 120 p.

Commission canadienne de pédologie, 1972. Clasification canadienne des sols. Ministère de l'agriculture du Canada, Ottawa (Ont.), 220 p.

Commission canadienne de pédologie, 1978. Le système canadien de classification des sols. Pub. 1646, Ministère de l'agriculture du Canada, Ottawa (Ont.), 170 p.

Couillard, L. 1978. Etude de la végétation de deux tourbières de l'île à Samuel, archipel de Mingan. B.Sc. thesis, Université Laval, Québec, 153 p.

Crum. H. A. & L. E. Anderson. 1981. Mosses of eastern North America. Columbia University Press, New York, vol. 1, 2: 1328 p.

Dansereau, P. & M. Raymond. 1948. Botanical excursions in Québec province; Montréal–Québec–Gaspé peninsula. Bull. Serv. Biogéographie, Université de Montréal, no 2: 13–19.

Dearden, P. 1975. The biogeography of the Table Mountain, Bonne Bay, Newfoundland. An investigation of plant community composition and distribution on a serpentine bedrock. M.Sc. thesis, Memorial Univ., St. John's, Newfoundland, 151 p.

Dearden, P. 1979. Some factors influencing the composition and location of plant communities on a serpentine bedrock in western Newfoundland. J. Biogeog. 6: 93–104.

Fernald, M. L. 1907. The soil preferences of certain alpine and subalpine plants. Rhodora 9(105): 149–193.

Fernald, M. L. 1925. Persistence of plants in unglaciated areas of Boreal America. Mem. Amer. Acad. Arts & Sci. 15: 239–342.

Frontier, S. 1983. Stratégies d'échantillonnage en écologie. Masson, Paris–Presses Univ. Laval, Québec, 494 p.

Gagnon, R. M. 1970. Climat des Chic-Chocs. M.P.-36, Ministère des richesses naturelles du Québec, Direction générale des eaux, Service de la météorologie, 103 p.

Gauthier, R. 1980. La végétation des tourbières et les sphaignes du parc des Laurentides, Québec. Études écologiques, no. 3, 634 p.

Gervais, C. 1964. Etude de la flore de la région du mont Logan. M.Sc. thesis, Université de Montréal, Montréal, 310 p.

Gervais, C. 1982. La flore vasculaire de la région du mont Logan, Gaspésie, Québec. Provancheria, no. 13, 63 p.

Grandtner, M. M., 1966. La végétation forestière du Québec méridional. Presses Université Laval, Québec, 216 p.

Gray, J. T. & Brown, J. E., 1979. Permafrost existence and distribution in the Chic-Chocs mountains, Gaspésie, Québec. Géogr. phys. Quat., XXXIII (3–4): 299–316.

Hale, M. E. & W. L. Culberson. 1970. A fourth checklist of the lichens of the continental United States and Canada. The Bryologist 73(3): 499–543.

Hartmann, K. V. H. 1980. Beitrag zur Kenntnis der Pflanzengesellschaften Spitzbergens. Phytocoenologia, 8(1): 65–147.

Kallio, P. & S. Heinonen. 1973. Ecology of *Rhacomitrium lanuginosum* (Hedw.) Brid. Rep. Kevo subarctic res. stat., 10: 43–54.

Lebrun, C. & M. Thériault. 1979. Étude géomorphologique du Mont Albert. B.Sc. thesis, U.Q.A.R., Rimouski. 70 p. + 1 map.

Legault, A. & V. Blais. 1968. Le *Cheilanthes silicosa* Maxon dans le nord-est américain. Naturaliste can. 95: 307–316.

Legendre, L. & P. Legendre 1979. Ecologie numérique. T. II. La structure des données écologiques. Masson, Paris. Presses Univ. Québec, Montréal, 254 p.

Loew, O. & D. W. May. 1901. The relation of lime and magnesium to plant growth. Bull. Bur. Pl. Ind. U.S.D.A. 1, 1–53.

Marie-Victorin, Fr. 1964. Flore Laurentienne, 2nd edition, Presses Univ. Montréal, Montréal, 925 p.

McKeague, J. A. 1978. Manuel de méthodes d'échantillonnage et d'analyses des sols. Soil Research Inst., Can. Dep. Agric., Ottawa, Ontario, 223 p.

Moisan, G. 1956. Le caribou de Gaspé. II. Analyse de l'habitat hivernal. Naturaliste can., 83: 262–274.

Mueller-Dombois, D. & H. Ellenberg. 1974. Aims and methods of vegetation ecology. John Wiley, New York, 547 p.

Payette, S. 1974. Classification écologique des formes de croissance de *Picea glauca* (Moench.) Voss et de *Picea mariana* (Mill.) BSP. en milieux subarctiques et subalpins. Naturaliste can., 101: 893–903.

Payette, S. & F. Boudreau 1984. Evolution postglaciaire des hauts sommets alpins et subalpins de la Gaspésie. Can. J. Earth Sci., 21(3): 319–335.

Payette, S., F. Boudreau, & R. Ledoux. 1981. Paléoécologie du mont Jacques-Cartier. *In* Gray, J. T., (Coord.), Weathering zones and the problem of glacial limits, Canadian Quaternary Association and Association Québecoise pour l'Etude du Quaternaire, Papers and guidebook for excursion and conference in Gaspésie, Québec, p. 69–87.

Payette, S. & E. Lepage. 1977. La flore vasculaire du Golfe de Richmond, Baie d'Hudson, Nouveau-Québec. Provancheria, no. 7, 68 p.

Pédro, G. & K. E. Bitar. 1966. Contribution à l'étude de la genèse des sols hypermagnésiens: recherches expérimentales sur l'altération chimique des roches ultrabasiques (serpentinites). Ann. agron. 17(6): 611–651.

Proctor, J. 1971. The plant ecology of serpentine. III. The influence of a high Mg/Ca ratio and high nickel and chromium level in some British and Swedish serpentine soils. J. Ecol. 59: 827–842.

Proctor, J. & S. R. J. Woodell. 1975. The ecology of serpentine soils. Adv. Ecol. Res. 9: 256–366.

Raymond, M. 1950. Esquisse phytogéographique du Québec. Mémoires du Jardin Botanique de Montréal, no. 5, 147 p.

Raymond, M., R. Pomerleau, & J. Kucyniak. 1954. Guide d'excursion no 15 Gaspé-générale. Report of IXth International botany congress, Montréal, p. 10–14.

Rieger, S. 1983. The genesis and classification of cold soils. Academic Press, New York, 230 p.

Roberts, B. A. 1980. Some chemical and physical properties of serpentine soils from western Newfoundland, Canada. Can. J. Soil Sci. 60(2): 231–240.

Rousseau, C. 1974. Géographie floristique du Québec–Labrador. Presses Univ. Laval, Québec, 799 p.

Rowe, J. S. 1972. Forest regions of Canada. Can. Dep. Environ., Can. For. Serv., Publ. 1300, 172 p. + map.

Rune, O. 1953. Plant life on serpentine and related rocks in the north of Sweden. Acta phytogeographica Suecica, 31, 139 p.

Rune, O. 1954. Notes on the flora of the Gaspé peninsula. Svensk bot. tidskr., 48(1): 117–136.

Scoggan, H. J. 1950. The flora of Bic and the Gaspé peninsula, Québec. Nat. Mus. Can., Bull. 115, 399 p.

Scoggan, H. J. 1978. The flora of Canada. Vol. 1–4, Nat. Mus. Can., Botany pub. no. 7, Ottawa, 1711 p.

Sirois, L. 1984a. Le plateau du mont Albert. Etude phytoécologique. Final Report, Ministère de l'environnement du Québec, 152 p. + 15 tables + 4 maps.

Sirois, L. 1984b. Le plateau du mont Albert. Etude phytoécologique. M.S. thesis, Université Laval, Québec, 152 p. + 15 tables + 1 map.

Sirois, L. & M. M. Grandtner. 1982. Aperçu des problèmes phyto-écologiques du plateau du mont Albert, parc de la Gaspésie. Abstracts of 50th ACFAS congress, p. 31 and unpublished report, 12 p.

Spence, D. H. N. 1957. Studies on the vegetation of Shetland. I. The serpentine debris vegetation in Unst. J. Ecol. 45: 917–945.

Stotler, R. & Crandall-Stotler, B. 1977. A checklist of the liverworts and hornworts of North America. The Bryologist 80(3): 405–428.

Tallis, J. H. 1964. Growth studies on *Rhacomitrium lanuginosum*. The Bryologist 67: 417–422.

Wagner, W. H. & E. Rouleau. 1984. A western holly fern,

Polystichum x scopulinum in Newfoundland. American fern Journal 74(2): 33.

Walter, H., E. Harnickell & D. Mueller-Dombois. 1975. Climate-Diagram Maps. 9 maps and text with 14 figures, Springer-Verlag, Berlin, 36 p.

Wanless, R. K., R. D. Stevens, G. R. Lachance & R. N. Debibio. 1973. Age determination and geological studies K-Ar Isotopic ages. Canadian geological survey, report 11, Ottawa, 139 p.

Authors' Addresses:

L. Sirois
Dép. de Biologie
Université du Québec à Rimovski
300 allée des Vrsutines
Rimovski
Québec
Canada 45C 3A1

M. M. Grandtner
Dép. Sciences Forestières
Faculté de Foresterie
Université Laval
Québec
Canada 41K 7P4

VI

Chemical and ecological studies on the vegetation

of ultramafic sites in Britain

J. PROCTOR

Abstract. The vegetation of British outcrops of ultramafic rocks occurs under a wide range of climatic conditions from the southernmost area of the English mainland to the northernmost Scottish islands. The rocks themselves vary in composition and are covered to a varying extent by foreign material. As a result, generalisations about this vegetation are difficult and it ranges in type from debris with the less than 1% plant cover to mature deciduous woodland with a species-rich ground flora. In this paper the physical environment and the vegetation of several British ultramafic outcrops are described. In most cases the vegetation is not as dramatically unusual as that described for many other countries. The reasons for this are discussed and those special floristic features which do occur are described. A summary of plant analyses for the British ultramafic outcrops is included and shows some important features. There are no nickel accumulators although there are reports of plants with high foliar iron concentrations. Some analysed species preferentially absorb magnesium, others calcium. Experimental work on plants of British ultramafic soils began using crop plants and provided an early demonstration of acute nickel toxicity in oats (*Avena sativa*) grown in a farm soil containing drift derived from ultramafic material. Acute nickel toxicity has rarely been demonstrated since, for any plant in an ultramafic soil in Britain. Later work with crops in sand and water culture demonstrated some important interactions between supplied nickel and magnesium. Experiments involving native species have demonstrated the existence of plant races, collected from ultramafic soils, which are tolerant to relatively high concentrations of nickel and magnesium and have shown the effects on plant growth of interactions between supplied calcium, magnesium, nickel and micronutrients. Some field experiments have been made on barren sites and these have demonstrated a substantial increase in plant cover on fertilisation with major nutrients. Nutrient limitation (in some cases it has been shown to be phosphorus specifically) is no doubt a factor which prevents full cover developing on some sites. The most recent work involving a range of approaches has suggested the importance of drought in maintaining sites in an open state and this may interact with nutrient limitation. The role of metal toxicity in maintaining the openness of sites is not clear, but seems generally less important than shortage of water or nutrients.

B. A. Roberts and J. Proctor (eds), The ecology of areas with serpentinized rocks. A world view, 135–167.
© *1992 Kluwer Academic Publishers.*

Introduction

The vegetation of British outcrops of ultramafic rock is not as dramatically unusual as that reported for many other countries. There are several reasons why this should be so. First, Britain was under ice or at least severely affected by periglacial conditions during the last Ice Age (until the glaciers disappeared from there about 10,000 years ago). This means that the flora of British ultramafic areas has a relatively recent origin and partly explains the absence on them of clear speciation. Secondly, there is extensive contamination by glacial material, loess, or sea spray. These contaminants reduce the effects of bedrock on the vegetation. Thirdly, the high precipitation/ evaporation quotient of the climate means that many of the soils are highly leached and the influence of the rock is reduced, particularly where it is massive and slow weathering. Fourthly, several of the outcrops have unusually calcareous rocks and the calcium ameliorates the effects of toxic elements in the soil. Fifthly, the outcrops have a small area. The largest contiguous area of ultramafic bedrock is the Lizard Peninsula where it is about 50 km^2. The area of soils derived from the bedrock is very much smaller than this and residual ultramafic soils occupy no more than a few km^2 at any site and usually much less. Finally, physiognomic contrasts are reduced because of the general removal of forest cover.

However the vegetation of British ultramafic outcrops is often distinctive in some way and in several cases it has important rarities. The earliest written reference is that of Traill (in Neill 1806) in 1803 who observed that the hills north of Balta-sound were 'in many places totally divested of vegetation (even of Lichens), presenting to the wearied eye a naked waste of an iron-brown colour'. Another early reference is that of MacGillivray (1855) who made brief remarks on the green appearance of the Coyles of Muick. The first detailed account of any site was by West (1912) who described several of the rare plants, including unusual varieties, on Unst, Shetland. He made the first attempt at an ecological explanation of the barren areas and showed an early appreciation of the possible roles of toxic metals and their interaction with the environment: 'The Serpentine yields chromic iron, and it is possible that in some parts of the area there may be some harmful constituents of the weathered rock which inhibit the growth of many plants, and also hinder the ground from being fully occupied by those that attempt to cover it, and if this be so, it may also serve as an ally to the bleakness of the place in augmenting the dwarfing of the plants.'

Although taxonomic work and species recording continued there was little ecological work until the 1950's. Interest was revived following the experimental work of Hunter & Vergnano (1952) and the account of Rune (1953) for Swedish ultramafic areas, and the synthesis of work on ultramafic areas in North America (Kruckeberg 1954, Walker 1954, and Whittaker 1954). There followed the classic descriptive papers by Coombe & Frost (1956a, b) and Spence (1957) which laid the foundation for modern ecological work on British ultramafics and the subject matter of this review.

I describe here the locations of the British ultramafic areas and their physical environment including the soils. The vegetation is then discussed, along with special features of the flora, and several analyses of the plants are collated. I go on to describe the relevant experimental work and conclude by assessing how far we can explain the important features of the vegetation.

The location of British ultramafic outcrops

In England the outcrops are restricted to the south west, with those on the Lizard Peninsula by far the most important; in Wales there is one small outcrop in Anglesey; in Scotland there are a large number of outcrops. Many sites are little known botanically. For several sites there is a limited amount of information: Clicker Tor and Polyphant in south-west England (Proctor & Woodell 1971); the solitary Welsh site (Slingsby & Brown 1977); Glendaruel in west Scotland (Proctor & Woodell 1971); Corrycharmaig in central Scotland (Steele 1955, Ferreira 1963, Roberts & Stirling 1974); and the following in northern Scotland, Beauty Hill (Proctor & Woodell 1971), Bridgend (Roberts & Stirling 1974), Brown Hill (Ferreira 1963), Glen Urquart (Steele 1955, Proctor & Woodell 1971, Roberts & Stirling 1974), and Portsoy (Smith 1968).

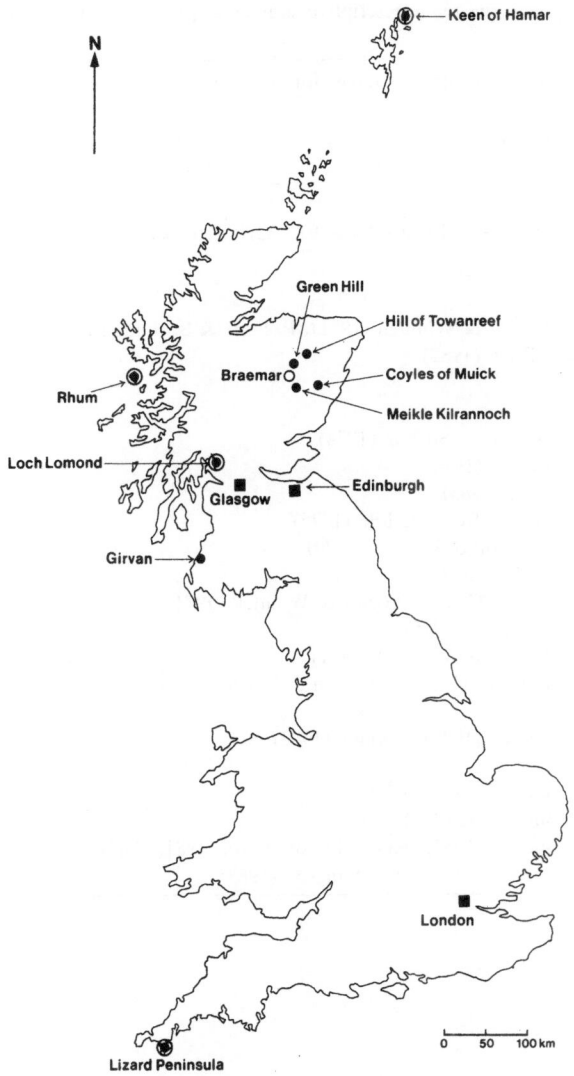

Fig. 1. The approximate location of the Britisch ultramafic areas where ecological work has concerned (●), and the sites of the meteorological measurements (○) summarised in Tables 3 and 4.

Ecological work has concentrated (not equally) in six areas (Fig. 1): the Lizard Peninsula (henceforth called the Lizard) in south-west England; outcrops a few km south of Girvan in south-west Scotland; several outcrops (including the Islands of Creinch and Inchcailloch in Loch Lomond, and the nearby Conic Hill and Lime Hill) along the western part of the Highland Boundary Fault; a group of sites in north-east Scotland, the Coyles of Muick, Meikle Kilrannoch, Strathdon (particularly Green Hill), and the Hill of Towanreef; the Island of Rhum in the Inner Hebrides of

western Scotland; and the Shetland Islands, mainly Unst (where the Keen of Hamar is the most intensively studied site) and Fetlar. Some features of these areas together with the main descriptive sources are given in Table 1.

Geology

Ultramafic rocks cover about 0.1% of Britain's land area and the nature and origins of their main occurrences are here briefly described.

The ultramafic rocks of the Lizard are an excellent example of a slice of Upper Mantle associated with a major crustal fault zone. The rocks are among the oldest in England, probably Pre-Cambrian in age (older than 600×10^6 years). The original rock was a peridotite, i.e. an olivine-rich ultramafic variety, but is now largely altered to serpentinite. The boundaries of the serpentinite, which occupies about 50 km^2 are faulted and near vertical, and its 3-dimensional shape resembles that of a vertical cylinder. Within the serpentinite body, a number of smaller bodies of quartz-rich red granite-gneiss (up to 1 km across) have been intruded at a later stage. Minor intrusions intermediate in composition between granite and ultramafic rocks, such as gabbro and dolerite, also cut the Lizard.

The thin slices of serpentinite exposed along the Highland Boundary Fault in Scotland include the outcrops near and in Loch Lomond. The Highland Boundary Fault is a deep crustal fault and the easily deformed serpentinite appears to have lubricated movement along it. The chain of bodies of ultramafic rock (mainly serpentinites) which includes the Coyles of Muick, Green Hill and the Hill of Towanreef are associated with a large intrusion of metamorphosed gabbro and probably originated in a magma chamber within the continental crust. The Cowal serpentinites (including Glendaruel) are of similar origin. All the Central Highland serpentinites are of Caledonian age.

Rhum has the largest area of unaltered peridotite in Britain. Of Tertiary age (60×10^6 years old), the peridotite formed in a magma chamber just below the surface of the crust and was subsequently exposed by deep erosion. The peridotite is interlayered with anorthosite, a rock type rich

137

Table 1. The areas (km²) of ultramafic bedrock, their altitude (a.s.l.), and the main descriptive sources of their vegetation, for the British sites on which ecological research has concentrated

Site	Approximate area of ultramafic bedrock* (km²)	Altitude (m)	Main descriptive sources for vegetation
England			
Lizard Peninsula	50	c. 90 (average)	Coombe & Frost (1956a, b); Marrs & Proctor (1978, 1979)
Scotland			
South-west			
Girvan	20	0–170	Proctor & Woodell (1971), Roberts & Stirling (1974), Birse (1982)
Western (Highland Boundary Fault)			
Ben Bowie	<1	280	Roberts & Stirling (1974)
Conic Hill	<1	15–300	Stirling (1974)
Creinch	<1	15–30	Nicol (1983)
Inchcailloch	<1	15–85	Horrill, Sykes & Idle (1975)
Lime Hill	<1	170–180	Johnston & Proctor (1979)
North-east			
Coyles of Muick	2	440–600	Spence (1970), Proctor & Woodell (1971)
Green Hill	6	460–570	Sedgwick (1975)
Hill of Towanreef	4	280–430	Proctor & Woodell (1971)
Meikle Kilrannoch	<1	870	Marshall (1959), Johnston & Proctor (1980)
Inner Hebrides			
Rhum	30	0–810	N.C.C. (1974), Looney (1982)
Shetland			
Mainland	<1	150–240	Spence (1970, 1979)
Fetlar	20	15–160	Slingsby *et al.* (1983)
Unst	40	15–150	Spence (1957, 1958); Slingsby (1979, 1981, 1982); Carter, Proctor & Slingsby (1987)

*Areas with ultramafic soil are much smaller.

in feldspar and thus not ultramafic. The peridotite on the mountain of Ruinsival and the western part of the mountain of Trallval is, however, free of anorthosite, although later intrusions of gabbro and dykes of dolerite do occur. The anorthosite weathers to a distinctive grey-white, while the peridotite interlayered with anorthosite forms a smaller expanse in the Cuillin Hills of neighbouring Skye. The two volcanoes were part of a tentative, aborted rift, later followed by a successful rift to the west which gave rise to the Atlantic Ocean.

In the Unst ultramafics, as in the Lizard Complex, the original peridotite has been altered to serpentinite which weathers to a characteristic ochre-brown colour. An additional episode of alteration has resulted in the formation of talc

along the margins and along certain faults. Chromite, a mineral which occurs in small amounts in most peridotites, is locally extremely abundant in the Unst serpentinite and was formerly quarried. The Unst serpentinite is a possible example of a sheet or nappe of oceanic crust which was emplaced onto a continental crust during the Caledonian episode of mountain building about 400 million years ago. The small mass of serpentinite at Ballantrae, on the south-west coast of Scotland, near Girvan, is more definitely a product of such a mechanism.

Rock analyses from several British ultramafic sites are shown in Table 2. In general, such analyses have not been helpful in understanding the differences that exist between the vegetation of the sites. An exception is the dolomitized rock

Table 2. Chemical analyses (%) of various British ultramafic rocks

	Tremolite-serpentine, Lizard[1]	Antigorite-serpentine, Unst[2]	Peridotite-serpentine, Unst[2]	Dunite-serpentine, Unst[2]	Antigorite-serpentine, Corrycharmaig[2]	Bastite-serpentine, Ayrshire[2]	Dolomitized-serpentine, Highland Boundary Fault rock[2]	Olivine adcumulate, Rhum[3]	Harrisitic cumulate, Rhum[3]
SiO_2	40.87	42.67	38.00	33.22	40.61	38.58	39.1	38.66	39.20
Al_2O_3	3.93	0.80	0.64	0.72	2.30	1.65	1.8	1.95	5.11
Fe_2O_3	6.17	1.98	3.85	4.64	3.67	3.94	5.4	3.33	3.28
FeO	2.37	4.23	3.39	3.64	3.23	2.49	–	9.52	8.24
MgO	32.86	37.86	39.58	40.85	37.52	37.84	12.4	41.99	36.38
CaO	2.36	0.25	0.93	0.27	0.15	0.04	15.0	0.64	2.94
Na_2O	0.39	0.03	0.08	0.09	0.02	0.62	–	0.12	0.38
K_2O	0.04	n.d.	n.d.	0.02	tr.	0.11	–	0.05	0.05
$H_2O > 105°C$	–	11.46	12.25	13.97	11.61	12.68	–	2.91	3.39
$H_2O < 105°C$	–	0.14	0.64	0.96	0.10	1.49	–	0.0	0.0
TiO_2	0.16	tr.	n.d.	0.07	0.03	0.04	tr.	0.15	0.26
P_2O_5	–	0.02	0.02	0.01	tr.	0.04	tr.	0.03	0.03
MnO	0.29	0.04	0.11	0.11	0.08	–	0.1	0.20	0.10
Cr_2O_3	0.25	0.37	0.39	1.23	1.07	0.24	0.3	0.37	0.45
NiO	–	0.21	0.23	0.19	0.15	0.06	–	(0.18)	(0.21)

Abbreviations: – = not analyzed; n.d. = not detected; tr. = trace.
[1]Butler (1953).
[2]Guppy (1956).
[3]Wadsworth (1961). Values in parentheses refer to % Ni^{2+}.

Table 3. Mean daily maximum and minimum temperatures (°C) for each month at five stations (see Fig. 1 for location). The data are for at least ten years at each site

Station	Altitude (m)	Jan	Feb	Mar	Apr	May	Jun	Jul	Aug	Sept	Oct	Nov	Dec
Lizard[1]	73	8.6	8.2	9.7	11.8	13.9	16.6	18.2	18.4	16.9	14.5	11.5	9.8
		5.0	4.4	5.4	6.6	8.6	11.2	12.6	13.1	12.3	10.4	7.5	6.1
Baltasound[1]	24	5.3	5.5	6.6	8.4	10.8	13.5	14.5	14.8	12.8	10.5	7.3	5.9
(Keen of Hamar)		1.0	0.9	1.6	2.7	5.1	7.5	8.9	9.2	7.8	5.8	2.7	1.5
Braemar[1]	339	3.5	3.8	6.4	9.4	12.9	16.3	17.1	16.6	14.3	11.0	6.5	4.5
		−2.7	−2.8	−0.9	1.0	3.5	6.7	8.3	8.0	6.4	4.0	0.3	−1.1
Kinloch[2]	10	6.9	7.3	8.8	11.2	14.2	16.4	17.1	17.4	15.6	12.5	9.7	8.1
(Rhum)		1.9	1.9	3.3	4.3	6.3	8.9	10.5	10.5	9.5	7.3	4.6	3.1
Arrochymore[3]	30	6.2	6.1	8.0	11.3	14.5	16.9	18.7	18.3	15.5	12.4	8.8	7.3
(Loch Lomond)		1.0	0.6	1.5	3.0	5.5	8.2	10.3	10.4	8.6	6.1	2.8	1.7

[1]Data from British Meteorological Office.
[2]Data from Nature Conservancy Council (1974).
[3]Data from Nicol (1983).

from the Loch Lomond area which supports a distinctive vegetation which depends to some extent on the high calcium concentrations in the rock. The geological history of these calcareous serpentinites is described by Henderson & Fortey (1982) and their use for limestone fertilizer has been documented by Mitchell & Mitchell (1983).

Climate

Some features of the climate near the major British ultramafic outcrops are summarised in Tables 3 and 4. The climatic differences explain some of their vegetational contrasts. For example, the uniqueness of the Lizard is partly dependent on its mild winters; high rainfall causes the coarse-textured soils on Rhum to be severely leached and hence bear more acidophilous vegetation. Recently, closer inspection of climatic data has shown that, surprisingly in view of the pervading cool moist climate, periodic droughts may be an important factor for plants of shallow skeletal soils of low water-retaining capacity. Carter, Proctor & Slingsby (1987) found that absolute summer droughts of 7–21 days have occurred on Unst, Shetland, in most of the 25 years between 1959 and 1983. J. H. H. Looney & J. Proctor (unpublished) have demonstrated regular summer droughts of up to 32 days on Rhum. On the Scottish sites, winters are usually sufficiently cold to cause frost heaving which has been experimentally demonstrated on Rhum (Ferreira & Wormell 1970) and on the Keen of Hamar (Carter, Proctor & Slingsby 1987, Fig. 5). The microclimate of open areas on skeletal soils is

Table 4. Mean monthly rainfall (mm) for five stations (Fig. 1). The data are for at least ten years at each site

Station	Altitude (m)	Jan	Feb	Mar	Apr	May	Jun	Jul	Aug	Sept	Oct	Nov	Dec	Total
Lizard[1]	73	100	72	69	50	59	49	58	67	73	80	100	103	880
Baltasound[1]	24	128	87	107	77	58	62	63	68	115	130	147	141	1185
Braemar[1]	339	82	67	52	55	69	55	61	83	81	87	89	98	879
Coire Dubh[2] (Rhum)	270	285	137	236	171	190	209	232	271	322	362	249	297	2960
Arrochymore[3]	30	175	113	135	65	79	88	94	97	159	144	202	158	1525

[1]Data from British Meteorological Office.
[2]Data from Nature Conservancy Council (1974).
[3]Data from Nicol (1983).

relatively harsh. Spence (1957) recognised this for the Keen of Hamar but his emphasis on the key role of wind was probably misplaced (Proctor & Woodell 1971). Carter, Proctor & Slingsby (1987) have made temperature measurements on skeletal soils and heathlands on ultramafic rocks on Unst. They showed that the diurnal temperature range was much greater on the former.

Soils

Classification

The most recent attempt to classify British soils with an ultramafic component in the parent material has been in Scotland where the soils have been named as magnesian variants of sub-groups used by the Soil Survey of Scotland (Heslop & Bown 1969, Birse 1982). Four main categories have been recognised by these workers and these are (with the proposed (Birse 1980) synonyms to relate them to the World Soil Map (F.A.O./U-NESCO 1974):

a) magnesian skeletal soils (magnesian variants of eutric regosols);

b) brown magnesian soils (freely-drained versions are magnesian variants of eutric cambisols and imperfectly drained ones are magnesian variants of eutro-gleyic cambisols);

c) magnesian gleys (magnesian variants of eutric gleysols or humoeutric gleysols); and

d) alpine magnesian soils (magnesian variants of cryo-eutric cambisols and cryo-eutric gleysols).

Birse (1982) has described as 'hemicryic,' subalpine brown magnesian soils which are subject to frost heave to a limited degree. The above classification seems to account for most of the British ultramafic soils and the category 'brown magnesian soils' encompasses the 'eutrophic braunerde' described by Coombe & Frost (1956b) from the Lizard. The 'brown rankers' recognized by Kubiena (in Coombe & Frost 1956b) perhaps fall into the shallow phase of the brown magnesian soils noted by Birse (1980) but they might be the magnesian equivalent of a brown ranker (a type not yet unequivocally classified) (C. J. Bown personal communication).

Some ultramafic outcrops bear exceptional soils: for example soils on the ultramafics of the Loch Lomond area are so calcareous that they can hardly be described as 'magnesian variants' and this term is not apt for the skeletal soils on Rhum where the magnesium is often largely removed by leaching. On Rhum, Ragg & Ball (1964) described some soils as unusual variants of peat podzols.

Soil parent materials and pedogenesis

Coombe & Frost (1956b) showed that the most extensive soils overlying the serpentinites of the Lizard were developed on granite-derived loess which was deposited under periglacial conditions. Bedrock-derived soils on the Lizard are largely confined to the coastal areas where there is an important maritime influence on the soils which has been quantified by Malloch (1972). Butler (1953) described some weathering products in soil on tremolite serpentine on the Lizard. He found that the clay (excluding free iron oxides) consisted principally of chlorite, illite and kaolin and that there were distinctive if relatively small amounts of montmorillonite and talc.

On the Scottish ultramafics it seems that nowhere is there a simple developmental series from magnesian skeletal soils through to brown magnesian soils and magnesian gleys of increasing depth and surface acidity. There are likely to have been a number of depositional and erosional events. Vegetation succession and soil development on magnesian skeletal soils seem to be very slow as will be discussed in detail later. Much of the Scottish ultramafic outcrops are covered with glacial drift which may be partly or almost wholly derived from ultramafic rocks. Such drift is the parent material of the deeper soils. Magnesian skeletal soils occur on the more rapidly weathering ultramafic rocks where the drift has been eroded or perhaps locally (as on parts of Unst and much of Rhum) where it was never deposited.

Proctor & Woodell (1971) and Carter, Proctor & Slingsby (1987) have discussed how ultramafic rocks differ in the extent to which they weather. Some remain as massive blocks, others are jointed and produce skeletal soils. A striking example is that in Unst, Shetland where the unique Keen of Hamar has by far the largest area of dry skeletal soils and has a bedrock which shatters to give characteristic fragments. The Keen of Hamar rocks (and the very local outcrops which weather similarly elsewhere on Unst) are included in the 'dunite-serpentine' of Amin (1954) and are distinguished from the more widespread 'peridotite-serpentine' which weathers more slowly and never gives rise to dry skeletal soils. Although the 'dunite-serpentine' is distinctively low in silica (Table 2 and Amin 1954) the precise causes of the different types of weathering are unknown. Differences in weathering have also been observed in Scandinavia where Proctor (1969) pointed out the strikingly similar appearance of the skeletal soils of the sites at Bunnerviken and Kittlefjäll to those of the Keen of Hamar and their contrast with the smooth soil-free surface of the massive little-weathered ultramafic rocks at Atoklinten and Junsterklumpen.

The magnesian skeletal soils have been called 'debris' (e.g., by Spence 1957 and Proctor & Woodell 1971) and they consist of surface stones overlying (at least on the Keen of Hamar) a very stony sandy loam which grades into the bedrock. Carter, Proctor & Slingsby (1987) have pointed out that 'debris' is confusing as a pedological category. On the Keen of Hamar and probably elsewhere the erosion of magnesian gleys (developed from drift) can produce a surface stone line as a lag deposit (Ball 1967) or re-expose pre-existing lag deposits to give rise to soils superficially similar to magnesian skeletal soils but where the stones overlie a strongly-gleyed B horizon. There are important floristic differences between these two stony soil types. For example on Unst the 'associated debris' (*sensu* Spence 1957) is a true magnesian skeletal soil and is the habitat of most of the rare plants whilst the 'unassociated debris' is an eroded soil (most frequently a former gley) which generally lacks rarities.

Wilson & Berrow (1978) have studied pedogenesis in four Scottish soils developed at Green Hill on glacial drift partly derived from serpentinite. In the silt and clay fraction they found ferruginous saponite and chloritised primary vermiculite as products of pedogenesis from the serpentinite parent material. Wilson & Berrow (1978) stressed the importance of the non-serpentinite fraction in the soil development. A feature of this work was the emphasis on the changing distribution of chromium, cobalt and nickel during pedogenesis.

They found that chromium remained largely in the sand fraction and remained insoluble whilst cobalt was evenly distributed amongst the soil particle size fractions. By contrast, nickel was accumulated in the clay where, after its release following decomposition of the serpentine minerals it was fixed during the chloritisation of the vermiculite and the crystallisation of the saponite.

The weathering of a Scottish serpentinite by the lichen *Lecanora atra* has been studied by Wilson *et al.* (1980, 1981). The lichen apparently produces oxalic acid which causes the decomposition of minerals, particularly the fibrous magnesium silicate, chrysotile. The magnesium released is precipitated extracellularly in the form of magnesium oxalate dihydrate which accumulates in the lichen thallus. (Structurally similar oxalates of nickel, cobalt and iron are also insoluble and this explains how substantial quantities of these toxic metals may occur within the thallus and yet are excluded from the interior of cells). Lichens are usually sparse on ultramafic outcrops (Brodo 1974) and oxalic acid is not produced by all lichens (oxalates were not associated with *Ochrolechia posella* nor *Xanthoria parietina* on the serpentinite studied by Wilson *et al.* (1981). Nevertheless these observations on *Lecanora atra* do point to some of the likely chemistry of the weathering of ultramafic rocks since oxalic acid is a common constituent of soil solutions (Vedy & Bruckert 1979).

Particle-size composition of soils

Soils on British ultramafics are often coarse-textured with a high proportion of sand (Table 5). Many of the soils on Rhum are extreme in this respect and consist largely of coarse sand.

Soil chemical analyses

The results of chemical analyses of soils are given in Tables 6 and 7, soil solution analyses are given in Table 8. The problem of the most suitable soil extractant for nickel and chromium has been investigated in detail by Shewry & Peterson (1976). They showed that nickel was relatively easily soluble (on soils from Green Hill and the Keen of Hamar) and since they found that at least part was in the form of Ni^{2+} it seems that

Table 5. Particle size composition[1] of a range of soils on British ultramafic rocks

	Sample depth (cm)	Sand (%)	Silt (%)	Clay (%)
Lizard[2]				
'Rock Heath'	0–7	24	21	19
'Mixed Heath'	0–7	46	21	15
Green Hill[3] (a)	10–20	57	28	15
(b)	20–30	74	18	8
(c)	7–15	67	21	12
(d)	0–5	57	21	22
Keen of Hamar[4]	0–15	82	14	4
Rhum[5]				
Barkeval		95	4	1

[1]Sand is defined as particles of 20–2000 μ except for Rhum where it is 50–2000 μ. Silt is 2–20 μ except for Rhum where it is 2–50 μ. Clay is <2 μ throughout.
[2]From Coombe & Frost (1956).
[3]From Wilson & Berrow (1978). Data are for mineral horizons with little or no organic matter.
[4]From J. Proctor (unpublished). Data are for skeletal soils.
[5]From Ragg & Ball (1964). Data are for a skeletal soil.

exchangeable nickel is a reasonable measure of this element. They claimed that all methods of extraction and measurement of available chromium in the soils were unsatisfactory since plants growing on the soils contained appreciable quantities of chromium but little was removed from the soils by a range of extractants.

Proctor & Woodell (1971) described how analyses of exchangeable ions might underestimate the Mg/Ca quotient of the soil solution in which plant roots were bathed. Johnston & Proctor (1981) showed the efficacy of soil solution analyses as a basis for water culture experiments and now soil solution analyses have been made for a range of soils by centrifugation of water-saturated samples (Proctor *et al.* 1981, Looney 1982). They reveal several features (Table 8). Mg/Ca quotients are generally higher in solutions compared with those of exchangeable quantities of magnesium and calcium. In some cases the solution calcium concentrations are very low. The Meikle Kilrannoch soil solution has much higher concentrations of ions than the other sites. In the Keen of Hamar soils Cl^- is the main anion whilst in the inland sites it is NO_3^-. In fact, soil solutions from the Keen of Hamar resemble, with some enrichment by magnesium and nickel, c. 1% sea water. Soil solution nickel is at low concentrations

throughout. It is highest at Meikle Kilrannoch where its mean concentration $(0.7\,mg^{-1})$ has been shown to be toxic to plants (Johnston & Proctor 1981).

Vegetation

The vegetation of unenclosed uncultivated areas overlying the Lizard serpentinites was described by Coombe & Frost (1956a). They recognised four 'noda': 'Rock Heath,' 'Mixed Heath,' 'Tall Heath' and 'Short Heath.' The last two occur on soils in which granite-derived loess forms a large proportion of, or the whole of, the parent material. Marrs & Proctor (1979) showed that the vegetation of the enclosed, formerly-farmed areas has many features in common with the 'Tall Heath' and 'Short Heath.' I discuss here only 'Rock Heath' and 'Mixed Heath' because they occur (but not exclusively) on soils with ultramafic parent materials. 'Rock Heath' is the more local and coastal (Fig. 2). It has an abundance or co-dominance of *Erica vagans* and *Ulex europacus* and is species-rich (Coombe & Frost 1956a list seventy-three species). 'Mixed Heath' (Fig. 3) usually occurs on deeper soils, a little further inland from 'Rock Heath' and is even more species-rich (Coombe & Frost 1956a list seventy-seven species from the typical variant). It has a co-dominance of *Erica vagans* and *Ulex europaeus*. Both 'Rock Heath' and 'Mixed Heath' also occur over gabbro on the Lizard and Marrs & Proctor (1978) have investigated them chemically on both substrata. Few 'Rock Heath' and 'Mixed Heath' species are restricted to either ultramafic substrata or gabbro and it is clear that the mild winters (Table 3) are probably of prime importance in accounting for the large number of rarities. The Lizard heaths have affinities with those of western France (Coombe & Frost 1956a; Duvigneaud 1966). Open communities on skeletal soils, so much a feature of many Scottish ultramafic rock outcrops, are very local on the Lizard. I have seen one such small stony area (near Kynance). This was a flush and had abundant *Allium schoenoprasum* and *Radiola linoides* (neither of which are recorded from Scottish ultramafics).

For Scotland, Spence (1970) and Birse (1982) have made useful albeit incomplete attempts to

Table 6. Analyses of pH; loss-on-ignition; exchangeable potassium, sodium calcium, magnesium and the Mg/Ca quotient; and exchangeable nickel for a range of soils over British ultramafic rocks

Site	n	pH	Loss-on-ignition (%)	K (meq $100\,g^{-1}$)	Na (meq $100\,g^{-1}$)	Ca (meq $100\,g^{-1}$)	Mg (meq $100\,g^{-1}$)	Mg/Ca (meq $100\,g^{-1}$)	Ni ($\mu g\,g^{-1}$)
Lizard									
'Rock Heath'[a]	20	6.7	29	0.90	3.2	7.5	18.9	2.52	1–2
'Mixed Heath'[a]	20	6.2	16	0.49	1.6	3.1	4.2	2.98	1–2
Coyles of Muick									
Skeletal soil[b]	2	6.7	7.7	0.23	0.35	3.8	15.7	4.13	(10.6)
Grassland[b]	2	5.8	23.6	0.35	0.45	6.2	16.2	2.61	–
Green Hill[c]	16	7.2	–	0.20	0.14	1.6	17.0	10.6	8.7
Hill of Towanreef[c]	24	6.8	–	0.15	0.20	1.6	10.4	6.5	8.6
Inchcailloch[d]	1	4.6	22.9	0.55	0.42	1.8	2.1	1.18	–
Keen of Hamar									
Skeletal soil[e]	24	6.6	6.3	0.11	0.34	1.2	5.5	4.8	11.5
Skeletal soil[e]	31	6.7	4.8	0.09	0.31	0.9	4.8	5.3	8.3
Heath[e]	10	6.2	23.6	0.41	0.56	3.9	16.3	4.0	5.4
Heath[e]	51	6.3	31.4	0.63	0.93	4.5	20.2	4.7	15.8
Sedge-rich[e]	15	6.6	8.3	0.24	0.67	2.1	7.3	4.2	8.4
Lime Hill[f]	20	6.6	–	0.21	–	2.6	6.2	2.4	9.6
Meikle Kilrannoch									
MK1 skeletal soil[g]	25	6.9	–	0.08	0.05	0.85	10.4	12.2	14.9
MK1 skeletal soil[h]	2	6.8	3.0	0.05	0.35	0.4	11.0	13.8	–
MK1 heath[h]	2	4.9	15.4	0.04	0.20	0.2	2.4	12.0	–
MK2 skeletal soil[g]	12	6.6	–	0.06	0.10	0.52	14.8	28.5	20.3
MK2 skeletal soil[h]	2	6.3	4.8	0.04	0.20	1.0	13.4	13.4	–
Rhum									
Hallival-Barkeval skeletal soil[i]	10	5.1	2.6	0.03	0.02	0.05	0.10	2.00	0.7
Ruinsival skeletal soil[i]	10	6.1	5.1	0.10	0.20	1.56	2.07	1.33	2.1
Trallval skeletal soil[i]	10	6.2	2.8	0.02	0.04	0.24	0.53	2.21	3.0
Trallval heath[i]	10	6.0	29.5	0.39	1.30	4.06	6.94	1.71	8.6

[a]From Marrs & Proctor (1978). Sample depth 0–10 cm (approximately) for 'Rock Heath', 0–20 cm for 'Mixed Heath'.

[b]From Proctor & Woodell (1971). Value in parentheses is the mean for ten samples analysed by D. A. Cottam & J. Proctor (unpublished). Sample depth 0–15 cm.

[c]From Johnston & Proctor (1977) and W. R. Johnston (unpublished). Overall mean of four samples collected from under each of several species. Sample depth 0–10 cm (approximately). All skeletal soils.

[d]From Hornung & Mew (1970). Sample depth 0–9 cm.

[e]From Carter, Proctor & Slingsby (1987). Sample depth 0–15 cm.

[f]From Johnston & Proctor (1979). Sample depth 0–15 cm.

[g]From Johnston & Proctor (1980). Sampling as for c above.

[h]From Proctor & Woodell (1971). Sample depth 0–10 cm.

[i]From Looney (1982). Sample depth 0–10 cm.

describe the vegetation of ultramafic areas and relate it to vegetation units recognised by continental workers. Birse (1982) tried to fit his descriptions within the framework of the widely-used Braun-Blanquet system and his classification is summarized in Table 9. The most distinctive vegetation is the open communities on skeletal soils which occur more or less extensively on many ultramafic outcrops. Birse (1982) placed such communities in three associations within the Class *Violetea calaminariae* (Table 9) (Figs 4, 5). Table 10 lists species recorded from these skeletal soils by Proctor & Woodell (1971).

It must be emphasised however that the largest areas of British ultramafics (including those with open communities) are covered with grassland, heath or mire. These usually, but not always, show some influence of the underlying rock but none of the associations described from them (by Birse 1982) is restricted to ultramafics. Dry moorland and acid grassland are often extensive and, with the exception of the moorland on Unst and Fetlar, are included in the herb-rich subassociations of the general communities on other rocks. The Unst and Fetlar heathland has a distinctive combination of species which Birse (1982) has named as a separate community, the *Plantago maritima–Erica cinerea* Association (Table 9). It is not confined to ultramafics but extends locally on to metagabbro. On crest positions, where

Table 7. Nickel, cobalt & chromium analysed by a range of methods for soils over British ultramafic rocks

Site	n	Ni (bioassay) $\mu g\,g^{-1}$	Ni (total) $\mu g\,g^{-1}$	Ni (EDTA) $\mu g\,g^{-1}$	Ni (acetic acid) $\mu g\,g^{-1}$	Ni (exchangeable) $\mu g\,g^{-1}$	Co (total) $\mu g\,g^{-1}$	Cr (total) μg^{-1}
Lizard[a]								
'Rock Heath'[a]	5	14	2500	58	15	2	160	2600
'Rock Heath'[a]	5	13	2200	25	6	1	180	3000
Coyles of Muick								
Skeletal soil[a]	4	27	4400	146	52	7	320	5000
Grassland[a]	1	9	3200	113	19	4	250	5500
Green Hill								
Skeletal soil[a]	3	26	5800	172	43	6	250	4500
Grassland[a]	2	30	5100	286	34	11	290	4900
Skeletal soil[b]	3	–	2440	321	127	5.7	–	1600
Hill of Towanreef								
Skeletal soil[c]	2	–	1500	–	–	–	–	2750
Heath[c]	2	–	1500	–	–	–	–	4750
Keen of Hamar								
Skeletal soil[a]	4	19	8700	140	71	1	260	8000
Skeletal soil[a]	5	21	9700	155	85	3	390	12000
Heath[a]	3	19	4300	96	15	7	350	19100
Skeletal soil[b]	3	–	3330	390	222	3	–	1800
Meikle Kilrannoch								
Skeletal soil[d]	5	–	1100	–	–	10.3	c. 60	8200
Nikkavord (Unst)								
Skeletal soil[a]	3	43	6200	127	102	11	260	6800
Heath[a]	2	43	4200	113	36	20	220	6400
Rhum								
Hallival-Barkeval (skeletal soil)[e]	10	–	2170	–	–	0.7	110	60
Ruinsval (skeletal soil)[e]	10	–	1410	–	–	2.1	120	100
Trallval (skeletal soil)[e]	10	–	2740	–	–	3.0	160	50
Trallval (heath)[e]	10	–	1400	–	–	8.6	160	120

[a]Data of Slingsby & Brown (1977). Sample depth 0–15 cm. Bioassays refer to leaf nickel concentrations in oats. Total nickel, cobalt and chromium determined by X-ray fluorescence spectroscopy. EDTA is 10% wv disodium EDTA.
[b]From Shewry & Peterson (1976, Table 7) who include nickel and chromium determinations for several other extractants, sample depth 0–10 cm. Total nickel and chromium measured by DC arc spectroscopy. EDTA was 0.05 M disodium EDTA.
[c]From Proctor & Woodell (1971). Total nickel and chromium measured by X-ray fluorescence. Sample depth 0–15 cm.
[d]From Proctor & Johnston (1977). Sample depth 0–10 cm (approximately). All samples from around roots of *Lychnis alpina*. Total nickel, cobalt and chromium analyses by DC arc spectroscopy.
[e]From Looney (1982). Sample depth 0–10 cm. Total nickel, cobalt and chromium analyses in nitric acid digests.

Table 8. The means of ion concentrations ($mg\,l^{-1}$), pH and Mg/Ca quotients in soil solutions in skeletal soils from ultramafic areas in Scotland (from Proctor *et al.* 1981 and Looney 1982)

Sites	n	Ni^{2+}	Ca^{2+}	Mg^{2+}	K^+	Na^+	Cl^-	NO_3^-	PO_4^{3-}	pH	Mg/Ca
Coyles of Muick	10	<0.1	0.80	9.1	0.30	4.5	3.1	52	1.3	6.8	11.4
Green Hill	8	<0.1	0.45	9.4	0.36	5.1	4.2	50	0.95	6.5	20.9
Hill of Towanreef	10	<0.1	0.80	12	0.58	9.6	8.3	57	3.2	6.6	15.0
Keen of Hamar	11	0.13	5.4	28	5.0	110	200	29	0.97	5.8	5.2
Meikle Kilrannoch	11	0.67	11	180	5.4	15	–	830	13	–	16.4
Rhum	17	<0.1	1.5	5.6	4.0	10	17	–	1.1	5.4	3.7

Fig. 2. 'Rock Heath' with exposed serpentinite, Lizard Peninsula.

Fig. 3. 'Mixed Heath' with *Ulex galii* (foreground) and the rarity *Erica vagans* (midground), Lizard Peninsula.

Table 9. Plant communities on Scottish ultramafic rocks (from Birse 1982)

Communities	Sites of occurrence and underlying soils
Class: *Violetea calaminariae*	
Associations:	
Sileno-Armerietum maritimae metallicolae	
Sub-association with *Cochlearia officinalis*	Green Hill. Sub-alpine brown magnesium soil.
Cerastium nigrescens – Armeria maritima Ass.[a]	Keen of Hamar. Magnesium skeletal soil.
Lychnis alpina – Armeria maritima Ass.	Meikle Kilrannoch. Alpine brown magnesium soil and alpine magnesian gley.
Class: *Molinio – Arrhenatheretea*	
Associations:	
Helictotrichon pratense – Deschampsia cespitosa Ass.[b]	Green Hill. Gleyed humic brown magnesian soil and humic magnesium gley
Lolium – Cynosuretum	
Typical subassociation[c]	Gleyed brown magnesian soil and brown magnesian soil.
Subassociation with *Luzula campestris*	
Typical variant[d]	Gleyed brown magnesian soil and brown magnesian soil.
Variant with *Scilla verna*[e]	Brown magnesian soil.
Class: *Nardo – Callunetea*	
Associations:	
Plantago maritima – Erica cinerea Ass.[f]	
Variant with *Rhinanthus minor*	Unst. Gleyed brown magnesian soil and brown magnesian soil.
Variant with *Thalictrum alpinum*	Unst. Gleyed brown magnesian soil.
Variant with *Molinia caerulea*	Unst. Magnesian gley.
Vaccinio – Ericetum cinereae	
Subassociation with *Viola riviniana*	
Variant with *Helictotrichon pratense*[g]	Strathdon. Humic brown magnesian soil, gleyed humic brown magnesian soil, and brown magnesian soil.
Subassociation with *Cladonia arbuscula*	
Typical variant[h]	Strathdon. Peaty brown magnesian soil.
Achilleo – Festucetum tenuifoliae	
Subassociation with *Thymus drucei*	
Variant with *Scilla verna*	
Sub-variant with *Dactylorrhiza ericetorum*[i] ssp. *maculata*	Unst. Brown magnesian soil.
Variant with *Koeleria cristata*	
Sub-variant with *Plantago maritima*	Girvan. Brown magnesian soil.
Junco squarrosi – Festucetum tenuifoliae	
Subassociation with *Sieglingia decumbens*	
Variant with *Festuca vivipara*	Fetlar. Gleyed magnesian ranker.
Class: *Caricetea nigrae*	
Associations:	
Antennaria dioica – Carex pulicaris Ass.[j]	Unst and Fetlar. Gleyed humic brown magnesian soil, humic magnesian gley, gleyed brown magnesian soil
Caricetum hostiano – pulicaris	
Typical subassociation	
Variant with *Ranunculus acris*	Girvan. Magnesian gley and humic magnesian gley.
Schoenus nigricans Comm.[k]	Unst and Fetlar. Humic magnesian gley.
Carici dioicae – Eleocharitetum quinqueflorae	
Typical subassociation	Unst. Humic magnesian gley.

Table 9. Continued

Caricetum echinato – paniceae	
Typical subassociation	
Variant with *Nardus stricta*	Unst. Humic magnesian gley.
Carex panicea – C. demissa Comm.	Unst and Fetlar. Humic magnesian gley and magnesian gley.
Class: *Montio-Cardaminetea*	
Association:	
Montio-Philonolidetum fontanae	
Typical subassociation	
Variant with *Cratoneuron filicinum*	Unst. Subaqueous peat.
Class: *Asteretea tripolii*	
Associations:	
Scilla verna – Festuca rubra Community	
Subcommunity with *Poa pratensis*	
Typical variant[l]	Unst. Saline gleyed brown magnesian soil.
Variant with *Anthoxanthum odoratum*[m]	Unst. Saline gleyed humic brown magnesian soil.
Subcommunity with *Empetrum nigrum*	
Typical variant	
Subvariant with *Frullania tamarisci*[m]	Unst. Saline gleyed humic brown magnesian soil.
Class: *Oxycocco-sphagnetea*	
Association:	
Narthecio-Ericetum tetralicis	
Subassociation with *Cladonia uncialis*	
Variant with *Campylopus atrovirens*[n]	Unst. Peaty magnesian gley.
Class: *Vaccinio-Piceetea*	
Association:	
Trientali – Juniperetum communis	
Typical subassociation[o]	Strathdon. Brown magnesian soil.

[a]Placed in the *Arenaria norvegica – Cardaminopsis petraea* sociation by Spence (1970).
[b]Occurs in flushed situations and regarded by Birse as a semi-natural community derived from woodland or scrub. It is atypical of the class in not being confined to heavily grazed and formerly ploughed land.
[c]Rotational pasture.
[d]Formerly ploughed land.
[e]Permanent grazing of long standing.
[f]Equivalent to the herb-rich *Calluna – Erica cinerea* heath of Spence (1979).
[g]Referred to as the *Calluna – Erica cinerea – Juniperus* sociation by Spence (1970).
[h]On hill crests with highly leached soils.
[i]Also occurs on non-ultramafic soils.
[j]Included within the *Carex flacca – C. pulicaris – Festuca* sociation of Spence (1970).
[k]Probably equivalent to the *Schoenus – Molinia* sociation of Spence (1970).
[l]Equivalent to the *Carex flacca – C. demissa* sociation of Spence (1970).
[m]Subject to seaspray and wave action during storms.
[n]Unusual (for ultramafics) peaty soil. No chemical analyses available.
[o]Described as 'the nearest approach to climax vegetation' on ultramafics by Birse (1982).

flushing is absent and the influence of leaching is greater, there may be little or no floristic evidence of the ultramafic substratum although deep peats are characteristically lacking. Mires, flushed with drainage water from ultramafic rocks, sometimes have locally distinct communities, but are usually indistinguishable from mires on non-ultramafic rocks.

A number of authors have described communities which fit poorly or not at all into Birse's

Fig. 4. Sileno – Armerietum maritimae metallicolae, sub-association with *Cochlearia officinalis* surrounding patches of eroding heath, Green Hill.

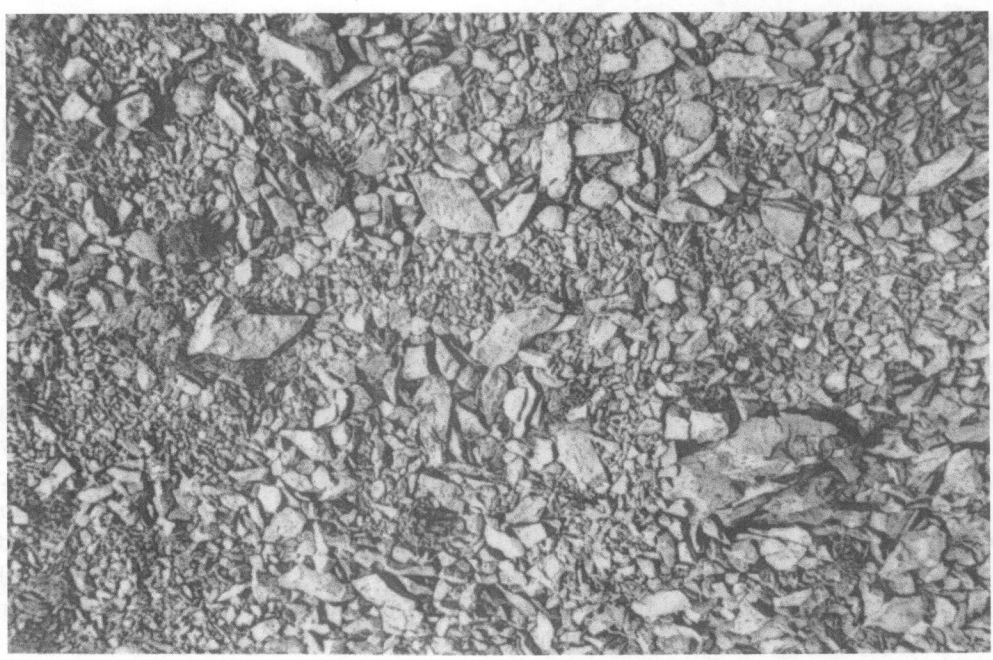

Fig. 5. Cerastium nigrescens – Armeria maritima Ass., Keen of Hamar. This area has very little plant cover and clearly shows stone sorting.

Fig. 6. Open, depauperate version of the *Plantago maritima* – *Erica cinerea*. Ass. near the summit of Hallival, Rhum. This is unusual since elsewhere in Scotland, open communities on residual ultramafic soils are in Associations within the Class *Violetea Calaminariae*.

classification (which he admits is incomplete). There is, for example an ultramafic-rock crevice community which was recognised by Spence (1970). In the Loch Lomond area of the Highland Boundary Fault there is semi-natural deciduous woodland which is not stunted, and which has a rich ground flora (Tittensor & Steele 1971, Horrill, Sykes & Idle 1975, and Nicol 1983).

The flora

Proctor & Woodell (1971) discussed the features of the British higher plant and fern flora on ultramafics in relation to six generalisations made by Rune (1953) from observations in Scandinavia

and North America. This remains a useful approach and I here re-examine these six points.

1. '*The serpentine flora is relatively poor in individuals as well as species.*' This is clearly not true of the species-rich heaths and grasslands but could be applied to the open vegetation of skeletal soils (as Rune presumably intended). However, where skeletal soils occur on other rocks (for example the quartzites of Ben Arkle and Foinaven in Sutherland, Scotland), the flora may be much more impoverished than that on ultramafics.

2. '*On serpentine, several species are represented by particular races differing ecologically and sometimes morphologically from the type races of the species.*' Ecologically different races are probably the rule on ultramafic areas and have been shown to differ in growth rate and metal tolerance from races collected on other soils (e.g. Proctor 1971a, b, and Johnston & Proctor 1981). Morphological races are frequent and perhaps most spectacularly developed on the Keen of Hamar, where there are unusual races of several species including *Plantago maritima*, *Rubus saxatilis* and *Rumex acetosa*. The *Plantago maritima* is very distinctive and has unusually short, broad leaves and is covered with silvery hairs (Fig. 7). However the race is not restricted to ultramafics and similar plants occur on skeletal soils on granite-diorite and quartzite elsewhere in Scotland. Several morphological variants occur on the ultramafic rocks of the Lizard Peninsula but many occur there on other substrata too. However the distinctive races of *Minuartia verna* ssp. *verna* and *Juniperus communis* ssp. *communis* are restricted to the 'Rock Heath' on ultramafics

There is an unusual race of the fern *Asplenium adiantum-nigrum* which occurs on many British ultramafic outcrops. It shows many morphological resemblances to the true serpentine species *Asplenium cuneifolium* which is not recorded from Britain. *A. adiantum-nigrum* and *A. cuneifolium* can be distinguished cytologically (the former is tetraploid, the latter diploid) and, despite one claim to have discovered the diploid in Britain (Roberts & Stirling 1974), it is now generally agreed that all British plants are referable to *A.*

Table 10. The species recorded from skeletal soils (including rock crevices*) during limited line transect sampling of six important Scottish ultramafic outcrops (Proctor & Woodell 1971)

	Coyles of Muick	Green Hill	Hill of Towanreef	Keen of Hamar	Meikle Kilrannoch	Rhum (Hallival-Barkeval ridge)
Asplenium adiantum-nigrum*	+	+	+			
A. viride*	+	+	+			
Deschampsia cespitosa	+	+			+	
Empetrum nigrum	+	+	+			
Minuartia verna ssp. verna		+	+			
Sagina procumbens	+	+	+			
Campanula rotundifolia	+	+				
Cerastium fontanum		+	+		+	
Cochlearia officinalis (sensu lato)		+	+	+	+	
Saxifraga hypnoides	+	+				
Antennaria dioica				+		+
Leontodon autumnalis				+		+
Plantago maritima				+		+
Silene acaulis				+		+
Rubus saxatilis				+		+
Carex flacca	+			+		+
Cardaminopsis petraea	+			+		+
Thymus praecox ssp. arcticum	+			+		+
Agrostis vinealis	+	+	+	+	+	+
America maritima	+	+	+	+	+	+
Festuca ovina/rubra/vivipara	+	+	+	+	+	+
Viola riviniana	+	+	+			
Euphrasia spp.	+	+		+	+	+
Anthyllis vulneraria	+			+		
Calluna vulgaris			+			+
Minuartia sedoides	+			+		
Silene maritima	+			+		
Rumex acetosa			+	+		
Racomitrium lanuginosum	+		+		+	+

Species recorded in one site only:
Coyles of Muick: Bryum caespiticium, Carex demissa, Carex pulicaris, Cerastium alpinum ssp. lanatum, Cladonia spp. Trichophorum cespitosum.
Green Hill: Polypodium vulgare, Viola tricolor.
Hill of Towanreef: Grimmia alpestris.
Keen of Hamar: Agrostis stolonifera, Arenaria norvegica ssp. norvegica, Cerastium arcticum ssp. edmondstonii, Coeloglossum viride, Rhinanthus spadiceus, Sagina nodosa, Scilla verna.
Meikle Kilrannoch: Bryum pseudotriquetrum, Campylopus subulatus, Carex bigelowii, Ceratodon purpureus, Lycopodium selago, Lychnis alpina, Racomitrium heterostichum var gracilescens.
Rhum: Campylopus atrovirens, Deschampsia flexuosa, Linum catharticum, Molinia caerulea, Polytrichum aloides, Potentilla erecta, Solidago virgaurea.

adiantum-nigrum (Sleep et al. 1978, Sleep 1980, 1985).

The causes of the odd races which occur on ultramafics must vary. The isolation of the outcrops must contribute to the development of morphological differences and in some cases, e.g. the hairiness of the Plantago maritima from the dry areas of the Keen of Hamar (Carter, Proctor & Slingsby 1987), the unusual features may be part of a mechanism of restricting water loss. The case of Asplenium adiantum-nigrum remains mysterious. Why are the same racial differences (which hardly seem adaptive) shown by many isolated populations on ultramafics of differing character? As a result of detailed cytological and experimental work on the ultramafic races of A. adiantum-nigrum, Sleep (1985) considers that they illustrate 'speciation in the act of happening'

Fig. 7. Plantago maritima with unusually short broad leaves, which are covered with silvery hairs, Keen of Hamar, Unst, Shetland.

and the situation 'represents one of the most striking examples of evolutionary processes in action that has yet been worked out among ferns.'

Ultramafics are not unique in their capacity to elicit morphological variants. Pigott (1956) has described unusual races (with some characters similar to those developed on ultramafics) on limestone in Upper Teesdale.

3. *'Many plants occur very disjunctively on serpentine.'* The following species occurring on soils with a largely ultramafic parent material have disjunctions of 80 km or more at one site at least (Proctor & Woodell 1971): *Arenaria norvegica* ssp. *norvegica*, *Asplenium viride*, *Cardaminopsis petraea*, *Cerastium arcticum* ssp. *edmondstonii*, *Lychnis alpina*, *Minuartia verna* ssp. *verna*, and *Juniperus communis* ssp. *communis*.

The most outstanding of these is *Cerastium arcticum* ssp. *edmondstonii* (Fig. 8) which is endemic to two sites on Unst, Shetland, the Keen of Hamar and a site about 1 km from the Keen, at Nikkavord. It differs from the type in having a more compact habit, obtuse sepals and darker green leaves which are densely glandular like the pedicels. Slingsby (1982) and Slingsby & Carter

(1986a) have shown that the Keen of Hamar and Nikkavord races differ morphologically and ecologically. The Nikkavord population has much fewer individuals and these tend to have longer, narrower leaves and longer internodes than those on the Keen. On Nikkavord the species is confined to stony eroded gley soils whereas on the Keen it is found only on the dry skeletal soils. Lusby (1985) has investigated the history and taxonomy of the Unst *Cerastium arcticum* whilst Slingsby & Carter (1986b) have provided counts of individuals and demonstrated its fluctuating numbers. Another important rarity is *Lychnis alpina*, a widespread arctic-alpine species which has its sole Scottish location on the two outcrops at Meikle Kilrannoch where it is common on the larger. Its only other British station is in the English Lake District about 300 km to the south. Proctor & Johnston (1977) have shown that there are different physiological races of *Lychnis alpina* at these two sites. *Lychnis alpina* is known as a metallophyte species outside Britain and it is characteristic of ultramafics in Scandinavia (Rune 1953). *Minuartia verna* is another metallophyte which shows disjunctions at three localities. The Lizard plants and some individuals at Girvan

Fig. 8. Cerastium arcticum ssp. *edmondstonii*, Keen of Hamar, Unst, Shetland.

grow in closed vegetation, a feature shown by many individuals of *Lychnis alpina* at Meikle Kilrannoch.

There are many smaller-scale disjunctions which involve intriguing differences between nearby ultramafic outcrops. These are perhaps best seen in Central Scotland where *Lychnis alpina* is common on one outcrop at Meikle Kilrannoch but is restricted to a few individuals at a site 1 km distant (Proctor & Johnston 1977). It has been suggested (Johnston & Proctor 1980) that the cause of this odd distribution is subtle soil differences between the sites but this idea needs testing further. Another curious example concerns the Coyles of Muick which has many floristic differences from other ultramafics in northeast Scotland. It is for example the only British ultramafic site with a population of *Cerastium alpinum* ssp. *lanatum*. Proctor & Woodell (1971)

suggested that the Coyles are relatively calcareous but again this is an idea which needs testing further.

Finally, it must be pointed out that disjunctions on ultramafics are by no means unique and there are many other habitats which harbour species with impressive disjunctions (e.g., the sugar limestone of Upper Teesdale (Pigott 1956). The ultramafic areas can be viewed as an example of open habitats with soils of relatively high pH which often have rare species (Pigott & Walters 1954).

4. '*The serpentine flora contains basicolous as well as acidicolous elements which grow together.*' This is an assertion which, whilst broadly true for many British soils on ultramafics, involves many complexities, not the least of which is the definition of basicoles and acidicoles. Many sites have a mosaic of base-rich skeletal soils and more acidic mature soils derived from drift. The interpretation of the occurrence of adjacent basicoles and acidicoles in these mosaics is straightforward. However there are many instances of more intimate occurrences of basicoles and acidicoles on the same soil type. Part of the explanation for this concerns the soil chemistry. Ultramafic soils are unusual amongst base-rich soils because their relatively high pH is combined with a relatively low concentration of calcium. This means that basicoles which do not require high calcium and acidicoles which are normally limited by high calcium (Jefferies & Willis 1964) might be expected to occur together if they are able to tolerate other toxicities on ultramafic soils. Well known major departures from calcifugy of such species as *Calluna vulgaris* occur in high rainfall low evaporation areas (e.g., McVean & Ratcliffe 1962, Pigott 1962) and these are the areas where most British ultramafic outcrops occur. 'Exacting calcicoles' (Steele 1955) and 'calciphiles' (Ferreira 1963, 1964) are absent from ultramafics except where they are enriched with calcium. Etherington (1981) has produced an index of calcifugy and it is striking that the two species with the highest values (*Galium saxatile* and *Deschampsia flexuosa*) are uncommon on skeletal soils on British ultramafics.

The joint occurrence of basicoles and acidicoles is by no means restricted to ultramafics. There are well-documented 'Chalk Heaths' and 'Limestone

Heaths' which have been discussed by Grubb, Green & Merrifield (1969) and Etherington (1981). The latter considers that calcifuges in the Limestone Heaths of south-west Britain may be relics from deeper loessic soils now eroded and survive largely by vegetative propagation in the shallow basic soils which are unsuitable for their seedlings. It is possible that erosion of drift cover may have resulted in a similar situation for calcifuges in the species-rich heaths on ultramafic rocks.

5. '*The serpentine flora has a relatively xerophytic character*.' Proctor & Woodell (1971) pointed out that ultramafics are not unusual amongst many British habitats in having species with xeromorphic characters (e.g., those in *Calluna vulgaris* and *Festuca* spp.). They suggested that this xeromorphism was probably related more to low tissue production under conditions of nutrient stress than to drought.

However, Carter, Proctor & Slingsby (1987) have recently emphasised the importance of droughts on the Keen of Hamar, Shetland, and have pointed out that a number of features of the species there (e.g. the hairiness of *Cerastium arcticum* ssp. *edmondstonii* and *Plantago maritima*) may be related to restricting water loss. It is likely that both xeromorphism and true xerophytism occur on British ultramafics.

6. '*The serpentine flora is often dominated by a certain family or certain genera, e.g., the Caryophyllaceae in northern Europe*.' This was presumably intended to apply to the flora of skeletal soils. Proctor & Woodell (1971) have shown that the Caryophyllaceae are usually conspicuous rather than dominant on British ultramafics. Several notable Caryophyllaceae including the rarities *Arenaria norvegica* ssp. *norvegica*, *Cerastium alpinum* ssp. *lanatum*, *Cerastium arcticum* ssp. *edmondstonii*, *Lychnis alpina*, and *Minuartia verna* occur abundantly on at least one British ultramafic outcrop.

Bryophytes, fungi and lichens

Neither bryophytes (D. F. Chamberlain, personal communication) nor fungi (R. Watling, personal communication) have species which are known to

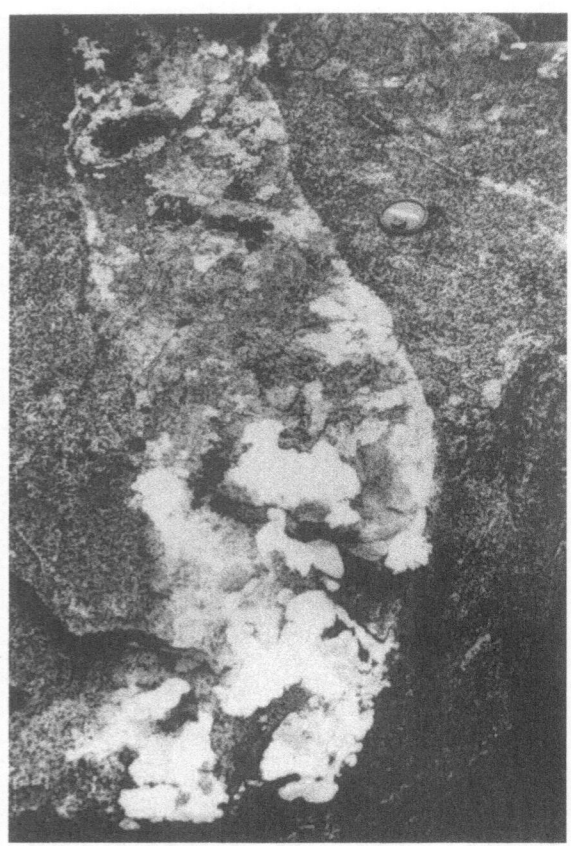

Fig. 9. Lichens growing profusely on gabbro but sparsely on the peridotite on each site of it, Rhum. (Photo by J. H. H. Looney)

be restricted to British ultramafics nor are there distinct species assemblages. Bryophytes are usually included in species lists for these areas but data on fungi are scarce. Watling (1969) included the ultramafic areas in his check list of fungi for Rhum. He commented that 'carpophores growing directly on ultrabasic substrates were found to be rare or absent; special climatic conditions may be required for carpophore initiation on these soils.'

A detailed study of the saxicolous cryptogams (including bryophytes and lichens) on Rhum was made by Bates (1978). He found that saxicolous cryptogams decreased in cover in the order: basalt, sandstone, limestone, ultramafic (Fig. 9). There was no evidence however of a distinctive assemblage of species on the ultramafic rock. Gilbert (1983) commented on the low lichen cover of the Rhum ultramafics and noted that 'a markedly basiphilous element appears to be absent in the lichen vegetation which is reminiscent of that on acidic spoil heaps associated with lead and zinc mines in the Pennines.' On the Unst, Shetland, ultramafics there are some noteworthy features of the lichen flora and P. W. James & O. W. Purvis (personal communication) have provided the following information.

The ultramafic rocks on Unst are characterised, in general, by their scant lichen growth and the predominance of a restricted range of crustose species, some characteristic of maritime habitats. However, whilst the lichen communities of the various ultramafic lithologies are distinctive, and include certain taxa which are otherwise rare, there appears to be no species which is indicative of a particular rock type in this series. The olivine-rich rock dunite is the most poorly colonized by lichens, but with the addition of pyroxene to the rock matrix, as in for example, harzburgite, there is an increase in the diversity, and cover of lichen species. Serpentinites and clinopyroxenites usually support luxuriant lichen growths, including more foliose species.

An unusual feature of these rocks, particularly well demonstrated by the peridotites, dunite and harzburgite, as well as soapstone, is a tendency to include a range of species usually considered typical of calcareous environments which are intermixed with those more characteristic of siliceous substrata. For example, on dunite, the calcicoles *Agonimia tristicula* (associated soil), *Caloplaca holocarpa*, *Catinaria grossa* (on mosses), *Lecania erysibe*, *Lecidea hypnorum* (on mosses), *Lecidella stigmatea*, *Microglaena moscorum* (on mosses), *Polyblastia cupularis* and *Verrucaria nigrescens* are often associated with *Candelariella vitellina*, *Fuscidea tenebrica*, *F*. c.f. *mollis* and *Huilia tuberculosa*, species characteristic of nutrient impoverished rocks with low contents of calcium carbonate. However, lichens characteristic of highly siliceous substrata e.g. *Rhizocarpon geographicum* and those restricted to rocks with high contents of calcium carbonate, e.g., *Placynthium nigrum*, *Verrucaria baldensis*, *V. caerulea* and *V. glaucina* are absent from these surfaces.

Where dunite boulders and those of other ultramafic lithologies are subjected to the influence of bird droppings, a characteristic luxuriant lichen flora (Xanthorion) tends to develop irregardless of rock type, dominated by *Xanthoria parietina*, *Physcia* spp. and *Rinodina confragosa*; these species are characteristic of nutrient enrichment on a wide range of rock substrata.

The observation of James & Purvis of mixtures of species of contrasting ecological preferences parallels that discussed for higher plants on soils derived from ultramafic parent materials.

Plant chemical analyses

The most detailed analyses of plants from British ultramafic areas have been made on Unst by Shewry & Peterson (1975, 1976) for several separate individual plants of a range of species growing on a relatively uniform soil on the Keen of Hamar. Their data (Table 11) show large differences in concentrations between individuals of the same species and some species are more variable than others. There are consistent interspecific differences which are clear evidence of different responses to calcium, magnesium, chromium and nickel. Some species had higher, others lower, Mg/Ca quotients than those (calculated for exchangeable magnesium and calcium) of their underlying soils. Further evidence of interspecific differences are shown by the different distribution of elements between roots and shoots. Roots of *Agrostis stolonifera* almost invariably had higher Mg/Ca quotients than shoots (Table 11). Roots of the same species and those of *Armeria maritima* and *Silene maritima* had higher concentrations of nickel than those of the shoots. Root chromium was much higher than that in the shoots in *Armeria maritima*, slightly higher in the case of *Agrostis stolonifera* and slightly less than that in the shoots in *Silene maritima*.

Analyses by other workers have not included roots and have been less intensive than those of Shewry & Peterson (1975, 1976) and have been expressed on a dry weight basis. The following points are noteworthy (Table 12): the large interspecific and inter-site differences in Mg/Ca quotients; the high iron concentrations in species from Green Hill, Hill of Towanreef and Rhum;

155

Table 11. The calcium, magnesium, nickel and chromium concentrations in the ash of individual plants growing on the northwest slope of the Keen of Hamar (Shewry & Peterson 1975, 1976). Values are means with ranges in parentheses

	n	Ca (%)	Mg (%)	Mg/Ca	Ni ($\mu g\,g^{-1}$)	Cr ($\mu g\,g^{-1}$)
Agrostis stolonifera (shoots)	20	1.79 (1.17–3.38)	3.94 (3.17–4.39)	2.17 (1.09–3.70)	756 (404–1077)	39 (22–65)
Agrostis stolonifera (roots)	20	1.38 (0.81–2.29)	5.12 (3.68–6.42)	3.70 (2.17–6.67)	3536 (2449–5147)	71 (45–95)
Anthyllis vulneraria	7	6.04 (5.08–7.33)	9.34 (8.42–10.3)	1.47 (1.22–1.85)	1653 (1112–2003)	130 (55–231)
Armeria maritima	20	7.5 (4.03–12.82)	10.4 (6.86–13.1)	1.35 (0.85–3.33)	987 (409–2424)	89 (24–224)
Cerastium arcticum ssp., edmondstonii	5	2.38 (2.00–3.26)	12.0 (10.5–13.7)	5.00 (3.85–6.67)	1564 (1347–1947)	147 (137–154)
Plantago maritima	10	5.72 (4.69–7.33)	12.6 (10.5–14.0)	2.22 (1.92–2.56)	719 (341–1166)	75 (25–129)
Silene acaulis	15	5.30 (3.23–8.83)	19.7 (13.9–30.8)	3.70 (2.44–6.25)	1781 (573–4480)	26 (14–43)
Silene maritima	20	9.44 (6.22–13.81)	7.08 (4.06–10.3)	0.71 (0.45–1.45)	328 (205–599)	22 (10–43)
Thymus praecox ssp. *arcticum*	10	9.97 (8.32–14.12)	10.4 (8.21–14.2)	1.02 (0.70–1.67)	1098 (643–1696)	68 (53–96)

the high potassium concentrations in several species from Hill of Towanreef, Lime Hill and Meikle Kilrannoch; high aluminium in *Minuartia sedoides* from Meikle Kilrannoch; high zinc in *Cochlearia officinalis* from all its sites; high sodium concentrations in several species, particularly those with maritime affinities (e.g., *Armeria maritima* and *Cochlearia officinalis*). No British species has been discovered which is an accumulator (*sensu* Baker 1981) of chromium or nickel. The highest concentrations of these elements known from British ultramafics occur in the above-ground parts of the bryophyte, *Racomitrium lanuginosum* (Johnston & Proctor 1977). They found dry matter concentrations of 235 $\mu g\,g^{-1}$ chromium in this species from the Hill of Towanreef and 470 $\mu g\,g^{-1}$ nickel in individuals from Green Hill. The Green Hill plants also had very high (21000 $\mu g\,g^{-1}$) iron concentrations. The uptake of metals by bryophytes merits further investigation.

Experimental work

Crop species

The first attempts to grow crops experimentally on a British ultramafic soil were made for a drift-derived farm soil at Whitecairns, Aberdeenshire (Hunter & Vergnano 1952). This soil produced nickel toxicity symptoms in a range of species which showed substantial interspecific differences in response. Barley (*Hordeum vulgare*) was least affected; wheat (*Triticum aestivum*), ryegrass (*Lolium perenne*) and bean (*Vicia faba*) were little affected; oats (*Avena sativa*), clover (*Trifolium* spp.), potato (*Solanum tuberosum*), turnip

Table 12. Element concentrations (μg g^{-1} dry matter) for a range of species on several British ultramafic sites

Site	Species	Plant part	n	P	K	Na	Ca	Mg	Mg/Ca	Ni	Cr	Fe	Authors
Lizard													
'Rock Heath'	Calluna vulgaris	Shoots	4	1430	3910	2100	2790	1630	0.58	31.6	26.3	220	Marrs & Proctor (1978)
'Mixed Heath'	Calluna vulgaris	Shoots	4	1680	4260	1320	2590	1730	0.67	17.2	26.3	380	
'Mixed Heath'	Erica vagans	Shoots	4	1350	5900	1900	1880	1560	0.83	11.5	16.9	190	
Creinch	Quercus sp.	Leaves	6	–	11000		6450	1580	0.24	12	–	44	Nicol (1983)
Green Hill	Agrostis vinealis	Leaf blades	3	–	12000	126	300	5730	19.1	151	83	2780	W. R. Johnston (unpublished) and Johnston & Proctor (1977)
	Calluna vulgaris	Green shoots	3	–	3500	337	5480	7030	1.28	147	72	4380	
	Cochlearia officinalis	Leaves	3	–	40000	1880	4500	19000	4.22	110	110	1400	
	Empetrum nigrum	Green shoots	3	–	3500	107	4160	5410	1.30	126	93	1230	
	Minuartia verna spp. verna	Leaves	3	–	18000	162	2340	16300	6.96	204	144	2830	
	Viola lutea	Leaves	2	–	24000	99	1840	15500	8.42	319	173	8500	
Green Hill	Calluna vulgaris	Green shoots	3	810	7470	1670	1940	5320	2.74	125	50	2380	Marrs & Bannister (1978)
Hill of Towanreef	Agrostis vinealis	Leaf blades	3	–	36000	1550	1000	6250	6.25	130	45	3880	W. R. Johnston (unpublished) and Johnston & Proctor (1977)
	Calluna vulgaris	green shoots	3	–	3600	1180	5360	3730	0.70	86	28	1180	
	Cerastium fontanum ssp. glabrescens	Leaves	3	–	58500	5090	1040	12800	12.3	148	37	4350	
	Cochlearia officinalis	Leaves	3	–	123000	3950	3280	8950	2.73	253	60	6320	
	Empetrum nigrum	Green shoots	3	–	2920	620	4160	5600	1.35	69	16	1240	
	Festuca rubra	Leaf blades	3	–	20700	810	900	4700	5.22	91	49	2130	
	Juniperus communis ssp. communis	Leaves	3	–	3020	417	10600	2790	0.06	32	15	310	
	Minuartia verna ssp. verna	Leaves	3	–	8800	1580	1660	9040	5.45	158	67	5600	
Hill of Towanreef	Calluna vulgaris	Green shoots	3	–	5400	2450	1370	7800	5.7	171	76	4300	Marrs & Bannister (1978)

Table 12. (Continued)

Site	Species	Plant part	n	P	K	Na	Ca	Mg	Mg/Ca	Ni	Cr	Fe	Authors
Keen of Hamar	Anthyllis vulneraria	Shoots	7	–	–	–	5420	8130	1.50	143	11	–	P. R. Shewry (personal communication)
	Armeria maritima	Shoots	20	–	–	–	5130	7420	1.45	76	7	–	
	Cerastium arcticum ssp. edmondstonii	Shoots	5	–	–	–	2490	12720	5.11	169	15	–	
	Plantago maritima	Shoots	10	–	–	–	4740	10370	2.19	58	6	–	
	Silene acaulis	Shoots	5	–	–	–	5170	21820	4.22	240	2	–	
	Silene acaulis	Shoots	10	–	–	–	4430	15280	3.45	113	2	–	
	Silene maritima	Shoots	20	–	–	–	14090	10340	0.73	48	3	–	
	Thymus praecox ssp. arcticum	Shoots	10	–	–	–	6700	6930	1.03	74	5	–	
Lime Hill	Calluna vulgaris	Green shoots	4	800	21000	390	17800	6400	0.36	73	60	400	Johnston (1976) and Johnston & Proctor (1979)
	Juniperus communis ssp. communis	Leaves	4	1300	22000	115	17000	9100	0.54	43	68	71	
	Plantago lanceolata	Leaves	4	1200	35000	9250	27000	20000	0.74	80	70	260	
	Teucrium scorodonia	Leaves	4	1100	24000	116	17000	11000	0.65	24	49	130	
Meikle Kilrannoch MK1	Agrostis vinealis	Leaf blades	4	2100	44000	240	500	9500	19.0	59	89	240	Johnston & Proctor (1980)
	Armeria maritima	Leaves	4	2800	75000	48000	9200	18000	1.96	27	61	150	
	Cochlearia officinalis	Leaves	4	2700	94000	8900	1800	46000	25.6	98	127	200	
	Festuca rubra	Leaf blades	4	1800	4200	160	450	7200	16.0	63	76	160	
	Lychnis alpina	Leaves	3	2200	66000	4000	11000	20000	1.82	74	83	250	
	Minuartia sedoides	Leaves	4	2000	31000	61	13000	13000	1.00	92	69	1200	
MK2	Cochlearia officinalis	Leaves	4	3500	80000	10000	2100	28000	13.2	72	63	290	
	Festuca rubra	Leaf blades	4	1900	27000	150	531	6570	12.4	52	53	210	
	Minuartia sedoides	Leaves	4	2600	31000	120	11000	14000	1.27	90	58	950	
Rhum	Agrostis vinealis	Leaf blades	3	–	6780	440	1120	7310	7.53	88	16	4400	Looney (1982)
	Calluna vulgaris	Leaves	3	–	5010	660	2940	4940	1.68	48	16	2500	
	Festuca vivipara	Leaf blades	3	–	3740	330	1360	10500	7.72	98	16	5290	
	Plantago maritima	Leaves	3	–	6540	5160	4220	28400	6.73	240	25	10400	

(*Brassica rapa*), cabbage and kale (*Brassica oleracea*) were badly affected but responded well to the application of lime and fertilizers; beet (*Beta vulgaris*) was worst affected and did not respond to the fertilisation treatment. The symptoms produced in oats were chlorotic leaves and necrosis as white longitudinal stripes. Since oats are relatively susceptible and the symptoms are characteristic, this species has since been used to assay nickel toxicities in ultramafic soils. Although positive results have been obtained elsewhere, e.g., in Zimbabwe (Proctor, Craig & Burrow 1980), the symptoms were not observed again for oats on British soils until recently. In glasshouse experiments on British ultramafics, oats either grew well or showed symptoms of nutrient deficiency or magnesium toxicity (e.g., Spence & Millar 1963, Proctor 1971a, Slingsby 1974, Proctor & Cottam 1982). Nickel toxicity symptoms were observed (Carter, Proctor & Slingsby 1987) in oats grown in a soil collected from a mire (with closed vegetation of common species not usually thought to have metalliferous affinities) which was known to have high soil nickel concentrations on the Keen of Hamar. It is noteworthy that both this demonstration of nickel toxicity in oats and the earlier one by Hunter & Vergnano (1952) did not involve skeletal soils.

It is known that the development of nickel toxicity symptoms in oats depends on a number of factors, e.g., the light regime (Anderson, Meyer & Mayer 1973); pH (Crooke 1956); nitrogen, potassium, calcium and magnesium (Crooke & Inkson 1955). Proctor & McGowan (1976) described the amelioration of nickel toxicity by high solution magnesium concentrations and emphasised that interactions between these two elements might be an important feature in soils over ultramafics.

There is a substantial amount of unpublished work at the University of Aberdeen on several pasture grass species and *Trifolium repens* grown on soils (not skeletal) from Green Hill (e.g., Fraser (1975), Gooch (1973), Khalid (1974), Munir (1970), Riley (1977), Soon (1971), Tassoulas (1970), Willett (1975)). This research has usually involved field and pot experiments and has usually (but not always) demonstrated increases in yields with fertilisation by major nutrients and calcium. There were marked differences in the efficacy of each element for different species.

Other Aberdeen workers have researched the growth of coniferous trees at Green Hill (Arnold 1973, Kilic 1976). They showed *Picea sitchensis* was more adversely affected than *Pinus sylvestris* or *P. contorta* and as a result of fertilizer trials and foliar analyses they concluded that phosphorus supply was the major factor which limited conifer growth. Blyth & MacLeod (1981) observed that the yield of *Picea sitchensis* was much less on an ultramafic site compared with that on soils with acidic parent material. By contrast, annual increment measurements on several deciduous tree species in the semi-natural woodlands on the calcareous ultramafics near Loch Lomond have shown no evidence of slower growth compared with that on other soil types in the same area (J. Proctor, unpublished).

Some biochemical work has been attempted on a range of agricultural grasses which grew well or badly on soil from Green Hill (Willett & Batey 1977). It was found that the species which grew best had root-surface phosphatases which were insensitive to changes in calcium concentrations (over the range $0-6 \, \text{mg} \, l^{-1}$). It was implied that root-surface phosphatases which have optimal activity without calcium are part of a 'fundamental mechanism by which plants may be adapted to growth in environments extremely deficient in Ca.' I return to a discussion of root-surface phosphatases later.

Native species

The first experiments involving native species of British ultramafics were made by Steele (1955) and Ferreira (1963) who were concerned with the varying degrees of calcicoly shown by British calcicoles. Steele (1955) grew a range of calcicole and calcifuge species on experimental plots of 'poor gravel soil' to which salts of calcium and magnesium were added. He demonstrated a range of responses to the additions and his work shows a number of examples of magnesium toxicity and an instance of a high magnesium requirement in the Lizard race of *Minuartia verna* ssp. *verna* which showed greatly improved growth on the addition of magnesium carbonate (but not with added magnesium chloride or sulphate). Ferreira (1963) grew a range of calcicoles and two species of ultramafics (*Lychnis alpina* and *Minuartia verna* ssp. *verna*) in pot experiments. The calcicoles were susceptible to magnesium toxicity (in the Hill of Towanreef soil or in treatments involving the addition of magnesium carbonate to a quartzite soil) whilst the *Lychnis* and *Minuartia* were tolerant.

Proctor (1971b) grew several species in a range of soils and demonstrated adaptation to ultramafics in *Agrostis vinealis*, *Agrostis stolonifera*, *Armeria maritima*, *Rumex acetosa* and *Silene maritima*. This work confirmed earlier experiments on *Avena sativa* (Proctor 1971a) in showing that ultramafic soils differ in their toxicity. He further demonstrated that adaptation to one ultramafic site does not necessarily confer adaptation to another. When grown on the high-magnesium Meikle Kilrannoch soil, races of *Agrostis vinealis* from Meikle Kilrannoch and Rhum and of *Agrostis stolonifera* from the Lizard and the Keen of Hamar grew at rates which were in the same rank order as the Mg/Ca quotients at their sites of collection. Thus, *A. vinealis* from Meikle Kilrannoch grew well but the same species from the low-magnesium soils on Rhum produced virtually no roots and the two *A. stolonifera* races (from moderately high magnesium soils) showed an intermediate response. Proctor (1971a) found that a non-ultramafic maritime race of *Armeria maritima* was able to grow fairly well in the Meikle Kilrannoch soil and suggested that maritime plants are to some extent pre-adapted since seawater has a moderately high Mg/Ca quotient (ca. 5 on a milli-equivalent basis).

In a further series of experiments Proctor (1971b), using the simple root elongation test of Wilkins (1957), demonstrated tolerance of both magnesium and nickel in *A. vinealis* and *A. stolonifera* races from a range of ultramafics. Races of *A. vinealis* from Rhum were tolerant of nickel but not magnesium. Proctor (1971b) also found that there was a tendency for ultramafic races to be more tolerant of chromium (Cr^{3+}) than races collected from other soils. The tolerance was less striking than that to nickel however and some ultramafic races were less tolerant of chromium than those from non-ultramafic soils. Evidence of adaptation to ultramafic soils under heathland was obtained by Marrs & Bannister (1978) who found that *Calluna vulgaris* from Green Hill grew

better in glasshouse experiments on soils from Green Hill and Hill of Towanreef than other races of *Calluna* investigated.

Marrs & Proctor (1976) studied the growth of tillers of *Agrostis stolonifera* in water culture. They used three races: a non-ultramafic commercial race; a race from a soil with a moderately high Mg/Ca quotient (from the Keen of Hamar); and a race from a soil with a very high Mg/Ca quotient (from Kittelfjäll, Sweden). They found evidence of magnesium tolerance and a higher magnesium requirement for maximum growth for both ultramafic races. Marrs & Proctor (1976) found that over much of the range of magnesium and calcium concentrations used in the culture solutions the ultramafic races had, compared with the non-ultramafic race, a higher concentration of magnesium in the roots and shoots and lower concentrations of calcium in the roots. These results along with those of some species' tissue analyses (Table 11) show that mechanisms of magnesium exclusion and avid calcium uptake are not essential features of plants of ultramafic areas. Such features have been emphasized by North American workers (e.g., Madhok & Walker 1969).

Many aspects of the earlier water culture experiments were criticised by Johnston & Proctor (1981). They pointed out that: the concentrations of toxic ions were arbitrarily chosen; the test solutions were very different in chemical composition from any likely to be found in soils in nature; and the degree of tolerance was often judged from some small feature such as the growth of a single adventitious root during a few days. In order to overcome these objections Johnston & Proctor (1981) carried out water culture experiments in which the media were based on analyses of soil solutions (Table 8). They grew, for several weeks, clones of *Festuca rubra* from Meikle Kilrannoch and from a non-ultramafic acid brown earth in four experiments with different combinations of Ca^{2+} and micronutrient concentrations. In each, the effects of different concentrations of Mg^{2+} and Ni^{2+} on plant growth were measured. They analysed the plants grown in the test solutions and found that their chemical composition was reasonably similar to those grown on soil in the glasshouse but less so to those occurring naturally in the field. The greatest growth of the Meikle Kilrannoch clone of *F. rubra* was in the water

160

culture with concentrations most similar to soil solutions from the same site. The acid-soil clone plants grew very slowly in this solution and were best in solutions containing higher concentrations of micronutrients and lower ones of Mg^{2+}, Ni^{2+} and Ca^{2+}. High Mg^{2+} concentration appeared to be the main cause of the slow growth of the acid-soil clone in Meikle Kilrannoch soils. However Ni^{2+} was shown to be toxic also. Both Mg^{2+} and Ni^{2+} toxicity were ameliorated by a higher concentration of Ca^{2+} and to some extent of micronutrients. The amelioration was associated with a reduction in tissue concentration of nearly all the elements analysed and may involve some restoration of membrane function. Johnston & Proctor (1981) found no indication of an amelioration of nickel toxicity by magnesium and their results contrast with those of Proctor & Mac-Gowan (1976) for oats, suggesting interspecific differences in response to combined magnesium and nickel.

Similar experimental work to that carried out by Johnston & Proctor (1981) has been made for *Agrostis* spp. and soils from the Rhum ultramafics (Looney 1982); from the Keen of Hamar, Shetland (Y. Bowen & J. Proctor, unpublished); and from Green Hill, Hill of Towanreef and Coyles of Muick (D. A. Cottam & J. Proctor, unpublished). In all these cases soil solutions (Table 8) were much more dilute than those at Meikle Kilrannoch and evidence of marked toxicity was restricted to the Green Hill solutions.

One of the possible mechanisms of adaptation of *Festuca rubra* to the Meikle Kilrannoch soil was investigated by Johnston & Proctor (1984). They tested the responses of root-surface phosphatases to ranges of concentrations of calcium, magnesium, nickel, and micronutrients which had been shown to cause different growth responses of the clones in solution culture (Johnston & Proctor 1981). It was shown that the phosphatases from the ultramafic race had greatest activity in a solution similar to that in the soil at its site of collection. The phosphatases from the non-ultramafic race were partially inhibited in the same solution. The results were taken to imply subtle enzymic adaptations to high concentrations of the potentially toxic ions. The ecological significance of these adapted phosphatases however remains unclear. The phosphatases of non-tolerant species investigated by Johnston & Proctor

(1984) and Willett & Batey (1977) never had any activity, even in the least favourable conditions, of less than 50% of that of the tolerant species in the best conditions. Greater differences might have been expected had these enzymes a fundamental role in the plants' survival on ultramafics since in the case of the Meikle Kilrannoch soil at least, non-tolerant plants are greatly stunted and moribund.

Field nutrient-addition experiments

A spectacular field nutrient-addition experiment was made by Ferreira & Wormell (1971) on a skeletal soil overlying ultramafic rock at an altitude of about 660 m on the Island of Rhum. They added major nutrients in 1965 and 1967 and major nutrients and calcium in 1968. The vegetation increased from 5% to 60% cover up to 1969. 'Overall, the vegetation had changed from Herb-rich *Calluna* heath to Species-rich *Festuca* grassland' (Ferreira & Wormell 1970). In 1982 I visited the site and found that the vegetation cover had increased to nearly 100%. The species-rich *Festuca* grassland had persisted but there was a high cover of mixed bryophytes in the lower part of the plot. It is of great interest that the effect of the fertilizers has remained in an area of high annual rainfall (ca. 3100 mm) on a coarse-textured soil for at least 14 years after their last application.

A closed 'Species-rich *Festuca* grassland' does occur frequently on Rhum in association with burrows of the Manx Shearwater (*Puffinus puffinus*) of which there are about 250,000 pairs on the island (Wormell 1976). These burrows are more or less restricted to the ultramafic rocks and their manuring effect provides a 'natural fertilisation experiment' which complements the results of Ferreira & Wormell (1971).

Slingsby & Carter (1986b) have recently summarised the results of a fertilisation experiment (with N, P, NP, NPK, Ca and CaNPK) observed over four years on the Keen of Hamar, Shetland. They found that phosphorus was the key element in causing vegetation change. The principal effect of the application of this element alone or in combination with others was (in this ungrazed site): an increase in grasses and a reduction of *Calluna* cover in herb-rich heath; an increase in grasses (particularly *Festuca* spp.) and other spe-

Fig. 10. Dramatic response of open sedge-rich vegetation to fertilization, Keen of Homar, Unst, Shetland.

cies on skeletal soils that were already moderately well covered with vegetation (Fig. 10); and a relatively small increase in cover of species on very bare skeletal soils.

Conclusions: successions on ultramafic skeletal soils

The barren skeletal soils which have rare species or varieties and on which succession is retarded are a feature of major interest. Much of the previous sections has related to them and the various lines of evidence concerning the maintenance of their open communities are now brought together.

There is a strong impression on all the sites with extensive areas of skeletal soils that these are extending as a result of erosion of deeper, drift-derived soils. On Green Hill (Wilson & Berrow 1978), the Keen of Hamar (Carter, Proctor & Slingsby 1987) and Rhum (Ragg & Ball 1964, Looney 1982) this has been confirmed by detailed soil studies. The rate of the erosion seems to vary from fairly rapid to very slow. Looney (1982) reported erosion rates of 10 cm per year on soil edges on Rhum. On the Keen of

Hamar no detectable (i.e, <1 cm) erosion occurred on the edge of an island of fully vegetated drift soil over 11 years. A comparison of aerial photographs taken over the Keen of Hamar in 1946 and 1975 showed no erosion between these dates although extensive erosion has occurred in the past. The causes of this widespread erosion on ultramafics are likely to be complex and to differ from site to site. For Rhum, Ragg & Ball (1964) suggested that the cause was climatic change or over-grazing and burning following changes in the Island's land-use. Carter, Slingsby & Proctor (1987) have considered that grazing is unlikely to account for the erosion on the Keen of Hamar.

Spence (1957, 1959) envisaged a cyclical succession on skeletal soils on the Keen of Hamar (although he recognised that the building phase might be very slow). Long-term field observations have shown no evidence of increased cover however and it seems that succession is very limited on the skeletal soils (Carter, Proctor & Slingsby 1987). Closed heath or grassland seems to be developed on the drift soils which once removed cannot be replaced. These conclusions possibly apply to most if not all British ultramafic skeletal soils. Four main causes for this slow or negligible

162

succession have been suggested: (a) low soil nutrient concentrations, (b) high soil Mg/Ca quotients; (c) high soil nickel concentrations; (d) adverse physical factors and (e) interactions between one or all these factors. I shall discuss these in turn.

a) *Low soil nutrient concentrations.* The results of the field nutrient addition experiments leave no doubt that at least on some parts of the Keen of Hamar (Carter, Proctor & Slingsby 1987) and Rhum (Ferreira & Wormell 1971) a shortage of nutrients prevents full plant cover developing. Phosphorus appeared to be the key limiting nutrient on the Keen whilst the experiment on Rhum did not separate the effects of nitrogen, phosphorus, potassium or calcium.

Slingsby & Carter (1986b) have described how accidental eutrophication by farmers is threatening many of the rare species on a substantial area of the Keen of Hamar. It would be interesting to see a nutrient addition experiment on Meikle Kilrannoch since Proctor (1971a) showed that nitrogen fertilisation increased the toxicity to oats in a pot experiment using soil from this site.

b) *High soil Mg/Ca quotients.* Magnesium toxicity has been demonstrated on the Meikle Kilrannoch soil (Proctor 1971a, Johnston & Proctor 1981) which has very high soil solution magnesium concentrations. Magnesium toxicity is probable but less acute on Green Hill and Hill of Towanreef (Proctor & Cottam 1982). The existence of magnesium-tolerant individuals from the Keen of Hamar (Proctor 1971b, Marrs & Proctor 1976) implies that high magnesium has some effect there, although it falls short of acute toxicity. It must be questioned if any soil toxicity can ever account for retarded colonisation since, when tolerant plants have emerged, then they should fully colonize the site unless another factor is limiting.

Proctor (1970) and Proctor & Woodell (1975) have emphasised magnesium toxicity rather than calcium deficiency as a key factor. They stressed the role of calcium in ameliorating toxic metal concentrations. Wyn Jones & Lunt (1967) showed that in the absence of toxic ions calcium was required in very low concentrations. More recently, soil solution analyses from Green Hill (Proctor *et al.* 1981) have shown that calcium may indeed be at such low concentrations that true calcium deficiency might operate. Experimental work is required to confirm this possibility. Proctor

(1971a) and Johnston & Proctor (1981) showed that relatively more calcium was required to ameliorate magnesium toxicity as magnesium concentrations increased. High Mg/Ca quotients are not invariably associated with the retarded colonisation since the skeletal soils of Rhum have low Mg/Ca quotients (Proctor & Woodell 1971, Looney 1982) and the plants there are not magnesium tolerant (Proctor 1971b).

c) *High soil nickel concentrations.* Nickel toxicity has been demonstrated for the Meikle Kilrannoch skeletal soil (Johnston & Proctor 1981) where it apparently acts interactively with magnesium toxicity. Acute nickel toxicity has never been observed on other skeletal soils although nickel tolerance of species on ultramafics is widespread (Proctor 1971b). In view of the recent demonstration of acute nickel toxicity in a completely vegetated mire soil on the Keen of Hamar (Carter, Proctor & Slingsby 1987), I question if nickel toxicity can account for retarded colonisation even in sites where it is at relatively high concentrations. It is relevant that Thompson & Proctor (1983) observed very high soil solution copper concentrations under vegetated soils on a spoil heap in Central Scotland and this appears to confirm the view expressed for magnesium, that species can adapt to metal toxicity which then no longer limits colonisation.

d) *Adverse physical factors.* It has been shown earlier how all the skeletal soils are likely to be subject to wide diurnal temperature fluctuations, frost heaving and serious droughts. All these factors will tend to retard plant colonisation particularly at the sensitive seedling stage.

e) *Interactions involving factors (a)–(d) and others.* Several interactive effects are likely in the skeletal soils including those just discussed. Low soil nutrient concentrations might interact with drought. Grime & Curtis (1976) carried out an addition experiment on shallow limestone soils in northern England. They showed that addition of distilled water or phosphate resulted in greatly increased growth of *Arrhenatherum elatius*. The phosphate reduced the effects of drought by stimulating the growth of a more extensive root system. Such an effect might explain the increased growth caused by nutrient addition on Rhum and on the Keen of Hamar where regular droughts occur. A further possibility is that soil solution concentrations of toxic elements could be greatly increased as they dry out, although we lack analyses of solutions extracted from dry soils.

Interactions involving lead and phosphorus have been shown to be important by Jeffrey (1971) in maintaining low phosphorus concentrations on soils under *Kobresia simpliciuscula* in Teesdale. Similar effects might occur with nickel or the abundant iron compounds (Dry & Robertson 1982) in the ultramafic soils. Soil-solution phosphorus concentrations are not unusually low (Proctor *et al.* 1981) although there are no data on the sizes of the labile pools of this element.

The results of Johnston & Proctor (1981) showed the possible interactive effects of low micronutrients in exacerbating the combined toxicity of nickel and magnesium toxicity.

In conclusion it seems that the causes of the slow or negligible succession on British skeletal soils derived from ultramafic substrata are complex and only partially explained. Different factors and their interactions are likely to be involved at different sites.

Note added in proof

An appraisal of the status of the Lizard rarities in a European context has been provided by Hopkins (1983) and further information about them is included in Margetts (1988).

The work of Looney (1982) on the vegetation of the Rhum ultramafics has been superseded by Looney & Proctor (1989a, b). Proctor *et al.* (1991) have discussed the ecology of Meikle Kilrannoch including the status of the sites' recently described endemic, *Cerastium fontanum* ssp. *scoticum.* Further disjunctions are the metallophyte crucifer, *Thlaspi caerulescens*, (Ingrouille & Smirnoff 1986) and the basicole *Filipendula vulgaris* (A.McG. Stirling personal communication) near Girvan. Carter *et al.* (1987) have described the results of their fertilization experiment on the Keen of Hamar which supersede the account in Slingsby & Carter (1986a). Alexander & Hardy (1981) found relatively high surface phosphatase activity in mycorrhizas of *Picea sitchensis* from a plantation of Green Hill although they have no evidence about the importance of the enzyme to the trees nutrition in the field. Information on the ultramafics of the west of Ireland is given by Dyos *et al.* (1991).

Acknowledgements

The following provided helpful comments on early drafts of this paper: Dr A. J. M. Baker, Mr C. J. Bown, Dr E. L. Birse, Mr S. P. Carter, Dr J. H. H. Looney, Dr N. T. Livesey and Dr D. R. Slingsby. Dr J. Volker wrote most of the geology section and Dr P. W. James and Mr O. W. Purvis are thanked for their information on lichens.

References

Alexander, I. J. & K. Hardy. 1981. Surface phosphatase activity of Sitka spruce mycorrhizas from a serpentine site. Soil Biology and Biochemistry 13: 301–305.

Amin, M. S. 1954. Notes on the ultrabasic body of Unst Geological Magazine 91: 399–406.

Anderson, A. J., D. R. Meyer, & F. K. Mayer. 1973. Heavy metal toxicities: levels of nickel, cobalt, and chromium in the soil and plants associated with visual symptoms and variation in growth of an oat crop. Australian Journal of Agricultural Research 24: 557–571.

Arnold, P. J. 1973. Coniferous tree nutrition on serpentine soils. M.Sc. thesis, University of Aberdeen.

Baker, A. J. M. 1981. Accumulators and excluders – strategies in the response of plants to heavy metals. Journal of Plant Nutrition, 3: 643–654.

Ball, D. F. 1967. Stone pavements in soils of Caernarvonshire, North Wales. *Journal of Soil Science* 18, 103–108.

Bates, J. W. 1978. The influence of metal availability on the bryophyte and macrolichen vegetation of four rock types on Skye and Rhum. Journal of Ecology 66: 457–482.

Birse, E. L. 1980. Suggested amendments to the world soil classification to accommodate Scottish mountain and aeolian soils. Journal of Soil Science 31: 117–124.

Birse, E. L. 1982. Plant communities on serpentine in Scotland. Vegetatio 49: 141–162.

Brodo, I. 1974. Substrate ecology. In The Lichens (V. Ahmadjian & M. E. Hale, eds), pp. 401–441, Academic Press, London.

Blyth, J. F. & D. A. MacLeod. 1981. Sitka Spruce (*Picea sitchensis*) in north-east Scotland I. The relationships between site factors and growth. Forestry 54: 41–62.

Butler, J. R. 1953. The geochemistry and mineralogy of rock weathering (1) The Lizard area, Cornwall. Geochimica et Cosmochimica Acta 4: 157–178.

Carter, S. P., D. R. Slingsby, & J. Proctor. 1987. Ecological studies on the Keen of Hamar serpentine, Shetland. Journal of Ecology 75: 21–42.

Carter, S. P., J. Proctor, & D. R. Slingsby. 1987. The effects of fertilization on part of the Keen of Hamar serpentine,

Shetland. Transactions of the Botanical Society of Edinburgh 45: 97–105.

Coombe, D. E. & L. C. Frost. 1956a. The heaths of the Cornish serpentine. Journal of Ecology 44: 226–256.

Coombe, D. E. & L. C. Frost. (1956b). The nature and origin of the soils over the Cornish serpentine. Journal of Ecology 44: 605–615

Cooper, A. & Etherington, J. R. 1974. The vegetation of carboniferous limestone in South Wales. I. Dolomitization, soil magnesium status and plant growth. Journal of Ecology 62: 179–190.

Crooke, W. M. 1956. Effect of soil reaction on uptake of nickel from a serpentine soil. Soil Science 81: 269–276.

Crooke, W. M. & R. H. E. Inkson. 1955. The relationship between nickel toxicity and major nutrient supply. Plant and Soil 6: 1–15.

Dry, F. T. & J. S. Robertson. 1982. Soil and Land Capability for Agriculture, Orkney and Shetland. 1 : 250,000 map and handbook. The Macaulay Institute for Soil Research, Aberdeen.

Duvigneaud, P. 1966. Note sur la biogéochemie des serpentines du sud-ouest de la France. Bulletin r. Société de Botanique de Belgique 99: 271–329.

Dyos, H., J. Proctor, & M. Sheehy Skeffington. 1991. Notes on the vegetation of the Dawros and other ultrabasic rock outcrops in Connemara. Irish Naturalists Journal (in press).

Etherington, J. R. 1981. Limestone heaths in south-west Britain: their soils and the maintenance of their calcicole-calcifuge mixtures. Journal of Ecology 69: 277–294.

F.A.O./U.N.E.S.C.O. 1974. Soil map of the world V. 1: Legend U.N.E.S.C.O., Paris.

Ferreira, R. E. C. 1963. Some distinctions between calciphilous and basiphilous plants I. Field data. Transactions of the Botanical Society of Edinburgh 39: 399–413.

Ferreira, R. E. C. 1964. Some distinctions between calciphilous and basiphilous plants II. Experimental data. Transactions of the Botanical Society of Edinburgh 39: 512–524.

Ferreira, R. E. C. & P. Wormell. 1971. Fertiliser response of vegetation on ultrabasic terraces on Rhum. Transactions of the Botanical Society of Edinburgh 41: 149–154.

Fraser, J. 1975. Responses of the vegetation of serpentine, Greenhill, Strathdon to soil conditions, physiography, and ameliorative treatments. M.Sc. thesis, University of Aberdeen.

Gilbert, O. L. 1983. The lichens of Rhum. Transactions of the Botanical Society of Edinburgh 44: 139–149.

Gooch, F. 1973. The growth of pasture species on the serpentine soils of Greenhill, Strathdon. M.Sc. thesis, University of Aberdeen.

Grime, J. P. & A. V. Curtis. 1976. The interaction of drought and mineral nutrient stress in calcareous grassland. Journal of Ecology 64: 975–988.

Grubb, P. J., H. E. Green, & R. J. C. Merrifield, 1969. The ecology of chalk heath: its relevance to the calcicole-calcifuge and soil acidification problem. Journal of Ecology 57: 175–212.

Guppy, E. M. 1956. Chemical Analyses of Igneous Rocks, Metamorphic Rocks and Minerals 1931–1954. London, H.M.S.O.

Henderson, W. G. & N. J. Fortey. 1982. Highland Border rocks at Loch Lomond and Aberfoyle. Scottish Journal of Geology 18: 227–245.

Heslop, R. E. F. & C. J. Bown. 1969. The Soils of Candacraig and Glenbuchat. Macaulay Institute for Soil Research, Aberdeen.

Hopkins, J.J. 1983. Studies of the historical ecology, vegetation, and flora of the Lizard district, Cornwall, with particular reference to heathland. Ph.D. thesis, University of Bristol, U.K.

Horrill, A. D., J. M. Sykes & E. T. Idle. 1975. The woodland vegetation of Inchcailloch, Loch Lomond. Transactions of the Botanical Society of Edinburgh 42: 307–334.

Hornung, M. & G. Mew. 1970. Report on the soils of the island of Inchcailloch, Loch Lomond National Nature Reserve. Nature Conservancy Council, Bangor, unpublished report.

Hunter, J. G. & O. Vergnano. 1952. Nickel toxicity in plants. Annals of Applied Biology 39: 279–284.

Ingrouille, M.J. & N. Smirnoff. 1986. *Thlaspi caerulescens* J. & C. Resl. (*T. alpestre* L.) in Britain. *New Phytologist* 102: 219–233.

Jeffrey, D. W. 1971. The experimental alteration of a Kobresia-rich sward in Upper Teesdale. The Scientific Management of Animal and Plant Communities for Conservation (Ed. by E. Duffey & A. S. Watt), pp. 79–89. Blackwell Scientific Publications, Oxford.

Jefferies, R. L. & A. J. Willis. 1964. Studies on the calcicole-calcifuge habit II. The influence of calcium on the growth and establishment of four species in soil and sand cultures. Journal of Ecology 52: 691–707.

Johnston, W. R. 1976. Studies in the plant ecology of two Scottish serpentine areas. M.Sc. thesis, University of Stirling.

Johnston, W. R. & J. Proctor. 1977. Metal concentrations in plants and soils from two British serpentine sites. Plant and Soil 46: 275–278.

Johnston, W. R. & J. Proctor. 1979. Ecological studies on the Lime Hill serpentine. Transactions of the Botanical Society of Edinburgh 43: 145–150.

Johnston, W. R. & J. Proctor. 1980. Ecological studies on Meikle Kilrannoch serpentines. Transactions of the Botanical Society of Edinburgh 43: 207–215.

Johnston, W. R. & J. Proctor. 1981. Growth of serpentine and non-serpentine races of *Festuca rubra* in solutions simulating the chemical conditions in a toxic serpentine soil. Journal of Ecology 69: 855–869.

Johnston, W. R. & J. Proctor. 1984. The effects of magnesium, nickel, calcium and micronutrients on the root surface phosphatase activity of a serpentine and non-serpentine clone of *Festuca rubra* L. New Phytologist 96: 95–101.

Jowett, D. 1958. Populations of *Agrostis* spp. tolerant of heavy metals. Nature, London 182: 816–817.

Khalid, R. B. Y. 1974. Effect of chemical treatments on plant growth in serpentine soils. Ph.D. thesis, University of Aberdeen.

Kilic, N. (1976). Factors affecting the growth of *Picea sitchensis* (Bong) Carr. and other conifers on serpentinic soils near Strathdon, Aberdeenshire. Ph.D. thesis, University of Aberdeen, U.K.

Kruckeberg, A. R. 1954. Plant species in relation to serpentine soils. Ecology 35: 267–274.

Looney, J. H. H. 1982. Chemical and ecological studies on plants and soils of ultrabasic and other areas on the Island of Rhum, Scotland. Ph.D. thesis, University of Stirling, U.K.

Looney, J. H. H. & J. Proctor. 1989a. The vegetation of ultrabasic soils on the Isle of Rhum I. Physical environment, plant associations and soil chemistry. Transactions of the Botanical Society of Edinburgh 45: 335–350.

Looney, J. H. H. & J. Proctor. 1989b. The vegetation of ultrabasic soils on the Isle of Rhum II. The causes of the debris. Transactions of the Botanical Society of Edinburgh 45: 351–364.

Lusby, P. S. 1985. The history and taxonomy of the Shetland taxon of *Cerastium arcticum* Lange. B.Sc. thesis, University of Aberdeen.

MacGillivray, W. 1855. The Natural History of Deeside and Braemar. London.

Madhok, O. P. & R. B. Walker. 1969. Magnesium nutrition of two species of sunflower. Plant Physiology, Lancaster 44: 1016–1022.

Malloch, A. J. C. 1972. Salt spray deposition on the maritime cliffs of the Lizard Peninsula. Journal of Ecology 60: 103–112.

Margetts, L. J. 1988. The difficult and critical plants of the Lizard District of Cornwall. Grenfell Publications, Bristol.

Marrs, R. H. & P. Bannister. 1978. The adaptation of *Calluna vulgaris* (L.) Hull to contrasting soil types. New Phytologist 81: 753–761.

Marrs, R. H. & J. Proctor. 1976. The response of serpentine and non-serpentine *Agrostis stolonifera* to magnesium and calcium. Journal of Ecology 64: 953–964.

Marrs, R. H. & J. Proctor. 1978. Chemical and ecological studies of heath plants and soils of the Lizard Peninsula, Cornwall. Journal of Ecology 66: 417–432.

Marrs, R. H. & J. Proctor. 1979. Vegetation and soil studies of the enclosed heathlands of the Lizard Peninsula, Cornwall. Vegetatio 41: 121–128.

Marshall, J. K. 1959. Studies on the serpentine vegetation at Meikle Kilrannoch, Clova, Angus. Hons. thesis, University of St. Andrews, U.K.

McVean, D. N. & D. A. Ratcliffe. 1962. Plant Communities of the Scottish Highlands. H.M.S.O., London.

Mitchell, J. & A. Mitchell. 1983. Limestone burning for Agricultural Fertilizer in the Parishes of Kilmaronock and Buchanan, Loch Lomondside, from the mid-18th to the early 19th Century. Unpublished report to the Drymen and District Local History Society.

Munir, Y. H. 1970. Comparative studies of the base status of magnesium soils. M.Sc. thesis, University of Aberdeen.

Nature Conservancy Council. 1974. Isle of Rhum National Nature Reserve: Reserve Handbook. Nature Conservancy Council, Inverness.

Neill, P. A. M. 1806. A tour through some of the islands of Orkney and Shetland. Edinburgh.

Nicol, K. A. J. 1983. A plant ecological study of the Island of Creinch, Loch Lomond. B.Sc. thesis, University of Stirling, U.K.

Pigott, C. D. 1956. The vegetation of Upper Teesdale in the North Pennines. Journal of Ecology 44: 545–586.

Pigott, C. D. 1962. Soil formation and development on the Carboniferous Limestone of Derbyshire I. Parent materials. Journal of Ecology 50: 145–156.

Pigott, C. D. & S. M. Walters. 1954. On the interpretation of the discontinuous distributions shown by certain British species of open habitats. Journal of Ecology 42: 95–116.

Proctor, J. 1969. Studies in serpentine plant ecology. D.Phil. thesis, University of Oxford, U.K.

Proctor, J. 1970. Magnesium as a toxic element. Nature, London 227: 742–743.

Proctor, J. 1971a. The plant ecology of serpentine II. Plant response to serpentine soils. Journal of Ecology 59: 397–410.

Proctor, J. 1971b. The plant ecology of serpentine II. The influence of a high magnesium/calcium ratio and high nickel and chromium levels in some British and Swedish serpentine soils. Journal of Ecology 59: 827–842.

Proctor, J., K. Bartlem, S. P. Carter, D. A. Dare, S. B. Jarvis, & D. R. Slingsby. 1991. Vegetation and soils of the Meikle Kilrannoch ultrabasic sites. Transactions of the Botanical Society of Edinburgh (in press).

Proctor, J., J. Burrow, & G. C. Craig. 1980. Plant and soil chemical analyses from a range of Zimbabwean serpentine sites. Kirkia 12: 127–139.

Proctor, J. & D. A. Cottam. 1982. Growth of oats, beet and rape in four contrasting serpentine soils. Transactions of the Botanical Society of Edinburgh 44: 19–25.

Proctor, J. & G. C. Craig. 1978. The occurrence of woodland and riverine forest on the serpentine of the Great Dyke. Kirkia 11: 129–132.

Proctor, J. & W. R. Johnston. 1977. *Lychnis alpina* L. in Britain. Watsonia 11: 199–204.

Proctor, J., W. R. Johnston, D. A. Cottam, & A. B. Wilson. 1981. Field-capacity water extracts from serpentine soils. Nature, London 294: 245–246.

Proctor, J. & I. D. McGowan. 1976. Influence of magnesium on nickel toxicity. Nature, London 134: 260.

Proctor, J. & S. R. J. Woodell. 1971. The plant ecology of serpentine I. Serpentine vegetation of England and Scotland. Journal of Ecology 59: 375–395.

Proctor, J. & S. R. J. Woodell. 1975. The ecology of serpentine soils. Advances in Ecological Research 9: 255–366.

Ragg, J. M. & D. F. Ball. 1964. Soils of the ultra-basic rocks of the island of Rhum. Journal of Soil Science 15: 123–133.

Riley, H. C. F. 1977. Land evaluation studies for grass production systems in the uplands of north-east Scotland. Ph.D. thesis, University of Aberdeen.

Roberts, R. H. & A. McG. Stirling. 1974. *Asplenium cuneifolium* Viv. in Scotland, Fern Gazette 11: 7–14.

Rune, O. 1953. Plant life on serpentines and related rocks in the north of Sweden. Acta Phytogeographica Suecica 31: 1–139.

Sedgwick, E. L. D. 1975. Soil-plant associations on serpentine at Greenhill, West Aberdeenshire. M.Sc. thesis, University of Aberdeen.

Shewry, P. R. & P. J. Peterson. 1975. Calcium and magnesium in plants and soil from a serpentine area on Unst, Shetland. Journal of Applied Ecology 12: 381–391.

Shewry, P. R. & P. J. Peterson. 1976. Distribution of chromium and nickel in plants and soil from serpentine and other sites. Journal of Ecology 64: 195–212.

Sleep, A. 1980. On the reported occurrence of *Asplenium*

cuneifolium and *A. adiantum-nigrum* in the British Isles. Fern Gazette 12: 103–107.

Sleep, A., R. H. Roberts, J. I. Souter, & A. McG. Stirling. 1978. Further investigations on *Asplenium cuneifolium* in the British Isles. Fern Gazette 11: 345–348.

Sleep, A. 1985. Speciation in relation to edaphic factors in the *Asplenium adiantum-nigrum* group. Proceedings of the Royal Society of Edinburgh 86B, 325–334.

Slingsby, D. R. 1974 The role of nickel in the ecology of British serpentine sites and in the physiology of *Avena sativa*. Ph.D. thesis, University of Bristol.

Slingsby, D. R. 1979. The Keen of Hamar, Shetland NNR and SSSI. A survey and monitoring scheme. Nature Conservancy Council, Aberdeen.

Slingsby, D. R. 1981. The Keen of Hamar, Shetland: A general survey and a census of some of the rarer plant taxa. Transactions of the Botanical Society of Edinburgh 43: 297–306.

Slingsby, D. R. 1982. The Vegetation of Crussa Field, Nikkavord and the Heogs, Unst, Shetland. Nature Conservancy Council, Aberdeen.

Slingsby, D. R. & D. H. Brown. 1977. Nickel in British serpentine soils. Journal of Ecology 65: 597–618.

Slingsby, D. R., S. P. Carter, R. J. King & C. Jenner. 1983. The Serpentine Vegetation of Fetlar and Haaf Gruney. Nature Conservancy Council, Aberdeen.

Slingsby, D. R. & S. P. Carter. 1986a. A comparative study of the morphology and ecology of *Cerastium arcticum* in Shetland and Faroe. Bulletin of the British Ecological Society 17: 25–28.

Slingsby, D. R. & S. P. Carter. 1986b. The ecological effects of eutrophication on the Keen of Hamar SSSI, Shetland. Nature Conservancy Council, Aberdeen.

Smith, R. A. H. 1968. An investigation into the general characteristics of the vegetation of the serpentines of NE Scotland, and an examination of the taxonomic status of some of the species characteristic of these rocks, and the factors determining the infertility of some of the soils derived from them. B.Sc. thesis, University of Aberdeen.

Soon, Y. K. 1971. Chemical factors affecting plant growth on serpentine soils, especially calcium supply. M.Sc. thesis, Univeristy of Aberdeen, U.K.

Spence, D. H. N. 1957. Studies on the vegetation of Shetland I. The serpentine debris vegetation in Unst. Journal of Ecology 45: 917–945.

Spence, D. H. N. 1958. The flora of Unst, Shetland, in relation to geology. Transactions of the Botanical Society of Edinburgh 37: 163–173.

Spence, D. H. N. 1959. Studies on the vegetation of Shetland II. Reasons for the restriction of the exclusive pioneers to serpentine debris. Journal of Ecology 47: 641–649.

Spence, D. H. N. 1970. Scottish serpentine vegetation. Oikos 21: 22–31.

Spence. D. H. N. & E. A. Millar. 1963 An experimental study of the infertility of a Shetland serpentine soil. Journal of Ecology 51: 333–343.

Spence. D. H. N. 1979. Shetland's living landscape. The Thule Press, Lerwick.

Steele, B. 1955. Soil pH and base status as factors in the distribution of calcicoles. Journal of Ecology 43: 120–132.

Stirling, A. McG. 1974. Conic Hill, Balmaha. Glasgow Naturalist 19: 140.

Tassoulas, J. A. 1970. Comparative studies of the composition of magnesian soils with particular reference to the heavy metal (Cr, Fe, Mn, Ni) status. M.Sc. thesis, University of Aberdeen, U.K.

Tittensor, R. M. & R. C. Steele. 1971. Plant communities of the Loch Lomond oakwoods. Journal of Ecology 59: 561–582.

Thompson, J. & J. Proctor. 1983. Vegetation and soil factors on a heavy metal mine spoil heap. New Phytologist 94: 297–308.

Vedy, J. C. & S. Bruckert. 1979. Les solutions du sol. Composition et signification pédogénetique. Constituants et Proprieźés du Sol (M. Borneau & B. Souchier eds) 2: 161–186. Paris, Masson.

Wadsworth, W. J. 1961. The layered ultrabasic rocks of southwest Rhum, Inner Hebrides. Philosophical Transactions of the Royal Society of London B, 244: 21–64.

Walker, R. B. 1954. Factors affecting plant growth on serpentine soils. Ecology 35: 267–274.

Watling, R. 1969. Check list of the plants of Rhum, Inner Hebrides (V.C. 104 North Ebudes) Part III: Fungi. Transactions of the Botanical Society of Edinburgh 40: 497–535.

West, W. 1912. Notes of the flora of Shetland, with some ecological observations. Journal of Botany, London 50: 265–275, 297–306.

Whittaker, R. H. 1954. The vegetational response to serpentine soils. Ecology 35: 275–288.

Willett, I. R. & T. Batey. 1977. The effects of metal ions on the root surface phosphatase activity of grasses differing in tolerance to serpentine soil. Plant and Soil 48: 213–221.

Wilkins, D. A. 1957. A technique for the measurement of lead tolerance in plants. Nature, London 214: 628.

Wilson, M. J. & M. L. Berrow. 1978. The mineralogy and heavy metal content of some serpentinite soils in northeast Scotland. Chemie der Erde B, 37: 181–205.

Wilson, M. J., D. Jones, & W. J. McHardy. 1981. The weathering of serpentinite by *Lecanora atra*. Lichenologist 13: 167–176.

Wilson, M. J., D. Jones, & J. D. Russel. 1980. Glushkinsite, a naturally occurring magnesium oxalate. Mineralogical Magazine 43: 837–840.

Wormell, P. 1976. The Manx Shearwaters of Rhum. Scottish Birds 9: 103–118.

Wyn, J. R. G. & O. R. Lunt. 1967. The function of calcium in plants. Botanical Reviews 33: 407–426.

Author's Address:
J. Proctor
Dept. of Biological and Molecular Sciences
University of Stirling
Stirling
Scotland FK9 4LA
U.K.

VII

Ecology of serpentinized areas

of north-east Portugal

E. MENEZES DE SEQUEIRA & A. R. PINTO DA SILVA

Abstract. After a brief description of the geology, climate and main soil types derived from serpentinized ultramafic rocks in north-east Portugal, detailed information is given regarding the behavior of some elements in relation to the weathering and soil forming processes.

The flora and vegetation and aspects of the autecology of some serpentinophytes are discussed regarding the most important ecological factors in the area: summer drought, nickel toxicity, and unbalanced Mg/Ca quotient. The serpentinomorphoses suggest an adaptation to drought, and to other soil toxicity characteristics.

The main adverse factors are probably the high Mg/Ca quotient, the high Ni and the low N, P, K and Ca concentrations. These may eliminate strongly competitive serpentinofuges and allow the establishment of peculiar serpentine flora and vegetation. When no erosion occurs, deeper and more fertile soils are developed (with higher water holding capacity, increased organic matter, and nutrient levels), tolerant shrubs and trees may grow, the serpentine character of the flora and vegetation decreases, but the Mediterranean character remains, not only because of summer drought, but especially because the toxic effects are not eliminated. A high concentration of nickel remains during the soil forming process – a dissolution – in spite of evolution diminishing its toxic effects.

In deep soils, nickel behavior and toxicity are probably also controlled by the relative stability of Cu and Ni organic complexes.

B. A. Roberts and J. Proctor (eds), The ecology of areas with serpentinized rocks. A world view, 169–197.
© *1992 Kluwer Academic Publishers.*

Introduction

The serpentinized areas of north-east Portugal cover some 8,000 ha and are subject to a mediterranean-type climate. The chemical composition of the substrates shows considerable variation, and this together with the variation in weathering and leaching of the soil cause important differences in the hydrological and chemical characteristics of the serpentinized sites. These aspects are addressed in the next paragraphs.

As in most other serpentinized areas in the world, the flora of the north-east Portuguese serpentinized areas is relatively poor. Nevertheless this flora contains a number of characteristic species and the plants possess the typical morphological features found in many serpentine plants. The flora, its morphological features, and the plant geographical characteristics are outlined. It is ecologically very interesting that species differ considerably in their ability to translocate various elements from their roots to their shoots. We pay some attention to this and present data.

The north-east Portuguese serpentine vegetation is characterized by four associations and two typical variants of wider-spread associations. These community types show interesting successional relationships, determined by soil development and anthropogenic disturbances. We discuss this in detail.

Geology

Unaltered peridotitic rocks are rare in north-east Portugal. They are usually metamorphosed. In the original olivine there is more forsterite (Mg_2SiO_4) than fayelite (Fe_2SiO_4), explaining the relatively high concentrations of Mg, Co and Ni which are often associated with forsterite. High content of Na as compared with many other serpentinic rocks (Robinson *et al.* 1935, Mhor & Van Baren 1954, Pedro & Bitar 1966) probably reflects the composition of the original magma. The abnormal quantities of CaO observed in some cases are input to assimilation from invaded media (Cotelo Neiva 1948). Detritic formations from the Cenozoic with coarse serpentinized fragments are occasionally present (Vasconcelos Ferreira 1965).

On the whole there are few serpentinized rocks with more than 1% Cr_2O_3; however, there is chromite accumulation (Cotelo Neiva 1948) in veins and pockets, from the intrusion of residual magma. The chemical composition of the north-east Portuguese ultramafic rocks varies within the following limits (%): SiO_2 – 27 to 50, Al_2O_3 – 0.05 to 7.6, Fe_2O_3 (including Fe_2O) – 6 to 13, TiO_2 – traces to 0.6, MgO – 14 to 39, CaO – 0.05 to 25, K_2O – 0.02 to 0.17, Na_2O – 0.25 to 1.17, Cr_2O_3 – traces to 12.40, MnO – 0.06 to 0.20, and NiO – 0.07 to 0.75 (Cotelo Neiva 1948, Vasconcelos Ferreira 1965, Menezes de Sequeira 1969). Lherzolite seems to be one of the most frequent occurrences, with the following composition (mean of 14 samples, ± standard deviation) (%): SiO_2 – 40 ± 5, Al_2O_3 – 2 ± 1, Fe_2O_3 – 8 ± 2, TiO_2 – 0.05 ± 0.02, MgO – 35 ± 4, CaO – 1.5 ± 1.5, K_2O – 0.05 ± 0.01, Na_2O – 0.40 ± 0.20, Cr_2O_3 – 0.5 ± 0.2, MnO – 0.15 ± 0.05 and NiO – 0.40 ± 0.20.

The serpentinized rocks in north-east Portugal are massive, dense, with a homogeneous structure and very often with chromite crystals. In more weathered superficial rocks (supergenic serpentinization) their large mesh structure can be easily seen (Vasconcelos Ferreira 1965). The microscopic structure is reticulate with a lattice of thin fillets of chrysotile or serpophite surrounding serpophite and, or, antigorite, sometimes also chrysotile, with or without chlorite and frequently with bastite. The brownish yellow color of these serpentinic rocks is due to oxide exuded during the serpentinization process (Cotelo Neiva 1948) (Photo 1, 2).

The ultramafic outcrops of north-east Portugal are irregularly distributed between latitudes 45°25′ and 41°54′ N (Pinto da Silva 1970). A similar latitudinal distribution can be found in Albania (Markgraf 1932), Alto Tibre (Pichi-Sermolli 1948) and Euboea (Krause, Ludwig & Seidel 1963). At Vinhais, the serpentinic outcrops extend from 500 to 830 m, in Bragança from 600 to 1060 m, and at Morais from 300 to 900 m a.s.l.

Climate

In the interglacial periods, the climate was of the Mediterranean type but with lower temperatures than present. During the last glaciation, Portugal

Photo 1. Unweathered serpentine rock thin section in plain light. Note possible grains of antigorite, chromite, etc.

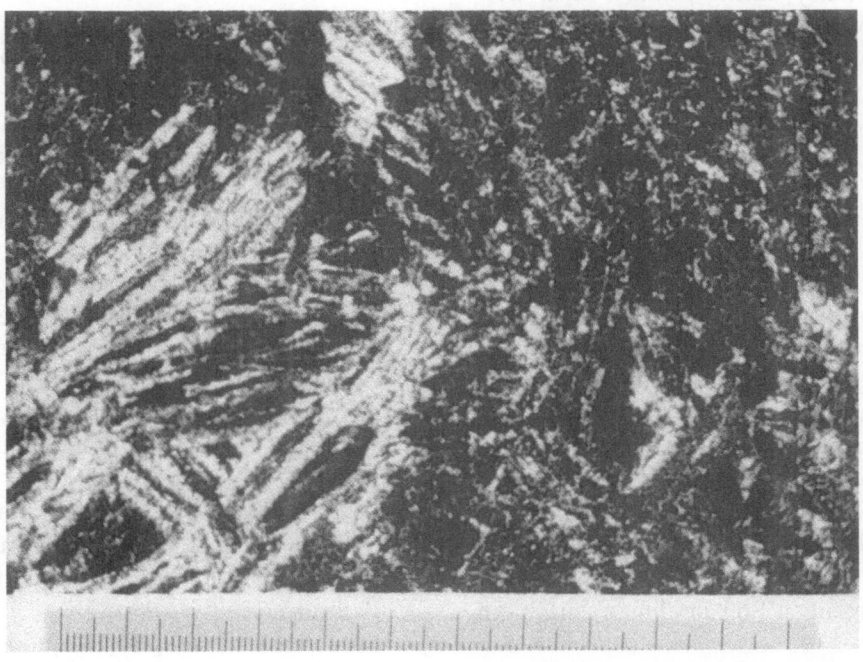

Photo 2. Weakly weathered serpentine rock thin section in plain light. Note the exudated iron oxides surrounding antigorite, enstatite, etc. grains.

had a humid climate with heavy rainfall in summer (Zbyszewski 1958). The lower limit of the glaciers was 1650 m a.s.l., and in the north of Portugal there were no trees above 850 m. It is accepted, because of the presence of fragmented shales and 'richter sides' (A. Ribeiro, pers. commun.), that periglaciation occurred. However, the serpentine areas offered refuge to Tertiary relics that have persisted, like in Mohelno, Kraubath, and Gulsen and in Italy (Pichi-Sermolli 1948), even if at more than 850 m above sea level (Löve 1963, Löve & Löve 1964). Later on serpentine endemics emerged and during interglacial times some sarmatic species migrated in (Pichi-Sermolli 1948, Demoulin 1966).

The present climate of these Portuguese areas can be classified:

a) In the Emberger classification (1955) as Mediterranean, varying from Mediterranean humid near Bragança and Vinhais to more arid variants at Monte de Morais (Macedo de Cavaleiros).

b) In the Thornthwaite classification (Mendes & Bettencourt 1980) it varies from $B_4B_1'sb_4'$ very humid, first mesothermal, with moderate water deficit in summer and with a moderate thermal efficiency during summer near Bragança and Vinhais to, $C_1B_2db_4'$ dry humid, second mesothermal, without or with little water surplus during the year and with a moderate thermal efficiency during the summer near Monte Morais.

Soils

From the different conditions of soil genesis, i.e. topography (erosion and infiltration), macromorphology of the rock, climate (alternate surplus and deficit of water), and vegetation cover, the following soil successions can be considered:

a) Fine earth accumulation in rock crevices.

b) Lithosols, mostly on steep slopes (Photo 3).

c) Humic litholic soils, mostly on gentle slopes (Photo 4).

d) Red Mediterranean Soil in climax conditions (Photo 5).

In addition, in depressions with poor drainage conditions, as a result of colluviation, there are hydromorphic soils with no illuvial horizon: in

Photo 3. Lithosol (profile no. 1). Note rock blocks and *Astragalus macrorrhizus* root system.

small depressions, with drainage and no erosion, there are Humic Litholic Soils. In both these soils there are no stones and gravel, and the vegetation presents less serpentine characteristics.

The morphological description of one example of the three most important soils in the succession is shown (Appendix I) with the analytical data (Table 1).

The general characteristics of these soils (Sequeira 1969; Appendix I and Table 1) are:

1. High level of heavy metals especially extractable nickel.

2. High level of Fe_2O_3 and MgO.

3. Unbalanced Mg/Ca quotient decreasing from rock to the surface.

4. Low levels of calcium and potassium.

5. Low level of soluble phosphorus.

6. Decrease of Fe_2O_3/Al_2O_3 quotient from rock to surface.

7. Decrease of SiO_2/R_2O_3 quotient from rock to sub-surface horizon.

8. The value of the pH in water varying between 5.4 to 7.2; there is a wide range of variation

Photo 4. Humic Litholic Soil (profile no. 2). Note foliated weathered serpentinic rock at the bottom of the profile and thin roots in the rock fissures.

of the pH within the same horizon, according to the size distribution of soil constituents.

9. High level of available copper in A horizons.
10. High stoniness and low depth.

Sequeira (1969) found that with soil and vegetation evolution the pH and Mg/Ca quotient decrease significantly. If the soil is ploughed, however, the pH and Mg/Ca increase.

Weathering and soil genesis

Under natural conditions of north-east Portugal, the weathering process is practically a dissolution. The bulk of the soil is always less than 20% of the original rock, in spite of local variations related to the interaction of climatic and physical aspects of the environment (Pedro & Bitar 1966, Menezes de Sequeira 1969, Menezes de Sequeira *et al.* in press).

On almost level surfaces, when the rocks are dense compact and with vertical joints, there is no erosion; percolation depends only on rainfall, and deep soils may be formed. On steep slopes, foliated rocks lying parallel to the surface will probably give rise to shallow soils, with many stones and gravel, because fine particles will be removed by runoff and seepage-flow.

The amount and type of weathering are also correlated with the hydrological balance, since the water surplus in winter and early spring determines leaching, and in late spring and summer, with high temperatures and evapotranspiration of accumulated moisture, rubefaction occurs.

As an example of the climatic range, the following diagrams (Figs 1, 2, 3, 4) represent the hydrological balance in the top 20 cm and in the whole profile of the two soils in extreme climatic conditions:

a) Bragança near Carrazeda (profile 2) $B_4B_1'sb_4$, very humid, 1st Mesothermal, with both moderate water deficit and moderate thermal efficiency during the summer.

b) At Monte de Morais (profile 3) $C_1B_2db_4'$ dry humid, 2nd Mesothermal, without or with little water surplus during the year and with a moderate thermal efficiency during summer.

Considering the topsoil, there is always a big water surplus in winter, a moderate surplus or a deficit in spring (April–June), and a great deficit

173

Table 1. Chemical and physical analytical data of serpentinized soils from north-east Portugal

Profile	1. Lithosol		2. Humic Litholic Soil			3. Red Mediterranean Soil					
Horizon	A_1	A_1/C	A_{11}	A_{12}	C/R	O	A_{11}	A_{12}	B	C_1	C_2
Depth (cm)	0–2.5/5	2.5/5–10/40	0–15/20	15/20–20/40	20/40–35/60	−1–0	0–20	20–50	50–90	90–150	150–250
Material > 2 mm (%)	41.0	62.8	22.2	31.3	56.9	6.6	13.3	2.4	3.0	9.2	5.9
Coarse sand (%)	n.d.	n.d.	18.1	42.8	49.4	n.d.	13.1	4.2	4.4	5.9	14.3
Fine sand (%)	n.d.	n.d.	62.3	41.4	36.5	n.d.	47.7	45.1	39.8	40.7	36.6
Silt (%)	n.d.	n.d.	18.5	13.4	12.1	n.d.	24.4	20.5	15.8	13.3	11.7
Clay (%)	n.d.	n.d.	1.1	2.4	2.0	n.d.	14.8	30.2	40.2	39.7	36.9
Moisture % at pF 2	n.d.	n.d.	39.42	18.82	17.20	n.d.	29.97	26.58	42.18	44.86	47.42
Moisture % at pF 2.7	n.d.	n.d.	28.85	13.61	15.43	n.d.	24.44	22.46	39.67	41.81	43.07
Moisture % at pF 4.2	n.d.	n.d.	9.74	9.11	10.37	n.d.	14.78	16.45	28.74	30.40	33.57
H_2O% at pF 2.7–H_2O% at pF 4.2	n.d.	n.d.	29.68	9.71	6.83	n.d.	15.19	10.13	13.34	14.46	13.85
pH in H_2O	6.80	6.90	6.20	6.80	6.95	5.45	6.15	6.70	7.00	7.00	6.90
Organic C%	0.98	0.48	1.22	0.29	0.17	9.06	1.69	0.43	0.50	0.14	0.06
Organic matter %	1.70	0.83	2.10	0.50	0.30	15.6	2.90	0.80	0.90	0.20	0.10
Total nitrogen %	0.16	0.09	0.14	0.05	0.03	0.56	0.19	0.07	0.04	0.03	0.01
Exchangeable Ca (meq 100 g^{-1})	2.99	3.35	0.81	0.29	0.35	6.46	2.08	0.81	0.43	0.23	0.30
Exchangeable Mg (meq 100 g^{-1})	17.00	29.50	12.0	12.5	20.0	15.5	19.0	25.5	38.5	41.0	43.5
Exchangeable K (meq 100 g^{-1})	0.27	0.33	0.07	0.02	0.02	0.69	0.10	0.09	0.10	0.06	0.04
Exchangeable Na (meq 100 g^{-1})	0.03	0.03	0.04	0.02	0.03	0.07	0.05	0.05	0.05	0.06	0.05
Mg/Ca (meq)	5.7	8.8	14.8	43.0	57.1	2.4	9.1	31.5	89.5	178.2	145.0
Exchangeable capacity (meq 100 g^{-1})	20.8	27.8	11.8	13.8	18.5	31.5	21.8	24.8	33.5	33.0	33.8
Soluble P (Bray I) (μg g^{-1})	7.3	3.3	0.6	0.4	0.4	95	0.8	0.2	0.1	0.2	0.3
Soluble K (Bray I) (μg g^{-1})	105	129	27	8	8	269	39	35	39	23	16
Available Cu (EDTA) (μg g^{-1})	2.6	1.6	2.8	2.2	2.7	1.6	0.5	0.2	0.2	0.2	0.2
Available Zn (EDTA) (μg g^{-1})	1.7	0.7	0.8	0.3	0.3	9.1	1.5	0.2	0.2	0.4	0.5
Available Mn (EDTA) (μg g^{-1})	150	240	110	140	120	1050	1050	700	420	340	250
Available Ni (EDTA) (μg g^{-1})	110	65	89	56	40	220	290	230	138	190	95
Available Cr (EDTA) (μg g^{-1})	0.3	0.4	0.2	0.4	0.4	0.9	0.8	0.4	0.5	0.4	0.4
Available Fe (EDTA) (μg g^{-1})	230	160	180	250	220	540	260	120	160	160	180

n.d. – not determined.

Photo 5. Serpentinic Red Mediterranean Soil (profile no. 3). Note deep soil (2.5 metres) between large blocks of serpentine.

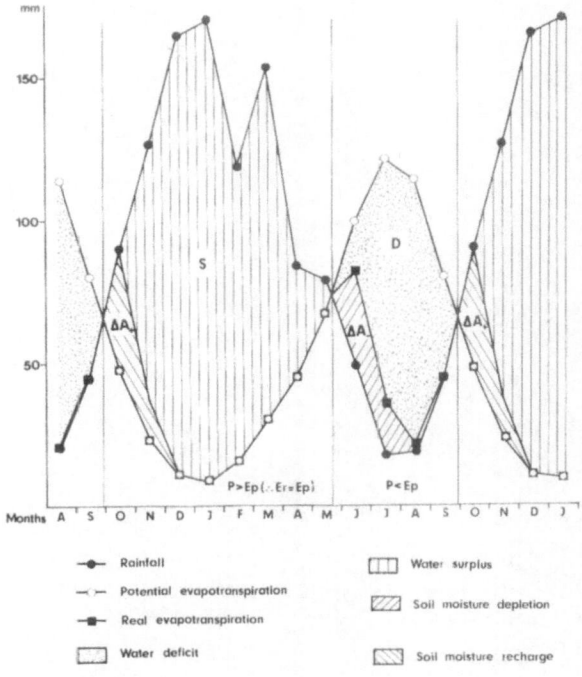

Fig. 1. Hydrological balance at Carrazeda Humic Litholic Soil Rainfall and evapotranspiration (Thornthwaite) mean of data from 1931 to 1960. Available water % moisture at pF 2.0–% moisture at pF 4.2. Depth 20 cm.

from July till October, after which there is a moderate surplus. The considerable water surplus from November till March (from 250 mm at Monte de Morais to 700 mm at Carrazeda) determines high leaching, which is conditioned by the amount of organic matter (CO_2 and free fulvic acids), and also by the soil and rain water temperatures (mean air temperatures ranging from 3–8°C at Carrazeda to 5–10°C at Monte de Morais).

In October and during spring (April–June) there is enough moisture and favourable temperatures (9–17°C at Carrazeda and 12–20°C at Monte de Morais) for great biological activity and sub-aerial weathering, but there is no leaching. In summer, drought, high temperatures and oxidation conditions probably stimulate rubefaction.

In the lower horizons, especially in deep soils (Red Mediterranean Soils) under driest conditions, leaching occurs only in February and March (the water surplus is reduced to 40 mm at the bottom of the profile at Monte de Morais),

while sub-aerial weathering lasts from October till February, and from April till August, and there are no conditions for rubefaction (no summer drought, lower temperatures and therefore no oxidation).

The original rock is mainly composed of olivine, serpentine, enstatite and analcime. In some cases there are traces of chlorite, talc etc., so there is a clear predominance of Mg, Si and Fe. Since the titanium containing minerals ilmenite, and likewise anatase and rutile, also present as traces, are extremely resistant to weathering (Mitchell 1964), TiO_2 can be considered as a standard oxide and the variation of the other oxides can be studied in relation to it (Mohr & van Baren 1954). So, admitting the stability of TiO_2, the oxide levels (MO) in the different soil horizons of a profile can be expressed as a quotient (titanic units MO % /TiO_2 %) in the original rock (less weathered) (Table 2), and in the preceding soil horizon (Table 3).

These results agree with the fact that olivine,

175

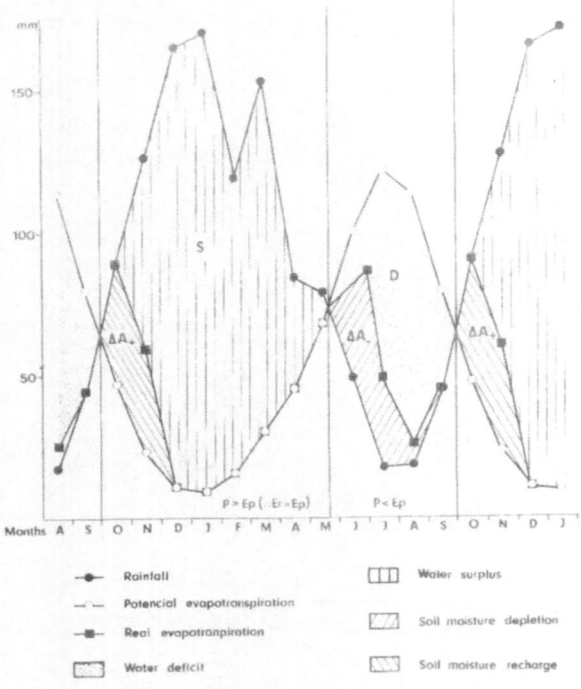

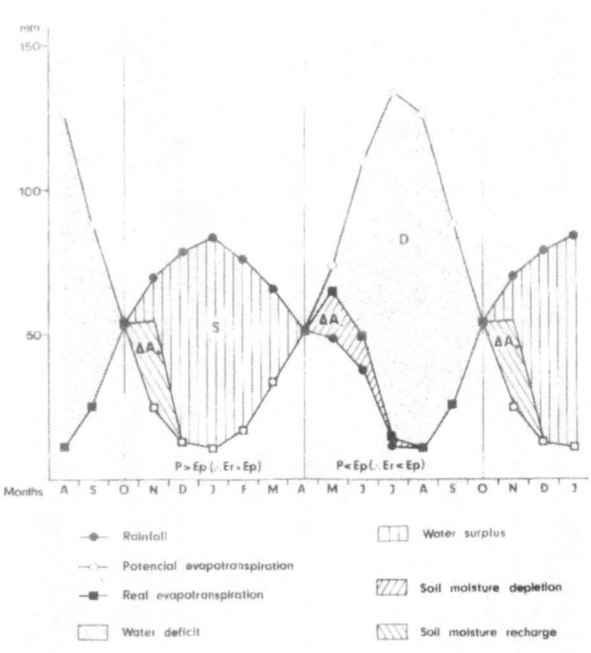

Fig. 2. Hydrological balance at Carrazeda Humic Litholic Soil Rainfall and evapotranspiration (Thornthwaite) mean of data from 1931 to 1960. Available water % moisture at pF 2.0-% moisture at pF 4.2. Depth 40 cm.

Fig. 3. Hydrological balance at Morais Red Mediterranean Soil. Rainfall and evapotranspiration (Thornthwaite) mean of data from 1931 to 1960. Available water % moisture at pF 2.0-% moisture at pF 4.2. Depth 20 cm.

serpentine and pyroxenes are the most easily weathered minerals (Mitchell 1964) and concur with those found by Pedro & Bitar (1966) in the decomposition of serpentinized rocks by carbonated water 'que dans la pratique correspond sensiblement à une dissolution' and with those verified by Menezes de Sequeira (1969) in similar soils using the same method. The weathering process (and not the resistance of the rock minerals to weathering) would thus explain the shallowness of most ultramafic soils in temperate climates (usually also subject to a severe erosion) in contrast with the deep ultramafic soils often formed in tropical climates where lateritization occurs (Robinson *et al.* 1935, Mohr & van Baren 1954).

In the shallow soils the order of leaching expressed in oxides, is $MgO > NiO > SiO_2 > Na_2O > Cr_2O_3 > Fe_2O_3 > Al_2O_3$ (Pedro & Bitar 1966). In the upper horizons of deeper soils Fe_2O_3 is leached more rapidly and probably corresponds to a 'podzolization' (Pedro & Bitar 1966). In the lower horizons of deep soils, with higher pH values and reduced leaching, there is an accumu-

lation of sesquioxides (Al_2O_3, Fe_2O_3 and Cr_2O_3 illuviation, MO/TiO > 100. Tables 2, 3). In uneroded soils with a good vegetation cover there is biological accumulation of K, Ca, Cu and Zn in the topsoil.

This interpretation must be considered *sensu lato*, since serpentinized rocks in temperate climates are very irregularly weathered, breaking down in decreasingly smaller different-size blocks. Very different phases of soil formation can thus be found simultaneously (particularly when there is a high percentage of coarse fragments; Table 1), as shown by variations of pH values within the same horizon (Menezes de Sequeira 1969).

Analyzing the data we may safely assume that the original olivine, enstatite, etc. are rapidly hydrolyzed to antigorite, chrysotile and lizardite (bastite and serpophite), with exudation of iron oxides. With the increasing loss of Mg, as well as some Fe, the serpentine group minerals are completely leached. Where more leaching occurs, such as at Carrazeda (Figs 1, 2), with the rapid

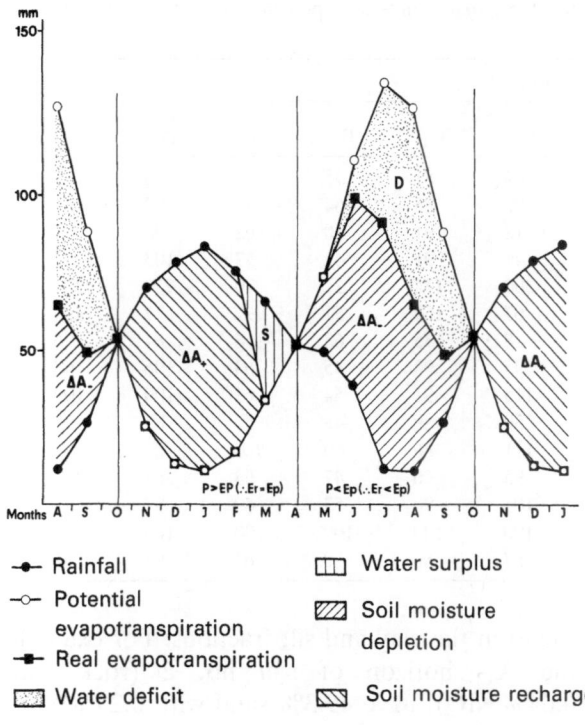

- ●— Rainfall
- ○— Potential evapotranspiration
- ■— Real evapotranspiration
- ▨ Water deficit
- ▭ Water surplus
- ▨ Soil moisture depletion
- ▨ Soil moisture recharge

Fig. 4. Hydrological balance at Morais Red Mediterranean Soil. Rainfall and evapotranspiration (Thornthwaite) mean of data from 1932 to 1960. Available water % moisture at pF 2.0–% moisture at pF 4.2. Depth 250 cm.

loss of Mg, talc becomes the most important sand mineral. Less leached soils (such as that of Monte de Morais) have more chlorite and smectite, and thus a higher clay content (Tables 1, 4).

The weathering of analcime $(Na(AlSi_2O_6))$, though slower, releases Na which is leached (Table 2) and produces pyrophilite $((Al_4(Si_8O_{20})(OH)_4)$ in all climatic conditions (Table 4). Some Al and SiO_2 are also released and, especially in deep illuviated horizons (Table 2), associate with Fe, Si and Mg in the formation of chlorite and smectite, clay minerals which will influence cation exchange capacity, swelling and structure of the soils. The slow leaching of Al and its illuviation in the B and C horizons of deep soils, as well as the Fe and Cr accumulation (Table 2), increase the formation of the clays chlorite and smectite (Table 1). Fe leached from olivine, chromite, enstatite, etc., may crystalize as goethite or hematite, or move in the soil solution as amorphous ferrous oxides until it is immobilized by oxidation (in summer) especially in the upper horizons. The amorphous ferric oxides form diffusion cutans (Photo 6) that give to the soil a characteristic dark red colour (rubefaction). Considering the coarser mineral fraction of the soil, it is interesting to notice that the dissolution of SiO_2, Mg and Ni in the weaker spots of the crystal lattice of olivines, serpentines, enstatites,

Table 2. Chemical composition of the profiles 2 and 3, expressed in titanic oxide units, as a percentage of the original values in the parent rock

Profile Horizon or weathered stage	Humic Litholic Soil				Red Mediterranean Soil							
	R_2	C/R	A_{12}	A_{11}	R_2	R_1	C_2	C_1	B	A_{21}	A_{11}	O
SiO_2	100	31	31	15	100	70	45	36	20	12	11	10
Al_2O_3	100	52	57	35	100	150	139	118	75	57	52	52
Fe_2O_3	100	40	53	32	100	120	133	120	61	37	34	28
MgO	100	30	36	16	100	63	26	23	10	6	6	5
CaO	100	73	76	32	100	166	500	650	217	151	186	225
K_2O	100	40	44	44	100	75	50	75	118	143	110	136
Na_2O	100	48	46	23	100	65	12	16	11	17	14	14
Cr_2O_3	100	36	52	33	100	139	132	125	56	36	38	26
MnO	100	45	58	31	100	93	106	97	53	32	32	32
NiO	100	28	34	18	100	144	85	75	36	20	18	12
CuO	100	67	53	23	100	113	125	107	64	65	55	82
ZnO	100	39	53	30	100	115	100	99	46	29	29	34
CoO	100	33	46	27	100	119	125	112	60	36	34	28
TiO	100	100	100	100	100	100	100	100	100	100	100	100
Total	100	33	36	18	100	75	50	43	21	13	12	13

Table 3. Chemical composition of the profiles 2 and 3, expressed in titanic oxide units as a percentage of the preceeding soil horizon

Profile Horizon	Humic Litholic Soil				Red Mediterranean Soil							
	R_2	C/R	A_{12}	A_{11}	R_2	R_1	C_2	C_1	B	A_{12}	A_{11}	O
SiO_2	100	31	19	50	100	70	64	81	55	63	88	95
Al_2O_3	100	52	109	61	100	150	90	85	64	76	92	99
Fe_2O_3	100	40	134	60	100	120	112	91	50	62	90	84
MgO	100	30	118	46	100	63	40	91	42	57	103	82
CaO	100	73	104	43	100	116	300	134	32	70	*124*	*122*
K_2O	100	40	109	100	100	75	67	150	157	*122*	77	*124*
Na_2O	100	48	97	50	100	65	18	135	68	158	84	96
Cr_2O_3	100	36	145	63	100	139	95	95	45	64	106	68
MnO	100	35	127	54	100	93	115	91	54	61	100	100
NiO	100	28	125	51	100	144	59	89	48	57	83	73
CuO	100	67	79	43	100	113	111	86	60	*101*	84	*120*
ZnO	100	39	136	58	100	115	95	90	47	64	*101*	*116*
CoO	100	33	140	58	100	119	105	90	53	60	97	81
TiO	100	100	100	100	100	100	100	100	100	100	100	100
Total	100	33	109	50	100	75	67	86	50	61	94	111

etc., results in porous sponge-like sand grains (Photos 7, 8).

Element behaviour in soil genesis

Soluble and exchangeable forms of magnesium (Mg^{2+}) increase with depth, but most of them (80–95%) are leached out of the profile; the remaining is part of the crystal lattices of talc, chlorite and smectite. The parent rock is very poor in calcium containing minerals, although this element may be locally concentrated in veins and pockets. The plants absorb Ca in considerable quantities and translocate it to the top; so whenever there are deeper soils, richer in organic matter, i.e. Red Mediterranean Soils, there is surface deposition of Ca. Biological accumulation by itself cannot explain the total Ca enrichment; clay minerals (smectite and chlorite) must retain Ca^{2+} more strongly than Mg^{2+}.

In the north-east Portuguese serpentinized rocks, nickel exists as chloanthite ($NiAs_3$), and also as isomorphic substitutions of magnesium in the crystal lattices of olivine, serpentine, enstatite and anthophyllite (Mitchell 1964). Nickel is held more weakly than Mg^{2+}, so it is more easily dislodged by H^+ entering into the crystal through weathering (Mitchell 1964; Tables 2, 3). In soils with large clay content (Red Mediterranean Soils and deep Humic Litholic Soils) part of the Ni that has been washed out of primary minerals can be found in the clay and silt fractions: For example the A_{12} horizon of soil no. 2 (rock with 0.64% NiO), have 49.3% sand with 0.23% NiO; 20.5% silt with 0.69% NiO; 24.6% clay minerals without free iron oxides with 0.67% NiO; 5.63% free iron oxides in clay fraction with 1.65% NiO (Menezes de Sequeira *et al.* in press). In these deep soils extractable nickel must be associated with Fe in amorphous oxides, and its toxic action is restricted to those plants with the capacity to solubilize iron oxides (Menezes de Sequeira *et al.* in press). In shallow soils, with low levels of clay, more than 90% of the total nickel occurs in the coarser fractions, as part of talc and other minerals, which will supply most of the extractable Ni released continuously by weathering of sand grains.

Whenever organic matter is present in the topsoil, organic complexes are formed, and there is an increase of toxicity. However, in deep soils with much organic matter there is also biological accumulation of copper; Cu organic complexes being more stable than Ni complexes (Mellor & Maley 1947), released Ni associates with Fe in amorphous oxides, and so becomes less toxic to plants. Nickel toxicity increases with cultivation of the soil: the plough, changing topsoil conditions, mixing stones and gravel with fine particles and organic matter, activates the weathering process, and consequently more nickel remain free to combine with organic acids as an organometal complex. For example, a non-dis-

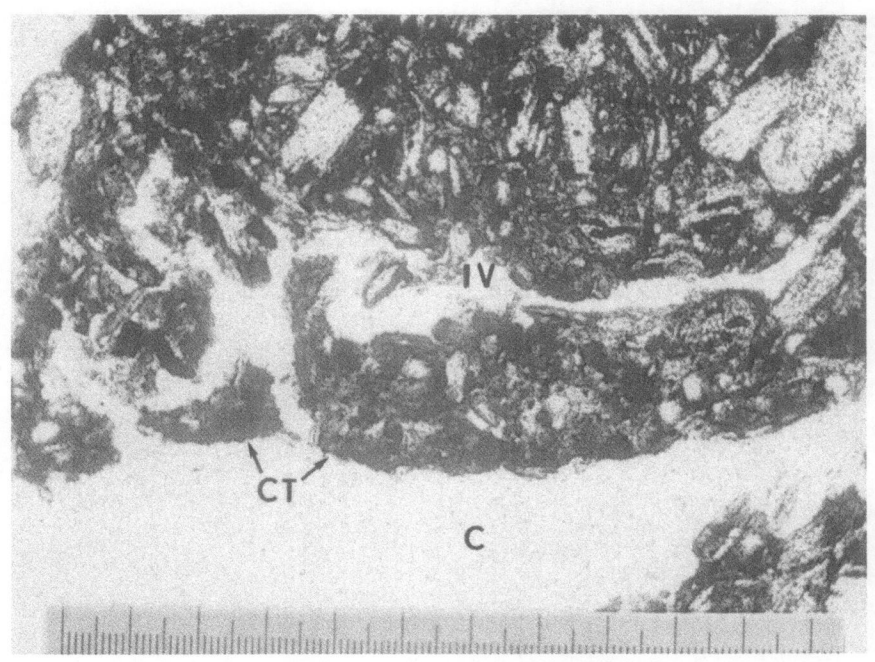

Photo 6. A/B Horizon of a serpentinic Red Mediterranean Soil (profile no. 3). Thin section in plain light. Note channels (C) among soil aggregates, interconnected vughs (IV), and iron oxide diffusion and illuviation cutans (CT).

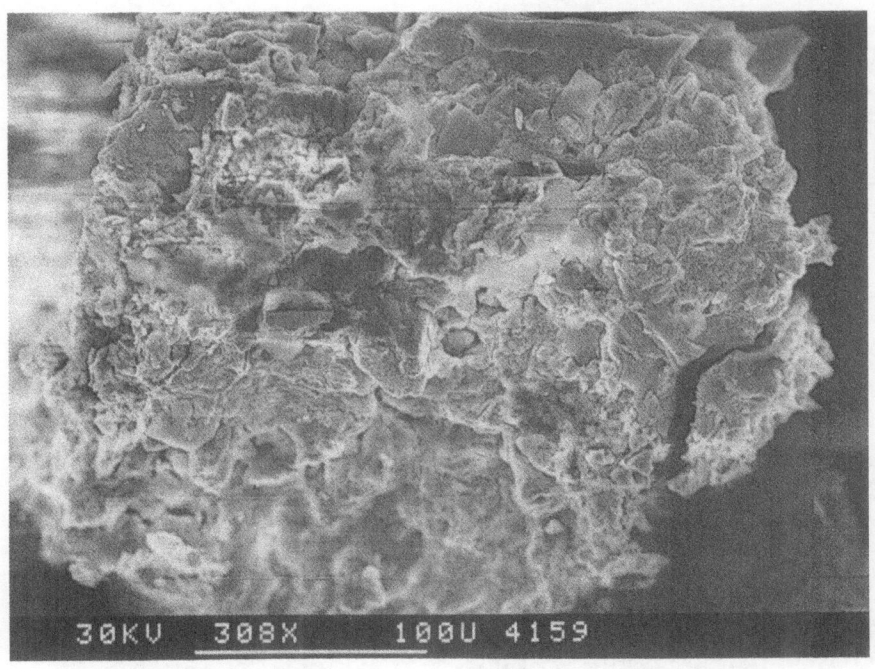

Photo 7(a).

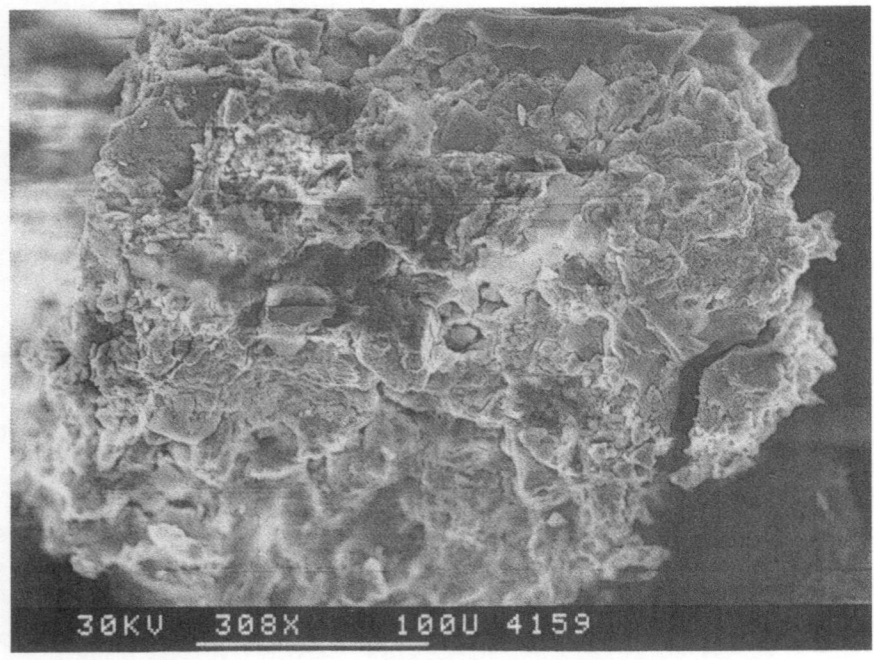

Photo 7a, b. Scanning electron micrographs of a sand antigorite grain collected in profile no. 3, B horizon. Note the weathering of weak spots in the crystal lattice. Material with a sponge-like appearance.

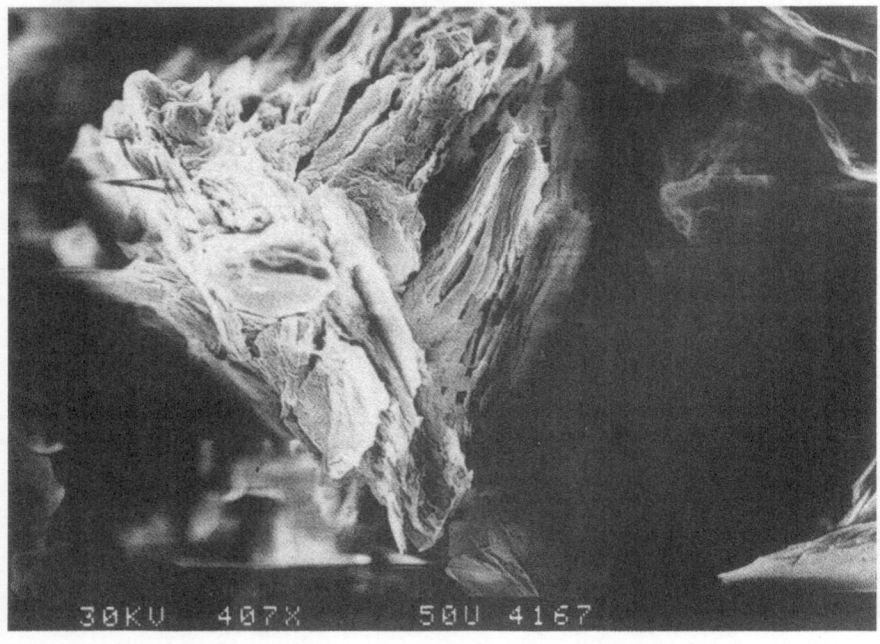

Photo 8(a).

180

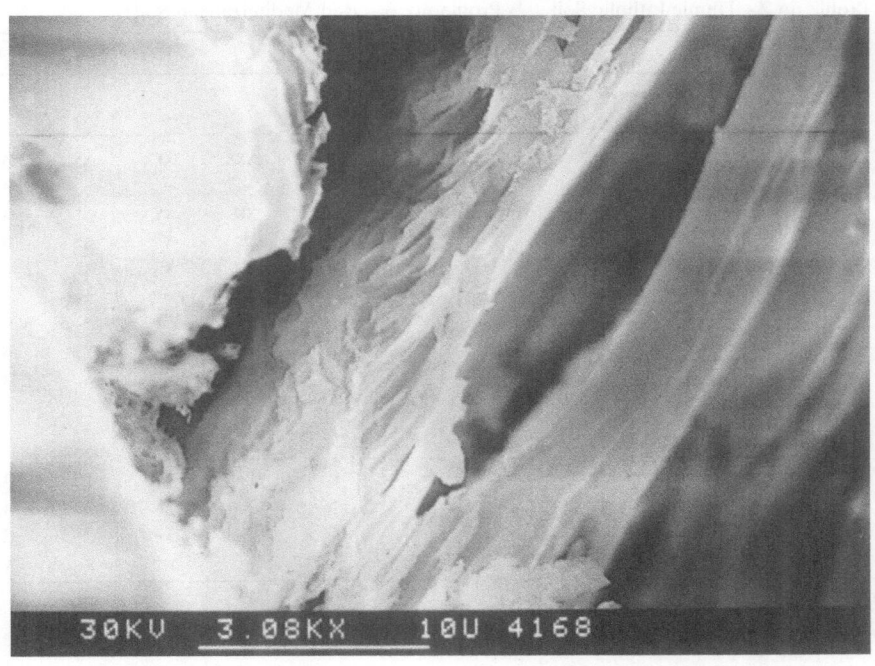

Photo 8(b).

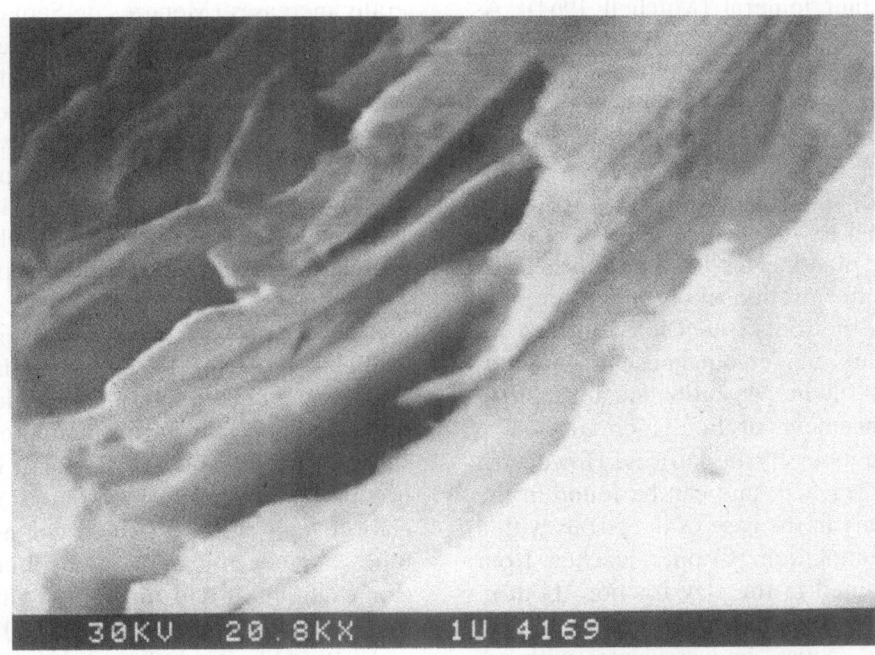

Photo 8(c).

Photo 8a,b,c. Scanning electron micrographs of a sand chrysotile grain collected in profile no. 3, C_1 horizon. Note the holes created by the weathering process in the weak spots of the crystal lattice.

Table 4. Semi-quantitative mineral composition of profiles 2 and 3 (X-ray diffraction analysis)

Minerals	Profile no.2 – Humic Litholic Soil				Profile no. 3 – Red Mediterranean Soil							
	R_2	C/R	A_{12}	A_{11}	R_2	R_1	C_2	C_1	B	A_{12}	A_{11}	O
(a) Sand fraction												
Olivine	x	v	x	x	v	v	–	–	–	–	–	n.d.
Antigorite	x	xx	xx	xx	xxx	v	xxx	xxx	xxx	xxx	xxx	n.d.
Lizardite	–	x	x	x	–	–	–	–	–	–	–	–
Enstatite	x	x	v	x	x	x	x	xx	xx	xx	xx	n.d.
Analcime	x	xx	xx	xx	x	x	x	x	x	x	x	n.d.
Pyrophyllite	x	xx	x	xx	–	–	x	x	x	x	x	n.d.
Anthophyllite	x	–	–	–	–	–	–	–	–	–	–	–
Chromite	v	–	–	–	–	v	–	–	–	–	–	n.d.
Goethite	v	v	x	v	–	v	v	v	v	v	x	n.d.
Talc	x	xx	xx	x	–	v	v	v	v	x	x	n.d.
Micas	v	x	v	x	–	–	–	–	–	–	–	–
(b) Clay fraction												
Antigorite		v	1	1–2	v	1	1	1		2	2	
Chlorite		2	3	4	1	2–1	2	3–2		4–3	4	
Smectite		4	2	1	4	4–3	3–4	2–3		2	1	
Micas		–	–	–			v	v		v	v	

v, Traces; x, Low level; xx, Medium level; xxx, High level.
1 < 10%; 2 < 20%; 3 < 40%; 4 < 60%; 5 > 60%.
n.d. – not determined.

turbed soil has 89 μg g^{-1} of extractable Ni and 245 μg g^{-1} after ploughing.

Chromium is mostly present as chromite, and is a very resistant mineral (Mitchell 1964). As pointed out by Goldschmidt (1958), the mobilized chromium (less than 20% of total) becomes part of the composition of chlorite and smectite, which is shown by the Cr concentration in the clay (crystalline) fraction (crystalline clay, between 2000 and 4000 μg g^{-1} clay, amorphous iron oxides in clay fraction 200 to 400 μg g^{-1}); thus it remains unavailable for plants. This is also confirmed by the very low concentration of chromium in plant tissues, even in the roots (see Table 9).

Copper occurs in ferromagnesium minerals (Mitchell 1964) (olivine, enstatite and anthophyllite) as a replacement of Fe^{++}, creating, like nickel, weak spots in crystal lattices. However it is not so easily leached, and can be found in the topsoil, especially in the case of deep soils with a high litter accumulation. Copper leached from the sand is retained in the clay fraction, both in crystal lattices of smectite and chlorite, and in amorphous iron oxides, and associates with organic matter in the upper horizons (soil no. 2: total Cu 25 μg g^{-1} in rock, 2.8 μg g^{-1} extractable Al horizon; soil no. 3: total Cu 3 μg g^{-1} in rock,

1.6 μg g^{-1} extractable in 0 horizon, 0.1 μg g^{-1} in B horizon – Table 1). So, it seems that, with soil evolution, the availability of this element generally increases (Menezes de Sequeira 1969).

Organic matter associates preferentially with copper, forming complexes (Mellor & Maley 1947). So, there is an accumulation of this element in the organic phase as well as in the amorphous iron oxides. As copper displaces nickel, the latter is pushed into the lattice of clays, and to the inactive iron oxides. Nickel is normally more or less 1/3 of chelated copper (McLean 1966).

Zinc occurs also in ferromagnesian minerals, probably replacing Fe^{++}, but it is more stable than copper and biological accumulation has been observed especially in soils with great litter deposition (Tables 1 and 2). Zinc as it is leached out of sand, becomes part of the crystal lattice of chlorite and smectite which contains about ten times as much zinc as the amorphous iron oxides (for example at soil no. 3 – C_1 horizon clay – 9% amorphous iron contains 150 μg g^{-1} ZnO; and 91% chlorite and smectite contains 144 μg g^{-1}). The lower stability of zinc organic complexes as compared with copper, nickel and cobalt complexes (Mellor & Maley 1947) probably explains the low level of zinc in organic matter (except in litter and in the surface horizons) and its higher proportion in the clay crystal lattice.

Cobalt occurs also in ferromagnesian minerals, possibly replacing Fe^{2+} and, in the same way as copper and nickel, determining weak spots in the mineral lattice. It also accumulates in the clay fraction, mostly associated with Fe in amorphous oxides (for example at soil no. 3 the C_2 horizon clay 3.7% amorphous iron oxide contains 666 μg g^{-1} CoO and 36% chlorite and smectite contain only 135 μg g^{-1}).

Flora

Only 409 species and subspecies (28.7% of all the taxa except hybrids and formae indicated by Rozeira in 1944 for the main territory of the province of Trás-os-Montes e Alto Douro) have been found growing on the ultramafic soils of north-east Portugal. Besides its poorness, this serpentine flora exhibits many of the peculiarities already recorded in other serpentine areas:

1. The high number of taxa considered as serpentinophytes and serpentinicolous relics as defined by Pichi-Sermolli in 1948 (Table 5) concurs with those found in the unglaciated "old" serpentine floras of Central and Southern Europe.
2. The occurrence of some cases of disjunction.
3. The presence of both acidicolous and basicolous plants of Mediterranean pasture communities of the *Therocistetea guttatae* and *Thero-Brachypodietea* classes respectively. However, the calcicoles found on limestone in the vicinity do not grow within the serpentinic areas.
4. The xerophytic character indicated by the high number of Mediterranean taxa and the existence of a high percentage of therophytes (Table 6).
5. The typical (Raunkiaer 1934) Mediterranean biological spectrum (Table 6).
6. The relatively high representation of some families, such as the *Caryophyllaceae*, *Leguminosae*, *Compositae*, *Gramineae* and of ferns, in contrast with the peculiar behaviour of *Ericaceae* and *Pinus*, which are very scarce or absent.
7. Many families and genera seem to be relatively less represented on the serpentinic sites than in the province as a whole (Table 7).
8. The number of endemics found in the serpentine areas of north-east Portugal is rather high. Some of them are vicariads of taxa found on nearly non-ultramafic soils such as: *Dianthus marizii* for *Dianthus laricifolius*, *Armeria eriophylla*, *A. langei* and *A. daveaui* for *A. transmontana*; or in rather distant and higher granitic mountains, such as *Arenaria tetraquetra* ssp. *fontiqueri* for *A. tetraquetra* ssp. *querioides*, *Jasione crispa* ssp. *serpentinica* for *J. crispa* ssp. *crispa* and *Plantago radicata* var. *radicata* for *P. radicata* var. *monticola*.
9. The occurrence of a high number of more or less combined serpentinomorphoses (Table 8) including small-leaved forms of the sclerophyll trees, glaucescence frequently occurring in grasses, and nanism in *Trifolium* and *Allium* species.

Serpentinophytes occur more frequently in a number of genera (Pichi Sermolli 1948, Pinto da Silva 1970). The following genera, in Europe, have more than one serpentinophyte (in brackets the actual number of serpentinophytes and the corresponding Portuguese taxa):

Alyssum (8, *A. pintodasilvae*), *Anthyllis* (3, *A. sampaiana*), *Arenaria* (5, *A. tetraquetra* spp. *fontiqueri*), *Armeria* (8, *A. eriophylla*, *A. langei*, *A. daveaui*), *Asplenium* (2, *A. cuneifolium*), *Centaurea* (4), *Cerastium* (8), *Cytisus* (2), *Dianthus* (6, *D. marizii*), *Euphorbia* (4), *Festuca* (4, *F. brigantina*), *Fumana* (2), *Ionopsidium* (2, *I. abulense* fo.), *Koeleria* (3, *K. crassipes*), *Minuartia* (3), *Onosma* (3), *Plantago* (3, *P. radicata* var. *radicata*), *Potentilla* (6), *Scorzonera* and *Podospermum* (4, *S. hispanica* var. *asphodeloides*, *P. tenuifolium*), *Sempervivum* (3), *Seseli* (3, *S. peixotianum*), *Silene* (incl. *Melandrium*) (8, *S. legionensis* var.), *Stachys* (3), *Stipa* (3), *Thymus* (3), *Veronica* (2).

Of these twenty-eight genera, fourteen are present on Portuguese ultramafic soils. Of the 105 European serpentinophytes included in those genera, sixteen (more than 15%) belong to the Portuguese serpentine flora (Pinto da Silva 1970).

Glaucescence (and purpurescence) with 59%, followed by plagiotropism (48%), stenophylly (46%) and nanism (43%) are the main types of serpentinomorphosis. Perennial macrorhizy, then

Table 5. Serpentinophytes and serpentinicole relics of north-east Portugal

Cheilanthes marantae (L.) Domin
Pteridium aquilinum (L.) Kuhn fo. *congestum* Pinto da Silva
Asplenium cuneifolium Viv. (4n)
Quercus rotundifolia Lam. A small leaved forma.
Quercus suber L. A small leaved forma.
Bucephalophora aculeata (L.) Pau ssp. *hispanica* (Steinh.) Löve & Kapoor, fo. *plagiotropica* Pinto da Silva
Arenaria tetraquetra L. ssp. *fontiqueri* Pinto da Silva
Spergularia purpurea (Pers.) G. Don fo. *congesta* Pinto da Silva
Agrostemma githago L. fo. *nana* (Hartman) Lundström
Silene legionensis Lag. var.?
Dianthus marizii (Samp.) Samp.
Alyssum pintodasilvae Dudley (*A. serphyllifolium* Desf. ssp. *lusitanicum* Dudley & Pinto da Silva)
Ionopsidium abulense (Pau) Rothm. fo.
Iberis linifolia Loefl. fo. *serpentinicola* Pinto da Silva
Reseda virgata Bss. & Reut. fo.
Sesamoides canescens (L.) O. Kze. fo.
Umbilicus rupestris (Salisb.) Dandy fo. *violaceus* Pinto da Silva
Genista hystrix Lange var. *villosa* Lange
Trifolium bocconei Savi fo.
Trifolium cherleri L. cf. var. *perpusillum* Briq.
Trifolium strictum L. var. *minus* (Rouy) Asch. & Graebn.
Anthyllis sampaiana Rothm.
Lotus tenuis Waldst. & Kit. ex Willd. var. *serpentinicus* Pinto da Silva
Astragalus macrorhizus Cav.
Eryngium tenue Lam. fo. *pumilum* Pinto da Silva
Seseli peixotianum Samp.
Armeria eriopylla Wk. ex P. Cout.
Armeria langei Bss. ex Lange
Armeria daveaui (P. Cout.) Pinto da Silva
Convolvulus arvensis L. cf. fo. *parviflorus* Lange
Linaria aeruginea (Gouan) Loscos & Pardo var. *simplex* Pinto da Silva
Plantago radicata Hoffgg. & Lk. var. *radicata*
Crucianella angustifolia L. fo. *plagiotropica* Pinto da Silva
Jasione crispa (Pourr.) Samp. ssp. *serpentinica* Pinto da Silva
Santolina semidentata Hoffgg. & Lk.
Tragopogon crocifolius L. fo. *serpentinicola* Pinto da Silva
Scorzonera hispanica L. var. *asphodeloides* (Wallr.) Samp.
Podospermum tenuifolium Hoffgg. & Lk.
Phleum bertolonii DC. fo.
Agrostis castellana Bss. & Reut. fo.
Molineria laevis (Brot.) Hack. fo. *violacea* Pinto da Silva
Periballia involucrata (Cav.) Janka fo.
Trisetaria ovata (Cav.) Paunero fo.
Gaudinia fragilis (L.) P. Beauv. fo. *violacea* Pinto da Silva
Koeleria crassipes Lange
Dactylis glomerata L. spp. *hispanica* (Roth) Nym. var. *microstachya* (Webb) P. Cout. fo. *glauca* Pinto da Silva and fo. *violacea* Pinto da Silva
Festuca brigantina (Mgf-Dgb.) Mgf-Dbg.
Ctenopsis delicatula (Lag.) Paunero fo. *quinqueflora* Pinto da Silva
Micropyrum tenellum (L.) Lk. var. *tenellum* fo.
Phleum phleoides (L.) Karsten fo.
Elytrigia repens (L.) Nevski var. *arvensis* (Rchb.) fo.
Aegilops geniculata Roth var. *hirsuta* (Eig) fo. *nana* (Pinto da Silva) Pinto da Silva
Taeniatherum caput-medusae (L.) Nevski ssp. *crinitum* (Schreb.) Pinto da Silva var. *serpentinicola* Pinto da Silva
Hordeum hystrix Roth fo. *decumbens* Pinto da Silva
Carex muricata L. ssp. *lamprocarpa* Čelak fo. *arcuata* Pinto da Silva
Allium gaditanum P. Lara fo. *exiguum* Pinto da silva
Allium sphaerocephalon L. var. *pallidum* Pinto da Silva
Allium vineale L. fo. *minus* Pinto da Silva
Scilla autumnalis L. var. *deflexo-scaposa* Pinto da Silva & Q. Pinto da Silva

Table 6. Comparison of the biological spectrum of the serpentine flora of north-east Portugal with that of serpentine flora of Upper Tiber valley, and that of non-serpentine soils, both in Argentario (Italy) (%)

Territories	Total of species	Th	Cr	H	Ch	Ph
Ultramafic soils of north-east Portugal	409	41	10	33	8	8
Argentario (Raunkiaer 1934)	866	42	11	29	6	12
Ultramafic soils of the Upper Tiber Valley	405	26	14	40	9	11

Th = Therophytes, Cr = Cryptophytes (mainly Geophytes and Helophytes), H = Hemicryptophytes, Ch = Chamaephytes, Ph = Phanerophytes.

excluding therophytes, cryptophytes and Pteridophytes (30) will reach 54%. The serpentinomorphoses observed in the flora may reflect an adaptation to drought, but seem mainly be caused by soil chemical factors. Congeneric species of *Armeria*, *Quercus* and *Trifolium*, with distinct ranges are similarly affected by the same serpentinomorphosis (Table 8).

Table 9 shows the Mediterranean character of the serpentine flora of north-east Portugal, though the climate can be classified (according to Thornthwaite) as very humid to dry subhumid. The Mediterranean floral element, if local endemics and cosmopolitan and widespread species were excluded, will attain 66%, and rise to more than 73% if, because of its Mediterranean character, the Hispanic-Mauritanian species and the Hispanic endemics are included in this element. For the counties of Vinhais, Bragança and Ma-

cedo de Cavaleiros, in which the ultramafic rocks and soils occur, Rozeira (1944) indicates more than 52% of the taxa as belonging to the Centro-European element and thus, the serpentine condition reverses the phytogeographic spectrum of the flora.

Plant behaviour and chemistry

For a better characterization of serpentinic habitat the behaviour and chemistry of some serpentinophytes are described. The selected taxa include some serpentine endemics, including a Ni hyperaccumulator, and some ubiquists.

Asplenium cuneifolium Viv. (4 n) is a serpentinophyte, intolerant of direct sunshine, that in north-east Portugal grows typically in shaded crevices of 1 to 7 cm width with a northern exposure, similar to its habitat in Italy (Pichi-Serm-

Table 7. Number and percentage of species of some families represented in the ultramafic areas of north-east Portugal and in continental Portugal as a whole

Families	Portugal (P. Coutinho 1920)		Trás-os-Montes e Alto Douro (Rozeira 1944)		Ultramafic areas of north-eastern Portugal	
	No.	%	No.	%	No.	%
Polypodiaceae (s. lato)	28	1.0	16	1.1[1]	11	2.7
Caryophyllaceae	137	5.0	64	4.4	32	7.8
Ranunculaceae	61	2.2	36	2.5	6	1.5
Cruciferae	116	4.3	55	3.8	12	2.9
Rosaceae	78	2.8	52	3.6	9	2.2
Leguminosae	281	10.4	123	8.6	48	11.7
Umbelliferae	118	4.3	52	3.6	12	2.9
Ericaceae	15	0.6	10	0.7	3	0.7
Labiatae	100	3.7	40	2.7	10	2.4
Scrophulariaceae	103	3.8	56	3.9	15	3.7
Compositae	287	10.6	140	9.8	52	12.7
Gramineae	217	8.0	125	8.8	68	16.6
Cyperaceae	74	2.7	26	1.8	11	2.7
Liliaceae	83	3.0	42	2.9	15	3.7
Other families	998	36.9	585	41.1	105	25.6
Totals	2696	99.3	1422	99.3	409	99.8

[1]This percentage would attain 1.3 if *Asplenium cuneifolium* Viv. and *Cheilantes marantae* (L.) Domin were added to the list of the flora of the province Trás-os-Montes e Alto Douro. For continental Portugal its value would then be 1.1.

Table 8. Incidence of serpentinomorphoses in the serpentine plants of north-east Portugal (Pinto da Silva 1970, revised)

Taxa	Glaucescence (incl. purpurescence)	Stenophylly (incl. microphylly)	Nanism	Plagiotropism	Macrorhizy	Glabrescence
Jasione crispa spp. *serpentinica*	×	×	×	×	×	×
Podospermum tenuifolium	×	×	×	×	×	×
Astragalus macrorhizus	×	×	×	×	×	
Plantago radicata var. *radicata*	×	×	×	×	×	
Dianthus marizii	×	×	×	×	×	
Armeria eriophylla	×	×	×	×	×	
Armeria langei	×	×	×		×	
Seseli peixotianum	×	×		×	×	×
Reseda virgata fo.	×	×		×	×	×
Alyssum pintodasilvae	×	×		×	×	
Santolina semidentata	×	×		×	×	
Koeleria crassipes	×	×	×	×		
Festuca brigantina	×	×		×		×
Arenaria tetraquetra spp. *fontiqueri*	×	×		×		
Linaria aeruginea var. *simplex*	×	×	×			
Quercus rotundifolia (a small leaved form)	×	×	×			
Gaudinia fragilis fo. *violacea*	×	×	×			
Aegilops geniculata var. *hirsutas* fo. *nana*	×		×	×		
Scorzonera hispanica var. *asphodeloides*	×	×			×	
Ionopsidium abulense fo.		×	×	×		
Genista hystrix var. *villosa*		×		×		×
Asplenium cuneifolium Viv.	×			×		
Umbilicus rupestris fo. *violaceus*	×		×			
Allium sphaerocephalon var. *pallidum*	×		×			
Taeniatherum caput-medusae var. *serpentinicola*	×	×				
Armeria daveaui	×				×	
Quercus suber (a small leaved form)		×	×			
Spergularia purpurea fo. *congesta*			×	×		
Sesamoides canescens fo.		×		×		
Silene legionensis var.?				×	×	
Phleum bertolonii fo.	×					×
Agrostis castellana fo.	×					
Molineria laevis fo. *violacea*	×					
Periballia involucrata fo.	×					
Elytrigia repens fo.	×					
Phleum phleoides fo.	×					
Trisetaria ovata fo.	×					
Dactylis glomerata fo. *glauca* & fo. *violacea*	×					
Micropyrum tenellum fo.	×					
Ctenopsis delicatula fo. *quinqueflora*	×					
Tragopogon crocifolius fo. *serpentinicola*	×					
Iberis linifolia fo. *serpentinicola*		×				
Lotus tenuis var. *serpentinicus*		×				
Convolvulus arvensis cf. fo. *parviflorus*		×				
Agrostemma githago fo. *nana*			×			
Trifolium bocconei fo.			×			
Trifolium cherleri cf. var. *perpusillum*			×			
Trifolium strictum var. *minus*			×			
Erungium tenue fo. *pumilum*			×			
Allium gaditanum fo. *exiguum*			×			
Allium vineale fo. *minus*			×			
Pteridium aquilinum fo. *congestum*				×		
Anthyllis sampaiana				×		
Hordeum hystrix fo. *decumbens*				×		
Carex muricata fo. *arcuata*				×		
Bucephalaphora aculeata fo. *plagiotropica*				×		
Scilla autumanalis var. *deflexo-scaposa*				×		
Crucinnella augustifolia fo. *plagiotropica*				×		
Cheilanthes marantae						×
Totals	35	26	25	27	14	8
Percentages	59	44	42	46	24	14

Table 9. Phytogeographic spectrum of the serpentine flora of north-east Portugal and a comparison with that of the Upper Tiber Valley

Phytogeographical elements and subelements	Species and subspecies %	
	North-east Portugal	Upper Tiber Valley (Pichi-Sermolli 1948)
Eurosiberian-boreal-american	21	38
Mediterranean	43	42
(Main) endemic taxa	22	3
Pluriregional and cosmopolitan species	14	17

olli 1948). It is a character species of the *Umbilico-o-Asplenietum cuneifolii* pioneer association (Photo 9).

The fine earth of the crevices in ultramafic rocks has an organic matter content of 5%, a pH from 6.5 to 6.9, with 3 to 20 me $100 \, g^{-1}$ exchangeable Mg, 1 to 3 me $100 \, g^{-1}$ exchangeable Ca, 0.1 to 0.2 me $100 \, g^{-1}$ exchangeable K, very low soluble P (about $1 \, \mu g \, g^{-1}$ P by Bray I method) and extractable Ni $> 100 \, \mu g \, g^{-1}$ (EDTA). The analysis of plant tissues (Table 10) shows a low concentration of toxic elements Mg, Co, Cr, Fe, Mn and especially Ni in the leaves, compared with the roots. This indicates a retarded translocation of these elements from roots to shoots. The Mg/Ca quotient, which is 12 in the soil (<2 mm), is reduced to 2.5 in the leaves, and while there are $140 \, \mu g \, g^{-1}$ extractable Ni in the soil, the roots have $930 \, \mu g \, g^{-1}$, and only $30 \, \mu g \, g^{-1}$ can be found in the leaves.

Arenaria tetraquetra L. ssp. *fontiqueri* Pinto da Silva is a characteristic and frequent endemic plant of ultramafic lithosols and gravelly and stony humic litholic soils, with deep roots descending to the weathered rock (Fig. 5). This typical serpentinophyte, with a high fidelity to the serpentinic habitat, shows high serpentinomorphoses (glaucescense, stenophylly and plagiotropism). It has a deep root system and a relatively high absorption of Ca, Fe and Zn, while it is also able to resist an adverse Mg/Ca quotient.

Growing in a soil with high Mg/Ca ratio, high Ni ($1000 \, \mu g \, g^{-1}$ total NiO) and Cr levels (4000 to 8000 $\mu g \, g^{-1}$ total Cr_2O_3) (Table 1 soil no. 2), the concentration of Mg, Ni and Cr in the roots and shoots (very similar) is clearly lower than that of the soil (Table 10).

Reseda virgata Bss. & Reut. fo. This plant, with a root system attaining the weathered rock (Fig. 6), is considered a preferent serpentinophyte (Pinto da Silva 1970) with evident serpentinomorphoses (glaucescence, stenophylly, plagiotropism, macrorhizy and glabrescence). It is probably a distinct taxon. Growing in shallow serpentinic soils with high levels of heavy metals (Table 1, soil no. 2) it appears to be able to restrict absorption (Table 10).

Anthyllis sampaiana Rothm. This endemic species is a typical serpentinophyte, in spite of exhibiting only plagiotropism and a deep root system that penetrates between the plates of weathered ultramafic rock (Fig. 7). Among all the analyzed serpentinophytes this species has the highest Co and Mn concentrations in the shoots, though the whole plant has a medium level of Ni, and the flowers alone have low Ni, Co and Mn contents (Table 10).

Astragalus macrorhizus Cav. Although this taxon in Spain and Algeria grows in non-serpentinic soils (Willkomm in Willkomm & Lange 1877) in north-east Portugal it shows a high serpentinic fidelity, growing only in stony and gravelly shallow litholic serpentinic soils (Table 1, soil 1). It shows high serpentinomorphoses especially macrorhizy (Fig. 8, Photo 3), besides plagiotropism and stenophylly. It shows good adaptation to prolonged drought conditions with roots penetrating more than 1 m into the rock crevices at the bottom of the soil profile. Belonging to a large genus, the Portuguese plant may prove to be a different taxon, of specific or subspecific level. Though it has low rates of absorption for Ni, Cr and Co, there is a high translocation of Cr from the roots to the shoots (Table 10).

Armeria eriophylla Wk. ex P. Cout. This plant, with a high fidelity to serpentinic habitats is well represented in the *Armerio-Arenarietum fontiqueri* association, especially in shallow, stony, gravelly lithosol or humic litholic soils with good exposure to light. It shows strong serpentinomorphoses – macrorhizy (Fig. 9), stenophylly, glaucescence, plagiotropism and nanism – all pointing to a good adaptation to summer drought. The analysis of plant tissues (Table 10) shows low absorption of toxic elements, especially Ni, Cr and Mg, but a high Zn absorption. This plant is not found in the surrounding *Genisto-Quercetum rotundifoliae* and in the *Thaeniathero-Alyssetum*

Table 10. Elemental concentration of plant parts of some serpentinophytes (dry matter)

	Ni $\mu g\,g^{-1}$	Ca %	Mg %	Mg/Ca me/me	Cr $\mu g\,g^{-1}$	Cu $\mu g\,g^{-1}$	Zn $\mu g\,g^{-1}$	Mn $\mu g\,g^{-1}$	Fe $\mu g\,g^{-1}$	Co $\mu g\,g^{-1}$
Taxa with low absorption of Ni										
*Armeria eriophylla**										
Roots	31	0.26	0.43	2.8	2.4	4.0	55	24	150	1.5
Stems and leaves	38	0.26	0.55	3.5	2.5	4.3	60	31	490	0.5
Flowers	24	0.16	0.41	4.3	3.8	5.2	39	31	350	2.0
*Astragalus macrorhizus**										
Roots	32	0.21	1.00	8.0	1.8	2.5	4.4	15	235	2
Stems	66	0.29	0.74	4.3	2.7	2.5	4.0	23	290	2
Leaves	43	0.93	1.72	3.0	4.2	4.2	8.8	52	450	1
Seeds	18	0.21	0.35	2.8	0.5	2.7	10.9	15	70	1
Plantago radicata var. *radicata***										
Roots	27	0.16	0.17	1.8	1.0	8.5	22.9	18	140	1.5
Leaves	38	0.58	0.57	1.7	2.5	10.0	23.5	47	348	traces
Flowers	41	0.43	0.62	2.3	4.6	12.3	25.9	38	742	traces
Reseda virgata fo.*										
Roots	32	0.11	0.19	2.8	3.0	3.0	7.0	18	265	2.0
Stems	14	0.15	0.55	6.2	0.6	4.0	8.0	16	50	0.5
Leaves	55	0.65	1.67	4.3	1.9	5.5	25.0	88	234	traces
Taxa with medium levels of Ni										
Arenaria tetraquetra ssp. *fontiqueri**										
Roots	78	0.62	0.78	2.2	3.7	6.3	29.6	43	347	2.0
Tops	66	0.65	0.60	1.5	4.3	3.1	30.0	37	750	2.5
Jasione crispa ssp. *serpentinica**										
Roots	52	0.22	0.36	2.6	6.1	4.2	23.9	23	543	4.6
Shoots	81	0.50	1.24	4.2	7.4	8.3	40.5	71	1153	1.2
*Santolina semidentata**										
Roots	89	0.17	0.33	2.2	9.5	8.5	12.4	52	1200	6.5
Stems	26	0.20	0.26	2.2	3.2	7.5	17.5	19	310	1.5
Leaves	69	0.63	0.67	1.8	3.8	22.5	51.5	51	380	2
Taxa with low translocation of Ni										
*Anthyllis sampaiana**										
Roots	104	0.34	0.50	2.5	4.9	6.2	13.7	261	689	13.8
Shoots	90	0.54	1.74	5.3	2.2	3.9	30.0	560	320	19.0
Flowers	50	0.65	0.98	2.5	2.0	3.7	24.5	146	260	5.5
Genista hystrix var. *villosa**										
Roots	160	0.16	0.48	5.0	4.0	9.5	26.8	66	380	2.0
Shoots	17	0.16	0.33	3.5	1.7	5.0	17.2	150	275	0.5
*Festuca brigantina**										
Roots	129	0.32	0.31	1.7	7.7	5.2	70.2	91	820	8.2
Shoots	24	0.19	0.17	1.5	2.3	5.3	22.1	32	200	1.0
Agrostis castellana fo.*										
Roots	66	0.12	0.11	1.5	1.8	3.1	16.0	34	225	1.5
Shoots	19	0.19	0.31	2.6	1.9	2.6	11.4	32	211	1.0
Asplenium cuneifolium (4 n)*										
Roots	930	0.54	1.36	4.2	22.8	7.8	13.8	188	4000	16.9
Rhizome and basal part of petioles	152	0.35	1.14	5.5	13.7	4.0	14.0	67	2000	6.5
Leaves (laminae)	30	0.40	0.60	2.5	1.1	4.3	15.9	18	146	traces
Ni hyperaccumulator										
*Alyssum pintodasilvae**										
Roots	800	0.30	0.13	0.7	0.2	1.7	17.0	13	49	traces
Stems	2700	0.72	0.27	0.7	1.3	2.5	41.5	23	260	2.0
Leaves	8650	2.79	0.83	0.5	2.2	3.1	40.8	67	315	1.5

*Plants growing on soil no. 2 (Table 1).
**Plants growing on soil no. 1 (Table 1).

188

Photo 9. Asplenium cuneifolium Viv. in the *Umbilico-Asplenietum cuneifolii* association.

pintodasilvae associations, which suggests a very low competitive character.

Plantago radicata Hoffgg. & Lk. var. *radicata*. Though normally present in serpentinic soils and with a high level of serpentinomorphosis this species grows also outside the serpentinic areas. Normally it is frequent in the *Armerio-Arenarietum fontiqueri*, *Cisto-Genistetum hystricis*, *Genisto-Quercetum rotundifoliae* and *Taeniathero-Alyssetum pintodasilvae* associations, suggesting a strong competitive character and also a good adaptation to the serpentinic habitats. The serpentinomorphoses of this plant are very similar to those of the former species (root system represented in Fig. 10); data show that it is able to resist toxic levels of Ni, Co and Mg by slowing down their absorption, but there is a relative high level of Cr in the flowers (Table 10).

Jasione crispa (Pourr.) Samp. ssp. *serpentinica* Pinto da Silva has a high fidelity to serpentinized habitats. It occurs in all typical serpentinic associations, thereby showing a high competitive character. Its serpentinomorphoses (glaucescence, stenophylly, nanism, plagiotropism, glaucescence and macrorhizy; Fig. 11), reveal its good adaptation to the adverse serpentinic habitat. It has a high level of Ni and the highest level of Cr found in plant shoots, besides the highest Fe absorption and translocation to the shoots (Table 10).

Santolina semidentata Hoffgg. & Lk. This species is only a preferent serpentinophyte common in most serpentinized soils, and it grows also outside the serpentinic habitat (Pinto da Silva 1970). It has distinct serpentinomorphoses (glaucescence, stenophylly, plagiotropism and macrorhizy). Its deep roots penetrate in the soil profile down to the crevices and even between the plates of the weathered rock (Fig. 12). It has high levels of Ni and Cr in the shoots (Table 10).

Agrostis castellana Bss. & Reut. A form of this polymorphic and common species occurs in all the serpentinic associations. It has a very deep root system (Fig. 13) and violet colour which may be due to adaptation to the adverse serpentinic habitat. It has a very strong competitive character. The elemental analysis of plant tissues shows that it has a high capacity to retard Ni translocation from roots to shoots (Table 10). Like the hyperaccumulator *Alyssum*, this perennial grass is able to survive tillage operations.

Festuca brigantina (Mgf-Dbg.) Mgf-Dbg. This endemic grass of north-east Portugal is one of the most interesting typical serpentinophytes. It is present only in shallow soils in the *Armerio-Arenarietum fontiqueri* association. It shows nanism and glaucescence combined with a typical violet colour, as well as plagiotropism. It is one of the few plants growing in serpentinic soils that has a superficial root system (Fig. 14) and shows no

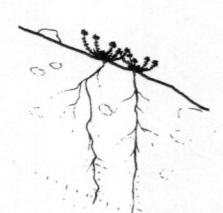

Fig. 5. *Arenaria tetraquetra* ssp. *fontiqueri* root system (profile 2).

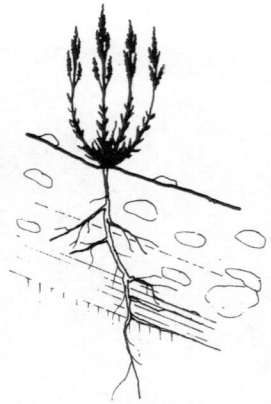

Fig 6. *Reseda virgata* fo. root system (profile 2).

Fig. 7. *Anthyllis sampaiana* root system (profile 2).

Fig. 8. *Astragalus macrorhizus* root system (profile 2).

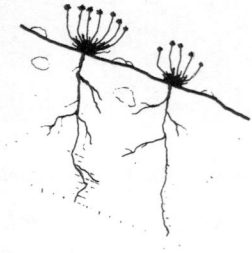

Fig. 9. *Armeria eriophylla* root system (profile 2).

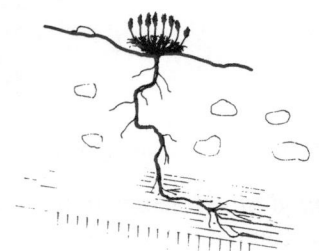

Fig. 10. *Plantago radicata* var. *radicata* root system (profile 2).

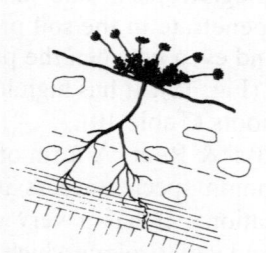

Fig. 11. *Jasione crispa* ssp. *serpentinica* root system (profile 2).

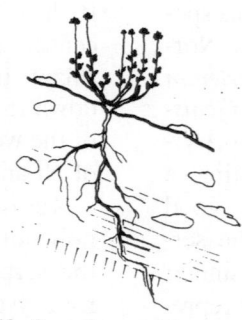

Fig. 12. *Santolina semidentata* root system (profile 2).

Fig. 13. *Agrostis castellana* fo. root system (profile 2).

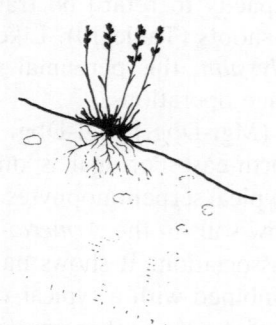

Fig. 14. *Festuca brigantina* root system (profile 2).

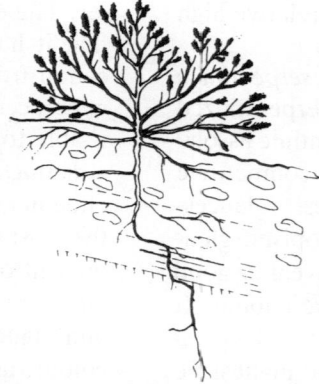

Fig. 15. *Genista hystrix* var. *villosa* root system (profile 2).

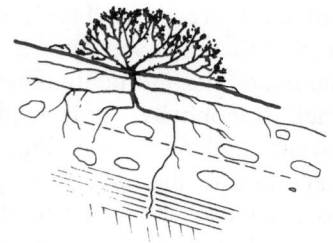

Fig. 16. *Alyssum pintodasilvae* root system (profile 2).

190

competitive vigour. The analysis of plant tissues (Table 10) indicates that there is a strong mechanism to inhibit the translocation of heavy metals from roots to shoots, especially Ni, Cr and Co. The very low level of Mg in the shoots is remarkable.

Genista hystrix Lge. var. *villosa* Lge. is an ubiquist plant growing well in serpentinic habitats. The roots deeply penetrate the soil profile but explore also the upper part of the soil (Fig. 15). It shows also a high capacity to retard toxic heavy metal translocation from roots to shoots (Table 10) but the Mn and Fe levels are not reduced.

Alyssum pintodasilvae Dudley is the only Ni hyperaccumulator in north-east Portugal and one of the 14 hyperaccumulating European species (all in the Section Odontarrhena) in the 168 species of *Alyssum* L. recognised (Brooks *et al.* 1979). This interesting endemic species occurs in all the typical serpentinic associations, but especially in the *Taeniathero-Alyssetum pintodasilvae* that covers the soils which have been ploughed and left uncropped (fallows). These communities are very similar to the '*Alyssum-Fluren*', with other *Alyssum* spp., pointed out by Krause, Ludwieg & Seidel (1963) for the Balkans. The roots are thin and not very long, but very abundant and ramified, occupying mainly the upper horizon of the soil profile (Fig. 16). This cruciferous species has a high capacity to translocate all elements from roots to shoots, especially Ni, Ca, Cr and Co (Table 10). It is the only Portuguese plant that has more than $1000 \, \mu g \, g^{-1}$ Ni (hyperaccumulation), and more than 1% Ca and $10 \, \mu g \, g^{-1}$ Co in the dry matter of the leaves, and needs a high level of heavy metals in the soil for its good development (Brooks *et al.* 1981). In contrast to all other serpentinophytes, it shows the best development in soils disturbed by ploughing, that is soils where there is a mixture of weathered and non-weathered material with organic matter. As a result it probably increases Ni availability and at the same time this hyperaccumulating plant also increases the availability of Ni in the soil. For example in one place, the soil had $89 \, \mu g \, g^{-1}$ extractable Ni in the A horizon, while in the rhizosphere of *Alyssum* it was $110 \, \mu g \, g^{-1}$, and after ploughing Ni increased to $245 \, \mu g \, g^{-1}$.

Vegetation

The main associations found in the vegetation of the serpentinized ultramafic soils of north-east Portugal are (Pinto da Silva 1970):

1. *Cheilanthetum marantae transmontanum*, a community of the dry warm crevices of ultramafic outcrops with southern exposure.
2. *Umbilico-Asplenietum cuneifolii*, a typical community of the crevices of ultramafic rocks with shading and cold exposure (Photo 9).
3. *Armerio-Arenarietum fontiqueri*, the common pioneer community of lithosols and shallow humic litholic soils in serpentinized ultramafic areas (Photo 10).
4. The *serpentinicola* variant of the *Genisto-Quercetum rotundifoliae*, that is the regional climax vegetation under xerophytic conditions, which usually occupies the deep humic litholic soil and the Red Mediterranean Soil of serpentinized ultramafic rocks (Photo 11).
5. *Taeniathero-Alyssetum pintodasilvae*, the vegetation of fallows in serpentinic soils that are occasionally cultivated (Photo 12).
6. The *serpentinicum* variant of the scrubby *Cisto-Genistetum hystricis* typical of litholic soils of serpentinized areas under xerophytic conditions.

The succession of the serpentine vegetation of north-east Portugal is presented in Fig. 17.

Weathering of the rock outcrops alters the peculiar habitat of the crevice vegetation, allowing the establishment of a considerable number of aggressive species of the *Armerio-Arenarietum fontiqueri typicum*. However successional change is often very slow, both initially as well as from the *Armerio-Arenarietum fontiqueri* to the *Genisto-Quercetum rotundifoliae*. The reason is that, although serpentinized rocks are very easily weathered, the weathering process is practically a dissolution (very little of the parent material remaining) and shallow soils result. Thus the *Armerio-Arenarietum fontiqueri typicum* behaves as a rather stable community, especially on steep slopes where the foliated rock lies parallel to the surface.

Where there are favourable weathering conditions and no erosion, resulting in deeper soils, more competitive ubiquist shrubs appear. This leads to the development of the *genistetosum* and

Photo 10. *Armerio-Arenarietum fontiqueri* association by the profile no. 2 at Carrazeda (Bragança). Note the contrasting aspects of the foreground serpentinic habitat and the normal cultivated landscape on amphibolithic soil with large-leaved trees and rye fields.

santolinetosum subassociations of the *Armerio-Arenarietum fontiqueri*.

In the deepest soils (Humic Litholic Soils and Red Mediterranean Soils), with a higher water-holding capacity and a higher supply of nutrients, acidophile shrubs and evergreen oaks (*Quercus rotundifolia*) may grow, which exclude all the serpentinophytes as well as the Caryophyllaceae. The development of the *serpentinicola* variant of the climax forest association (*Genisto-Quercetum rotundifoliae*) is the result.

When the forest or shrubby vegetation cover is destroyed, the *Armerio-Arenarietum fontiqueri* can return; however, if destruction is caused by farming, the *Taeniathero-Alyssetum pintodasilvae* occurs and occupies the fallows. Whenever cultivation is abandoned, in deep non-eroded soils there is a slow but usually direct development from the *Taeniathero-Alyssetum pintodasilvae* to

the *Genisto-Quercetum rotundifoliae* var. *serpentinicola* facies *fruticosum* (occasionally through the *Cisto-Genistetum hystricis* var. *serpentinicum*). Somewhat later on, the facies *arboreum* of the climax association develops. On shallow gravelly eroded soils, the fallow association develops into the subassociations of the *Armerio-Arenarietum fontiqueri*.

The analysis of serpentine vegetation of northeast Portugal suggests that ultramafic soils induce also the serpentinomorphosis of the vegetation (sinserpentinomorphosis) expressed by characteristic communities growing on rock crevices, rocky-gravelly soils and fallow land. However in the vegetation of grain fields, meadows, scrub and in the climax woodland only variants of more widespread communities induced by serpentine can be observed, characterized only by a few serpentinophytes, and by nanism and stenophylly of trees and shrubs.

The parallelism between the *Armerio-Arenarietum fontiqueri* and the dolomiticole *Armerietum junceae* Braun-Blanquet, Roussine & Nègre (1952) can contribute to explain the ecology of the serpentine soils:

Armerietum junceae		—	*Armerio-Arenarietum fontiqueri*	
V^1	*Armeria girardii* (*A. juncea*)	—	III^1	*Armeria eryophylla*
V^{1-2}	*Arenaria tetraquerta* ssp. *capitata*	—	IV^{+2}	*Arenaria tetraquerta* ssp. *fontiqueri*
I^+	*Seseli glaucum*	—	III^+	*Seseli peixotianum*
II^+	*Alyssum serpyllifolium* ssp.	—	IV^{1-2}	*Alyssum pintodasilvae* (*A. serpyllifollium* ssp. *lusitanicum*)
IV^+	*Anthyllis vulneraria* ssp.	—	II^+	*Anthyllis sampaiana*
I^+	*Dianthus virgineus*	—	V^{+2}	*Dianthus marizii*
III^+	*Thymus serpyllum* ssp. *angustifolius*	—	II^+	*Thymus zygis*

Presence Degree: (I) rare (1 to 20% of the stands); (II) seldom present (20 to 40% of the stands); (III) often present (40 to 60% of the stands); (IV) mostly present (60 to 80% of the stands); (V) constantly present (80 to 100% of the stands). Suffixes (−) mean abundance and cover in the stands: (+) sparcely or very sparcely present; (1) plentiful but small cover value; (2) very numerous but cover below 1/20 of the area; (3) any number of individuals but cover between 1/4 to 1/2 of the area (Braun-Blanquet, J. 1932. *Plant sociology. The study of plant communities*. Translated, revised and edited by Fuller G.D. & Conard H.S., McGraw-Hill Book Company Inc. N.Y. and London.)

Photo 11. Genisto-Quercetum rotundifoliae var. *serpentinicola*, facies *arboreum* association near profile no. 3 at Monte de Morais.

Photo 12. Taeniathero-Alyssetum pintodasilvae association, near Bragança. Note the contrast with the landscape of non-serpentinic soils.

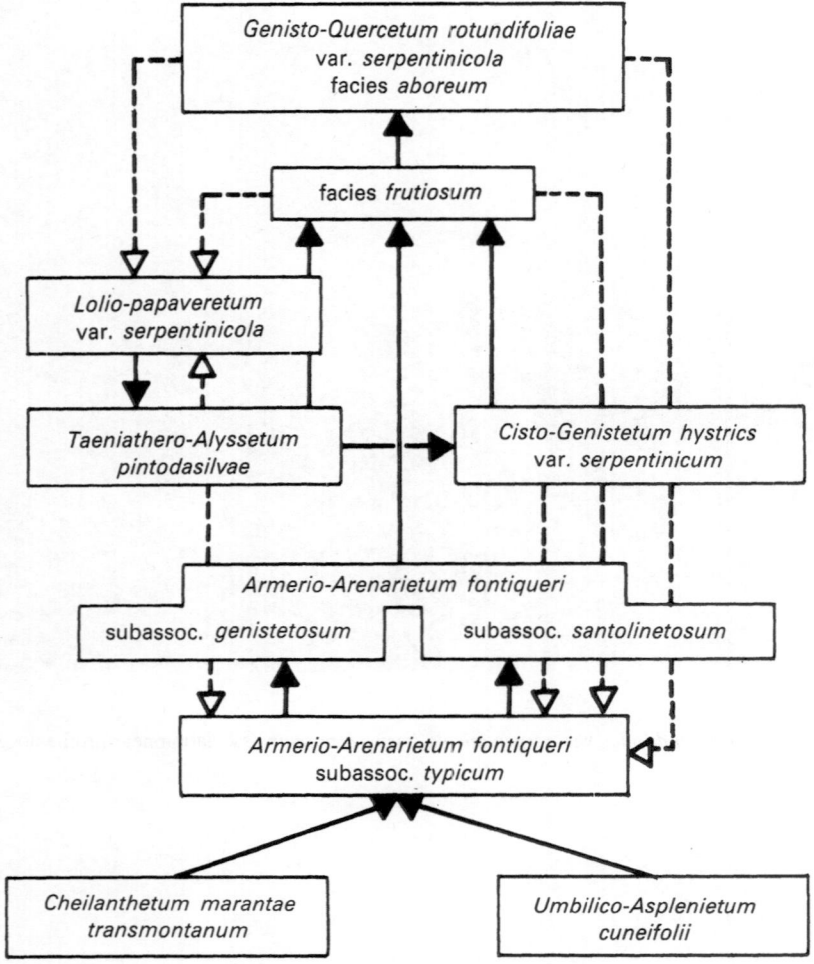

Fig. 17. Scheme of the succession of the serpentine vegetation in north-east Portugal (see the text). Solid lines and arrows mean succession towards the climax; broken lines express possible degradings. *Lolio-Papaverctum* var. *serpentinicola* is the vegetation of grain fields on serpentine soils.

Asperula cynanchica (III⁺, II⁺), *Allium sphaero-cephalon* (III⁺, I⁺), *Cerastium pumilum* (I⁺,III⁺), *Trifolium scabrum* (I⁺, I¹), *Helichry-sum stoechas* (V⁺, I⁺) and *Fumana procumbens* (III⁺, I⁺), which are characteristic species of the order or class, or merely high-presence companions of the *Armerietum junceae*, all occur in both associations.

The Mediterannean character and xeromorphism of the serpentine vegetation of north-east Portugal with a sclerophyll, small-leaved climax forest cannot be explained by the climate only, as it contrasts with the mesophilous climax vegetation of the non-ultramafic soils in the surrounding areas (Photo 11). This Mediterranean character, both of the flora and vegetation, is

probably a consequence of a high summer water deficit in the soils which have a low water holding capacity. In addition, high evaporation rates due to the sparse vegetation cover and dark soil colour, as well as conditions due to the soil forming process, and also to the soil toxicity and low soil nutrients have to be considered. The main toxic factors are the high Mg/Ca quotient and the high Ni concentration.

The unusual soil chemistry excludes widespread, strongly competitive species and allows the establishment of a peculiar serpentine flora and vegetation. Where on level sites with no erosion, deeper and more fertile soils develop (with higher water holding capacity, increased organic matter, and nutrients) resistant shrubs and trees may grow, and the serpentine character of the flora and vegetation is diminished, but the Mediterranean character remains. The high concentration of nickel remains during the soil forming process – a dissolution – in spite of soil evolution and helps to diminish its toxic effects.

Appendix I

Soil morphological description

Profile 1. Lithosol. Bragança, north of Samil, 750 m a.s.l. on a hillside with 10% slope, S.E. exposure, uncultivated soil, with vegetation of *Armerio-Arenarietum fontiqueri*, very stony, high risk of erosion, excessive external drainage and good internal drainage (Photo 3).

Horizon	Depth (cm)	
Al	0–2.5/5	Dark brown (10 YR 3/3) (dry), dark yellowish brown (10 YR 3/4) (moist), silt loam with many small stones and gravel, friable to very friable, soft non sticky, slightly plastic, moderate fine crumb structure, with many thin roots, humid. Gradual boundary to Al/C.
Al/C	2.5/5–40	Brown (10 YR 4/3) (dry), dark brown (10 YR 3/3) (moist), silt loam with many stones and gravel, firm non sticky, with many medium and thin roots between stones. Irregular and abrupt boundary to C/R.
C/R	>40	Greenish gray hard weathered serpentine rocks. Some fine earth between the stones; interlaced thin and medium roots between the contacting faces.

Profile 2. Humic Litholic Soil. Bragança, 1 km south of Carrazeda, 860 m a.s.l. on a hillside with 23% slope, Northern exposure. Uncultivated, with vegetation of *Armerio-Arenarietum fontiqueri*. High stoniness, high risk of erosion, excessive external drainage and good internal drainage (Photo 4).

Horizon	Depth (cm)	
A11	0–15/20	Brown (7.5 YR 5/4) (dry), dark brown (7.5 YR 3/2) (moist), loamy with many small stones and gravel, friable to very friable, soft, non sticky, slightly plastic, moderate medium crumb structure, with many medium and thin roots, humid. Gradual boundary to A12.
A12	15/20–20/40	Brown (10 YR 5/4) (dry), dark brown (10 YR 3/3) (moist) with greenish and yellow spots, loamy with many gravel and stones from weathered serpentinic rocks, friable, soft, non sticky, moderate medium crumb structure, with many medium and thin roots, humid. Irregular and abrupt boundary to C/R1.
C/R1	20/40–35/65	Greenish gray serpentine rock, foliated with plane faces parallel to the soil surface, interlaced thin and medium roots between the contacting faces. Some fine earth between the plates.

R2	35/65–?	Dark serpentine rock in hard blocks. In the fissures, among the blocks there are some roots decreasing in number with depth, but there are still a few at 70 cm.

Profile 3. Red Mediterranean soil from ultrabasic rock. Macedo de Cavaleiros at 3 km S.E. Monte de Morais, 630 m a.s.l., close to a hill top, S.E. exposure, 10% slope. Uncultivated soil with vegetation of *Genisto-Quercetum rotundifoliae*. High stoniness, no risk of erosion, good external and internal drainage (Photo 5).

Horizon	Depth (cm)	
O	−1–0	Litter from *Genista hystrix*, *Quercus rotundifolia* etc.
A1	0–20	Yellowish red (5 YR 4/6) (dry), dark red (2.5 YR 3/6) (moist), silt loam with many large serpentinized stones and gravel, friable to very friable, soft, non sticky, slightly plastic, moderate medium crumb structure, with many large medium and thin roots, humid. Gradual boundary to A/B.
A/B	20–50	Yellowish red (5 YR 4/8) (dry), dark reddish brown (5 YR 3/4) (moist) silt loam with many gravel, great quantity of voids, slightly loose, sticky, slightly plastic, blocklike composed by medium granular structure with large medium and thin roots, humid. Gradual boundary to B.
B	50–90	Dark reddish brown (5 YR 3/4) (dry), yellowish red (5 YR 4/6) (moist), clay loam with many gravel, firm, sticky, plastic, prismlike composed by medium blocklike structure, with slickenside, with some large medium and thin roots, humid. Gradual boundary to C1.
C1	90–150	Dark yellowish brown (10 YR 4/4) (dry), dark grayish brown (2.5 YR 4/2) (moist), with little greenish spots, clay loam, with many gravel, firm, sticky, plastic, blocklike structure, with many large and medium roots. Gradual boundary to C2.
C2	150–250	Yellowish brown (10 YR 5/4) (dry), dark yellowish brown (10 YR 3/4) (moist), with many little greenish spots, clay loam with many stones and gravel, very firm, sticky, non plastic, blocklike structure, with some roots, humid. Abrupt boundary to R1.
R	250–?	Greenish gray, very weathered serpentinized rock.

References

Braun-Blanquet, J., N. Roussine, & R. Nègre. 1952. *Les groupements végétaux de la France Méditerranéenne*. Centre Nat. Rech. Sci., Service de la Carte des Groupments Végétaux, Montpellier.

Brooks, R. R., R. S. Morrison, R. D. Reeves, T. R. Dudley & Y. Akman. 1979. Hyperaccumulation of nickel by *Alyssum* Linnaeus (Cruciferae). Proc. R. Soc. Lond. Serie B 203: 387–403.

Brooks, R. R., S. Shaws, & A. Asensi Marfil. 1981. Some observation on the ecology, metal uptake and nickel tolerance of *Alyssum serpyllifolium* subspecies from the Iberian peninsula. Vegetatio 45: 183–188.

Cotelo Neiva, J. M. 1948. Rochas e minérios da região de Bragança-Vinhais. Relat. Serv. Fom. Mineiro 14: 1–251.

Demoulin, V. 1966. L'Espagne méditerranéenne et continentale. (L'Ibérie sèche). Natura mosana 19 (2–3): 33–48.

Emberger, L. 1955. Une classification biogéographique des climats. Recl. Trav. Labs. Bot. Géol. Zool. Univ. Montpellier, sér. Bot. 7: 3–43.

Ferreira, M. R. Portugal Vasconcelos. 1965. Geologia e petrologia da região de Robordelo-Vinhais. Revta Fac. Ciênc. Univ. Coimbra 36: 1–287.

Goldschmidt, V. M. & A. Muir (eds.). 1958. *Geochemistry*. Clarendon Press, Oxford.

Krause, W., W. Ludwig, & F. Seidel. 1963. Zur Kenntnis der Flora und Vegetation auf Serpentinstandorten des Balkans.

6. Vegetationsstudien in der Umgebung von Mantoudi (Euböa). Bot. Jb. 82: 337–403.

Löve, A. 1963. Conclusion. In: *North Atlantic biota and their history*: 391–397. Pergamon Press, Oxford.

Löve, A. & D. Löve. 1964. The North Atlantic flora – its history and late evolution. Tenth Intern. Bot. Congr., Abstracts of Papers: 139–140.

Markgraf, F. 1932. Pflanzengeographie van Albanien. Bibl. Bot. 105: 29–89.

McLean, G. W. 1966. Retention and release of nickel by clays and soils. Ph.D. Thesis, Univ. Calif., Berkeley.

Mellor, D. P. & L. Maley. 1947. Stability constant of internal complexes. Nature, Lond. 159: 370.

Mendes, J. C. & M. L. Bettencourt. 1980. *O clima de Portugal*. Fascículo XXVI. Contribuição para o estudo climatológico de água no solo e classificação climática de Portugal Continental. Inst. Nacional de Meteorologia e Geofísica.

Menezes de Sequeira, E. 1969. Toxicity and movement of heavy metals in serpentinic soils. (North-Eastern Portugal). Agronomia lusit. 30: 115–154.

Menezes de Sequeira, E., A. M. Soares da Silva, & J. M. A. Vieira e Silva. (in press). Génese dos solos ultrabásicos transmontanos. Pedologia, Oeiras.

Mitchell, R. L. 1964. Trace elements in soils. In: F. E. Bear, *Chemistry of the soil*. 2. ed., Reinhold Publishing Co., New York.

Mohr, E. G. J. & F. A. van Baren. 1954. Tropical soils. A critical study of soil genesis as related to climate, rock and vegetation. N. V. Uitgeverij W. van Hoeve – The Hague and Bandung. Interscience Publishers, Ltd. London. Interscience Publishers, Inc., N.Y.

Pedro, G. & K. E. Bitar. 1966. Contribution à l'étude de la génese des sols hypermagnésiennes: recherches experimentals sur l'alteration chimique des roches ultrabasiques (serpentines). Annls Agron. 17: 661–651.

Pereira Coutinho, A. X. 1920. Breves considerações estatísticas acerca da flora portuguesa. Bolm Soc. broteriana 28: 95–121.

Pichi-Sermolli, R. 1948. Flora and vegetatione delle serpentine e delle altre ofioliti dell'Alta Valle del Tevere (Toscana). Webbia 6: 1–380.

Pinto da Silva, A. R. 1965. Os hábitats serpentinicos e o seu racional aproveitamento agrário. Primeiras achegas aceica do caso português. Comun. Colóq. Aportación de las investigaciones ecológicas y agricolas a la lucha del mundo contra el hambre, Madrid. 20 al 25 de Octubre: Pinto da Silva 1–40.

Pinto da Silva, A. R. 1970. A flora e a vegetação das áreas ultrabásicas do Nordeste Transmontano. Subsídios para o seu estudo. Agronomia Lusit. 30(3–4): 175–364.

Pinto da Silva, A. R. 1981. Mais algumes plantas Serpentinicoles do Nordeste Trans montano. Bolm Soc. broteriana 2 ser. 54: 239–247.

Raunkiaer, C. 1934. The life forms of plants and plant statistical plant geography. Clarendon Press, Oxford.

Robinson, W. O., C. Edington, & H. C. Byers. 1935. Chemical studies of infertile soils derived from rocks high in magnesium and generally high in chromium and nickel. Tech. Bull. U.S. Agric. 471.

Rozeira, A. 1944. A flora da província de Trás-os-Montes e Alto Douro. Mems Soc. broteriana 3: 1–203.

Volker, A. 1973. Fieldbook for land and water management experts. Professional edition. Int. Inst. for Land reclamation and improvement. ILRI, Wageningen.

Willkomm, M. & J. Lange. 1877. *Prodromus florae hispanicae*. 3 E. Schweizerbart (E. Koch), Stuttgart.

Zbyszewski, W. 1958. Le Quaternaire du Portugal. Bolm. Soc. Gel. Portugal. 13: 3–227.

E. Menezes de Sequeira[1] & A. R. Pinto da Silva
[1]Departamento de Pedolosia
[2]Departamento de Fifossistemáticae geo botânica
Estação Agronómica Nacional
2780 Oeiras
Portugal

VIII

Distribution of serpentinized massives on the

Balkan peninsula and their ecology

B. TATIC & V. VELJOVIC

Abstract. In the serpentinized habitats of the Balkan Penninsula a specific type of soil is developed which appreciably affects flora and vegetation. A certain number of plant species (serpentinophytes) thrives exclusively in these habitats and animal species are scarce.

So far a considerable number of serpentinophytes has been described, but we are of the opinion that the relevant descriptions were rather arbitrary. The status of a certain number of species must be checked because true serpentinophytes may be considered only those plants which thrive on shallow soil or even on the barren rocks so that their root system is in a direct contact with the maternal rock.

Vegetation of serpentinized area is very degraded, but on the basis of abundant data it may be concluded that not long ago these terrains were forested. Even today some serpentinized terrains are covered with natural pine, oak and other trees. The pine stumps dug out in village dirt roads prove that these barren areas used to be covered with forests.

By applying the method of Braun–Blanquet for the study of serpentinized terrains of the Balkan Peninsula a certain number of communities was described. Some of them are presented in this work. We hope that future investigations with a more detailed approach including ecological and physiological aspects will allow the revision of these communities.

In some serpentinized areas soil is denuded to such an extent that natural afforestation is impossible. It must be carried out by seeding or planting.

B. A. Roberts and J. Proctor (eds), The ecology of areas with serpentinized rocks. A world view, 199–215.
© 1992 *Kluwer Academic Publishers.*

Introduction

The Balkan Peninsula is situated in the far southeast of Europe. It is separated from the Apennines Peninsula by the Adriatic and Ionian Seas and from Asia by the Aegean and Black Seas. In the north the Balkan Peninsula (as used in this paper) is bordered by the Danube and Sava Rivers.

The Balkan Peninsula is rugged (orographically and hydrographically very indented); it has the massives of the Dinaric Alps, Pindus Mountains, the Rodope and Balkan Ranges. These extend in different directions, have different altitudes (up to 2,800 m), and have a diverse geological composition all of which are factors which made the Peninsula an important refuge for plants and animals during the glaciations.

The hydrographic net of the Balkan Peninsula is complex since its rivers flow into the Adriatic, Aegean and Black Seas. Its numerous lakes are of different size, origin and ecological characteristics. Most important are glacial lakes, located at different altitudes and with a great number of endemic and relict species. The Peninsula has several areas of ultramafic rocks. Grubic (1956) states that in the Balkan Peninsula the zone of serpentinized rocks extends with interruptions from the Alps to the Aegean Sea. Data on the general geological composition of the Peninsula are given by Cvijic (1924) and are illustrated on his geological map (1926). In Yugoslavia large complexes of ultramafics occur: on the massives on both sides of the Ibar River on Zlatibor Mountain (about 14,000 km^2), on Tara Mountain around Doboj and Maglaj, in Albania in the Drim River basin and the region of Morava River, in Greece on Vurinos Mountain (i.e., on its northern part as well as on the Isle Eubŏa). In addition to these localities there are smaller areas of serpentinized rocks in Yugoslavia, Albania, Greece, and Bulgaria (Karatoglis, Babalonas & Kabasakalis 1982). One smaller area of these rocks occurs in the European part of Turkey, but it is of limited interest and is not discussed.

Climate

The climate of the Balkan Peninsula is diverse. The narrow region of the mainland along the Adriatic Coast, the whole of Greece and the littoral region of Bulgaria has a mediterranean climate. In the valleys of the rivers which flow into the Adriatic and Aegean seas, the climate is submediterranean. The northern parts of the Peninsula have a markedly continental climate which almost has the characteristics of a steppe climate. The territory which lies between these mentioned climatic zones has a moderately continental climate. The climatic zones correspond to three distinct vegetation zones or regions: Mediterranean (hard-leaf evergreen vegetation), steppes and forest-steppes, and the deciduous forests. Depending on the altitude and the distance from the sea, the zonation, regarding both climate and vegetation, is different so that some regions are subalpine and alpine. The complexity of the relief of the Balkan Peninsula creates a variety of habitats, including snow patches in the alpine zones.

The Peninsula has had a history of human disturbance since the ancient Greek and Roman cultures. This has resulted in great ecological changes including the impoverishment of the indigenous flora and fauna.

Geomorphology and soils

Ultramafic massives are different in appearance and on many localities they have virtually no soils or vegetation and appear as stony barren ground. Elsewhere they are covered with soil of various depths having different degrees of development of anthropogenic forest and meadow-pasture vegetation. To some extent the differences reflect those of the mineralogical and petrographic composition of the rock which in turn reflect characteristics of the magma, the date of ultramafic massive formation, and the degree of serpentinization.

The soil cover on serpentine massives has a specific genesis and its connection with the phenomena occuring in the Neogene is of importance. These are the regression of the Pannonian Sea and the formation of isolated lakes and their gradual regression. Ultramafic lake sediments, chemically different from autochtonous serpentine soils also occur.

Table 1 shows the chemical composition of different mafic and ultramafic rocks in the Peninsula. Though there are obvious differences be-

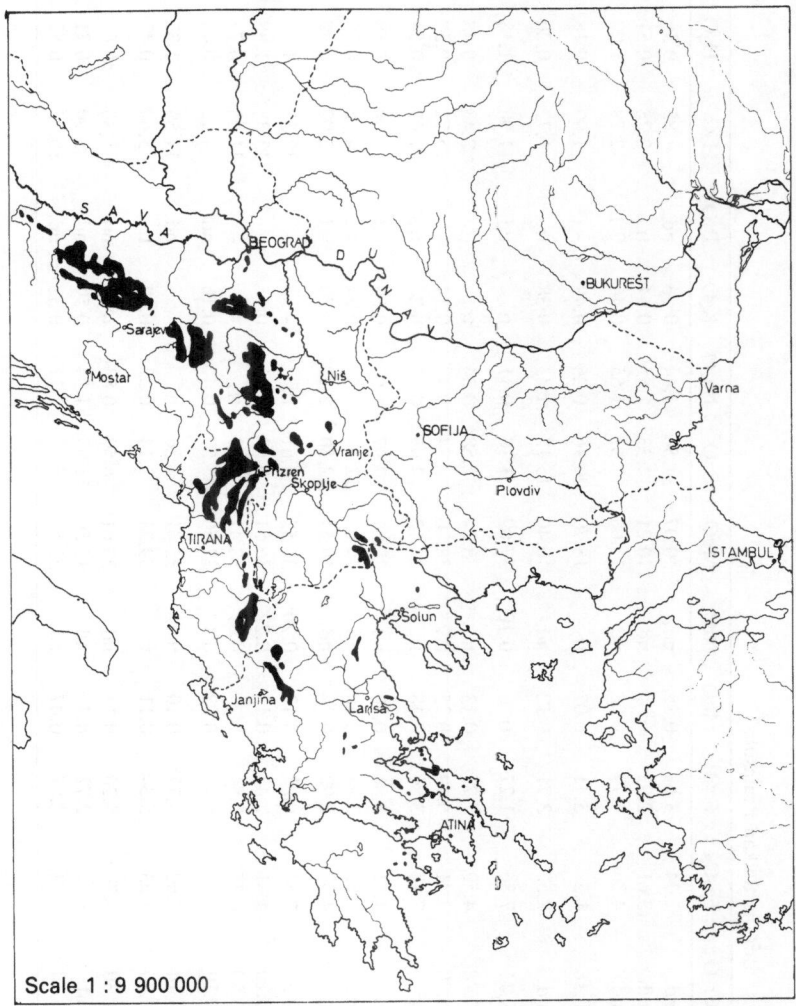

Fig. 1. Map of the Balkan Peninsula with designated localities of serpentine in black.

tween the habitats on barren ultramafic rocks and those covered with soil, there is some uniformity in their physical and chemical properties and their effect on organisms. The habitats on barren ultramafic massives are warm and dry, poor in major nutrients and have high amounts of magnesium, cobalt, chromium and nickel.

Flora and vegetation

There are characteristic species which generally inhabit or tolerate ultramafic soils. We recognize two types: *serpentinophytes* and *serpentino-morphoses*. Pichi-Sermolli (1948) regarded *serpentinomorphoses* as *nanism*, *stenophyllism*, *glau-* *cescence*, *purpuraescence*, *plagiotropism* and *macrorhizy*. The presence of chromium, cobalt and nickel in soils, together with unfavourable physical characteristics and scarcity of mineral nutrients contribute to the process of speciation. The serpentine massives are characterized by many relict and endemic species. True serpentin-ophytes – *obligate serpentinophytes* – occur only on the open spots of ultramafic rocks. The occurrence of dense forests on serpentine massives reflects the occasional presence of deeper soil. Obligate serpentinophytes develop in forest areas which have been degraded and in devasted forests where the soil cover has been removed. In these areas the natural re-establishment of forests does not occur.

Table 1. Chemical composition % of mafic and ultramafic rocks from the Balkan Peninsula

Rock	Locality	SiO_2	TiO_2	Al_2O_3	Cr_2O_3	Fe_2O_3	FeO	MnO	NiO	MgO	CaO	Na_2O	K_2O	P_2O_5	H_2O^+	H_2O^-
Coarse-grained gabbro	Kolnic	46.74	0.53	18.85	nd.	2.34	3.62	0.15	nd.	10.07	14.56	1.44	0.56	nd.	1.85	0.36
	Maljen	47.12	0.09	16.40	nd.	3.61	2.93	tr.	nd.	10.21	14.33	1.55	0.64	tr.	2.78	0.51
	Bogut. banja	47.47	0.07	14.25	nd.	4.36	4.79	0.08	nd.	11.00	11.67	1.55	0.46	0.10	2.79	1.00
Dialagitte	Brezovica	51.80	tr.	1.75	0.18	1.39	3.41	0.07	tr.	18.79	20.76	0.26	0.49	–	1.09	0.18
Gabbro	Kozh	46.51	0.29	17.80	nd.	2.52	3.71	0.12	nd.	10.40	15.12	1.43	0.38	tr.	1.72	0.20
Harzburgite	Bogut. banja	38.87	tr.	1.58	0.11	7.86	1.27	tr.	0.10	37.09	1.10	0.11	tr.	nd.	11.08	1.18
Lherzolite	Maljen	38.41	tr.	3.41	0.28	4.59	7.77	0.13	nd.	38.24	2.58	0.34	tr.	–	2.43	0.48
	Viseg. banja	40.70	tr.	2.81	0.20	2.21	6.19	0.25	0.21	38.20	2.74	0.42	0.14	–	5.72	0.58
	Kozuh	38.84	0.79	6.81	tr.	7.10	3.86	0.18	tr.	28.12	4.95	0.62	0.24	tr.	7.63	0.54
Olivine gabbro and peridotite	Maljen	42.89	tr.	7.11	tr.	4.31	3.86	0.10	nd.	27.57	7.07	1.04	0.80	nd.	5.09	0.58
	Viseg. banja	42.88	0.07	20.24	tr.	1.53	2.94	0.10	nd.	14.03	12.14	1.59	0.24	tr.	4.05	0.55
	Bogut. banja	47.26	tr.	18.62	tr.	4.07	2.92	0.14	nd.	10.51	12.30	2.12	0.30	nd.	2.02	0.29
Serpentine	K. Mitrovica	38.16	tr.	0.92	0.31	7.54	0.82	0.12	0.20	37.52	0.61	tr.	–	–	12.32	1.51
	Partiz. vode	38.12	tr.	1.21	0.20	4.81	2.42	0.13	0.22	36.51	0.63	0.61	0.13	–	13.38	1.88
	Dubostica	40.12	tr.	0.21	0.50	6.22	0.84	tr.	0.21	38.41	1.13	tr.	–	–	11.62	0.71
	Zlatibor	41.58	0.10	3.48	0.19	1.81	6.48	0.08	0.17	37.50	3.50	0.47	0.16	nd.	4.12	0.44
Serpentinized harzburgite	Priboj/Lim/	40.40	tr.	0.91	0.14	8.60	0.77	0.10	0.12	35.41	1.26	tr.	–	nd.	11.59	1.32
	Konjuh	41.32	tr.	1.61	0.20	3.93	5.34	0.11	tr.	39.21	2.14	0.32	0.09	tr.	5.64	0.42
Troctolite	Deli Jovan	39.13	0.10	13.51	1.30	3.79	5.20	0.13	nd.	23.61	6.50	0.83	tr.	nd.	5.30	0.65
	Viseg. banja	43.13	tr.	21.93	nd.	1.61	2.72	0.3	nd.	13.50	11.03	1.54	0.50	nd.	3.90	0.31
	Konjuh	37.58	0.03	10.65	0.10	7.08	2.87	0.07	nd.	23.50	5.49	1.11	0.32	nd.	10.75	0.82

Flora and vegetation on serpentine habitats of the Balkan peninsula

The early studies dealing with the flora of the Balkan Peninsula provided abundant data on plant species occurring on serpentine terrain: Grisebach (1843); Pančić (1859); Adamovic (1909); Janchen (1920); Markgraf (1928, 1931); Jávorka (1921) and others. Though in these early papers floristic rather than ecological observations were made, the emphasis was also put on the specificity of the flora and its substratum. More extensive studies on the serpentine flora were carried out later covering ecophysiological and cytogenetic aspects. Other authors dealing with the study of the Balkan Peninsula are Novak (1928), Rechinger (1957), Malý (1908, 1910 etc.), Markgraf (1905), Nyárády (1927–1929), Hayek (1923).

Pančić (1859) attributed the occurrence of serpentine taxa to the black soil that is heated very much during summer and is dry due to its coarse texture. He thought that serpentinophytes were almost exclusively field species. In later studies morphogenetic changes in plants were ascribed to the Mg/Ca quotient in the soil, e.g., Novak (1937). He concluded that a Mg/Ca quotient of > 1 and low concentrations of $< 1 \mu g \ g^{-1}$ for nitrate and phosphate in ultramafic soils were particularly important. More recently, attention has been directed towards the effect of cobalt, chromium, and nickel.

The floristic investigations were completed by Tutin *et al.* (1964–68) which covers also other relevant data for some taxa as regards habitat, endemism and disjunction. Tutin *et al.* (1964–80) list 45 serpentinophytes for the Balkan Peninsula as proper serpentinophytes (see Appendix I).

On the basis of our recent investigations and those of others over much of the Balkan Peninsula, we drew the conclusion that the number of obligate serpentinophytes is smaller than is reported by Tutin *et al.* Undoubtedly, obligate serpentinophytes are *Alyssum markgrafii, Asplenium adulterinum, A. balcanicum, A. cuneifolium, A. smolikanum, Bornmüllera dieckii, Fumana bonapartei, Gypsophila spergulifolia, Halacsya sendtneri, Haplophyllum boissierianum, Onosma elegantissima, Onosma euboica, Polygala dörfleri, Potentilla visiana, Saponaria sicula,* and *Sedum serpentini.* Further work on obligate serpentinophytes is required.

In addition to the floristic studies vegetation studies were made. Applying the methods of Braun-Blanquet (1928), a great number of papers were published and plant communities specific to the ultramafic habitats were described. Some communities on deeper ultramafic soils on gentle slopes and flat terrains are not restricted to ultramafic soils. Much more work is required to establish which communities are truly restricted to ultramafic soils.

The results of numerous investigations done so far concerning plant cover of serpentine habitats are discussed in detail by Horvat *et al.* (1974). Studies have continued and we list here the associations we recognize as occurring on serpentinized soils.

Forest and shrub phytocenoses
Abieto-Fagetum, Calluno-Quercetum serpentinicum, Cotynus coggria-Satureja thymifolia, Erica verticillata-Pinus halepensis, Erico-Abieti-Fagetum, Erico-Pinetum nigrae, Erico-Quercetum petreae, Fago-Abietetum serpentinicum, Myrtus communis-Pinus halepensis, Nerium oleander-Tamarix tetrandra, Ostryeto-Quercetum petreae serpentinicum, Piceetum excelsae, Piceetum omoricae, Pinetum heldreichii, Pinetum nigrae serpentinicum, Pinetum nigrae silvestris, Pinetum silvestris-nigrae seslerietosum latifoliae, Pinetum silvestris-nigrae seslerietosum rigidae, Potentilleto-Pinetum nigrae, Potentillo albae-Quercetum, Genisto-Callunetum croaticum, Querceto-Ostryetum, Quercetum montanum rubetosum hirsuti, Querco-Carpinetum serpentinicum, Querco-Pinetum, Quercus ilex-Arbutus andrachne, Seslerio serbicae-Pinetum, Vaccinio-Myrtilli-Fagetum.

Meadow, pasture and stony ground phytocenoses
Agropyretum juncei mediterraneum, Alyssum euboeum-Alyssum murale, Alyssum-Onosma heterophylla-Silene fabarioides, Alyssum-Psilurus incurvus-Aegilops neglecta, Artemisio-teucrietum montani, Bromo-Chrysopogonetum grylli, Centaureo-Bromion fibrosi, Convolvulo-Festucetum vallesiacae, Crepis neglecta-Aira elegans, Crytmo-Staticetum, Cynancho-Saponarietum intermediate, Danthonietum calycinae, Diantho-Seslerietum latifoliae, Erysimo-Sempervivetum heuffelii, Euphorbieto-Fumanetum bonapartei, Festuco-Carycetum laevis, Festuco-Nardetum strictae, Festuco-Plantaginetum serpentini, Halac-

sya sendtneri-Potentilla mollis, Halacsyo-Seslerietum rigidae, Hyperico-Euphorbietum glabriflorae, Linarietum concoloris, Malcolmia scyria-Alyssum praecox, Molinietum coeruleae, Onosomo-Scabietosum fumaroides, Pinguicula hirtiflora-Adianthetum, Poëto mollineri-Plantaginetum carinatea, Polygalo-Genistetum hassertianae, Potentillo-Fumanetum bonapartei, Sedo-Bornmüllerietum diecki, Selaginello-Thelypteretum dryopteris, Silenetum willdenowii serpentinae, cretica-Alyssum euboeum, Alyssum serpentini.

This survey of phytocenoses indicates the heterogeneity of the plant cover of the Balkan serpentinized areas and may be incomplete. The major groups are composed of phytocenoses of deciduous forests, evergreen forests, meadows and pastures, and of stony habitats. We emphassize that there is a lack of investigations on the most open habitats on ultramafic rocks.

Animals

Zoology of ultramafic habitats

The first published data on animals on ultramafic soils was by Pančić (1859) from Serbia (east Yugoslavia). He reported a tiny rodent as the only mammal found which, 'accidentally or because of necessity arrives and quickly leaves this hostile environment as soon as possible.' According to Pančić only the domesticated goat 'feels at home in this habitat because only this environment is suitable for its spontaneous jumps and cunning climbing.' The occurrence of rodents was also noted by Tatić (1969) who found a freshly swallowed field mouse in the stomach of a dissected snake on the slopes of Studena Mountain in the Ibar Gorge.

Pančić (1859) also reported about birds: 'On these habitats only two constant guests are found – grey *Saxicola oenanthe*, which often perches on projecting teeth of the rocks, and its dark feathered associate *Saxicola maura* Pall.' He noted that the great scarcity of warm-blooded animals in these habitats seems to make even birds of prey avoid serpentine hills. He also pointed out that he spent weeks and weeks searching through ultramafic terrain but failed to reveal even one

pellet which is otherwise frequently found on karst and other rocks.

On ultramafics, reptile species are much more numerous. Among lizards the most frequent are *Lacerta viridis* L. and *L. muralis* L., and the snakes *Coronella austriaca* L. and *L. Vipera amodytes* L. The snake *Eremias variabilis* Fiber is less common along with the blindworm, *Anguis fragilis* L.

Arthropods are very numerous on ultramafic soils. In addition to common Isopoda, Pančić reported two species of centipedes, *Intus communia* L. and *Scolopendra morsistans* L. According to Pančić insects are very abundant on the ultramafic hills of the Balkan Peninsula although their herbaceous cover is sparse.

At the meeting of the Zoological–Botanical Society at Vienna on 1st of June, 1859 Pančić's study dealing with the flora of serpentine hills, occupying the middle part of Serbia, reported on seven species of insects found on serpentine terrains: two Coleoptera (*Procrustus banaticus* Doll. and *Ludius ferrugineus* Latr.), two ants (*Hypoclinea quadripunctata* L. and *Monocumbus viaticus* Fabr.), one grasshopper (*Oedipoda germanica* Latr.) and two crickets (*Cicada orni* L. and *Lanterna* sp.). These were the first ever reported data on insects occurring in the territory of Serbia.

In 1883 Pančić published a paper on Orthoptera of Serbia. He described some new species including the grasshopper, *Pyrgomorphula serbica* Panč. endemic to Serbia. The very narrow range of distribution of this grasshopper is taken as a typical sample of stenoendemism. This species has been found on ultramafic habitats occupying the Tara Mountain and the southern part of Serbia on areas densely covered with *Erica carnea* L.

Zivadinovic & Ritter-Studnička (1970) found many species of Collembola on dolomite and ultramafic complexes of Bosnia and Hercegovina. Out of total number of 103 species of Collembola 21 are found only on dolomite habitats, while 43 species are found on both serpentine and dolomite. Among Collembola studied, 63 species have a broad distribution. In addition, the authors reported the following new species: *Neanura eburnea* Gisin, *Tullbergia novemspina* Gisin and *Tetracanthella specifica* Palissa.

Zivadinovic & Ritter-Studnička (1970) also stressed the unfavourable conditions for the development of Collembola and for fauna in general on dolomitic and particularly on ultramafic substrates. Investigations on Collembola showed that their distribution is related to (1) plant communities, (2) soil development, and (3) the soil water content. Hence, great differences in the collembola fauna have been observed between that under the pioneer communities of black pine (*Pinus nigra* Arn.) and oaks on serpentine substrata and that on mesophilious substrates. Zivadinovic and Ritter-Studnička concluded that old and rare species of Collembola survive on dolomite and serpentine as relict species. However, on habitats with more favourable conditions they lose in competition with other species.

Conservation

The greatest part of the Balkan Peninsula lies within the zones of deciduous forest and hard-leaved, evergreen, mediterranean vegetation. The considerably smaller remaining part in the north belongs to the steppe and forest-steppe zone. These complexes form a mosaic depending on the underlying geological composition.

Records dating back to the beginning of the 19th century testify to the great abundance of forests of the Peninsula at that time. Novaković (1884) reported on it.

The plant cover of ultramafic substrates is heterogeneous. Many ultramafic substrata were once under forest. However, man has destroyed much of the forest. The influence of human activity such as felling of trees, wildfire, clearcutting, resin bleeding and converting of forests to arable lands, even on very steep slopes, is very evident. In serpentine habitats forests have been replaced by meadows, pastures, arable lands, steppes, stony grounds and rock barrens.

Pančić (1859) wrote of middle Serbia: 'Once these hills were entirely covered with forest communities, which is reflected by the presence of single dwarfish, almost dried trees of *Quercus pubescens* W., *Esculus* L., *Q. cerris* L., *Q. pedunculata* W., *Pinus laricio* Pois., and by the occurrence of *Fagus sylvatica* L., found in the northern gentle slopes and rounded hillsides. Unfortunately, it is obvious that forests are disappearing posing great harm to our economy.'

Livestock, especially pigs, through grazing added to the removal of vegetation. The ultramafic substrates eroded rapidly after vegetation removal. Other woody species, present on ser-

Fig. 2. Spring picture of broad-leaved and mixed forest on Golija Mt. in the central part of Yugoslavia (Photo M. Janković).

Fig. 3. Devastated forest on ultramafics caused by excessive cutting (Photo M. Janković).

pentines, are scarce, e.g., *Euonymus verrucosus Fraxinus ornus* L., *Pyrus malus* L., *Prunus machaleb* L., *Rhamnus tinctoria* W.K., *S. domestica* L., *Sorbus torminalis* L., and *Viburnum lantana* L. However, *Rhus cotinus* L., *Acer tataricum* L. and *Juniperus oxycedrus* L. can form greater or smaller patches as for example near Brdjani, the Ibar Gorge and Kopaonik Mountain.

On some serpentine massives, there are still well preserved forests such as on Gostovic Mountain in Bosnia and Herzegovina (Krause *et al.* 1963), on Golija Mountain in West Serbia (Tatic *et al.* 1969), on Goc Mountain and Studena Mountain near Kraljevo, on Kodza Balkan Mountain near Prizren, on the mountains of Albania and also in some areas of Greece. However, in Yugoslavia few massives today are still covered with forest communities (Fig. 2).

The inaccessible and sparsely populated Tara Mountain region in the western part of Serbia has well preserved forests of the spruce, *Picea omorika* Panc., which is the most characteristic endemic species and tertiary relict species of the central part of the Balkan Peninsula. This is due mainly to the inaccessibility and sparse population of the region. Fairly well preserved forests on ultramafic massives are found in the regions of the northern and western part of the Balkan Peninsula which has more favourable climatic conditions and which were populated at a later date. Though it is well known that ultramafic soils have unfavourable physical and chemical properties and that forests on them have a protective role they are still over exploited. Recent deforestation on ultramafic substrata is shown in Fig. 3.

Forests are also damaged by the practice of bleeding of trees for resin. Some evergreen trees, especially some species of pine (*Pinus*) produce great quantities of resin. Resin from old trunks of the widespread Scot's pine (*Pinus silvestris* L.),

206

Fig. 4. Bleeding old pine trees for resin. (Photo A. Bayer).

black pine (*P. nigra* Arn.) and allepic pine (*P. halepensis* L.) is used by industry. Private and public owners of forests bleed these trunks for resin in the most uneconomic way. Even in the steep and inaccessible terrains where forest is valued for erosion control this practice of bleeding trees for resin is continuous. For collecting the dripping resin, flower pots or any type of containers are placed under trees and the bleeding procedure is repeated at various heights above the trunk (Fig. 4). Unfortunately, resin is abundant only in mature trees which, besides resins, produce much seed. Hence, the destruction of these trees decreases the number of seeds and weakens the reproductive capacity of the species.

Forests are also damaged by kindling stripping by farmers in montane areas who have houses near pine stands. Since the kindling is rich in resin it is easily lit. Unfortunately, even today with electricity in nearly all places, many villages still use kindling as a source of lighting.

The processes of degradation and devastation occurring under the effects of the many anthropogenic factors indicate that on serpentinized habitats with devastated plant cover, the revegation is very slow and is almost impossible under natural conditions. It is even more difficult on very steep slopes undergoing strong erosion processes due to which soil development is absent. On such habitats forest cover gradually diminishes and erosion processes become increasingly drastic (Fig. 5).

When forests disappear meadows, pastures, rocky grounds and barrens develop on the ultramafic soils. Meadows and pastures are characteristic of the serpentinized massives with a more favourable topography. In some localities grasslands with clumped tree growth are a stable climax community. These habitats have favourable hydrological characteristics, although they are in hilly regions. Floristically they are similar to meadows on flat lands and on gentle north-facing slopes. The main species are *Cynosurus cristatus*, *Agrostis vulgaris* and *Chrysopogon gryllus*. For south-facing meadows with deeper soils characteristic species are *Andropogon ischaemum*, species of *Festuca*, *Danthonia calycina*, *Poa mollineri*, *Sesleria rigida*, *S. serbica* and others. The total area of well-preserved meadows is small because they are now often ploughed as part of a policy of evermore intensive land management practices. Arable lands on ultramafic soils undergo erosion, especially on abandoned slopes leaving permanent shrubby vegetation or barrens. Such complex changes in plant cover on ultramafic soils are presented in Fig. 6.

After deforestation, especially on very steep areas, rocky ground and sparse remnant trees and shrubs remain. A few remnants of the forest floor are also visible. Though the cover of these terrains is sparse there is still a difference between particular slopes which results from differences in the degree of isolation and steepness (Fig. 7).

The vegetation of degraded rocky ultramafic grounds is similar on steep and dissected areas in hilly regions at low altitudes with slightly rounded relief and on gentle slopes of ultramafic areas in Greece, Albania and in the central part of Yugoslavia. For example, the massive of Vour-

Fig. 5. Erosion in devastated oak forest (Photo M. Janković).

Fig. 6. Serpentine massive, near Stragari, central part of Yugoslavia. Above: Virgin oak forest. Left: Area grown with herbaceous cover. Right and in the middle: Asbestos mine and flotation waste deposits (Photo V. Veljović).

Fig. 7. Serpentine massive of the Ibar gorge converted to stony grounds. Scattered remaining trees and shrubs of the forest indicate the degradation process (Photo D. Čolić).

inos Mountain in the region of Kozana is composed entirely of ultramafic rocks. It was deforested and degraded into rocky ground where some species of *Quercus coccifera*, *Q. pubescens*, *Fraxinus ornus* forests occur as well as many specis of shrubs and perennial plants.

Typical ultramafic open rocky habitats are distinct in their high number of obligate serpentinophytes which make up a substantial proportion of the plant cover. It seems that the more unfavourable the conditions, the greater the role of serpentinophytes. These open habitats pose the greatest problem in the renewal of plant cover and are subject of the re-afforrestation studies.

Ultramafic massives often have commercial deposits of chromium, nickel and other minerals. Mining was established in medieval times and has continued to the present causing great changes in the forest cover. Forests were cut to construct houses for miners and probably for ore process-

ing. Many deforested ultramafic areas have remained essentially uncolonized since the early mining period indicating the difficulties of the renewal of plant cover (Fig. 8).

On ultramafic substrates, especially on stony soils, plant cover increases slowly since erosive effects are severe. In the re-establishment of forest cover an important role is played by the remaining fragments of the forest floor. The remaining tree species carry on reproduction but the number and quality of the seeds is reduced and the fate of the seeds is influenced by many adverse factors. Remnant species of shrubs may play a pioneer role, as was already pointed out by Pančić (1859) for sumac (*Rhus continus*).

In forest management practices sumac is of particular interest. On gentle slopes and lee areas it provides a thick canopy in the shade of which a number of herbaceous plants can develop. In this way further scattering of loose rocks is hindered

Fig. 8. Vein of magnesite in the serpentine rock on Tara Mt. (Photo P. Stevanović).

and at a later stage soil humus is formed and oak seeds can germinate. Sumac is an inhabitant of serpentine massives where it has a high aesthetic value since its leaves are characterized by a variety of colours ranging from yellowish to orangered and markedly red. Thus, extensive regions covered with sumac show wonderful autumn colours (Fig. 9). On many ultramafics initially devoid of forest cover, sumac stands gradually diminish under the shade of larger oaks. Pančić (1859) described mummy-like trees of sumac which persist for many years.

Juniper (*Juniperus communis*) may play a similar role in re-establishment (Čolić 1957). In Serbia, Macedonia, Albania and Greece, on rocky open ultramafic habitats we found many red juniper (*J. oxycedrus*) stands and both this and *J. communis* are noted as important pioneer species on ultramafics here and elsewhere. Species of *Prunus, Malus, Cytisus, Rosa, Crataegus, Coronilla*, may also play a pioneer role on ultramafics. On burnt areas this role is attributable to species of *Salix, Sambucus*, and *Rubus*, among others. Natural re-establishment was studied by Vukićević (1965) who observed that colonization of burnt areas on ultramafics differs from that on karst, silicate and other substrates. It is interesting to notice that herbaceous plants colonize burnt areas on ultramafics at a slower rate than on other substrates.

Ground flora species such as *Erica carnea* and grasses such as *Sesleria, Andropogon, Nardus*, play an important role in the re-establishment of the vegetation cover. However, in areas where *Sesleria* and *Naruds* species have formed a dense cover, the re-establishment of forest is slow.

Fig. 9. Sumac (*Rhus cotinus*) on the steep slopes of serpentine stony grounds as a pioneer species in the succession towards forest (Photo M. Fischer).

These grasses form a thick cover where there are unfavourable conditions for seed germination of other species.

In open areas with serpentinophytes succession seems to be arrested. Serpentinophytes grow either in rosettes or lying on the ground or as tiny shrubs. Although their total cover is small they are of importance for erosion protection and natural reforestation on these areas does not occur.

Recently, protective measures are being undertaken to prevent erosion on the ultramafics. Cascade dams have been built and suitable tree spe-
cies are planted. Locust (*Robinia pseudoacacia*) is the first to be planted and then species of *Pinus*, *Larix* and others. The initial afforestation projects on denuded massives have shown a substantial increase in plant cover. *Pinus nigra* is particularly successful on ultramafics. Its survival is high, seedlings grow fast and quickly reach maturity. With planting, the dense cover of grass is destroyed and creates favourable microclimatic conditions for the pine. These afforestation projects were started about 30 years ago and continue extensively as a practicable way to bring the land back under forest cover.

Appendix I.

A list of 45 serpentinophytes for the Balkan Peninsula (Tutin *et al*. 1964–80)

Halacsa sendtneri (Boiss.) Dörfler – Al. Yu.
Stachys scardica (Griseb.) Hayek – Bu. Gr. Yu. Al.
Thymus longicaulis C. Presl. – Al. Yu.
Veronica scardica Griseb. – Al. Yu.
Verbascum serpentinicum Rech. fil. – Gr. Yu.
Anchusa serpentinicola Rech. fil. – Gr. Yu.
Viola beckiana Fiala – Al. Yu.
Viola ducadjinica W. Becker et Kosanin – Al. Gr. Yu.
Sanguisorba albanica Andrasovszky et Jav. – Al. Yu.
Potentilla visianii Pancic – Al. Yu.
Onosma elegantissima Rech. fil. et Goulimy. – Gr.
Haplophyllum boissierianum Vis. et Pancic – Al. Yu.
Verbascum bosnense K. Maly – Al. Gr. Yu.
Genista hassertiana (Bald.) Bald. et Buchegger – Al. Yu.
Fumana bonapartei Maire et Petitmengin – Al. Gr. Yu.
Euphorbia gregersenni K. Maly – Yu.
Euphorbia glabriflora Vis. – Al. Gr. Yu.
Eryngium serbicum Pancic – Al. Yu.
Cytisus pseudoprocumbens Markgraf – Al. Yu.
Linum tauricum Willd. Balkan and beyond
Aster albanicus Degen – Al. Yu.
Scorzonera serpentinica Rech. fil. – Gr.
Centaurea kosaninii Hayek – Al. Yu.
Centaurea candelabrum Hayek et Kosanin – Al.
Forsythia europaea Degen et Bald. – Al. Yu.
Anthyllis serpentinicola Rech. fil. et Goulimy – Al. Gr. Yu.
Narthecium scardicum Kosanin – Al. Gr. Yu.
Sesleria serbica (Adamovic) Ujhelyi – Gr. Yu. Bu.
Brachypodium serpentini C. E. Hubbard – Al.
Polygala dörfleri Hayek – Al. Yu.
Koeleria eriostachya Pancic – Balkan and beyond
Alyssum balcanicum Nyár. – Al. Gr. Yu.
Alyssum markgrafii O. E. Schulz. – Al. Yu.
Alyssum smolikanum Nyár. – Al. Gr.
Sedum serpentini Janchen – Al. Gr. Yu.
Alyssum densistellatum T. R. Dudley – Gr.
Bornmülera dieckii Degen – Yu.
Cardamine plumieri Vill. Balkan and beyond
Arenaria serpentini A. K. Jakson – Al.
Silene pindicola Hauskn. – Gr.
Gypsophyla spergulifolia Gris. – Al. Yu.
Saponaria sicula Rafin – Al. Gr. Yu.
Asplenium adulterinum Milde – Balkan and beyond
Asplenium cuneifolium Viv. – Balkan and beyond
Onosma auboica Rech. fil. – Gr.

Abbreviations: Al.-Albania, Gr.-Greece; Bu.-Bulgaria and Yu.-Yugoslavia.

References

Adamović, L. 1909. Die Vegetationsverhältnisse der Balkanländer (Mösische Länder). Die Vegetation der Erde, Leipzig.

Adamović, Z. 1976. Josif Pančić entomolog. Spomenica srpske Akademije Nauka, Beograd.

Babalonas, D. 1984. Zur Kenntnis der Flora und Vegetation auf Serpentin-Standorten Nordgrichenlands I. Feddes Repert. B. 95, Heft 9–10, Berlin.

Babalonas, D. 1984. Armeria maritima subsp. smolikana, ein neues Taxon aus NW-Grichenland. Willdenowia 14.

Babalonas, D. 1984. Floristisches aus dem Vourinos Gebierge (N. Grichenland). Willdenowia 14.

Bayer, A. 1928. Smolarźeni v západofrancouzaských Landech. Priroda XXI, 11.

Beck-Mannagetta, G. 1906, 7, 9, 11. Flora Bosne, Hercogoviae i Novpazarskog Sandžaka. Glasnik zemaljskog muzeja u Bosni i Hercegovini, Sarajevo.

Blaženčić, Z., R. Vučković. 1983. Kserofilna zajednica Convolvulo-Festucetum vallesiacae prov. u okolini Beograda. Acta Biologica Yugoslavica, Vol. 18, No. 2, Beograd.

Blečić, V., B. Tatić & F. Krasnići. 1968. Kratak prilog flori Jugoslavije. Bull. de l'Inst. et du Jardin bot. de l'Univ. de Beograd, Tom III nov. ser.

Blečić, V., B. Tatić & F. Krasnići. 1969. Tri endemične zajednice na serpentinskoj podlozi u Srbiji. Acta Botanica Croatica, Zagreb.

Bornmüller, J. 1925. Beiträge zur Flora Mazedoniens I. Engler's Bot. Jahrbücher, Bd. LIX, Leipzig.

Braun-Blanquet, J. 1928. Pflanzensoziologie. Grundzüge der Vegetationskunde, Berlin.

Cvijić, J. 1924. Geomorfologija I i II deo, Beograd.

Cvijić, J. 1928. Geološka karta Balkanskih zemalja, Beograd.

Čolić, D. 1957. Ispitivanje uloge kleke (Juniperus communis L.) na serpentinskom erozivnom području. Conseil des Academies de la. R. P. F. de Yougoslavie, Beograd.

Derganc, L. 1903. Geographische Verbreitung der Daphne blagayana Freyer. Allg. Bot. Zeitschr. Karlsruhe.

Dimitrijević, B. 1946. Agrogeologija. – Naučna kajiga, Beograd.

Djordjević, P. 1931. Pinus nigra Arn. var gočensis. Oesterr. Bot. Zeitschr. Bd. 80. Wien.

Djordjević, P. 1972. Petrohemijska korelacija peridotitskih i gabroidnih plutonskih asocijacija Jugoslavije. Manuskript, Beograd.

Gajić, M., M. Kojić & M. Ivanović 1954. Pregled šumskih fitocenoza planine Maljena. Glasnik šumarskog fakulteta, Beograd.

Gayer, G. 1928. Senecio serpentini. Annales musei conit. Castriferrei, sectio hist. natur. A. Szombathely.

Grisebach, A. 1843. Spicilegium Florae Rumelicae et Bithymicae. Brunsvigiae.

Grubić, A. 1956. Paleozoik i serpentin na Kopaoniku. Zapisnik Srpskog geološkog društva za 1954. godinu, Beograd.

Hayek, A. 1923. Zweiter Beitrag zur Kenntnis der Flora von Albanien. Denkschr. der K. ak. Wiss. Math. naturw. Klasse, Bd. 94, Wien.

Hayek, A. 1927–1933. Prodromus Florae peninsulae Balkanicae. Rep. spec. nov. Berlin.

Hegi, G. 1906. Illustrierte Flora von Mittel-Europa. Wien.

Heywood, H. 1964. Bornmüllera Hausknecht ap. T. G. et al. Flora Europaea 1. Cambridge.

Horvat, I. 1948. Biljne zajednice. Zagreb.

Horvat, I. 1954. Pflanzengeographische Gliederung Südosteuropas. Vegetatio 5/6, Den Haag.

Horvat, I. 1974. Vegetation Südosteuropas. Geobotanica selecta, Band IV, Gustv Fischer Verlag, Stuttgart.

Hunter, I., O. Vergnamno. 1952. Nickel toxity in plants. Ann. Appl. Biol. 39.

Janchen, E. 1920. Die systematische Gliederung der Gattung Fumana. Oesterr. Bot. Zeitschr. Wien.

Janchen, E. 1920. Vorarbeiten zu einer Flora der Umgebung von Škodra. Oesterr. Bot. Zeitschr. LXIX. Wien.

Janković, M. 1976. Josif Pančić et la protection du milieu naturel. Spomenica, Srpska akad. nauka i umetnosti, Beograd.

Jávorka, S. 1920–21. Új adatok Albánia flóráhájoz (Novitas florae Albaniae). Bot. Közlemények, XIX, Budapest.

Jovanović, B. 1954. Fitocenoza Quercetum confertae cerris kao biološki indikator. Glasnik šumar. fakulteta 8. Beograd.

Jovanović, B. 1954. O šumama Srbije početkom XIX veka. Geografski lik Srbije u doba prvog srpskog ustanka. Posebno izdanje SGD. sv. 32, Beograd.

Jovičić, M. 1891. Nekoliko serpentina iz Srbije. Geološki anali 3. Beograd.

Karatoglis, S., D. Babalonas & B. Kabasaklis. 1982. The ecology of plant populations growing on serpentine soils. Phytom, (Austria), 22.

Kišpatić, M. 1897. Kristalinsko kamenje serpentinske zone u Bosni. Rad Jugosl. Akad. Knj. 133. Zagreb.

Kišpatić, M. 1904. Petrografske bilješke iz Bosne. Ibid. 159.

Kojić, M. 1959. Zastupljemost, uloga i značaj djipovine (Chrysopogon gryllus) u livadskim fitocenozama Zapadne Srbije. Arh. za polj. nauke. Sv. 37, Beograd.

Kojić, M., M. Ivanović. 1953. Fitocenološka istraživanja livada na južnim padinama Maljena. Zbor. radova polj. fak. God. I, Beograd.

Košanin, N. 1911. Vegetacija planine Jakupice u Makedoniji. Glasnik Akad. 85. Beograd.

Košanin, N. 1913. Die Verbreitung von Forsythia europaea Deg. et Bald. in Nord Albanien. Magy. Bot. Lapok, Bd. 12, Budapest.

Košanin, N. 1926. Nove vrste u flori Južne Srbije. Glas Srpske Kralj. Akad. CXIX., prvi razred 54, Beograd.

Krause, W., W. Ludwig. 1957. Zur Kenntnis der Flora und Vegetation auf Serpentinstandorten des Balkans. Flora 145, Jena.

Krause, W., W. Ludwig, & F. Seidle. 1963. Zur Kenntnis der Flora und Vegetation auf Serpentinstandorten des Balkans 6. Vegetationsstudien in der Umgebung von Mantoudi (Euböa). Bot. Jahrb. 82.

Kretschmer, L. 1930. Die Pflanzengesellschaften auf Serpentin im Gurhofgraben bei Melk. Verh. Zool. Bot. Ges. Bd. 80, Wien.

Kruckeberg, A. 1951. Intraspecific variability in the response of certain native plant species to serpentine soils. Am. Jour. Bot. 38.

Kruckeberg, A. 1954. The ecology of serpentine Soils III.

Plant species in relation to serpentine soils. Ecology 35.

Kubiena, W. 1953. Bestimmungsbuch und Systematik der Böden Europas. Stuttgart.

Lämmermayr, L. 1926. Materialen zur Systematik und Ökologie der Serpentinflora. Sitzungs. Ak. Wiss. Math. nat. Klasse 133 Wien.

Lämmermayr, L. 1934. Übereinstimungen und Unterschiede in der Pflanzendecke über Serpentin und Magnesit. Mitteil. Nat. Vereins Steiermark Bd. 71. Graz.

Lovrić, Z. 1984. History and Genetics of Tolerance to the Natural Phyto-toxic Media of Western Balkans. Being Alive on Land. by N. S. Margaris. Junk Publ. Hague.

Maksimović, Z. 1957. Geohemija raspadanja ultrabazičnih stena u Srbiji. Manuskript. Beograd.

Malý, K. 1908, 10, 12, 19, 20, 23, 28. Prilozi za floru Bosne i Hercegovine. Glasnik Zem. Muz. Sarajevo.

Markgraf, F. 1928. Pflanzen aus Albanien. Denkschr. d. Akad. Wiss. Math.-nat. Klasse, Bd. 102, Wien.

Markgraf, F. 1931. Pflanzen aus Albanien. Ibid.

Markgraf, F. 1932. Pflanzengeographie von Albanien. Bibl. Bot. 105.

Messeri, A. 1936. Ricerche sulla vegetazione dei dintorni di Firenze. Nuovo Giorn. Bot. ital. Nuovo Serie. Vol. XLIII, Firenze.

Mikinčić, V. 1953. Geološka karta F. N. R. Jugoslavije, i:500.000, S. A. N. Beograd.

Milde, J. 1868. Filices criticae. Asplenium adulterinum. Botanische Zeitung, 26 Jahrg. Leipzig.

Nemec, Fr. 1937. Mineralogie, Petrografie a geologie okoli Mohelna (monografie hadce). Mohelno, Brao.

Novak, F. 1927. Ad florae Serbiae . . . Preslia, V, Praha.

Novak, F. 1928. Ekologizké úvahy o hadcovych rasách a hadcové vegetacie. Veda Prirodni, IX, Praha.

Novak, F. 1929. Ad florae Serbiae . . . VIII, Preslia, Praha.

Novak, F. 1937. Kvetena a vegetace hadovych pud. Mohelno, Ia, Brno.

Novaković, S. 1894. Srbija u godini 1834. Pisma grofa Boa le Konta grofu de Rinji o tadašnjem stanju u Srbiji. Spomenica, SKA, XXIV, Beograd.

Nyárády, E. 1927. Studiu preliminar asupra unor specii de Alyssum diu sectia Odontarrhena. Bull. Gradini Bot. Univ. Vol. V–IX, Cluj.

Pampanini, R. 1912. Flora dei serpentini di Montignoso . . . Nuovo Giorm. Bot. Ital. Vol. XIX, Firenze.

Pančić, J. 1859. Die Flora der Serpentinberge in Mittelserbien. Verh. d. K. Zool. bot. Ges. Bd. IX, Wien.

Pančić, J. 1870. Nešto o našim šumama. Prilikom zatvaranja prvog izloga zemaljskih proizvoda 4. oktobra 1870, Kragujevac.

Pančić, J. 1874. Flora Kneževine Srbije. Beograd.

Papanicolau, K., D. Babalonas & S. Kokkini. 1983. Distribution patterns of some Greek mountain endemic plants in relation to geologic substrate. Flora, 174.

Paribok, T., N. Alekseeva-Popova. 1966. Soderžanie himičeskih elementov v rastenijah Poljarnova Urala v svjazi s problemoj serpentinofitovoj rastiteljnosti. Bot. Zurnal T. Li, 3. Moskva-Leningrad.

Pavićević, N. 1962. Tipovi zemljišta Jugoslavije. Zadružna knjiga, Beograd.

Pavlović, Z. 1961. Vegetacija planine Zlatibora. Zbornik inst. ekol. i biogeogr. XI, Beograd.

Pavlović, Z. 1962. Karakteristični elementi serpentinske flore Srbije. Glasnik prirod. Muz., 18. Beograd.

Pichi-Sermolli, R. 1936. Osservazioni sulle principali morfosi delle piante del serpentino. Nuovo Gior. Bot. XLIII, No. 2, Firenze.

Pichi-Sermolli, R. 1948. Flora e vegetazione delle serpentine e della altri ofioliti dell alta valle del Tevere (Toscana). Webbia 6. Firenze.

Randjelović, N., & F. Rexhepi. 1979. Biljne zajednice severoistočnog Kosova. Poseban otisak, II kongres ekologa Jugosl. Zagreb.

Randjelović, N., & M. Ružić. 1983. Pašnjačka serpentinska vegetacija jugoistočne Srbije. Glas. prir. muz. B., knj. 38. Beograd.

Rechinger, K. 1957. Plantae novae graeco-macedonicae, imprimis serpentinicolae. Anz. Österr. Ak. Wiss. Math. nat. Kl. 2.

Rechinger, K. 1961. Die Flora von Euböa. Engler's Bot. Jahrb. Bd. 80, Heft 3, 4, Stuttgart.

Rexhepi, F. 1979. Prilog poznavanju flore na serpentinima Kosova. Prirodno-mat. fakultet, Priština.

Rexhepi, F. 1979. Endemic plant community Potentillo-Fumanetum bonapartei. Ibid.

Ritter-Studnička, H. 1963. Biljni pokrov na serpentinima u Bosni. Godišnjak Biol. Inst. Univ. XVI, Sarajevo.

Ritter-Studnička, H. 1964. Anatomske razlike izmedju biljaka sa serpentinske, dolomitske i krečnjačke podloge. Ibid. XVII, Sarajevo.

Ritter-Studnička, H. 1968. Die Serpentinomorphosen der Flora Bosniens. Bot. Jahrb. 88. Stuttgart.

Ritter-Studnička, H. 1978. Die Flora der Serpentinvorkommen in Bosnien. Bibl. Bot., 130.

Robinson, W., G. Edgington, & H. Byers. 1935. Chemical studies of infertile soils . . . Dept. Agr. Techn. Bull.

Rune, O. 1953. Plant life on serpentines and related rocks in nord Sweden. Acta phytogeographica Suecica. Upsala.

Školjnik, M., & J. Smirnov. 1970. O vazmožnih pričinah serpentinomorfozov . . . Botaničeskij Zurnal, T. 55, 12., Moskva-Leningrad.

Tatić, B. 1969. Flora i vegetacija Studene Planine kod Kraljeva. Bull. de l'Inst. et du Jard. Bot. de l'Univ, de Beograd.

Tatić, B., V. Veljović, A. Marković, & B. Petković. 1981. Prilog proučavanju serpentinske flore Jugoslavije. Biosistematika Vol. 7, No. 2, Beograd.

Tatić, B., & V. Veljović. 1982. Potreba za reviziju termina serpentinofit. Glas. rep. Zavoda za zašt. prir. 15. Titograd.

Tešić, Z., M. Ristanović, & H. Ritter-Studnička. 1967. Prilog poznavanju mikroflore u dolomitskim i serpentinskim staništima pod raznim stadijama biljne sukcesije u Bosni i Hercegovini. Mikrobiologija, Ser. B. 4. Beograd.

Tutin, T. G. et al. 1964–1980. Flora Europaea 1-5. Cambridge Univ. Press, Cambridge–London–New York–Melbourne.

Ujhelyi, J. 1959. Species Sesleriae Generis novae. Feddes Rep. Berlin.

Veljović, V. 1967. Vegetacija okoline Kragujevca. Glasnik prir. muzeja, Ser. B. kaj. 22. Beograd.

Vergnano, O. 1958. Ni, Cr e Co nel dinamismo nutritivo delle piante serpentinicole, Nuovo Diorn. Bot. Ital. 66.

Vukićević, E. 1965. Sukcesija vegetacije i prirodno obnavljanje šuma na šuzskim požarištima u Srbiji. Glasnik šum. fak. 29. Beograd.

Walker, R. 1954. The ecology of serpentinic Soils II. Ecology 35.

Wandelberger, G. 1974. Die Serpentinpflanzenvorkommen des Burgenlands in ihrer Pflanzengeographischen Stellung. Wiss. Arbeiten Bgld. 53, Österr.

Wettstein, R. 1918. Moltkea dörfleri Wett. und die Abgrenzung der Gattung Moltkea. Österr. Bot. Zeitschr. Wien.

Whittaker, R. 1954. The ecology of serpentine Soils IV. Ecology 35.

Živadinović, J. and H. Ritter-Studnička 1970. Vrste Colemballa u zajednicama Basne i Hercegovine. – Godišnjah muzeja u Sarajevu.

Author's Address:
B. Tatić
Faculty of Sciences
University of Belgrade
Belgrade
Yugoslavia

V. Veljović
Faculty of Science
Kragujevac
Yugoslavia

IX

The distribution and ecology of the vegetation of

ultramafic soils in Italy

O. VERGNANO GAMBI

Abstract. The vegetation, geology and distribution of serpentinized-ultramafic areas in Italy are described and the investigations on serpentine vegetation briefly outlined. The recent ecophysiological research on serpentine vegetation, including the endemic nickel accumulator *Alyssum bertolonii* Desv., is reported.

B. A. Roberts and J. Proctor (eds), The ecology of areas with serpentinized rocks. A world view, 217–247.
© *1992 Kluwer Academic Publishers.*

Introduction

The occurrence of plant species restricted to ophiolitic substrates was documented by the sixteenth century: Cesalpino (1583) described a '*Lunaria quarta alias Alysson*' found only on the 'black stones' of Montauto (Upper Tiber Valley). This plant was identified by Pier Antonio Micheli as *Alyssum perenne*, and is mentioned in his manuscripts describing the Tuscan flora. It was considered typical of the darkly coloured stony hills near Florence (Monte Ferrato, Impruneta, etc.). But it was only after a physician, G. Amidei (1841), brought the close relationship between some plant species and ultramafic outcrops in Tuscany to the attention of the 3rd Meeting of the Italian Scientists in Florence 1841, that a real botanical interest was aroused.

T. Caruel was the first to thoroughly investigate the relationship between plant composition and ultramafic areas. In a communication 'Sur la flore des gabbres de Toscane' delivered to the International Botanical Congress in Paris (1867) he stated that out of 200 species collected on these outcrops, the great majority was made up of xerophilous species commonly found on barren and rocky areas of various lithological composition. But some species (plants are named according to Tutin *et al.* (1964–1980) such as *Alyssum bertolonii* , *Armeria denticulata*, *Asplenium cuneifolium*, *Centaurea aplolepa* ssp. *carueliana*, *Cheilanthes marantae*, *Euphorbia nicaeensis* ssp. *prostrata* (Fiori) Arrigoni, *Festuca inops* and *Minuartia laricifolia* , seemed restricted to ultramafic soils. Only *Alyssum bertolonii* and *Cheilanthes marantae* were strictly endemics. Caruel concluded that these few species had a preference for this substrate and that the soil composition had caused the appearance of varieties or forms characterized by peculiar morphological changes (Caruel 1867, 1871).

There were many investigations on ultramafic vegetation at the beginning of the present century, both in the Alps and in the Apennines: in Piedmont by Vaccari (1903) and Mussa (1908), in the Ligurian Apennine by Gola (1912) and Pavarino (1918), in the Emilian Apennine by Fiori (1917, 1919) and Pavarino (1914, 1915, 1918), in Tuscany by Pampanini (1912) and Fiori (1914, 1920). Sometimes ultramafic vegetation

was compared with that of adjacent areas with different rocks such as limestone or gneiss. The xeric character of ultramafic vegetation was always stressed and the small number of species was observed both by Vaccari and Pavarino. The richness of ferns and hemicryptophytes was noted (Pavarino 1915). All these botanists, although generally cautious in relating the nature of the soil to the presence of peculiar species, agreed that this substrate induced morphological alterations in various plant organs (absence or reduction of pubescence, stunting, dwarfism, plagiotropism, greater root development and in some instances increased pubescence). They agreed that *Alyssum bertolonii*, *Alyssum argenteum* and possibly *Asplenium cuneifolium* are restricted to ultramafites. Species with a disjunct distribution such as *Cheilanthes marantae*, *Scorzonera austriaca* and *Stipa pennata*, or species such as *Minuartia laricifolia*, typical of more northern latitudes, are also present in these areas. The presence of vicariant endemisms (e.g. *Alyssum bertolonii* in Tuscany, *A. argenteum* in the northern Apennine) and of species with a disjunct distribution is considered evidence of the ancient origin of this flora. Before considering in detail the investigations on the vegetation of the Italian ultramafic areas, their distribution and geology will be briefly outlined.

Geology

The mafic and ultramafic complexes of the Alps and Apennines (Fig. 1) consist of Mesozoic rocks (including Jurassic ophiolites proper) and pre-Mesozoic rocks. The Mesozoic rocks are widespread in the Alps (Fig. 2) and in the northern (Fig. 3) and southern Apennines, the pre-Mesozoic rocks occur only in the metamorphic Palaeozoic basement of the Alps and in some of the higher tectonic units (overriding sheets) in the southern Apennines, which are probably equivalent to part of the Alpine basement.

Alps

Pre-Mesozoic rocks: These are included in the highly metamorphic basement exposed in the eastern (Austro-Alpine) district and in the exter

218

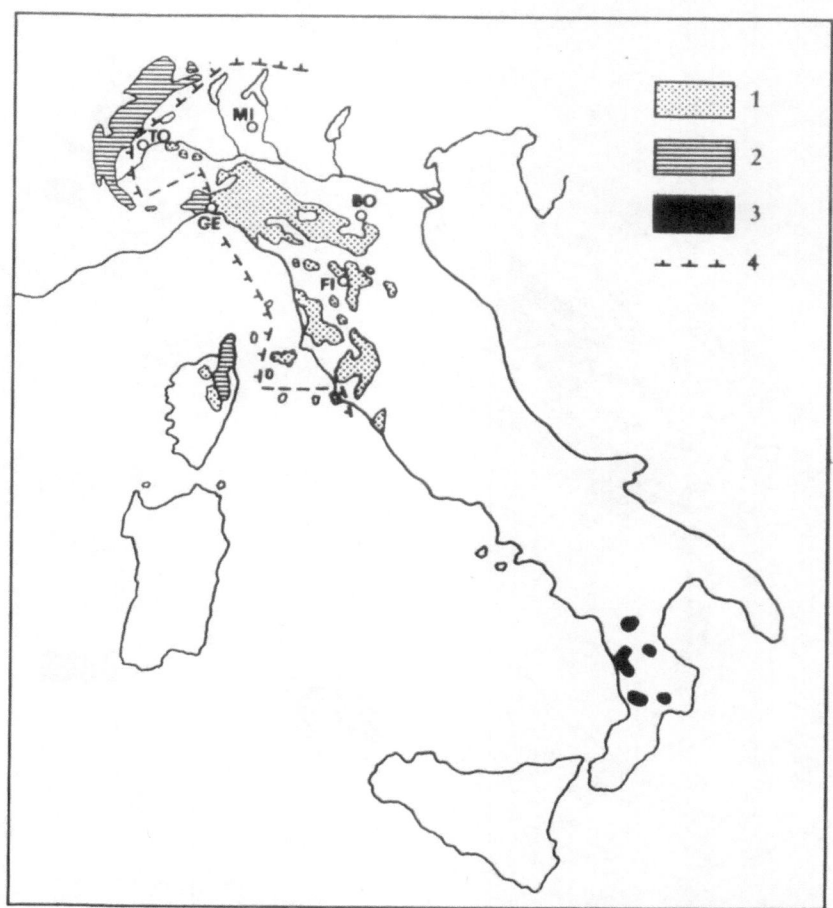

Fig. 1. Distribution of the Italian ophiolitic terrains. 1. Ligurids; 2. Schistes Lustrès of Piedmont, Corsica, Gorgona, Giglio and Argentario; 3. Main ophiolitic outcrops of Calabria; ⊥ Eastern boundary of the alpine orogenic metamorphism. (After Abbate, Bortolotti & Principi, 1980).

nal Hercynian massifs (Aar-Gotthard, Mont Blanc, Pelvoux, Argentera). They are also represented in the basement that forms part of the lower Penninic nappes and in the inner massifs of Sesia-Lanzo and Ivrea. The ultramafics are believed to have originated below the continental crust, i.e. they may represent part of the upper mantle, and were affected by high-grade metamorphism.

Mesozoic rocks: Ophiolites are believed to represent rocks extruded along the axis of a midocean ridge and are usually associated with a sedimentary cover of late Jurassic radiolarian cherts, pelitic rocks often rich in manganese, and early Cretaceous limestones and argillites. Ophiolitic

complexes are more or less metamorphosed. Ophiolites include serpentinites, gabbros and basalts. They are widely distributed in the southern 'Piedmont belt', largely formed of Mesozoic argillaceous, marly and calcareous sediments. Ophiolites were largely fragmented by tectonic movements and occur as discontinuous masses varying from several km to a few m or even less in size.

The Voltri massif of Liguria, probably the largest ophiolite outcrop in Italy, may be included in this unit (the geological boundary between the Alps and Apennines runs N–S parallel to the eastern border of this massif, slightly to the W of Genova).

In the 'Schistes Lustrés' belt of Valais, mafics

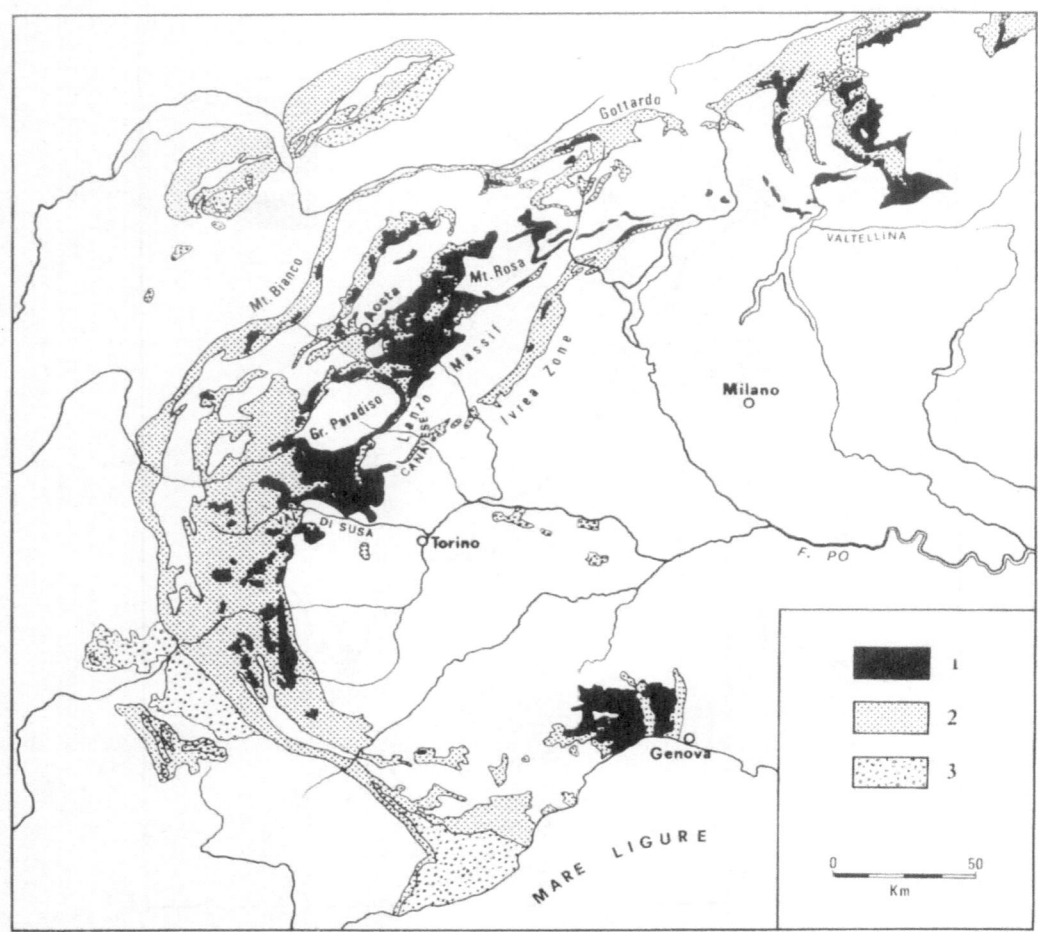

Fig. 2. Distribution of ophiolites in the Alps. 1. Ophiolites undifferentiated (Mesozoic, mainly Jurassic to Early Cretaceous) and mafic and ultramafic rocks of uncertain origin, mainly pre-Triassic (Ivrea zone, Lanzo Massif); 2. Ophiolitic belts, ophiolites and related oceanic sediments; 3. Ophiolitic detritus in flysch-type sediments. (After Dietrich, 1980).

are represented by volcanic and subvolcanic rocks and rare ultramafics and are believed to have been formed in a separate basin, to the north of the main ophiolitic belt.

Northern Apennines

The ophiolites comprise serpentinites, gabbros and basalts. Like the ophiolites of the Alps they are covered by radiolarian cherts, limestones and shales ranging from late Jurassic to early Cretaceous and seem to represent extrusions from a mid-ocean ridge.

According to their geological occurrence the ophiolites may be divided into three groups. To the West, the high-pressure, metamorphic ophiol-

ites of the Gorgona and Giglio islands and of the Argentario promontory. In an intermediate belt, which is mainly developed in eastern Liguria between the coast and the Vara valley, ophiolites and their sedimentary cover occur in a more or less coherent nappe structure. In an eastern belt, ophiolites are dispersed in a large number of small masses and ophiolite breccias in the external Ligurid nappes, an incompetent mass of clays, marls, marly limestones and sandstones which moved North-east as successive landslides, overriding Tertiary deep sea formations. Small masses of granite are occasionally associated with these ophiolites of the eastern Ligurids. The basin in which the oceanic crust was extruded was presumably of limited extent (about 100 km) and formed the south-west extension of the western Alpine basin. Ophiolites of both the internal and external Ligurids show low grade metamorphism and their original structures are well preserved.

Southern Apennine

In northern Calabria, an upper tectonic unit formed by Hercynian crystalline basement rocks, including metamorphics of western Alpine affinity (Dioritic-Kinzigitic Formation) includes metamorphic mafic and ultramafic rocks. This tectonic unit overrides ophiolite-bearing complexes of southern Lucania and northern Calabria and is comparable in age, origin and structure to those of the northern Apennine and of the Penninic belt of the Alps. Here too, ophiolites are fragmented into small masses and show metamorphism of various grades, from low to high pressure facies. Mafic rocks are relatively rare in this area, where ultramafics predominate. Radiolarian chert of late Jurassic age forms a sedimentary cover, as in the northern Apennine.

Vegetation

Western and central Alps

Although in the western and central Alps (Fig. 2) ophiolite outcrops are quite widespread covering a total area of 333.4 km² of which 104 km² in Susa Valley and Canavese, 135.8 km² in Aosta Valley and 35.7 km² in Valtellina, only restricted areas of Canavese, Susa Valley, Aosta Valley and Valtellina have interested botanists.

Canavese, Susa Valley

In the mountainous relief between the Dora Riparia and the Stura di Lanzo rivers, the Mount Musiné flora has attracted attention for its peculiar characters. This mountain (1150 m) is mainly formed by lherzolitic serpentinites associated with gabbros and is characterized by the presence of bare rocks and by shallow reddish soil. The vegetation is typically xerophilous, because the precipitation runs off or enters into the rock crevices (Rigotti 1930). Mussa (1908, 1937) reported on the scarcity of trees: conifers are represented by

a few stunted specimens of *Pinus sylvestris* and by *Juniperus communis*. Mussa also observed a predominance of xerophilous species belonging to the mediterranean element (e.g. *Heteropogon contortus*) or to a more eastern component (e.g. *Stipa pennata*). Among the species restricted to ultramafics he mentioned *Alyssum argenteum* and *Asplenium cuneifolium*. *Cheilanthes marantae* is also present.

In this mountainous relief which belongs to the Lanzo massif, the area Madonna della Neve-Monte Lera (1370 m) has been proposed as a nature park (Associazione Italiana Naturalisti 1979) as it includes also the *locus classicus* and the most important habitat of *Euphorbia gibelliana*, endemic to Piedmont ultramafics. This heliophilous species, closely related to *Euphorbia hyberna*, is restricted to the mountain ridges where soil pH is lower (4.5–5.3) and in an area with high summer rainfall ($\approx$315 mm) coinciding with the flowering stage.

The vegetation cover is mostly mesophilous; many species are typical of the order *Fagetalia sylvaticae* Br.Bl., although the thermophilous element is also present and characterized by species of the order *Quercetalia pubescentis-petraeae* Br. Bl. (e.g., *Tilia cordata* is dominant in the area). The acidity of the soil (pH range is 4.5–5.3) and the high leaching of the surface soil do not generally favour the development of soil lichens. The dominance of crustose forms on the bare rocks is an indication of an early stage of colonization.

Aosta Valley

Research in the Aosta Valley has been limited to two ultramafic areas, one included in the valleys of Ponton, Champdepraz and Champorcher and the other in the Valley of Challant-Ayas.

During his frequent excursions in the Aosta Valley, Vaccari (1903) could not miss the striking contrast between floras of different substrates: he noticed how ultramafics were barren and poor in number of species. He attributed both characters to the chemical nature of the bedrock and to its slow weathering. The vegetation cover seemed richer and better developed where the Pleistocene glaciers had carried downwards calcareous or amphibolitic deposits.

The rocks of the Ponton and Champdepraz valleys are mainly serpentinites: the soil is very stony with almost no trace of humus. In the Ponton valley the eastern side up to 2000 m is covered by pine woods (most probably *Pinus uncinata*) with a few firs (*Picea excelsa*, *Abies alba*) and more numerous larches (*Larix decidua*), associated with *Alnus viridis*, *Betula pendula*, *Juniperus communis*, *Populus alba* and *P. tremula*. It is interesting to note the absence of *Juniperus sabina*, which is very common in all nearby valleys. The grass cover is always scarce and similar in both valleys. Calciphilous species are not totally absent, but often the specimens are few and poorly developed. Several basiphilous species are very frequent among which is *Thlaspi rotundifolium*. Not many special forms were identified: Vaccari first mentioned the presence only on the Ponton ultramafics of a glabrous form of *Saxifraga moschata* (common in all the Aosta Valley); of a more pubescent form of *Saxifraga murithiana*, and described a new species of *Hieracium salassorum* Zahn & Besse. *Alyssum argenteum*, present on several ultramafic areas of the Aosta Valley, is not mentioned but it has been collected also in this area (Vaccari 1904–1911, Vergnano Gambi *et al.* 1979).

Recently in the Ayas area Verger (1979, 1982) has considered the pioneer vegetation of prasinites and serpentinites, the former belonging to the *Androsacion alpinae* Br.Bl. group and the latter to the *Thlaspeion rotundifolii* Br.Bl.; in both cases no particular forms were noted. In the *Pinus sylvestris* woods of the middle part of the Ayas valley the pine deformed branches are noted and the presence of *Alyssum alpestre* is mentioned. The association *Caricetum fimbriatae* is considered typical of the large ultramafic outcrops of the Pennine Alps at high altitudes (Verger 1983).

The extreme eastern branchings of the ophiolitic outcrops which surround Mount Rosa are found in the Vigezzo Valley, which extends from the Ossola Valley to the Canton Ticino (Switzerland). Becherer (1969) has noticed the ferns *Asplenium cuneifolium* and *Cheilanthes marantae* on these ultramafics.

Valtellina

There are only a few investigations on the ophiolitic areas of the central Alps. In Valtellina one of the main outcrops is included in the Val Malenco, which has been considered in a phytosociological survey of the Province of Sondrio (Credaro & Pirola 1975). On the ultramafic boulder terrain of the Val Malenco the presence in the Association *Galeopsido-Rumicetum* Br.Bl. of some species such as *Silene vulgaris* ssp. *prostrata*, *Thlaspi rotundifolium* var. *corymbosum* and *Trisetum distichophyllum* marks the transition to more basiphilous types (*Thlaspion rotundifolii* Br. Bl.).

Northern Apennines

Western Ligurian Apennines

The vegetation of the ultramafic outcrops which extend from Monte Beigua (North of Varazze) to the western side of the Polcevera valley and belong to the formation known as the Voltri Massif (Nos. 1 & 2 in Fig. 3, covering a total area of 79 km^2) has been studied by Martini & Orsino (1969) (Monte Beigua) and by Guido & Petroni (1975) (the hinterland of Arenzano). In both areas the soil is shallow, stony and skeletal. The clay fraction is dominant and organic matter scarce: the pH range is 6.4–6.9.

Hemicryptophytes are the main life forms. Geophytes and therophytes, although well represented, do not contribute appreciably to the plant cover. There is a low number of chamaephytes, but some such as *Daphne cneorum* and *Euphorbia spinosa* are characteristic of rocks and screes. Phanerophytes are not frequent, and the tree cover, with the exception of pine woods (*Pinus pinaster*, *P. sylvestris*), is scarce (Fig. 4).

The typical endemic taxa *Asplenium cuneifolium* and a special form of *Sesamoides pygmaea* var. *firma* (a western mediterranean element found in the driest parts of the ultramafic outcrops of western Liguria) are restricted to the open habitats (rocks, screes, boulders) where the soil is scarce and poor in nutrients. These pioneer species are both heliophilous and xerophilous and are able to withstand unfavourable local conditions, but not strong competition, and gradually disappear with soil evolution, being excluded by mesophilous and sciophilous species. In the area of Monte Beigua *Viola bertolonii* is also present and *Erica cinerea* and *Scorzonera austriaca* are preferentially located on ultramafics. On the

222

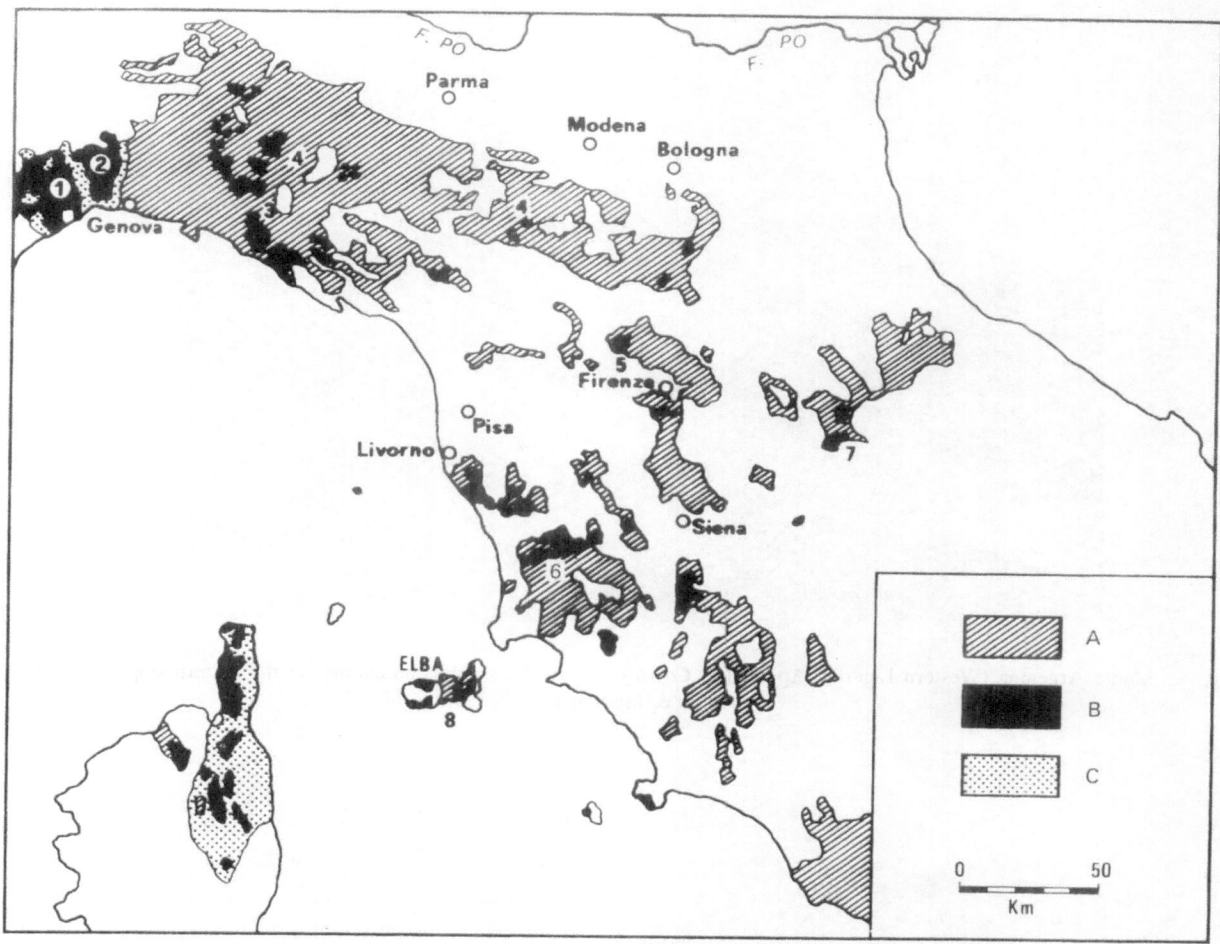

Fig. 3. Distribution of ophiolites in the Apennines. A. Ligurids; B. Ophiolites; C. Schistes Lustrès. Location of ophiolitic outcrops: 1 Voltri, 2 Sestri-Voltaggio, 3/4 Eastern Ligurian-Emilian Apennine, 5 Florence basin, 6 Southern Tuscany, 7 Tuscan Apennine, 8 Elba Island. (After Abbate, Bortolotti & Principi, 1980).

ultramafics there are relict species such as *Carex humilis, Cheilanthes marantae* (Pichi Sermolli & Chiarino-Maspes 1963), *Daphne cneorum, Plantago holosteum*, often located on soils with low calcium content.

On these outcrops are found co-existing species belonging to different geographical elements: species with eumediterranean or mideuropean affinities occur with eurosibiric and circumboreal species. Many alpine and mountain species occur at lower altitudes than usual and can be considered relicts of the last glaciation. Their survival depends both on climatic conditions and on a lack of competition. Tree communities are repre-

sented by pine, oak and beech woods, but the commonest is the *Pinus pinaster* wood of anthropogenic origin; stands or isolated specimens of *Pinus sylvestris* on the mountainous slopes in both areas are worth mentioning as they represent a relict of a colder and drier climate. At higher elevations, with shallow soils and bare rocks, there are oak stands of *Quercus petraea*, with densely branched specimens not higher than 4–5 m. In the more maritime areas, single specimens or small stands of *Fagus sylvatica* reach lower altitudes than usual.

On the Scogli neri (Black cliffs), in the area of Pegli, plant communities are characterized by the

Fig. 4. Monte Argentea (Western Ligurian Apennine): *Corylus avellana* L. shrub communities on the eastern slope. (Photo: Guido & Montanari).

Fig. 5. Rocca negra (Western Ligurian Apennine): Isolated specimens of *Pinus sylvestris* L. on rocky slopes. (Guido & Petroni 1975).

presence of *Asplenium cuneifolium, Euphorbia spinosa*, by a stenophyllic form of *Silene vulgaris* and by a glabrous form of *Galium rubrum*. Sometimes *Alyssum argenteum* is present (Pavarino 1918).

Eastern Ligurian and Emilian Apennines
This area (Nos. 3 & 4, Fig. 3) is characterized by a great fractionation of ophiolitic outcrops (total area 73.3 km², with single outcrops ranging from 0.5 to 15 km²), which are localized in the eastern part of the Ligurian Apennines (Cinque Terre, Val di Vara) and on the northern slope of the Apennines which overlooks the Po Plain.

Eastern Ligurian Apennines: The plant community which characterizes the coastal area of the Ligurian Apennines from Deiva (Savona) to Levanto (Furrer & Hofmann 1969) is the *Euphorbietum spinosae-ligusticae* A. Hofm. & Furrer, with *Euphorbia spinosa* ssp. *ligustica* (Fiori) Pignatti endemic to these ultramafics. This species forms large cushions (0.5–1 m) and shows a well developed taproot (1–2 m). Three other species are typical of this association: there are frequent *Centaurea aplolepa* f. *ligustica* Fiori, and *Potentilla hirta* var. *angustifolia*, and the less abundant *Minuartia laricifolia*. Morphological variants are often observed and even *Pinus pinaster* has shorter needles than usual and seeds of reduced size and weight.

Trebbia river basin: The vegetation of the mountainous chain (including Monte Aiona, 1700 m) in the upper valleys of the Trebbia and Aveto rivers, is of particular interest. The area had been glaciated (Würmian) and has the highest rainfall in the whole Apennines (≈3000 mm/y). The ultramafic rocks of Monte Aiona can be related to lherzolitic types with every degree of transformation into serpentinites; they are closely associated with outcrops of spilitic basalts whilst gabbros are lacking. The top of the mountain looks like a highland with a chaotic distribution of rocks with a rough grey surface; its vegetation has been studied by Guido & Montanari (1983).

Trees are represented by a beech wood which reaches 1600 m and on the western slopes by specimens of *Pinus uncinata* var. *rotundata*, which is indicative of an upper mountain horizon now completely absent following the climatic changes which took place in this area. Among the pioneer species of the scree vegetation: *Festuca spectabilis, Rumex scutatus, Silene vulgaris* var. *angustissima* Fiori and the endemic *Asplenium cuneifolium*. In the mesothermophilous meadow another typical endemic taxon *Minuartia laricifolia* ssp. *ophiolitica* Pignatti is present along with *Sesamoides pygmaea* var. *firma*, but some typical Tuscan endemics (*Alyssum bertolonii, Armeria denticulata*, etc.) are lacking (Fig. 6).

The analytical data compared with other Apennine ophiolites, show high calcium concentrations in rocks (CaO=3.2–3.5%) and soils (CaO=3.02–6.1); soil pH ranges from 5.0 to 6.6.

The name of the mountains, Sassi Neri (1000 m), Pietra Corva (1060 m), Roccabruna (1419 m) (Black Stones, Raven Stone, Dark coloured Rock), indicate in the Bobbio area the slowly weathered cliffs, peaks and slopes of serpentinites or gabbro debris. For such a relatively high altitude, the winter is rather mild and the summer particularly dry. This climatic condition is due to the influence of strong winds and to the nature of the substrate.

In the first botanical investigation Pavarino (1915) compared the ultramafic vegetation of the Bobbio area with that of adjacent calcareous formations.

On the lherzolitic serpentinites of Monte Sassi Neri and Pietra Corva, or di Corvo (the latter has been taken recently in consideration for the establishment of a nature park) plant life is xeric, poor and characterized by the presence of *Alyssum argenteum, Calluna vulgaris, Juniperus communis*, etc. On the rocky cliffs some xerophilous species faithful to ultramafics are growing (*Cheilanthes marantae, Cerastium arvense*, etc.). Although the ultramafic flora is always poorer in number of species, than that of similar calcareous areas, Pavarino remarked on the abundance of ferns and of species of the Caryophyllaceae, Crassulaceae, Labiatae and Compositae. Some species show characteristics which deserve the rank of varieties and many are represented by dwarf specimens. Some anatomical modifications to reduce transpiration, such as the thickened walls in the leaf epidermal cells in *Saxifraga cuneifolia*, the viscous coating in *Silene nutans*, and the thick indumentum in *Cheilanthes marantae*, confer an overall typical dark grey colour to the plant cover.

Fig. 6. Monte Aiona (Eastern Ligurian Apennine): Mesothermophilous pasture with populations of *Allium schoenoprasum* L. (Guido & Montanari 1983).

In spite of the fact that a number of plants are preferentially located on ultramafics, only two species are constantly linked with this substrate: *Alyssum argenteum* and *Asplenium cuneifolium*.

The rock and scree communities of this area can be referred to the *Thlaspetea rotundifolii* Br. Bl. (Pirola 1967). Pignatti Wikus & Pignatti (1977) also describe the various associations on ultramafics in the northern Apennine.

Secchia river basin: in the Modena Apennine, ophiolitic outcrops are rather common and have been thoroughly studied (Pellizzer 1961); analyses of the soil composition are given by Negodi (1943).

On the dark green or black lherzolitic serpentinite of Varana (400–450 m, soil pH=6.0–6.8), Negodi (1941) lists 216 species of which 195 are ubiquitous, a few are calcicolous and only two are exclusively linked to soils low in calcium: *Cheilantes marantae* and *Achillea tomentosa*. The vegetation of the Varana outcrop was also described by Bertolani Marchetti (1953), and can be ascribed to the submontane horizon with a tree community with *Ostrya carpinifolia* associated with *Acer monspessulanus*, *Fraxinus ornus*

and *Quercus pubescens*. Xerophilous elements are present in the barren areas. The glacial and postglacial periods have eliminated the typical elements of the mediterranean flora, but some mediterranean or continental elements survive on these arid and exposed soils, as *Cheilanthes marantae* or *Achillea tomentosa*, the latter being a western mediterranean steppic element which can be considered as a continental relict. The flora of this outcrop has eighty-one species common to Monte Ferrato (Tuscany) but, with the exception of the two examples mentioned and the addition of *Melica ciliata* ssp. *magnolii* (another species typical of dry habitats) no other plant often found on ultramafics has been recorded. The typical endemics are lacking here.

Negodi (1943) has compared the vegetation of the Varana outcrop with that of the nearby Sasso Tignoso (1430 m), which is formed by rocks of red basalt. Negodi, in this case, considered the effects of altitude, and although generally not inclined to give great importance to soil composition in relation to plant distribution, he tentatively explained the absence of the archaic endemisms typical of ultramafics or of species

226

Fig. 7. Monte Nero (upper Nure valley, Emilian Apennine): Stunted specimens of *Pinus mugo* Turra. (Photo: Guido & Montanari).

preferentially located on this substrate, to the presence of higher calcium concentrations in red basalt, which might exclude species more strictly linked to ultramafics.

Tuscany

Almost all the ultramafic outcrops in Tuscany have been throughly investigated, particularly in the Florence area (Monte Ferrato, Impruneta (No. 5, Fig. 3, total area 6.7 km²) and in the Tuscan Apennines (Upper Tiber Valley, No. 7, Fig. 3, total area 7.4 km²).

Florence area
Monte Ferrato (350–440 m): Lherzolitic serpentinites (with quarries of the 'green marble of Prato'), gabbro and basalt, form this complex of hills near Prato (Florence). The soil is often skeletal

or shallow and has a low water-holding capacity, and pH ranges from 6.6 to 7.4 on the serpentinite, and from 6.8 to 7.0 on gabbro.

This complex has been frequently visited by botanists because it is close to Florence and isolated from the surrounding formations. Messeri (1936) conducted an important floristic, ecological and phytogeographical study of Monte Ferrato and prepared a guide to this site for the 7th International Phytogeographical Excursion. The vegetational aspect of M. Ferrato is that of a thin pine wood with *Pinus pinaster* (of anthropogenic origin) and the remains of a population of *Quercus cerris* and *Q. pubescens*. The predominance of caespitose hemicryptophytes and suffrutescent chamaephytes in the most xeric habitats lends a more steppic than mediterranean character to the plant cover.

The vegetation can be distinguished according to different habitats; the ultramafic endemics and plants with the highest degree of morphological variants (Messeri 1936, Pichi Sermolli 1936), are located on rocks, screes and pebbles. Because of the scarce tree cover, the dryness of the substrate and the dark colour of the rocks, the local climatic conditions are more continental (temperature ranges are lower and rainfall higher and more uniformly distributed in comparison to the Florence basin) so that many continental relicts of various origin are located here. The endemics which confer a typical aspect to this vegetation are forms which belong to central, southern or eastern European species which are also found in Russian or Bohemian steppes, such as *Stachys recta* or rupicolous species of central and northern Europe such as *Asplenium cuneifolium*; or mediterranean species which here reach their most northern distribution (*Thymus striatus*). In the thicker wood stands, limited mostly to gabbro, some atlantic species are also able to survive (e.g., *Genista pilosa*, *Ulex europaeus*).

Arrigoni *et al.* (1983) distinguished two vegetation types: on the screes and debris, a pioneer type the *Armerio-Alyssetum bertolonii* which includes the Tuscan ultramafic endemics. On lithosols, or more highly evolved soils there is a type belonging to the *Bromion*, *Brometalia erecti* Br. Bl. related to a steppic meadow.

The lichens which are important weathering agents, have been studied by Sambo (1927) and

Fig. 8. Impruneta (Florence area): Pine wood with *Pinus pinaster* Aiton, *Pinus pinea* L. and *Juniperus oxycedrus* L. ssp. *macrocarpa* (Sibth. et Sm.) Ball.

Cengia Sambo (1937), who reported dwarf and narrower forms, particularly in *Parmelia*. A first contribution to the bryology of Monte Ferrato has been recently made by Cortini Pedrotti (1975).

Impruneta (300 m): Bargoni (1943) made a statistical-ecological survey of the vegetation of this well known ophiolitic outcrop which is a few km from Florence. This rather flat area is covered by a thin pinewood (*Pinus pinaster*) mixed with *Juniperus oxycedrus* ssp. *macrocarpa*. The plant species include the typical Tuscan endemics that are found on Monte Ferrato. These species are not particularly late in resuming growth or in flowering, whilst the ubiquitous species are markedly late. Many species show the persistence of dried up stems and branches for long periods, so seeds can take advantage of a longer dispersal period.

Metalliferous hills
Several limited outcrops of serpentinites, gabbro and basalt associated with formations of different constitution exist in the south-western Tuscan area which includes the 'Colline metallifere' (No.

6 Fig. 3, total area 45.4 km², with single outcrops ranging from 4 to 6 km²).

The flora of the hills in the province of Pisa (Monte Pelato, Gabbro, etc. 300–380 m), of Siena (Monte Gabbro 485 m) and of Grosseto (Monte Gabbro 560 m) were examined by Fiori (1920) and Vergnano (1953a). The flora includes the typical endemics of Monte Ferrato; these species are absent on gabbro and basalt.

Tuscan Apennines
Upper Tiber Valley (Monte Murlo, Monte Petroso, Monti Rognosi of Albano and Montauto): the well known work of Pichi Sermolli (1948) on the flora and vegetation of this area made an important contribution to the knowledge of the ecology of the flora and vegetation of ultramafics.

Soil analyses from samples of both serpentinite and gabbro show that the two substrates have important distinctive properties: soils derived from serpentinite are shallow, rich in stones and of dark brown or blackish colour. Gabbro is more easily weathered producing less stony, grey coloured soils, with deeper profiles, of 50–75 cm.

228

Fig. 9. Impruneta: *Cheilanthes marantae* (L.) Desv.

Generally, the ultramafic soils have a higher Mg/Ca quotient and a pH range between 6.4 and 7.4. This flora compared with that of other Tuscan areas shows the influence of a higher rainfall and of a tendency toward subcontinental temperatures; the biological spectrum indicates that it belongs to the submediterranean vegetation with some euro-centralasiatic affinities. Plant communities were analyzed in detail and divided according to different habitats and degree of soil evolution: the typical ultramafic vegetation is that found on rocks, screes and boulder terrain.

Pastures represent the less typical ultramafic communities.

The tree vegetation, mostly of *Quercus pubescens*, did not originate spontaneously, as it is the case of *Quercus cerris* on grabbo, but came from the neighbouring soils and did not affect to a great extent the existing typical ultramafic vegetation.

Pichi Sermolli considers as *typical serpentinophytes* those species exclusively restricted to ultramafic substrates and as *preferential serpentinophytes* those found also on other magnesium rich soils. *Serpentinicolous relicts* are those plants which once had a wider distribution but nowadays are almost exclusively or preferentially restricted to ophiolitic outcrops.

The morphological modifications observed in plants growing on ultramafic soils are also described in detail, adding to the ones previously quoted stenophyllism and glaucescence. They are mainly attributed to edaphic factors, particularly to drought.

On the basis of the geological and climatic events which occurred since the Upper Miocene, the possible causes of the establishment and survival of the various elements of the ultramafic flora are discussed, and the genesis of this flora outlined.

Comparing the vegetation of the Upper Tiber Valley with that of Monte Ferrato and Impruneta, one notes the absence of some typical endemics, such as *Centaurea carueliana* and *Euphorbia prostrata*. There are additional varieties restricted to ultramafics (*Campanula glomerata* var. *serpentinicola* Pic.Ser., *Silene cucubalus* var. *zlatiborensis* (Novák) Pic.Ser., *Stachys recta* var. *ophiolitica* Pic. Ser. and several serpentinicolous relics (*Cheilanthes marantae*, *Cirsium pannon-*

Fig. 10. Impruneta: *Alyssum bertolonii* Desv.

icum, Daphne cneorum, Plantago holosteum, Scorzonera austriaca, etc.).

Elba Island

The ophiolitic outcrops of Elba (No. 8 Fig. 3, 10.4 km²) have not received particular attention, probably because they also lack endemics. A short survey (Corti 1940) of the flora of Monte Orello, which is constituted mainly of red basalt, points out its unusual plants.

Southern Apennines

Investigations on the vegetation of the ophiolitic outcrops of Calabria, although particularly interesting for their origin and position, are almost completely lacking. They cover approximately an area of 26 km².

Conclusions

This survey aimed at giving a general outlook on the main botanical investigations of Italian serpentinized areas. Only a few have been studied compared to the great number of ophiolitic out-

crops distributed on the whole peninsula and on the islands.

The species which are generally recognized as typical endemics (serpentinophytes) are the following:

1. *Asplenium cuneifolium*, in the Alps and northern Apennine.
2. *Minuartia laricifolia* ssp. *ophiolitica* Pignatti, in the northern Apennine.
3. *Alyssum bertolonii*, almost exclusively Tuscan, and its vicariant *Alyssum argenteum* in the Alps and in the northern Apennine. The populations of the northern Apennine and of the Susa Valley belong to *A. argenteum* f. *bertolonioides* Nyárády. These two species are allopatric and have the same (2n=16) chromosome number (Arrigoni *et al.* 1983). *A. argenteum* seems restricted to the Alpine area (Aosta Valley).
4. *Euphorbia nicaeensis* var. *prostrata* (Fiori) Arrigoni, in Tuscany.
5. *Euphorbia spinosa* ssp. *ligustica* (Fiori) Pignatti, in the eastern Ligurian Apennine.
6. *Euphorbia gibelliana*, limited to Piedmont.
7. *Armeria denticulata*, with an almost exclus-

ive distribution in Tuscany; its vicariant *A. marginata* (Levier) Bianchini in the northern Apennine does not seem restricted to ultramafics.

8. *Stachys recta* ssp. *serpentini* (Fiori) Arrigoni, and the var. *ophiolitica* Pic. Ser., in Tuscany.

9. *Thymus acicularis* var. *ophioliticus* Lacaita, in Tuscany.

10. *Centaurea aplolepa* ssp. *carueliana* (Micheletti) Dostál in Tuscany.

In some other cases it is difficult to establish that species whose distribution or taxonomic relevance are still uncertain are restricted to ultramafic habitats. This is the case for: *Leucanthemum pachyphyllum* Marchi et Illuminati (*Chrysanthemum leucanthemum* var. *crassifolium* Fiori), *Festuca robustifolia* and *F. inops* (Arrigoni *et al.* 1983), *Biscutella coronopifolia* (Pignatti Wikus & Pignatti 1977), *Sesamoides pygmaea* var. *firma* (Martini & Orsino 1969), *Silene cucubalus* var. *zlatiborensis* (Novák) Pic. Ser., *Campanula glomerata* L. var. *serpentinicola* Pic. Ser. (Pichi Sermolli 1948), to mention only the most important.

The vegetation of the Tuscan outcrops is considered the most typical because it shows the highest number of endemisms and peculiar forms. It can be related with that of the Balkan peninsula and of Portugal which are both richer in endemics and numbers of species as a consequence of the much lesser influence of Pleistocene glaciations and possibly of a greater consistence of the original indigenous floristic stock (Pignatti Wikus & Pignatti 1977).

The genesis of the ultramafic flora probably goes back to the middle Tertiary, and a relict of this period is most probably *Cheilanthes marantae*. The strong climatic changes which occurred during Pleistocene differently affected the various elements of this flora: they certainly did not have the same climatic requirements, but the xeric habitat and the reduced competition of serpentinized areas enabled many of them to survive.

Ecophysiological Research

The peculiar characteristics of serpentine vegetation, particularly in Tuscany, have stimulated

many investigations. Attention has focused not only on vegetational aspects, but also on the ecology and physiology of plants living on the ultramafic substrates. The main lines of research can be summarized as follows:

1. Effects of ultramafic soils on:
   (a) plant morphology and anatomy
   (b) elemental composition of native vegetation
   (c) elemental composition of cultivated plants
   (d) other metabolic aspects.

2. Physiological aspects of nickel uptake and tolerance in *Alyssum bertolonii*
   (a) elemental composition
   (b) interaction between nickel and other ions, particularly calcium and magnesium
   (c) nickel localization
   (d) nickel tolerance and uptake
   (e) nickel complexing compounds
   (f) enzyme adaptation
   (g) seed germination.

I shall now discuss these lines of research in turn.

Effects of ultramafic soils

Plant morphology and anatomy

Bargoni (1940) investigated whether one of the morphological variants in serpentine plants (i.e., reduction in size of leaves) was maintained when plants (*Armeria denticulata*) were cultivated on normal soil. The reduction in leaf size as other xeromorphic characters (thick cuticle, multi-layered palisade tissue, etc.) were not maintained on normal soil, only the "denticulate" character which distinguishes the leaves of this endemism, was present in all the specimens.

Lisanti (1952, 1958) grew *Stachys recta* in Hoagland solution with the addition of Ni^{2+}, SO_4^{2-} and Mg^{2+}. He noted that the reduction in leaf size, dwarfism and stunting appeared only when nickel was added to the solution or when the plants were grown on ultramafic soil. Lisanti concluded that the narrower leaves, characteristic of *Stachys serpentini*, were induced by the presence of nickel, although more recent reports show that in addition to nickel, other elements such as boron, chromium and cobalt at high concentrations can produce similar effects (Shkolnik & Smirnov 1970). Narrower and weaker leaves were also observed in oat plants when Ni^{2+} or Co^{2+} was added to the nutrient solution (Vergnano & Hunter 1953, Vergnano 1953b). The anatomical alterations due to excess nickel were very similar, or identical, with those observed in a soil derived from ultramafic glacial drift in Aberdeenshire (Hunter & Vergnano 1952).

Elemental composition of serpentine plants
Research on the elemental composition of these plants (Vergnano 1945, Minguzzi & Vergnano 1953) began with the analysis of some species restricted to serpentinized-ultramafic soils (the so called serpentinophytes), *Alyssum bertolonii, Euphorbia prostrata, Centaurea carueliana* and of an ubiquitous species, *Helichrysum italicum*, collected on the ultramafic outcrop of Impruneta (Florence). Minguzzi & Vergnano (1948) reported the exceptional nickel concentration of *A. bertolonii*. The total rock and soil composition reflected the general characters of ophiolitic outcrops with high magnesium (14.7% MgO) and iron (Fe_2O_3 18.34%) concentrations. However, because iron is principally present as magnetite, it is mostly unavailable for the plants. Because the extent of the Impruneta outcrop is limited and surrounded by different types of rocks, its calcium (CaO 1.10%) and potassium (K_2O 0.8%) concentrations are higher than those found in other Tuscan ultramafites. Further, soil nickel (NiO 0.33%) and chromium (Cr_2O_3 1.66%) are particularly high.

The plants showed a high magnesium concentration (2,000–8,000 $\mu g\,g^{-1}$), but calcium concentration was higher than expected (5,000–12,000 $\mu g\,g^{-1}$). As a result the Mg/Ca quotient (on a dry mass basis) was generally lower than 1. Iron concentrations were very variable (800–2,500 $\mu g\,g^{-1}$) and in *Helichrysum italicum* not higher than those observed in samples of the same species collected on sandstone, so that a possible iron toxicity in the area seemed unlikely.

The concentration of nitrogen, phosphorus and potassium was low, but on the other hand the scarcity of chlorides and sulphates, considered by Novák (1928) a common feature of these soils, was not confirmed by plant analysis.

Chromium, cobalt and nickel were also ana-

lysed (Vergnano 1958a): an exceptional nickel concentration, at least an order of magnitude higher than for other plants in the same area, was found in leaves, flowers and seeds of *Alyssum bertolonii*, so that this species can be considered a nickel-accumulator. Among the other species particularly high nickel values were found in the dried up stems of *Centaurea* (290 $\mu g\,g^{-1}$) and *Helichrysum* (480 $\mu g\,g^{-1}$). Chromium values were higher than average in leaves and stems of *Euphorbia* (280 $\mu g\,g^{-1}$) and *Helichrysum* (178 $\mu g\,g^{-1}$). In all samples cobalt showed the lowest values, ranging from 2 to 40 $\mu g\,g^{-1}$. The conclusion of these investigations was that although many factors were involved in ultramafic soils, some of the peculiar morphological variants shown by the plants, could be ascribed to high concentrations of chromium, cobalt and nickel (Vergnano 1958b). No abnormal levels of other trace elements such as boron, copper or manganese, were found in the Impruneta plants (Vergnano Gambi 1966).

Plants from other Tuscan outcrops (Monte Ferrato and Cecina, near Leghorn), *Cheilanthes marantae* and *Plantago holosteum*, showed fairly high nickel concentrations (200–300 $\mu g\,g^{-1}$) in the roots, where the quotients of Cu/Fe and Mn/Fe showed constant and typical values (Pacetti 1969).

More recently Sasse has examined the composition of several plants and soils (total and extractable elements) from three Tuscan outcrops (Impruneta, Monte Ferrato and Pieve S. Stefano, in the Upper Tiber Valley). His data, which also include other European ultramafic outcrops, show that "a dense and richer vegetation is always and primarily correlated with a lower nickel content of these soils" (Sasse 1979a). Once again, the importance of this element in characterizing serpentine vegetation is stressed. With the exception of *A. bertolonii*, the highest nickel concentrations generally occur in the roots (Sasse 1979b).

A comparison between soil and plant concentrations of chromium, cobalt and nickel in ultramafic outcrops in Tuscany (Monte Ferrato, Impruneta), in the Emilian Apennine (Bobbio), and in a very restricted area of the Eastern Ligurian Apennine (Framura, near Deiva) was carried out by Vergnano Gambi *et al.* (1982). Tables 1 and 2 summarize the data; those referring to the Ligur-

Table 1. The total and extractable nickel, chromium and cobalt ($\mu g\,g^{-1}$ on air dry soil) and nickel in soil solution ($\mu g\,l^{-1}$), and the total calcium, magnesium and iron (%) of the soils from the Appennine collection sites. Average values on at least two replicates

	M. Ferrato	Impruneta	Bobbio	Framura
Nickel total	2700	2500	2300	2700
CH₃COOH 2.5%	65	88	88	111
HCl 0.1 N	35	58	42	76
H₂O dist.	<1	<1	<1	<1
Soil solution	100	170	-	-
Chromium total	1700	3900	1700	1300
CH₃COOH 2.5%	5	2	4	5
HCl 0.1 N	<1	<1	<1	<10
H₂O dist.	<1	<1	<1	<1
Cobalt total	141	312	188	177
CH₃COOH 2.5%	6	12	18	18
HCl 0.1 N	12	12	18	18
H₂O dist.	<1	<1	<1	<1
Calcium	0.60	0.81	2.06	-
Magnesium	14.70	14.25	20.50	-
Iron	5.36	6.40	3.05	-
pH	7.7	7.4	7.2	7.2

ian plants have been omitted as their chromium, cobalt and nickel concentration is within the range of the average values for other areas. If we consider the nickel concentrations of these different groups of plants with the exception of *A. bertolonii* in Tuscany and of its vicariant *A. argenteum* f. *bertolonioides* in the northern Apennine, the great majority has a nickel leaf concentration below 50 $\mu g\,g^{-1}$. Only leaves of plants with a well developed rosette as *Armeria* and *Plantago* show a higher nickel concentration (130–200 $\mu g\,g^{-1}$), but on the whole serpentine plants tend to limit nickel uptake and translocation. Of the three elements, nickel reaches the highest values in plant tissues; none of the species examined had such high concentrations of chromium or cobalt. Generally chromium shows higher values than cobalt, but it is generally concentrated in the roots and only occasionally reaches high leaf concentrations as in *Cerastium* (36 $\mu g\,g^{-1}$), *Plantago* (62 and 91 $\mu g\,g^{-1}$) and *Armeria* (70 $\mu g\,g^{-1}$). Nickel and cobalt are positively correlated in serpentinophytes, because nickel accumulation might also possibly favour cobalt uptake. Calcium, magnesium and iron concentrations were also examined (Table 3). Normally, the high magnesium of ultramafic soils is

reflected in high magnesium concentrations in plant organs, but this might also indicate an unusual magnesium and low calcium requirement observed in some experiments (cf. Proctor & Woodell 1975) and recently demonstrated by Main (1981) for the serpentine endemic *Poa curtifolia* Scribn. Nevertheless, there are species with unusually high concentrations of calcium, such as both species of *Alyssum*, or with high calcium concentrations not associated with exceptional nickel values, such as *Dianthus* and *Sedum*. In these cases the tissue Mg/Ca quotients are low, while other plants such as *Cerastium ligusticum*, *Euphorbia prostrata* and both *Plantago* spp. have a high Mg/Ca quotient. Iron values are occasionally high and sometimes comparable with those observed by Ritter-Studnička & Dursun-Grom (1973) in serpentine vegetation in Bosnia.

Although interesting correlations between elements are detectable, it is difficult to trace their possible metabolic importance.

The investigations were also extended to the vegetation of some alpine ophiolitic outcrops in the Aosta Valley (located in the valley of the Evançon river in its lower (Challant) and upper part (Ayas) (Vergnano Gambi *et al.* 1982). Chromium, cobalt and nickel concentration was as-

Table 2. Nickel, chromium and cobalt concentration ($\mu g\,g^{-1}$ on dry weight) of typical serpentinophytes and non-serpentinophytes collected at Impruneta, M. Ferrato and Bobbio

	Impruneta			M. Ferrato			Bobbio		
	Ni	Cr	Co	Ni	Cr	Co	Ni	Cr	Co
Serpentinophytes									
Asplenium cuneifolium				25	4	<1	40	7	<1
Minuartia ophiolitica									
leaves				36	13	8	12	3	<1
woody stems				266	83	18	65	<1	<1
Alyssum bertolonioides							5,620	21	6
Alyssum bertolonii	9,330	8	62	11,860	23	35			
Euphorbia prostrata	115	4	12	53	<1	6			
Armeria denticulata									
green leaves	20	4	3	23	5	3	12	6	<1
old leaves	158	70	12	106	54	12			
Stachys serpentini				23	5	7			
Thymus ophioliticus	37	7	2	30	7	<1			
Non-serpentinophytes									
Cheilanthes marantae	35	3	<1	43	2	<1			
Ceterach officinarum				73	11	6			
Cerastium ligusticum	56	38	<1	70	36	<1	48	32	<1
Silene italica									
leaves	35	14	<1	29	3	<1	6	<1	<1
roots	64	24	<1						
Dianthus sylvestris	27	4	6				35	6	<1
Sedum sp.	18	<1	<1	12	2	<1			
Sanguisorba minor									
leaves	47	11	<1	34	6	<1			
roots	167	43	<1	194	19	9			
Euphorbia cyparissias							88	13	<1
Plantago holosteum	217	91	13						
Plantago lanceolata				135	62	<1			
Senecio erucifolius							94	13	9
Centaurea rupestris				12	2	<1	18	3	<1

sessed in a number of species collected at different altitudes: among the plants sampled in the Challant Valley (1400–2000 m) none showed high nickel concentrations with the exception of *Alyssum argenteum* which, as *A. bertolonii* in Tuscany and *A. bertolonioides* in the northern Apennine, is a nickel accumulator (5000–9500 $\mu g\,g^{-1}$). Instead, almost all the species collected in the Ayas Valley (Table 4), and in particular those growing on the terminal moraine formed by antigoritic serpentinites of the Verra glacier (2500 m) had a higher than usual nickel concentration. In some cases (*Thlaspi rotundifolium*, *Cardamine resedifolia*, *Luzula lutea* and *Linaria alpina*) nickel concentrations exceeded 1000 $\mu g\,g^{-1}$ indicating that in this substrate the element was highly available.

From the soil analyses (Table 5) it is evident that total nickel concentration is not markedly different in the Apennine (Table 1) or in the Alpine soils, but nickel concentration in the extractable fractions, and particularly in the soil solution is more closely related to plant concentration, demonstrating the higher nickel availability of both alpine localities. In these plants the accumulation of nickel in non-permanent organs such as leaves can represent a mechanism of detoxification of the underground parts, which generally show lower values and which are in these habitats the only perennating organs.

To verify if nickel accumulation was related to seasonal and/or local conditions, samples of the most interesting species were collected in different years and in several localities of the Ayas Valley (Vergnano Gambi & Gabbrielli 1981):

Table 3. Calcium, magnesium and iron concentration. (% of dry weight) of typical serpentinophytes and non-serpentinophytes from Impruneta, M. Ferrato and Bobbio

	Impruneta			M. Ferrato			Bobbio		
	Ca	Mg	Fe	Ca	Mg	Fe	Ca	Mg	Fe
Serpentinophytes									
Asplenium cuneifolium				0.42	0.40	0.01			
Minuartia ophiolitica				1.17	0.54	0.09	0.63	0.31	0.01
Alyssum bertolonioides							4.84	0.23	0.02
Alyssum bertolonii	3.44	0.69	0.05						
Euphorbia prostrata	1.18	0.89	0.26	0.57	0.83	0.01			
Armeria denticulata	1.30	0.63	0.03	0.62	0.81	0.03	0.51	1.02	0.02
Stachys serpentini				0.43	0.31	0.02			
Thymus ophioliticus	1.17	0.57	0.05	0.83	0.56	0.03			
Non-serpentinophytes									
Cheilanthes marantae	0.37	0.13	0.04	0.19	0.15	0.05			
Ceterach officinarum				0.36	0.41	0.08			
Cerastium ligusticum	0.25	0.12	0.22	0.34	1.40	0.30	0.32	0.65	0.18
Silene italica	0.78	0.42	0.10	0.74	0.60	0.01	0.76	0.63	0.01
Dianthus sylvestris	1.02	0.45	0.02				2.36	0.31	0.02
Sedum sp.				1.55	0.64	0.02			
Sanguisorba minor				1.30	0.64	0.03			
Euphorbia cyparissias							1.55	0.85	0.03
Plantago holosteum	0.56	0.85	0.07						
Plantago lanceolata				0.81	1.37	0.40			
Senecio erucifolius							0.77	1.10	0.06
Centaurea rupestris				0.55	0.34	0.01			

Fig. 11. Ayas Valley (Aosta): Lateral moraine af the Verra glacier.

Table 4. Nickel, chromium and iron ($\mu g\,g^{-1}$ on dry weight), calcium and magnesium concentration (%) of plants collected in the western Alps (Ayas Valley, Aosta)

Species	Collection site	Ni	Cr	Co	Fe	Ca	Mg
Trisetum distichophyllum	Moraine	82	<1	4	218	0.22	0.28
Luzula lutea, red leaves	2500 m	1130	8	12	798	0.72	0.67
green leaves		2390	5	29	240	1.18	1.26
Salix myrsinites, leaves		522	95	29	4190	0.85	0.90
stems		446	46	18			
Cerastium latifolium		710	110	18	5350	0.46	1.24
Silene vulgaris		223	13	5	850	0.96	0.90
Cardamine resedifolia		1280	9	28	395	1.78	1.03
Thlaspi rotundifolium, leaves		7880	9	130	503	0.85	0.52
stems		5000	<1	24			
Saxifraga autumnalis		106	26	6	1520	1.62	1.79
Linaria alpina		2610	9	24	558	1.08	1.38
Plantago serpentina		210	5	<1	180	0.59	0.49
Cystopteris fragilis	Pian	6	<1	<1	350	0.23	0.25
Thlaspi alpestre	di Verra	4510	21	35	1120	0.58	0.40
Bupleurum ranunculoides	2000 m	165	<1	6	268	0.55	0.40
Gentiana amarella		64	9	12	580	0.24	0.49
Solidago virgaurea		730	15	6	880	0.41	1.09
Chrisanthemum leucanthemum		141	17	12	1240	1.0	1.57
Biscutella laevigata		18	2	12	173	5.35	0.53

Thlaspi rotundifolium was the only species which consistently, in every habitat and collection year, accumulated nickel to a great extent (6,600–

Table 5. The total and extractable nickel, chromium and cobalt ($\mu g\,g^{-1}$ on air dry soil) and nickel in soil solution ($\mu g\,l^{-1}$), and the total calcium, magnesium and iron (%) of two collection sites in the upper Ayas Valley (Aosta)

	Pian di Verra	Moraine
Nickel total	1180	1700
CH$_3$COOH 2.5%	70	164
HCl 0.1 N	82	129
H$_2$O dist.	<1	<1
Soil solution	1100	1300
Chromium total	1990	2200
CH$_3$COOH 2.5%	1	1
HCl 0.1 N	<1	<1
H$_2$O dist.	<1	<1
Cobalt total	80	159
CH$_3$COOH 2.5%	7	12
HCl 0.1 N	6	6
H$_2$O dist.	<1	<1
Calcium	1.50	3.00
Magnesium	11.40	16.01
Iron	6.60	6.00
pH	6.6	6.9

23,800 $\mu g\,g^{-1}$). The nickel accumulation of the other species (*Cardamine*, *Linaria*, *Minuartia*) was also influenced by the substrate composition and by the climatic characteristics of the collection year, but these species showed also a wide range of low values, from 200–2,300 $\mu g\,g^{-1}$ for *Cardamine*, from 150–1,990 $\mu g\,g^{-1}$ for *Linaria*, from 240–1,900 $\mu g\,g^{-1}$ for *Minuartia*, so that the unusual concentration above 1,000 $\mu g\,g^{-1}$ was not typical of the species as it was in *Thlaspi*.

A further indication of a marked tendency of *Cerastium* to accumulate chromium is the high concentration of this element in *C. latifolium* (55–190 $\mu g\,g^{-1}$). Also relevant is the chromium concentration in *Campanula scheuchzeri* (20–110 $\mu g\,g^{-1}$). Cobalt values do not generally exceed 40 $\mu g\,g^{-1}$, only *Thlaspi rotundifolium* showed higher concentrations (100–700 $\mu g\,g^{-1}$).

In the moraine plants with lower nickel concentrations there is a significant positive correlation between nickel and iron. The majority of the species shows an almost equal distribution of nickel between roots and shoots, with a tendency to foliar accumulation when concentrations are particularly high (Gabbrielli *et al.* 1987).

Elemental composition of cultivated plants
The elemental composition of some crops (wheat,

oats, barley and *Medicago*) grown on ultramafic soils in the area of the Monti Rognosi in the Upper Tiber Valley (Vergnano 1959a) was also investigated. The samples were collected in May and June and all the plants showed signs of chlorosis and stunting. Soil pH ranged between 7.2 and 7.4.

In all the crops nitrogen and phosphorus were always at concentrations which are associated with deficiency symptoms. Magnesium was generally higher than calcium uptake, so that the Mg/Ca quotient in leaf extracts was often higher than 1 in oats and wheat. Iron showed normal values (76–300 $\mu g\,g^{-1}$), chromium and nickel were both relatively low (4–15 $\mu g\,g^{-1}$). No nickel toxicity symptoms as those described for the Scottish ultramafic soil at Whitecairns, Aberdeen (Hunter & Vergnano 1952, 1953) were observed in the oat crops. Manganese (9–50 $\mu g\,g^{-1}$) and zinc (20–60 $\mu g\,g^{-1}$) had rather low values so that in plants at a later growth stage, deficiency symptoms could be expected; copper was often low (5–28 $\mu g\,g^{-1}$) and there were indications of a possible molybdenum deficiency (<0.3–1 $\mu g\,g^{-1}$). Particularly in wheat, oats and barley there was a negative correlation between phosphorus and chromium (Vergnano 1959b).

Other metabolic aspects
In oat plants (*Avena sativa* cv. Victory) grown for 60 days in pots with the Impruneta ultramafic soil and in sand cultures with the addition of 2.5 mg l^{-1} Ni^{2+} (Vergnano Gambi *et al.* 1976), high nickel concentrations occurred in the roots, but the element was not translocated to a great extent to the leaves. In the plants grown on the ultramafic soil, the leaves had a concentration of 15–20 $\mu g\,g^{-1}$ and showed a marked interveinal chlorosis, but no white striping, which actually occurs with higher nickel concentrations (Vergnano 1953b). It happened instead in the sand cultures, where leaf nickel concentration reached 100 $\mu g\,g^{-1}$.

A reduction in total chlorophyll and a gradual drop of the percentage of HCl-soluble iron along with a marked increase in total iron and phosphorus concentration were observed with increasing leaf age both in the ultramafic and in the sand cultures. This condition indicated an interference of nickel with iron metabolism. In addition, an abnormally high concentration of citric acid oc-

curred in the roots and leaves of all the plants. In *Helichrysum italicum*, high concentrations of citric acid were also observed in samples collected on ultramafics compared with samples collected on sandstone. The high concentrations of malic and malonic acids found in *Alyssum bertolonii* (Pelosi *et al.* 1976, Pancaro *et al.* 1978) and the occurrence of higher organic acid concentrations in plants collected on ultramafites (Ritter-Studnička 1971) or in metal-resistant ecotypes (Mathys 1977, Brookes *et al.* 1981, Thurman & Rankin 1982) suggested further investigations along this line.

In a preliminary general survey (Vergnano Gambi & Gabbrielli 1985) of species growing on ultramafic and on other substrates, it has been possible to observe that in some species the ultramafic soil induces a foliar accumulation of citric (*Helichrysum*) or malic acid (*Cistus*), while in others (*Festuca*) both acids are produced in significantly higher proportion. There are also plants which do not show any appreciable variation in acid concentration. Therefore, the production of higher quantities of these acids, besides being species specific, is not always strictly related to the presence of nickel or other metals in the soil.

Physiological aspects of nickel uptake and tolerance in Alyssum bertolonii

Elemental composition
A. bertolonii Desv., one of the fourteen European species of *Alyssum* which accumulates nickel (Brooks & Radford 1978) was the first nickel-accumulating plant recorded with a concentration of 12,000 $\mu g\,g^{-1}$ in the leaves, other organs showing lower but still relevant values (4,000–5,000 $\mu g\,g^{-1}$). This species was therefore suggested as a useful indicator plant in mineral prospecting for nickel (Minguzzi & Vergnano 1948), and it is recognized as an universal nickel indicator (Brooks 1983).

In this plant the quotients (Mg+Ni)/Mg and Ca/Ni are remarkably constant, indicating a possible interaction of nickel with calcium and magnesium.

The relationship between nickel and major elements was investigated at several stages during the life cycle of the plant, both at Monte Murlo (Upper Tiber Valley) and at Impruneta (Vergnano Gambi *et al.* 1977, 1980a). The particularly

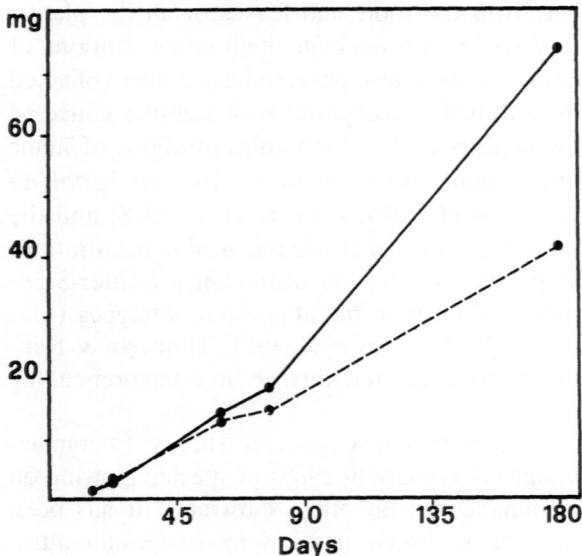

Fig. 12. Dry mass production (mg per seedling) of *A. berto-lonii* grown on ultramafic soil (dotted line) and on garden soil (solid line).

good development of the plants on the stony slopes of Monte Murlo did not seem related to a lower nickel concentration or to a lower Mg/Ca quotient, although in this plant magnesium is at relatively higher concentrations in the older stems and roots, so that its translocation to the leaves is more limited. Green stems and leaves at Monte Murlo show the lowest magnesium concentrations. In this locality phosphorus deficiency symptoms were occasionally evident, while nitrogen supply was apparently not limiting.

To better understand the metabolism of this species, plants have been grown from seeds in pots with normal soil to obtain specimens in which the nickel supply of the seeds is exhausted and the element brought to very low concentrations (Vergnano Gambi *et al.* 1980b). Seedlings developed equally well on ultramafic and normal soil till they were 60 days old; afterwards garden soil plants were larger and had a greater production of dry matter (Fig. 12).

Nickel distribution from cotyledons to leaves was followed (Pancaro & Cardini 1981): the high seed nickel content, of 3–5 μg per seed (4,000–7,000 μg g^{-1}) is distributed partially in the seed coats (ca. 10%) but mainly in the cotyledons and embryo. With the shedding of the cotyledons the greater part (ca. 60%) of the element is elimin-

ated and the remaining 40% is distributed in the leaves, particularly in the first ones; the periodical shedding of the leaves gradually lowers nickel concentrations and after 7 years (1978–1985) of cultivation in normal soil plants show a leaf nickel concentration of only 4 μg g^{-1}.

As nickel concentration decreases, that of calcium gradually increases as does dry-mass production. The plants show a good vegetative development although seed production has always been either nonexistent or scarce, but recently (1984) also viable seeds have been produced, from which new plants will be obtained. Such results would suggest that *A. bertolonii* does not have a special nickel requirement, but the development of nickel depleted plants will provide further indications on this problem.

The analysis of freeze-dried leaves (Table 6) from plants grown for 2 years on garden soil compared to that of the ultramafic samples, showed that such elements as nitrogen, potassium, sodium, iron and sulphur had higher and obviously magnesium lower concentrations. Only phosphorus concentration did not increase showing the frugality of plants adapted to soils of low phosphorus content. With an increase in calcium the magnesium and nickel concentration dropped. In all samples high chloride values were found: 250–315 μmol g^{-1} in ultramafic and 520 μmol g^{-1} in garden soil plants, this clearly being a specific feature of the plant.

Generally all the essential microelements (Table 7) have higher values in plants grown on ultramafic soil, but the possibility of boron toxicity, suggested by Shkolnik & Smirnov (1970) is to be excluded for *A. bertolonii*, which shows values common in natural vegetation. Molybdenum concentration is generally low: however scarce molybdenum availability has been considered to contribute to poor plant growth on some ultramafic soils (Walker 1948, Proctor & Woodell 1975) and it can certainly affect plants not adapted to this substrate as is the case of the cultivated plants in the Upper Tiber Valley. Copper and zinc are particularly scarce in seeds.

Total phenols and free amino acids were also evaluated. The prominent feature in the distribution of total phenols is the low value (25 μmol g^{-1}) in the nickel-depleted plants, in which their concentration is reduced to almost 50% com-

238

Table 6. Elemental composition of *A. bertolonii* leaves and seeds (% on freeze-dried tissue)

	N	P	K	Ca	Mg	S	Na	Fe	Ni
Serpentine soil									
leaves									
February	2.60	0.17	1.04	3.80	0.43	0.34	0.11	0.037	1.21
June	2.19	0.11	1.13	2.70	0.61	0.28	0.06	0.026	0.83
October	1.76	0.07	0.95	2.98	0.67	0.30	0.06	0.032	1.04
seeds		0.50		0.70	0.18				0.70
Garden soil									
leaves									
June	3.80	0.44	2.74	4.16	0.20	1.05	0.18	0.059	0.002

pared to the ultramafic samples (43–52 μmol), which show also an almost constant phenol concentration. These compounds are often implicated in complexing metals (Gomah & Davies 1974, Foy *et al.* 1978), and as they are involved in the synthesis of lignin, a certain amount of nickel could be chelated and held in insoluble form also in the conducting tissues. In addition they are also connected with the synthesis of anthocyanin, which often accumulates in plants under toxic metal stress. An increase in red pigmentation is typical of leaves and stems of *A. bertolonii*: this phenomenon is particularly evident when plants are grown in water cultures with addition of Ni^{2+}. The frequent appearance of red colouring (purpurescence or erithrism) is considered by Ritter-Studnička (1968) as a typical variant in plants growing on ultramafic soils.

The free amino acid composition of *A. bertolonii* shows that the differences between the plants on the two substrates are mainly quantitative: the free amino acid content is much higher in the garden plants as is their total nitrogen

Table 7. Microelement composition of leaves and seeds of *A. bertolonii* (μg g^{-1} on freeze-dried tissue)

	B	Mn	Cu	Zn	Mo	Co	Cr
Serpentine soil							
leaves							
February	13.4	190	3.8	29	0.6	35	4
June	33	360	5.1	70	0.4	70	3
October	36	280	14.0	100	0.8	88	16
seeds	8.8	40	2.0	22	0.9	7	1
Garden soil							
leaves							
June	12	36	4.6	55	6.3	2	1

value, but those amino acids like cystine and methionine, which might be involved in complexing metals, do not reach higher concentrations in plants grown on ultramafic soil: actually methionine was clearly detected only in the garden sample. Proline is the major amino acid of the leaves of *A. bertolonii* and in February and October accounts for more than 40% of the total free amino acids (Bick *et al.* 1982).

In the roots of a copper tolerant race of *Armeria maritima* Farago & Mullen (1979) observed high proline concentrations, but in this case proline was associated with water-soluble copper. The high proline values of *A. bertolonii* could be implicated in nickel tolerance mechanisms, but they seem more a characteristic of this species possibly connected with adaptation to xeric habitats as the nickel-poor garden plants are also rich (c.30%) in this amino acid.

Interaction between nickel and other ions, particularly calcium and magnesium

A relationship among these three elements has been previously noted: therefore plants of *A. bertolonii* were cultivated in Arnon dilute (1:10) solution modified as suggested by Main (1974, 1981), in which the Mg/Ca quotients had values included between 0.125 and 8 (L. Pancaro, unpublished). In this case the best plant development (evaluated as dry matter yield) was obtained, in absence of Ni^{2+}, when the Mg/Ca mol quotient was equal to 4 (2 mM Mg^{2+} and 0.5 mM Ca^{2+}), showing that this serpentinophyte too has a higher magnesium and lower calcium requirement.

When seedlings of *A. bertolonii* and *A. berto-*

lonioides were kept in a solution of Ca(N-O₃)₂(1 g l⁻¹), their nickel concentration had shown a highly significant negative correlation with calcium (Gabbrielli *et al*. 1982). Taking as an index, root length, Gabbrielli & Pandolfini (1984) demonstrated that in *A. bertolonii* root growth is stimulated by a solution of 0.1 mM Mg^{2+} and inhibited by one of 1 mM Mg^{2+} (c. 24 mg l⁻¹), and the latter has the same inhibiting effect as a solution of 0.1 mM Ni^{2+} (c. 6 mg l⁻¹). The presence of Ca^{2+} (1 mM) cancels the toxic effect of Mg^{2+} but is not sufficient to reduce Ni^{2+} toxicity, so that roots have a very limited growth. If both ions, Ca^{2+} and Mg^{2+}, are added to the solution, even if Ca^{2+} concentration is low, root growth is enhanced. In the presence of 0.1 mM Ni^{2+} the best root growth is obtained with the addition of both ions at the concentrations 0.1 mM Ca^{2+} and 1.5 mM Mg^{2+} to the solution. Although in these experiments seedlings were not grown in a complete nutrient solution, preventing therefore a comparison with the previous experiments, nevertheless it is clear that Mg^{2+} and Ca^{2+} internal concentrations counteract Ni^{2+} toxicity or in any case enhance nickel tolerance and that this species in the presence of Ni^{2+} requires Mg^{2+} at a concentration higher than that of Ca^{2+}.

The stimulating effect on root growth could be ascribed to a lower nickel tissue concentration: actually Ca^{2+} and Mg^{2+} depress Ni^{2+} uptake, but root growth cannot be related only to a lower nickel concentration, as the interaction between nickel and the other ions is more complex.

Nickel localization

A first attempt at nickel localization in *A. bertolonii* tissues was carried out by Vergnano Gambi (1967), who noticed a preferential deposition of the Ni-dimethylglyoxime complex in the epidermis and in the schlerenchyma around the vessels; Gabbrielli (1980) observed a strong accumulation of the complex not only in the stem and leaf epidermis, but also in the root cortex and stele. Differential centrifugation of leaf homogenates showed that nickel was distributed mainly in the vacuolar sap (i.e., in the water soluble fraction), followed by cell walls and chloroplasts (Table 8).

Nickel tolerance and uptake

Using root elongation and plasmatic resistance (Repp 1963) tests, the degree of nickel tolerance in some *Alyssum* species was examined (Gabbrielli *et al*. 1982). In addition to *A. bertolonii* and *A. bertolonioides*, *A. nebrodense*, endemic to Sicily, not growing on ultramafics, but belonging to the Section Odontarrhena in which are grouped all the nickel accumulating species of *Alyssum* (Brooks & Radford 1978), and the closely related *Alyssum saxatile* belonging to the Section Aurinia, were tested.

The degree of nickel tolerance was evaluated as ED50 using the probit method as suggested by Craig (1977): *A. bertolonii* was the most tolerant species, being four times more tolerant than *A. argenteum* f. *bertolonioides* and ten times more than *A. nebrodense*; it was also 100 times more tolerant than *A. saxatile*. Also the plasmatic resistance tests in epidermal cells confirmed this behaviour.

The high nickel tolerance threshold of *A. bertolonii* and *A. bertolonioides* is certainly connected with their ability to bind and isolate the element in non-toxic forms, considering its distribution in the whole cell; the higher nickel tolerance of *A. bertolonii* is probably related to its higher nickel accumulating capacity; none of the samples of the Italian populations of *A. argenteum* collected in the Alps and in the Apennines, including herbarium specimens, reaches the nickel concentrations observed in *A. bertolonii* (Brooks & Radford 1978, Vergnano Gambi *et al*. 1979).

At high internal nickel concentrations a saturation mechanism operates in both the tolerant species, whereas in *A. nebrodense* and *A. saxatile* nickel uptake is clearly linear, so that a toxic effect is observed even at relatively low internal nickel concentrations (Fig. 13).

Identification of nickel-complexing compounds

In a first attempt to identify the possible nickel-complexing compounds which would allow nickel accumulation but which would remove Ni^{2+} from sensitive sites within the cell, various solvents were used to extract the metal from the leaves of *A. bertolonii* (collected from plants growing on an ultramafic outcrop near Pisa). Eighty percent of the metal was water soluble and probably associated with organic acids (Pelosi *et al*. 1974).

Table 8. Nickel concentration on dry weight and percentage distribution in stem and root tissues and in subcellular fractions from leaves of *Alyssum bertolonii* Desv.

		% DW	Ni	
			$\mu g\,g^{-1}$	%
Entire stem			2,620	100
cortex		42.5	2,135	81.5
stele		57.5	485	18.5
Entire root			592	100
cortex		43.2	526	88.5
stele		56.8	66	11.5
Entire leaf			12,210	100
crude debris	muslin filtration	19.3	1,820	2.8
cell wall debris	500 g × 2'	24.4	1,710	3.4
chloroplasts, etc.	7,000 g × 30'	12.0	2,430	2.4
supernatant	96,000 g × 60'		11,000	91.2
pellet		4.7	350	0.1

Further research involved the separation and identification of malic and malonic acids, which were considered to be involved in complexing nickel (Pelosi *et al.* 1976). To verify if *A. bertolonii* grown on ultramafic soils actually had a greater production of these acids, the organic acid concentration of plants collected on these soils and also cultivated on normal soil, was assessed

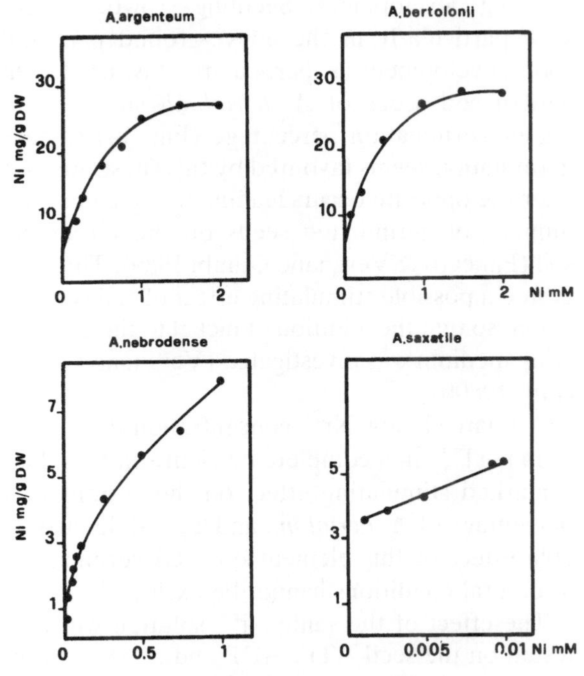

Fig. 13. Ni^{2+} uptake in various species of *Alyssum*.

(Pancaro *et al.* 1978). In the purified fractions obtained from the aqueous extracts of the ultramafic samples, malic and malonic acids were present almost in equimolecular ratios with nickel. In the garden soil plants, which had a low nickel concentration, these acids were scarce. These data were also confirmed by the analysis of freeze-dried leaves and seeds. Samples of leaves of *Alyssum serpyllifolium* ssp. *lusitanicum* Dudley et P. da Silva, endemic to Portugal ultramafics and cultivated in the Florence Botanical Garden on ultramafic soil, showed high levels particularly of malic acid. The predominance of malic and malonic acids in purified nickel-containing extracts of this species was observed also by Lee *et al.* (1978). Investigations on excised roots of *A. bertolonii* treated with a Ni^{2+} solution 0.1 mM for 24 or 48 h demonstrated an increase in malic acid production, but the same was observed in the presence of a K^+ 0.2 mM solution (Vergnano Gambi & Gabbrielli 1987). This would suggest that malic acid production is not a specific response to Ni^{2+}; the synthesis of malic acid could probably be the result of a higher proton efflux with consequent alteration of cytoplasmic pH, which controls PEP carboxylase activity.

On the other hand the selectivity in nickel uptake observed in *Alyssum* must be also taken in consideration as cobalt and zinc compete with nickel in uptake experiments with excised roots (Gabbrielli *et al.* 1988).

Therefore it seems plausible that the process of translocation to the shoots may be part of the mechanism which favours Ni^{2+} accumulation in *Alyssum*.

Enzyme adaptation

One of the aspects connected with tolerance mechanisms involves enzyme adaptations to high metal concentrations as those found in the soil solution and within the plant. The evidence for such an adaptation in cytoplasmic enzymes is so far lacking, but extracellular enzymes of the root system, or enzymes carried on the outer surface of the plasma membranes are more likely to show altered properties (Woolhouse 1983), and have received particular attention.

Research on soluble and cell-wall bound acid phosphatases in zinc-tolerant ecotypes (Wainwright & Woolhouse 1975, Cox & Hutchinson 1980) or in serpentine-tolerant clones (Willet & Batey 1977, Johnston & Proctor 1984) indicate that these enzymes might be involved in tolerance mechanisms.

Experiments on root surface acid phosphatases of seedlings of *Alyssum bertolonii* and *A. saxatile*, a species of wide occurence, cultivated in solutions to which various concentrations of Ni^{2+}, Ca^{2+} and Mg^{2+} were added (Gabbrielli *et al.* 1989), showed that Ni^{2+} (0.01 mM) stimulated *in vivo* phosphatase activity in *A. bertolonii*, but depressed that of *A. saxatile*. Ca^{2+} had a stimulating effect in both species, but the highest activity was obtained in *A. bertolonii* with low Ca^{2+} and in *A. saxatile* with high Ca^{2+} concentrations. The Mg^{2+} concentrations required by *A. bertolonii* for increased enzyme activity proved toxic to *A. saxatile*. In the ultramafic endemic favourable conditions for the increase of phosphatase activity were induced also by a high Mg^{2+}/Ca^{2+} quotient ($\simeq 10$), which improved also dry mass production and root growth.

These results indicate a higher tolerance to nickel and magnesium in the enzymes of plants typical of ultramafic soils, in accordance with the results reported by Johnston and Proctor (1984).

The role of root acid phosphatases still remains to be cleared but the suggestion that these enzymes should be involved in the turnover of phosphatases, which would otherwise be lost with cell lysis in the senescing tissues, seems plausible. In this case the extracellular phosphatases would be an important aspect of plant adaptation to these particular substrates with low phosphate concentrations.

Seed germination

Germination tests were carried out under controlled conditions to establish possible ecological differences among populations of *A. bertolonii* in Tuscany (Upper Tiber Valley, Impruneta, Monte Ferrato) and those of *A. argenteum* f. *bertolonioides* in the Emilian Apennine (Bobbio). The maximum germination percentage was obtained at 20°C for all the populations of *A. bertolonii* and at 15–17°C for *A. argenteum* f. *bertolonioides* adapted to colder areas (Pelagatti 1978). The two species do not show an absolute dormancy and the highest germination percentage occurs when seeds are 8–10 months old; seeds lose their viability rather early (20 months). Compared to *A. nebrodense*, the two ultramafic species show a lower germination percentage and a lower number of seeds germinated per day, i.e., a lower coefficient of velocity, calculated from the formula proposed by Kotowski (1926), which characterize species of habitats with little competition (Ernst 1965). In their natural conditions seeds start germinating between March and June as the germination capacity is at its highest and the climatic conditions are more favourable to seedling development. Seedling growth is very slow particularly in the above-ground part and root development is particularly favoured. On garden soil seeds of *A. bertolonii* start with a higher germination percentage (Fig. 14) so that germination seems favoured by this substrate, but later the opposite occurs leading to a greater final number of germinated seeds on the ultramafic soil (Pancaro & Vergnano Gambi 1980). This suggested a possible stimulating effect of soil composition, so that the addition of nickel to the germinating medium was investigated (Vergnano Gambi *et al.* 1980b).

A relatively low Ni^{2+} concentration (0.25 mM, c. 14 mg l^{-1}) in a complete nutrient solution, had a marked stimulating effect on the germination percentage of *A. bertolonii* and a possible favourable effect of this element on seed germination in natural conditions cannot be excluded.

The effect of the same Ni^{2+} solution was also tested on the seed ATP, ADP and AMP concentrations at different imbibition times (0, 24,

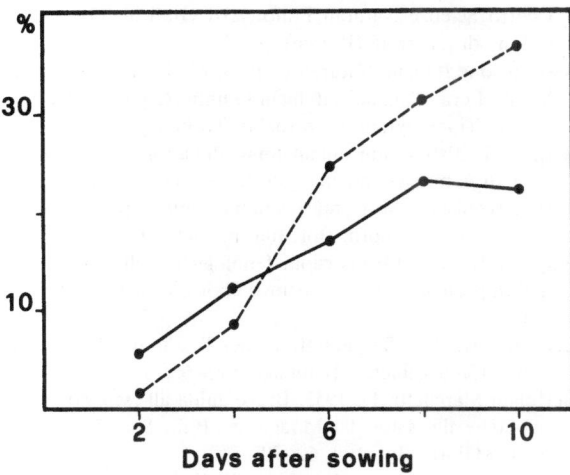

Fig. 14. Germination percentage of *A. bertolonii* on ultramafic soil (dotted line) and on garden soil (solid line).

48, 72, 96 h). No difference with the control was observed and the same occurred with a solution of 0.1 mM Co^{2+} (c. 0.5 mg l^{-1}), suggesting that Ni^{2+} uptake at low concentrations does not involve particular energy requirement. Instead higher concentrations (10 and 25 mM) of both ions markedly lowered the seed adenylate levels (Grossi 1985).

A more detailed investigation (Pancaro *et al.* 1981) on the germination of *A. bertolonii*, *A. argenteum* f. *bertolonioides* and *A. nebrodense* in the presence of Ni^{2+} (2.5–25 mM), Co^{2+} (2.5–20 mM) and Cr (VI) (0.2–1 mM) added to the nutrient solution, was carried out in controlled conditions, starting when seeds were harvested till they practically failed to germinate (after 20 months). The germination was evaluated with different parameters and the data statistically analysed. At the relatively high concentrations of the three ions, the effect per mMole is generally inhibitory: not significant for Co^{2+}, slight but significant for Ni^{2+} and marked and highly significant for Cr (VI).

The inhibition reaches its maximum when seeds are ripe and it is significantly related to the germination values of the control, suggesting a species-specific response. This specific inhibition is constant during the whole life span of the seeds and shows different values according to the species and ion used: it is greater in *A. argenteum* than in *A. bertolonii* and more marked for chromium than for nickel.

Therefore also these experiments confirm that *A. bertolonii* has a higher tolerance threshold for chromium and nickel than *A. argenteum* f. *bertolonioides*, a further indication of a possible less ancient evolution of metal tolerance in the latter species. This behaviour can be also connected with the wide geographical distribution, not always connected with ultramafites, of *A. argenteum*, from which originated the populations restricted to the ultramafic soils of the northern Apennine. Some investigations (unpublished) on the effect of Ni^{2+} on the germination of *A. argenteum* collected on an ultramafic outcrop in the Aosta Valley indicate that the nickel tolerance of this species is almost identical to that of *A. argenteum* f. *bertolonioides*.

Future research

The problems posed by the vegetation of ultramafic soils are manifold and complex and some have been almost totally ignored. For instance the presence of morphological variants in lichens and mosses of ultramafites (cf. Rune 1953) has not yet prompted further investigations in this mainly unexplored field. Also in higher plants morphological adaptations and their taxonomical relevance have not always received proper attention, probably because this problem involves a long and sometimes not easy experimentation. One question arises as to whether these adaptations imply the presence of toxic metals in the soil or are mainly a response to unfavourable xeric habitats. Almost lacking are anatomical and ultrastructural observations on ultramafic species. It would be of particular relevance to ascertain the occurrence of C_4-type of photosynthesis in these plants. Perhaps one should also look for intermediate anatomical characters between C_3- and C_4-type of photosynthesis.

Another point which should be stressed is the need for laboratory experimental work on nutritional requirements (ionic concentrations and optimum quotients), particularly for some elements as iron, phosphorus and nitrogen, of species adapted to ultramafic soils. The problem of the evolution of different tolerance mechanisms, exclusion and accumulation of chromium, cobalt and nickel are both well represented in serpentine races, opens a wide field of investigations. Accumulation of nickel considered limited to a few species, in some habitats such as those observed in the alpine area, seems to be of almost general occurrence even if seldom reaching extreme values. The edaphic condition of the moraine of the Verra glacier which has probably taken origin in the 'little ice age,' between the seventeenth and the beginning of the nineteenth century, has allowed the selection and the establishment of highly tolerant and metal-accumulating populations, belonging to different families, even if Crucifers are outstanding in their metal-accumulating capacity. One must also consider if the chemical form of some metals present in the soil and their solubility in addition to the presence of organic matter, might induce a greater metal uptake. Elsewhere plants more frequently show exclusion mechanisms: the characterization of these populations which seem to show a lower degree of metal tolerance would be very interesting and enzyme research would offer an important tool for future work.

In plants which accumulate metals (and an interesting genus would certainly be *Cerastium*), the identification of the biochemical pathways which lead to accumulation and compartmentation of the metal will be one of the major fields of interest and research.

Acknowledgements

I would like to express my thanks to Professor P. V. Arrigoni and Professor R. Pichi Sermolli for their very helpful comments on the manuscript, to Professor A. Azzaroli and Professor P. Passerini for their revision of the geological section, and to Dr. M. Guido and Dr. C. Montanari for kindly supplying some of the photos.

References

Abbate, E., V. Bortolotti & G. Principi. 1980. Apennine ophiolites: a peculiar oceanic crust. Ofioliti special issue 1: 59–96.

Amidei, G. 1841. Specie di piante osservate nei terreni serpentinosi. Atti 3a Riunione Scienziati Italiani, 523–524. Firenze.

Arrigoni, P. V., C. Ricceri, & A. Mazzanti. 1983. La vegetazione serpentinicola del Monte Ferrato di Prato in Toscana. Centro Scienze Naturali Prato. Arti Grafiche Prioreschi, Catena di Quarrata (Pistoia), pp. 27.

Associazione Italiana Naturalisti. 1979. Madonna della Neve-Monte Lera. Collana Cataloghi Giunta Regionale del Piemonte. Tipolitografia Scaravaglio, Torino, pp. 46.

Bargoni, I. 1940. Cenni sull'anatomia della foglia di due individui di Armeria denticulata Bert. cresciuti rispettivamente su serpentino e su terreno comune, non serpentinoso, in cultura. Nuovo Giorn. Bot. Ital. n.s. 47: 507–509.

Bargoni, I. 1943. Osservazioni fenologiche sulle serpentine dell'Impruneta (Firenze). Nuovo Giorn. Bot. Ital. 50: 232–251.

Becherer, A. 1969. Serpentinfarne des Tessin und des Italienischen Grenzgebietes. Bauhinia 4: 65–66.

Bertolani Marchetti, D. 1953. Breve guida all'escursione alle ofioliti e alle salse del Modenese. Bull. Soc. Bot. Ital., Nuovo Giorn. Bot. Ital. 60: 707–712.

Bick, W., O. Vergnano Gambi, & P. DeKock. 1982. A relationship between free amino acid and nickel contents in leaves and seeds of Alyssum bertolonii. Plant and Soil 66: 117–119.

Brookes, A., J. C. Collins, & D. Thurman. 1981. The mechanism of zinc tolerance in grasses. J. Plant Nutrition 3: 695–705.

Brooks, R. R. 1983. Biological methods of prospecting for minerals. John Wiley & Sons, New York, p. 36.

Brooks, R. R. & C. C. Radford. 1978. Nickel accumulation by European species of the genus Alyssum. Proc. R. Soc. London B200: 217–224.

Caruel, T. 1867. Sur la flore des gabbres de Toscane. Actes Congrès Intern. Botanique, Paris, 58–63.

Caruel, T. 1871. Flora dei Gabbri di Toscana. In "Statistica Botanica della Toscana", cap.10: 321–326. Pellas Firenze.

Cengia Sambo, M. 1937. Osservazioni lichenologiche sul gruppo del M. Ferrato. Nuovo Giorn. Bot. Ital. 44: 295–311.

Cesalpino, A. 1583. De Plantis Libri XVI. Marescotti, Firenze, p. 369.

Corti, R. 1940. Appunti sulla vegetazione dell'Isola d'Elba. I. Una gita a M. Orello e ai Monti tra Rio Alto e Portolongone con osservazioni sui distretti ofiolitici dell'Isola. Bull. Soc. Bot. Ital., Nuovo Giorn. Bot. Ital. n.s. 47: 494–504.

Cortini Pedrotti, C. 1975. La vegetazione pioniera del Monte Ferrato (Prato). Atti Soc. Tosc. Sci. Nat. Mem B. 81: 39–44.

Cox, M. & T. C. Hutchinson. 1980. The response of root acid phosphatase activity to heavy metal stress in tolerant and non-tolerant clones of two grass species. New Phytol. 86: 359–364.

Craig, G. C. 1977. A method of measuring heavy metal tolerance in grasses. Tr. Rhod. Sci. Ass. 58: 9–16.

Credaro, V. & A. Pirola. 1975. La vegetazione della Provincia di Sondrio. Tipolitografia Bonazzi, Sondrio pp. 104, Tav. 25, Tab. Fitosociologiche 45.

Dietrich, V. J. 1980. The distribution of ophiolites in the Alps. Ofioliti special issue 1: 7–51.

Ernst, W. 1965. Uber den Einfluss des Zinks auf die Keimung von Schwermetallpflanzen und auf die Entwicklung der Schwermetallpflanzengesellschaft. Ber. Deut. Bot. Ges. 78: 205–212.

Farago, M. E. & W. A. Mullen. 1979. Plants which accumulate metals. Part. IV. A possible copper-proline complex from the roots of Armeria maritima. Inorg. Ch. Acta 32: L.93–94.

Fiori, A. 1914. La flora dei serpentini della Toscana. II Confronto fra la flora del Monte Ferrato (serpentino) e quella della Calvana (calcare albarese). In Fiori A. e Pampanini R. La flora dei serpentini della Toscana. Nuovo Giorn. Bot. Ital. n.s. 21: 216–240.

Fiori, A. 1917. Piante raccolte sopra un masso inesplorato dell'Appennino Modenese. Bull. Soc. Bot. Ital. 9–11.

Fiori, A. 1919. Contribuzione alla flora dei serpentini del Pavese. Bull. Soc. Bot. Ital. 39–40.

Fiori, A. 1920. Rilievi geografici e forestali sulla flora del bacino della Cecina e località finitime. Ann. R. Ist. Sup. Forestale Naz. V: 149–186.

Foy, C. D., R. L. Chaney & M. C. White. 1978. The physiology of metal toxicity in plants. Ann. Rev. Plant Physiol. 29: 511–566.

Furrer, E. & A. Hofmann. 1969. Das Euphorbietum-spinosaeligusticae, eine Serpentingesellschaft in Ligurien. Acta Bot. Croat. XXVIII: 81–90.

Gabbrielli, R. 1980. Localizzazione del nichel in tessuti di A. bertolonii. Atti Congr. Soc. Bot. Ital. Palermo, Giorn. Bot. Ital. 114: 119.

Gabbrielli, R. & T. Pandolfini. 1984. Effect of Mg^{2+} and Ca^{2+} on the response to nickel toxicity in a serpentine endemic and nickel accumulating species. Physiol. Plant. 62: 540–544.

Gabbrielli, R., R. Birtolo, & O. Vergnano Gambi. 1982. Evaluation of nickel tolerance in Alyssum. Atti Soc. Tosc. Sci. Nat., Mem. Ser. B 88: 143–153.

Gabbrielli, R., F. Pedani & O. Vergnano Gambi. 1987. Ulteriori dati sulla composizione minerale della vegetazione degli affioramenti ofiolitici dell'alta valle di Ayas. Rev. Valdôt. Hist Nat. 41: 99–110.

Gabbrielli, R., C. Mattioni & O. Vergnano. 1988. Heavy metal tolerance in Alyssum bertolonii. VIth Congress FESPP, Split 4–10 Sept. 1988, Abstract VI/12.

Gabbrielli, R., L. Grossi & O. Vergnano. 1989. The effects of nickel, calcium and magnesium on the acid phosphatase activity of two Alyssum species. New Phytol. 111: 631–636.

Gola, G. 1912. La vegetazione dell'Appennino piemontese. Ann. Bot. (Roma) 10: 189–338.

Gomah, A. M. & R. I. Davies. 1974. Identification of the active ligands chelating Zn in some plant water extracts. Plant and Soil 40: 1–19.

Grossi, L. 1985. Changes in adenine nucleotide levels and energy charge in germinating seeds of Alyssum bertolonii in presence of heavy metals. Biochem. Physiol. Pflanzen 180: 277–283.

Guido, M. & C. Montanari. 1983. Studio e cartografia della vegetazione cacuminale del Monte Aiona (Appennino Ligure). Arch. Bot. Biogeogr. Ital. 59: 105–131.

Guido, M. A. & P. Petroni. 1975. Flora e vegetazione della valle del torrente Lerone (Appennino Ligure Occidentale). Webbia 29: 645–716.

Hunter, J. G. & O. Vergnano. 1952. Nickel toxicity in plants. Ann. Appl. Biol. 39: 279–284.

Hunter, J. G. & O. Vergnano. 1953. Trace-element toxicities in oat plants. Ann. Appl. Biol. 40: 761–777.

Kotowski, F. 1926. Temperature relations to germination of vegetable seeds. Proc. Amer. Soc. Hort. Sci. 23: 176–184.

Johnston, W. R. & J. Proctor. 1984. The effects of magnesium, nickel, calcium and micronutrients on the root surface phosphatase activity of a serpentine and non serpentine clone of Festuca rubra L. New Phytol. 96: 95–101.

Lee, J., R. D. Reeves, R. R. Brooks, & T. Jaffré. 1978. The relation between nickel and citric acid in some nickel accumulating plants. Phytochem. 17: 1033–1035.

Lisanti, E. L. 1952. Contributo allo studio delle morfosi che si riscontrano sui serpentini (possibilità di chemiomorfosi). Nuovo Giorn. Bot. Ital. n.s. 59: 349–360.

Lisanti, E. L. 1958. Sulle chemiomorfosi da serpentino. Esperimenti in ordine fattoriale con nichel, magnesio e ione solforico sulla Stachys recta tipica. Nuovo Giorn. Bot. Ital. n.s. 65: 452–459.

Main, J. L. 1974. Differential responses to magnesium and calcium by native populations of Agropyron spicatum. Am. J. Bot. 61: 931–937.

Main, J. L. 1981. Magnesium and calcium nutrition of a serpentine endemic grass. The Amer. Midland Nat. 105: 196–199.

Martini, E. & Orsino, F. 1969. Flora e vegetazione delle Valli dei torrenti Acquabona, Scorza e Lerca (gruppo del M. Beigua, Appennino Ligure). Webbia 23: 397–511.

Mathys, W. 1977. The role of malate, oxalate and mustard oil glucosides in the evolution of zinc-resistance in herbage plants. Physiol. Plant. 40: 130–136.

Messeri, A. 1936. Ricerche sulla vegetazione dei dintorni di Firenze. IV La vegetazione delle rocce ofiolitiche di Monte Ferrato presso Prato. Nuovo Giorn. Bot. Ital. n.s. 43: 277–372.

Minguzzi, C. & O. Vergnano. 1948. Il contenuto di nichel nelle ceneri di Alyssum bertolonii Desv. Atti Soc. Tosc. Sci. Nat. Mem B 55: 49–71.

Minguzzi, C. & O. Vergnano. 1953. Il contenuto di elementi inorganici delle piante della formazione ofiolitica dell'Impruneta (Firenze). Nota I: Determinazione degli elementi macronutritivi e ricerca degli elementi in tracce. Nuovo Giorn. Bot. Ital. n.s. 60: 287–319.

Mussa, E. 1908. Note floristiche delle Prealpi Torinesi fra la Dora Riparia e la Stura di Lanzo (Zona delle Pietre Verdi). Atti Soc. It. Sci. Nat. 47: 139–157.

Mussa, E. 1937. Sguardo al la vegetazione del M. Musinè (Val di Susa). Nuovo Giorn. Bot. Ital. n.s. 44: 715–730.

Negodi, G. 1941. La flora e la vegetazione dei serpentini di Varana. Ann. Bot. (Roma) 22: 117–142.

Negodi, G. 1943. Studi sulla vegetazione dell'Appennino Emiliano e della pianura adiacente. VI Caratteri delle associazioni del diabase rosso del Sasso Tignoso modenese (m. 1492). Ann. Bot. (Roma) 22: 133–152.

Novák, F. A. 1928. Quelques rémarques aux problèmes de la végétation sur les terrains serpentiniques. Preslia 6: 43–71.

Pacetti, A. 1969. Il contenuto in nichel, manganese, ferro, rame e fosforo in Plantago holosteum Scop. e in Notholaena marantae (L.) R. Br. ex Desv. degli affioramenti ofiolitici toscani. Inform. Bot. Ital. 1: 161–163.

Pampanini, R. 1912. La flora dei serpentini della Toscana. I. Montignoso. Nuovo Giorn. Bot. Ital. n.s. 19: 464–466.

Pancaro, L. & F. Cardini. 1981. Distribuzione di nichel, calcio, magnesio e potassio in Alyssum bertolonii coltivato in assenza di nichel. Atti Congr. Fisiol. veg. Pistoia, 18–20 VI, Giorn. Bot. Ital. 115: 254.

Pancaro, L. & O. Vergnano Gambi. 1980. Effetti del nichel sulla germinazione e sviluppo delle plantule di alcune specie di Alyssum. Atti Congr. Soc Bot. Ital., Palermo, Giorn. Bot. Ital. 114: 124–125.

Pancaro, L., M. Innamorati, O. Vergnano Gambi, & S. Occhiochiuso. 1981. Effetti del cobalto, nichel e cromo sulla germinazione di Alyssum durante il ciclo di postmaturazione e invecchiamento. Giorn. Bot. Ital. 115: 265–284.

Pancaro, L., P. Pelosi, O. Vergnano Gambi, & C. Galoppini. 1978. Ulteriori indagini sul rapporto tra nichel e acidi malico e malonico in Alyssum. Giorn. Bot. Ital. 112: 141–146.

Pavarino, G. L. 1914. Intorno alla flora del calcare e del serpentino nell'Appennino Bobbiese. Contribuzione seconda. Atti Ist. Bot. Univ. Pavia Ser. II, 14: 19–42.

Pavarino, G. L. 1915. Intorno alla flora del calcare e del serpentino nell'Appennino Bobbiese. Contribuzione prima. Atti Ist. Bot. Univ. Pavia Ser. XII, 12: 21–56.

Pavarino, G. L. 1918. Intorno alla flora del calcare e del serpentino. Contribuzione terza. Atti Ist. Bot. Univ. Pavia Ser. II, 15: 89–108.

Pelagatti, E. 1978. Influenza di vari fattori sulla germinazione di Alyssum bertolonii Desv. Dott. Nat. Sci. Thesis, University of Florence, Italy, pp. 85.

Pellizzer, R. 1961. Le ofioliti nell'Appennino Emiliano. Mem. Acc. Sci. Inst. Bologna ser. I, 8: 1–183.

Pelosi, P., C. Galoppini, & O. Vergnano Gambi. 1974. Sulla natura dei composti del nichel presenti in Alyssum bertolonii Desv. Nota I. Agric. Ital. 29: 1–5.

Pelosi, P., R. Fiorentini, & C. Galoppini. 1976. On the nature of nickel compounds in Alyssum bertolonii Desv. II. Agr. Biol. Chem. 40: 1641–1642.

Pichi Sermolli, R. 1936. Osservazioni sulle principali morfosi delle piante del serpentino. Bull. Soc. Bot. Ital., Nuovo Giorn. Bot. Ital. n.s. 43: 461–474.

Pichi Sermolli, R. 1948. Flora e vegetazione delle serpentine e delle altre ofioliti dell'alta valle del Tevere (Toscana). Webbia 6: 1–380.

Pichi Sermolli, R. E. G. & V. Chiarino-Maspes. 1963. Ricerche geobotaniche su Notholaena marantae in Italia. Webbia 17: 407–451.

Pignatti Wikus, E. & S. Pignatti. 1977. Die Vegetation auf Serpentinstandorten in den nördlichen Apenninen. Studia Phytol. honorem jub. A. O. Horvat: 113–124.

Pirola, A. 1967. Escursione sociale l967 (Guida alle escursioni). Giorn. Bot. Ital. 101: 395–400.

Proctor, J. & S. R. J. Woodell. 1975. The ecology of serpentine soils. Adv. Ecol. Res. 9: 256–366.

Repp, G. 1963. Die Kupferresistenz des Protoplasmas höherer Pflanzen auf Kupfererzböden. Protoplasma (Wien) 57: 643–659.

Rigotti, H. 1930. Significato fitogeografico delle florule dei serpentini submontani in Piemonte. Atti XI Congr. Geogr. Ital. 11: 72–74.

Ritter-Studnička, H. 1968. Die Serpentinomorphosen der Flora Bosniens. Bot. Jb. 88: 443–465.

Ritter-Studnička, H. 1971. Zellsaft Analysen zum Problem der Serpentinvegetation. Oesterr. Bot. Z. 119: 410–431.

Ritter-Studnička, H. & K. Dursun-Grom. 1973. Ueber den Eisen-Nickel und Chromgehalt in einigen Serpentinpflanzen Bosniens. Oesterr. Bot. Z. 121: 29–49.

Rune, O. 1953. Plant life on serpentines and related rocks in the North of Sweden. Acta Phytogeogr. Suec. 31: 1–139.

Sambo, E. 1927. I licheni del Monte Ferrato (Toscana). Nuovo Giorn. Bot. Ital. n.s. 34: 333–358.

Sasse, F. 1979a. Untersuchungen an Serpentinstandorten in Frankreich, Italien, Oesterreich und der Bundesrepublik Deutschland. I. Bodenanalysen. Flora 168: 379–395.

Sasse, F. 1979b. Untersuchungen an Serpentinstandorten in Frankreich, Italien, Oesterreich und der Bundesrepublik Deutschland. II Pflanzenanalysen. Flora 168: 578–594.

Shkolnik, M. J. & U. S. Smirnov. 1970. On possible causes of serpentinomorphosis and morphological variations in plants induced by high concentrations of boron. Bot. Zhurn. 55: 1764–1782 (in Russian).

Thurman, D. A. & J. L. Rankin. 1982. The role of organic acids in Zn tolerance in Deschampsia caespitosa. New Phytol. 91: 629–635.

Tutin, T. G. et al. 1964–1980. Flora Europaea. Cambridge Univ. Press.

Vaccari, L. 1903. La flore de la serpentine, du calcaire et du gneiss dans les Alpes Graies orientales. Bull. Soc. Flora Valdôt. 2: 32–75.

Vaccari, L. 1904–1911. Catalogue raisonné des plantes vasculaires de la Vallée d'Aoste. Soc. Flore Valdôt., vol. I, Aoste, 37.

Verger, J. P. 1979. Origine des sols sur prasinites et serpentinites sous végétation pionnière en climat alpin (Val d'Aoste). Doc. Cart. Ecol. XXI, 127–138.

Verger, J. P. 1982. L'étage montagnard sylvicole sur serpentinites en Vallée d'Ayas (Val d'Aoste). Doc. Cart. Ecol. XXV: 51–66.

Verger, J. P. 1983. Contribution à la connaissance d'un groupement alpin climacique original sur serpentines: le Caricetum fimbriatae. Phytosociologie et Pédologie. C.R. Acad. Sc. Paris 296, Ser. III: 775–778.

Vergnano, O. 1945. Ricerche comparative sulla presenza di alcuni elementi nelle ceneri di piante serpentinicole. Dott. Nat. Sc. Thesis, University of Florence, Italy, pp. 161.

Vergnano, O. 1953a. Erborizzazioni su alcune serpentine della Val di Cecina. Bull. Soc. Bot. Ital., Nuovo Giorn. Bot. Ital. n.s. 60: 330–332.

Vergnano, O. 1953b. L'azione fisiologica del nichel sulle piante di un terreno serpentinoso. Nuovo Giorn. Bot. Ital. n.s. 60: 109–183.

Vergnano, O. 1958a. Il contenuto di elementi inorganici delle piante della formazione ofiolitica dell'Impruneta (Firenze). Nota 2: Nichelio, Cromo e Cobalto nel dinamismo nutritivo delle piante serpentinicole. Nuovo Giorn. Bot. Ital. n.s. 65: 133–162.

Vergnano, O. 1958b. Sul determinismo delle morfosi della vegetazione sui terreni serpentinosi attraverso l'analisi della nutrizione minerale. Rend. Acc. Naz. Lincei, s. VIII, 24: 588–597.

Vergnano, O. 1959a. Metabolismo minerale di piante coltivate su terreni agrari d'origine ofiolitica nei Monti Rognosi (Alta Val Tiberina). Nuovo Giorn. Bot. Ital. n.s. 66: 100–150.

Vergnano, O. 1959b. La nutrizione fosforica in presenza di

246

elementi micro-nutritivi e tossici nel terreno. Agrochimica 3: 262–269.

Vergnano, O. & J. G. Hunter. 1953. Nickel and cobalt toxicities in oat plants. Ann. Bot. n.s. XVII: 317–328.

Vergnano Gambi, O. 1966. Il contenuto in manganese, rame e boro delle piante dell'affioramento ofiolitico dell'Impruneta (Firenze). Arch. Bot. Biogeogr. Ital. 42, 4a ser. XI: 1–13.

Vergnano Gambi, O. 1967. Primi dati sulla localizzazione del nichel in Alyssum bertolonii Desv.. Giorn. Bot. Ital. 101: 59–60.

Vergnano Gambi, O. & R. Gabbrielli. 1981. La composizione minerale della vegetazione degli affioramenti ofiolitici dell'alta valle di Ayas. Rev. Valdôt. Hist. Nat. 35: 51–61.

Vergnano Gambi, O. & R. Gabbrielli. 1985. Tolleranza ai metalli nelle piante superiori. Atti 2° Congr. Naz. SITE, Padova 25–29 giugno 1984, 1097–1099.

Vergnano Gambi, O. & R. Gabbrielli. 1987. The response of plants to heavy metals: organic acid production. Giorn. Bot. Ital. 121: 209–212.

Vergnano Gambi, O., R. R. Brooks, & C. C. Radford. 1979. L'accumulo di nichel nelle specie italiane del genere Alyssum. Webbia 33: 269–277.

Vergnano Gambi, O., R. Gabbrielli, & L. Pancaro. 1982. Nickel, chromium and cobalt in plants from Italian serpentine areas. Oecol. Plant. 3: 291–306.

Vergnano Gambi, O., L. Pancaro, & C. Formica. 1977. Investigations on a nickel accumulating plant: Alyssum bertolonii Desv. I. Nickel, calcium and magnesium content and distribution during growth. Webbia 32: 175–188.

Vergnano Gambi, O., L. Pancaro, & R. Gabbrielli. 1980a. Investigations on a nickel accumulating plant: Alyssum bertolonii Desv. II. Phosphorus, potassium, iron and trace element content and distribution during growth. Atti Soc. Tosc. Sci. Nat., Mem. B 86: 317–329.

Vergnano Gambi, O., L. Pancaro, F. Cardini, & R. Gabbrielli. 1980b. Nickel and serpentine effects on Alyssum bertolonii Desv.: seed germination and seedling development. 2nd FESPP Congr., Santiago de Compostela, Abstracts: 701–702.

Vergnano Gambi, O., F. Cardini, L. Pancaro, & R. Gabbrielli. 1976. Alcuni aspetti del metabolismo di piante coltivate sia su un terreno serpentinoso sia in presenza di nichel. Giorn. Bot. Ital. 110: 303–318.

Wainwright, S. J. & H. W. Woolhouse. 1975. Physiological mechanisms of heavy metal tolerance in plants. In "The Ecology of Resource Degradation and Renewal". Eds. M. J. Chadwick and G. T. Goodman, Blackwell, Oxford, pp. 231–257.

Walker, R. B. 1948. Molybdenum deficiency in serpentine barren soils. Science 108: 473–475.

Willet, I. R. & T. Batey. 1977. The effects of metal ions on the root surface phosphatase activity of grasses differing in tolerance to serpentine soil. Plant and Soil 48: 213–221.

Woolhouse, H. W. 1983. Toxicity and tolerance in the responses of plants to metals. Enc. Plant Physiol., New Series, 12C: 245–300.

O. Vergnano Gambi
Dipartimento di Biologia Vegetale
dell'Università di Firenze
Laboratorio di Fisiologia Vegetale
Via Micheli 1
50121 Firenze
Italy

X

The vegetation over ultramafic rocks

in the tropical far east

JOHN PROCTOR

Abstract. The tropical Far East has many outcrops of ultramafic rock including very large areas in Sulawesi and New Caledonia. The outcrops occur under many different climates and give rise to soils of a very wide range of chemical compositions. The vegetation of the ultramafic soils is poorly known but is very varied. At one extreme are stunted grassland and shrublands, at the other, species-rich large-stature rainforests. In general the causes of this variation remain unexplained. The best investigated area by far is in New Caledonia. Most of the ultramafic soils in that country are covered by a more or less bushy evergreen sclerophyllous vegetation, whilst about 10% of the soils bear evergreen rainforests. There are a very large number of species restricted to ultramafic soils in both these vegetation types. The vegetation overlying the ultramafic soils in New Caledonia is not necessarily lower nor more sparse than that on different substrates nearby. Foliar analyses have shown that a high proportion of the species on New Caledonian ultramafic soils have an unusual chemical composition. About 21% of the species had more than $1,000 \mu g\,g^{-1}$ manganese, 5% of the species had more than $1,000 \mu g\,g^{-1}$ nickel, including nine species with more than $10,000 \mu g\,g^{-1}$ nickel. Detailed information for vegetation on ultramafic soils elsewhere in the tropical Far East is available only for Gunung Silam, a small mountain in Sabah, Malaysia. The vegetation of Gunung Silam is species-rich and ranges from large-stature lowland evergreen rainforest to small-stature lower montane forest. Forest stature there seems to be determined more by altitudinal effects (which might include hydrological effects) rather than by any soil toxicities. In contrast to New Caledonia only two tree species on Gunung Silam are restricted to ultramafic soils. A further contrast is provided by studies of leaf chemical composition since Gunung Silam has only five species reported to have more than $1,000 \mu g\,g^{-1}$ manganese and only one with more than $1,000 \mu g\,g^{-1}$ nickel. Very little experimental work has been made on the plants of ultramafic soils of the tropical Far East.

B. A. Roberts and J. Proctor (eds), The ecology of areas with serpentinized rocks. A world view, 249–270.
© 1992 *Kluwer Academic Publishers.*

Introduction

Ultramafic rocks outcrop in many places in the Far East and their occurrence and geology over much of the area have been discussed by Hamilton (1979). Studies on their vegetation have been patchy but Whitmore (1984) has recognized the Formation 'Rainforest over ultrabasic rocks.' He has emphasized however that within the Formation there are large differences in stature and species richness. At one extreme are the extraordinarily stunted grassland and shrubland at about 600 m altitude in the Talaud Islands of Indonesia (Lam 1927) and at the other are species-rich large-stature rainforests such as those on the lower slopes of Gunung Silam, Sabah (Proctor *et al.* 1988). Part of the variation can be explained by climatic differences between areas but there are large unexplained differences in the vegetation within areas.

Only the island of New Caledonia in the southeast of the region has had its vegetation over ultramafic rocks intensively investigated and much of the work there has been discussed by Jaffré (1980). This island is floristically and climatically distinct and recent studies on the ultramafic Gunung Silam, Sabah (Proctor *et al.* 1988, 1989) have revealed many contrasts with New Caledonia so that no generalizations are possible for the region as a whole. Several recent initiatives have been made to study the vegetation on ultramafic rocks in the Far East and it is expected that this chapter will rapidly become out of date.

Climate

An overview of the climate is difficult because adequate information is rare and the outcrops are spread over such a large area and at such a wide range of altitudes. A useful summary of the rainfall over much of the region has been given by Whitmore (1984) and there are climatic details available, which are given below, for some sites.

Jaffré (1980) has described the climate of New Caledonia. Rainfall is markedly seasonal there and ranges from less than 1000 mm per year up to 5200 mm per year on the summits of high mountains. There is a great deal of variation between years and at one locality (on the west coast) a mere 168 mm annual rainfall was once recorded. Temperature records for a station at 170 m altitude in the south of New Caledonia show that mean monthly maximum temperatures range from 22.1°C (August) to 28.8°C (February); and ranges of minima are 14.8°C (August) to 21.6°C (February). An absolute minimum of 7.2°C has been recorded in August and an absolute maximum of 36.0°C in December. The lapse rate is about 0.57°C 100 m^{-1}, and above about 1000 m temperatures down to 0°C are not rare.

The climate around Gunung Silam has been discussed by Proctor *et al.* (1988). The annual rainfall at the base of the mountain at a station at 10 m altitude varied from 1405 mm to 3132 mm (mean 2011 mm) between 1972 and 1983. August tended to be the driest month but there is no well-defined dry season. There was little correspondence in the rainfall pattern between this station and another at the summit (884 m) when measurements were made in July to September for 1983 and 1984. The summit, which is often covered by a cloud cap, usually had less rainfall. The temperatures at 10 m had a mean daily minimum of 23.4°C and a mean daily maximum of 31.7°C in July to September 1984 and these are probably representative of the temperatures for the whole year. The lapse rate was 0.48°C 100 m^{-1} for the same period.

Whitmore (1969) has discussed some aspects of the climate of the Solomon Islands. These have a mean annual rainfall for coastal stations which varies from 3120 mm to 6250 mm except for rain shadow areas which receive from 1250 mm to 3120 mm. Rainfall inland is certainly higher than at the coast.

Soils

Ultramafic soils in the Far East have been studied most intensively in New Caledonia. Trescases (1969a, b, 1973, 1975, 1976) has discussed the chemistry of the weathering processes. There are two main phases: first a removal of silica and magnesium and a concentration of iron in well-drained profiles; secondly a recombination of silica and magnesium to form 2:1 type smectite clays at the base of slopes (where excess silica recrystallizes to form quartz or chalcedony, and excess magnesium is precipitated as a crust of magnesium carbonate (giobertite). The first phase occurs only in wet conditions and forms ferralitic (oxisol) soils (Latham 1975b) whilst the second phase occurs in dry conditions and leads to the formation of 'sols bruns eutrophes hyper-magnésiens' which include inceptisols and verti-sols. The oxisols of New Caledonia are much more extensive than the present climate would indicate and presently are forming only in parts of the south of the island (Trescases 1969a).

Latham (1975a, b, c) has provided a detailed account of soil formation at a range of altitudes on the Boulinda massif on the west of New Caledonia. This mountain (summit at 1330 m) is a mass of peridotite resting on serpentinite. There are 'sols bruns eutrophes hypermagnésiens' at the base of the mountain, oxisols above them up to 1200 m, above which there are various 'sols á accumulation humifère' with a layer of mor humus. There is much variation within each of these categories and soil zonation and morphology are often complicated by erosional events which move materials and truncate profiles.

Ultramafic oxisols are widespread in New Caledonia (Jaffré 1980) and have been recorded in Sabah (locally on Gunung Silam by Proctor *et al.* 1988 and near Ranau by Fox & Tan 1971), and are important in the Solomon Islands (Lee 1969) and Sulawesi (Parry 1985). 'Sols eutrophes bruns hypermagnésiens' occur in a number of places in New Caledonia in similar topographical situations to those on the Boulinda massif; chemically related ultramafic inceptisols occur in the plots investigated by Proctor *et al.* (1988) on Gunung Silam. Their highest plot (at 870 m) on Gunung Silam had a soil resembling the 'sols à accumulation humifère' of Latham (1975a) and similar soils with a marked organic horizon have been recorded elsewhere in New Caledonia (Virot 1956), from the Solomon Islands at low altitudes (350 m) by Lee (1969) and M. Latham (unpublished), and from high altitudes (2000–2800 m) on Gunung Kinabalu, Sabah (Burnham 1984). Chemical analyses for ultramafic soils on Gunung Kinabalu were made by W. R. C. Munro & J. Proctor (unpublished) and are included in Table 1. Latham (1975a) regarded the 'sols à accumulation humifère' as characteristic of ultramafic

substrata on mountains and notes that similar soils are lacking on Mont Panié (altitude 1600 m) which occurs on the east coast of New Caledonia and has a similar climate to Boulinda, but a non-ultramafic mother rock of mica schists and gneiss.

Ultramafic oxisols vary in depth according to their topographic position and on slight slopes are often several metres deep. They have coarse-textured surface horizons; the lower horizons have a sandy-clay texture with a high proportion of fine sand; they have a well-developed micro-porosity and a good water reserve. They are composed principally of oxides and hydroxides of iron which are often accompanied by chromium, especially in the gravelly horizons. Calcium and silica are low in all the profiles whilst magnesium and nickel increase with depth. The soils are acid and have a low cation exchange capacity. There are low exchangeable bases in all the horizons: Mg^{2+} is not abundant but is the dominant exchangeable base.

'Sols bruns eutrophes hypermagnésiens' are often shallow on slopes but on flat ground are deeper (>50 cm). They are stony and have numerous fragments of weathering rock at the surface. The clay content is much higher than in the oxisols but there is still a substantial proportion of sand. The soils have a pH around neutral and a high C.E.C. which is dominated by Mg^{2+} ions and there are relatively low concentrations of Ca^{2+} and K^+.

The organic layer of the montane 'sols à accumulation humifère' is at least 10 cm thick and often much thicker. It has a low pH (often c, 3 in a paste) and a low base saturation with Mg^{2+}, the most abundant base.

A summary of the chemical properties of some examples of the ultramafic soils is given in Table 1.

Vegation and flora

New Caledonia

The vegetation and flora of the ultramafic soils of New Caledonia is the best known for any country in the Far East thanks to the research of Jaffré and his co-workers. A vegetation type called 'maquis' occupies 80–90% of the ultramafic soils. The

Table 1. The mean pH, exchangeable potassium sodium, calcium, magnesium, the cation exchange capacity, the Mg/Ca quotient, exchangeable chromium and nickel and total phosphorus, cobalt, chromium and nickel concentrations in a range of ultramafic soils from the Far East (– means not determined)

Country	Soil type	Sample depth (cm) (where known)	n	pH	Exchangeable (m-equivs 100 g⁻¹) K	Na	Ca	Mg	CEC	Mg/Ca	Exchangeable (µg g⁻¹) Cr	Ni	P	Total (µg g⁻¹) Co	Cr	Ni	Author
New Caledonia	Sol brun eutrophe hypermagnésien		10	6.8	0.2	0.3	1.8	38.5	23.9	20.9	0.6	44	140	430	6,300	3,900	(1)
			10	6.9	0.2	0.3	1.5	32.7	33.6	22.5	0.2	66	230	570	12,700	4,700	
	Sol complexe ferrallitique sur alluvions serpentineuses		6	6.0	0.2	0.4	1.8	8.3	13.4	4.6	8.1	0.4	200	930	80,600	3,500	
	Sol complexe ferrallitique serpentineuse colluvial sur serpentinites		4	5.6	0.09	0.1	1.3	1.1	–	0.88	5.0	0.4	–	700	34,900	3,700	
	Sol ferrallitique érodé		17	6.1	0.04	0.06	0.7	1.5	6.2	2.3	1.4	8.7	150	800	25,500	10,400	
			1	5.2	0.05	0.03	0.5	0.8	10.4	1.8	–	–	–	1,000	13,700	7,200	
			4	5.9	0.04	0.07	0.9	1.9	4.1	2.3	1.6	27	200	750	28,700	6,500	
	Sol ferrallitique remanié colluvionné		25	5.3	0.05	0.07	0.7	0.6	6.2	0.82	11	6	300	1,300	24,300	7,100	
	Sol ferrallitique remanié colluvionné ± hydromorphe		13	5.2	0.03	0.03	0.2	0.4	2.8	2.3	3	8	–	700	25,900	5,200	
	Sol hydromorphe sur alluvions et colluvions ferrallitiques		13	5.4	0.05	0.2	0.5	2.2	17.8	5.0	1	35	–	400	48,900	4,100	
	Sol hydromorphe sur alluvions		2	5.2	0.05	0.2	0.3	0.9	5.5	3.0	–	–	–	500	24,100	4,400	
	Sol ferrallitique remanié colluvionné		1	6.1	0.02	0.01	0.01	0.3	1.2	3.2	–	–	–	500	34,100	5,800	
	Sol ferrallitique à gravillons et à cuirasse		12	4.9	0.03	0.06	0.6	0.3	5.5	0.55	2.0	0.8	310	230	55,700	1,900	
			1	4.4	0.02	0.02	0.2	0.2	–	1.1	–	–	–	200	30,000	1,900	
			12	4.9	0.02	0.02	0.6	0.3	1.6	0.51	3.2	0.2	–	270	35,500	2,400	
	Sol ferrallitique gravillonaire		5	5.5	0.1	0.1	0.7	0.5	3.6	0.76	1.2	0.5	–	350	28,600	1,300	
	Sol ferrallitique erodé		5	5.3	0.1	0.1	0.9	6.1	18.8	7.2	0.3	21	–	800	13,900	2,900	(2)
Sabah, Gunung Silam	Inceptisols at a range of altitudes																
	280 m	0–15	20	5.7	0.14	0.10	7.7	24.6	49	3.1	<2	13	–	–	–	–	
	330 m	0–15	20	5.8	0.17	0.13	2.3	15.7	61	6.7	<2	18	–	–	–	–	
	480 m	0–15	20	6.1	0.23	0.06	4.2	11.5	81	2.6	<2	16	–	–	–	–	
	610 m	0–15	20	6.0	0.42	0.17	12.4	10.6	102	0.8	<2	15	–	–	–	–	
	790 m	0–15	20	5.6	0.17	0.08	0.9	5.4	105	4.3	<2	12	–	–	–	–	
	870 m	0–15	18	4.0	0.53	0.41	1.2	5.6	105	3.7	<2	2	–	–	–	–	
Gunung Kinabalu	Oxisol	0–20	1	4.1	0.27	0.08	0.49	0.96	–	2.0	–	1	–	–	–	–	(3)
	?2600 m	0–15		4.8	0.39	0.08	2.5	2.2	117	0.9	–	0.7	–	–	–	–	(3)

252

Table 1. Continued

Country	Soil type	n	Sample depth (cm) (where known)	pH	Exchangeable (m-equivs 100 g⁻¹) K	Na	Ca	Mg	CEC	Mg/Ca	(μg g⁻¹) Cr	Ni	P	Total (μg g⁻¹) Co	Cr	Ni	Author
Ambun,	Orthic ferralsol	1	1–8	4.8	0.2	0.3	0.2	0.8	9.1	4.0	–	–	–	–	–	–	(4)
Sungai Takala;		1	1–30	5.8	0.3	0.2	0.2	52.9	51.9	265	–	–	–	–	–	–	
Entelben,	Orthic luvisol	1	3–33	6.5	0.1	0.4	3.9	10.8	16.4	2.8	–	–	–	–	–	–	
Kuala Binalik;		1	0–10	6.5	0.1	0.1	2.8	12.3	17.4	1.4	–	–	–	–	–	–	
Ulu Tingkayu	Chromic luvisol	1	3–23	6.5	0.1	0.2	11.6	30.7	46.9	2.6	–	–	–	–	–	–	
Solomon Islands, Santa Isabel	Acrorthox; Typic Haplohumox	1	2–10	5.2	0.2	0.2	0.1	0.4	8.6	4.0	–	–	200	240	21,280	3,680	(5)
Choiseul	Acrorthox;	1	0–8	4.8	1.4	2.3	9.1	3.2	56.4	0.35	–	–	300	140	4,430	1,160	(6)
	Typic Tropofluvent	1	0–13	6.7	0.1	0.0	3.9	3.7	11.3	0.95	–	–	380	170	1,500	1,170	
Sulawesi	Acrorthox;	1	0–12	5.3	0.05	0.04	13.3	11.6	34.8	0.87	1	40	–	–	9,900	4,050	(7)
	'Alluvial ultramafic soil';	1	0–13	6.3	0.5	0.3	5.5	11.1	10.0	2.0	1	4	–	–	975	825	
	'Colluvial ultramafic soil';	1	5–25	6.2	0.4	0.4	4.6	26.3	21.8	5.7	1	28	–	–	1,050	2,100	

1. Jaffré (1980).
2. Proctor et al. (1988).
3. W. R. C. Munro & J. Proctor unpublished.
4. Acres et al. (1975).
5. Hansell & Wall (1976).
6. Wall & Hansell (1976).
7. Parry (1985).

Fig. 1. Maquis on ultramafic soil with *Xanthostemon flavum*, New Caledonia (Photograph by Dr A. J. M. Baker).

term 'maquis' is used because the vegetation has a superficial resemblance to true maquis which is developed under a mediterranean climate in temperate zones. The New Caledonian maquis occurs from sea level up to 1600 m and under rainfalls varying from 900–4000 mm year^{-1}. It is an evergreen sclerophyllous vegetation (Fig. 1), more or less bushy or with a dense sedge layer, and has many variants including some transitional forms with forest. Jaffré (1980) classified the maquis into three main physiognomic types:

(a) 'Le maquis arbustif' which occurs at the base of mountains on 'sols bruns eutrophes hypermagnésiens'. It is a thicket of varying density with small shrubs and a low sedge understorey.

(b) 'Le maquis buissonant' is found on 'sols ferrallitiques gravillonaires ou cuirassés' on pla-

teau above 200 m altitude. It has a more or less continuous cover of much-branched shrubs and virtually no understorey herbs.

(c) 'Le maquis ligno-herbacé' occurs at a range of altitudes on 'sols ferrallitiques remaniés par erosion ou colluvionement' on slopes or flat ground. It is characterized by a herb layer of large sedges and by a shrub layer which is diversely bushy and discontinuous.

Each of the types of maquis has numerous physiognomic and structural variants in response to different altitudinal and edaphic conditions. Although maquis is diverse and may have different origins there are several unifying features. For example the leaves of most species are sclerophyllous, coriaceous, with a thick cuticle and are nanophylls to microphylls (i.e. 25–2,000 mm^2). Many species have hairy leaves; narrow leaves are common and so are scented leaves. Many have vertical leaves at the branch ends and this, in combination with the horizontal branching gives a striking 'candelabra' appearance which has evolved independently in many of the larger shrubs and small trees. Most of the maquis species (including those with big tap roots) have a well-developed superficial root system. Maquis species when tested on fertile soils have always been shown to be inherently slow growing although some of the smaller maquis species can become much larger under good conditions of moisture and fertility.

The contrast between the maquis vegetation on ultramafic soils and the vegetation on other soils varies. Jaffré (1980) has stressed that the vegetation overlying the ultramafic rock is not necessarily lower, more stunted or more sparse than that on different substrates nearby. Boundaries between basalts or siliceous rocks and ultramafics usually involve an abrupt change from a savanna or open maquis to a denser, higher, shrubby maquis on the ultramafic soils. Such changes involve big floristic differences and are most dramatic where 'sol bruns eutrophes hypermagnésiens' occur. The low vegetation on the non-ultramafic soils is caused by burning; fires are much less frequent in the maquis on ultramafic soils which is generally much less disturbed by man. In the south of the mainland at the boundary between ultramafic rocks and gabbro or granodiorite there is little change in the vegetation cover. On the

Fig. 2. Rainforest on ultramafic soil, mount Koghi, New Caledonia (Photograph by Dr A. J. M. Baker).

gabbro and granodiorite there are many hygrophilous species which are found on the ultramafics but only in hollows. Jaffré (1980) has suggested that this indicates that physical factors may be important in controlling the distribution since the non-ultramafic soils have more clay and hold water better.

Jaffré (1980) made an intensive study of 455 relevés in maquis from the south of the mainland and from the Boulinda massif. The results are not discussed in detail here but some of his conclusions are important. He established the principle of vegetation variation caused by differences within the ultramafic soil groups. He showed that the floristic richness and the specificity of the flora to ultramafic soils is highest in maquis on the 'sol bruns eutrophes hypermagnésiens' and much less in the oxisols and hydromorphic soils.

Evergreen rainforests (Fig. 2) are found on ultramafic soils usually between 1500 m and 3500 m under rainfalls ranging from 500 to 1000 mm per year. In the wetter east part of the island they go below 500 m but are restricted to valleys or humid sites sheltered from fire. The forests are most commonly found on steep slopes on eroded oxisols or skeletal rocky soils and are

rare on 'sols bruns eutrophes hypermagnésiens' or on deeper oxisols. This is partly because the 'sols bruns eutrophes hypermagnésiens' are found below the forest limit and partly because the deeper oxisols are in nickel-rich areas where the vegetation has been damaged by miners. The canopy is generally between 15 and 20 m high. The forests are often dominated by gymnosperms: *Agathis lanceolata* in the south and on the Boulinda massif and several different *Araucaria* species elsewhere. The herb layer consists mainly of ferns and orchids: grasses and sedges are rare. The forest understory has palms, *Prunus* spp. (Rosaceae), *Psychotria* spp. (Rubiaceae) and many species with big leaves. Lianes and semi-epiphytes are rare. Some species, by their abundance or distinctive appearance have been used to distinguish particular forest types, e.g., Jaffré (1980) recognised an *Arillastrum* (gum oak) (Myrtaceae) facies, an *Agathis ovata* facies, a *Nothofagus* (Fagaceae) facies, and an *Araucaria* facies.

Montane evergreen rainforests, with a canopy 8–15 m high, occur under high rainfall (>3,500 mm per year) above about 1,000 m on eroded oxisols. They are often rich in gymno-

255

sperms and in the south of the island are dominated by *Araucaria humboldtensis*. The individual trees have bent trunks with many epiphytes and a lot of species have small leaves and are sclerophyllous or hairy. On the 'sols à accumulation humifère' (Latham 1975a) there occurs a stunted montane forest which is only 6–10 m high, dominated by *Metrosideros dolichandra* (Myrtaceae), which has a rich flora of bryophytes, filmy ferns and lichens.

Forests which are on a boundary of ultramafic and other rock types usually show no obvious differences and often a detailed study is required before the line of separation can be recognized.

The flora of New Caledonia is remarkable in many ways. Its geographical affinities have recently been discussed by Morat *et al.* (1986) for maquis species and by Morat, Veillon & MacKee (1984) for forest species. There are about 3,000 species of phanerogams and 250 species of vascular cryptogam in a total land area of less than 20,000 km². The actual numbers of species must be greater than this since the flora is incompletely known. Seventy-six percent of the phanerogamic species are endemic. Several families which are common in many parts of the tropical Far East are absent, e.g., the Dipterocarpaceae, Ericaceae, Melastomataceae and Zingiberaceae, whilst others are poorly represented, e.g., the Compositae, Gramineae and Tiliaceae. Other families are relatively important, e.g., the Casuarinaceae, Cunoniaceae, Cyperaceae (which seem to occur often where grasses might be expected), Epacridaceae, and Myrtaceae. The Gymnosperms are very well represented and there are some primitive flowering plant families, e.g., the Winteraceae and Monimiaceae. The New Caledonian flora was regarded by Takhtajan (1969) as being one of five floristic subkingdoms of the Old World tropics and its history has been summarized by Morat, Veillon & MacKee (1984). Its relative richness must be viewed against a background of great climatic, geological, and topographic diversity on the island.

There are about 1500 species on the ultramafic soils (which occupy about one third (ca. 5,500 km²) of the land area of New Caledonia and virtually all are indigenous. The flora of the ultramafic soils of New Caledonia has more endemism than the rest of the territory. The maquis

has 944 species on ultramafic soils. Of these, 724 are restricted to ultramafic substrata and 97% of these restricted species are endemic to New Caledonia. For forests there are 1,511 species (including 77 epiphytes) of which 415 species (98% of which are endemic to New Caledonia) are restricted to ultramafic soils. A high proportion of the large tree species are not restricted to ultramafic soils. The ultramafic soils are also characterized by absences, the most notable of which are introduced species (often important on other substrata and including the Caryophyllaceae) and, of the indigenous species, many Compositae and Gramineae, one species of Myrtaceae and some of Verbenaceae. The families best represented in the flora of the forests on ultramafic rocks are (in order of their number of species): Orchidaceae (mainly epiphytes in the humid forests of the south), Rubiaceae, Myrtaceae, Euphorbiaceae, Apocynaceae, gymnosperms (of many families), Lauraceae, Sapotaceae, Proteaceae, Cunoniaceae, Flacourtiaceae, Araliaceae, Sapindaceae, Pandanaceae, Rutaceae, Myrsinaceae, Pittosporaceae, Leguminosae, Moraceae, Meliaceae and Celastraceae. In the maquis the Myrtaceae are the most numerous followed by the Rubiaceae, Euphorbiaceae, Apocynaceae, Orchidaceae, Cyperaceae, Cunoniaceae, Sapotaceae, Proteaceae, Rutaceae. Araliaceae, Leguminosae, Dilleniaceae, Flacourtiaceae and Epacridaceae.

The ultramafic rocks of New Caledonia have been dated at the Eocene/Oligocene boundary. Their emplacement was slow and did not involve the extinction of ancient taxa which were thus able to adapt to the new rock type. Several genera have shown an active speciation since the Oligocene and include 'neoendemic' species and there are several 'palaeoendemics' including the ancient gymnosperms *Dacrydium guillauminii*, *Decussocarpus minor* and *Neocallitropsis pancheri* which were almost certainly once more widespread and now survive on the ultramafics of the south of the island where competition is probably less severe.

New Guinea

Van Royen (1960, 1963) has described some aspects of the vegetation of the ultramafic outcrops which occur along the northern coast of Waigeo Island and also on a number of outcrops along the northern coast of the mainland of New Guinea. Much of the vegetation is unusually open and shrubby in an area where tall rainforest would be expected. It comprises such species as 'Dodonaea viscosa (Sapindaceae), Myrtella beccarii (Myrtaceae), Nepenthes ampullaria (Nepenthaceae), Schizoloma guerinianum (Lindsaeaceae), Schizoloma ensifolium, Cladium micranthes (Cyperaceae), Styphelia abnormis (Epacridaceae), Dillenia alata (Dilleniacae) and Alphitonia moluccana (Rhamnaceae) and an extensive undergrowth in many places of Gleichenia sp. (Gleicheniaceae).' In some places there are extensive stands of Casuarina (Casuarinaceae) species. This unusual vegetation occurs on 'purple-reddish soil' with high concentrations of chromium, cobalt and nickel and gives way to rainforest precisely at the boundary between this soil and others with lower concentrations of these elements. Van Royen (1963) commented that the unusual vegetation may be secondary and maintained by fire. Elsewhere on Waigeo, van Royen (1960) has described rainforest 'on lateritic loamy hills with an underlying limestone or ultrabasic mother-stone.' This forest is dominated with Vatica papuana (Dipterocarpaceae) and has Dillenia spp., Intsia bijuga (Leguminosae), Adina multiflora (Naucleaceae) and Myristica spp. (Myristicaceae) as co-dominants. Locally, Agathis labillardieri (Araucariaceae) dominates substantial areas of this forest.

Paijmans (1976) has commented that 'throughout Papua New Guinea ultrabasic rocks in particular and, less obviously, limestone appear to be correlated with the presence of dense, thin-stemmed, small-crowned and often low forest. Ultrabasic rocks often have a topography of steep, straight slopes with unstable, shallow and chemically poor soils, and hence have thin stemmed forest. However, forest tends to be denser and smaller crowned than average hill forest also on ultrabasics that have a general topography and deep soils.' He commented further, 'In Papua New Guinea there is no evidence that any tree species is confined to either ultrabasic rock or limestone. Casuarina papuana is commonly predominant in forest over ultrabasic rocks and occasionally in forest on limestone, but this

species has a wide ecological range and is not restricted to any particular rock type or soil.'

Philippines

There is little information on the substantial ultramafic areas of the Philippines although Podzorski (1985) includes some notes on the wide range of forest types which occur on the ultramafic rocks of Palawan island. An extreme type of stunted forest occurs on Mount Bloomfield with sclerophyllous trees of 2–3 m high at low altitude. Podzorski (1985) has commented: 'The heavy-metal indicators, Scaevola micrantha (Goodeniaceae), Brackenridgea palustris var. foxworthyi (Ochnaceae) and Exocarpus latifolius (Santalaceae) occurred here some 500 m below their normal altitude. Collections of Terminalia (Combretaceae), Gymnostoma (Casuarinaceae), Syzygium (Myrtaceae), Chionanthus (Oleaceae) and Guioa (Sapindaceae) do not appear to resemble any known species'.

Solomon Islands

The following account is based on Lee (1969) and Whitmore (1969). The total land area of this rugged and mountainous archipelago is about 30,000 km^2. Ultramafic rocks occur on several of the islands and have a combined area of about 1,040 km^2. The intact forest is poor in species and dominated over large areas by Casuarina papuana or Dillenia crenata (which is absent from the island of Choiseul), and a few other species all of which grow scattered through the other forests. Only four species are known to be restricted to the ultramafic soils: Galubia hombronii (Palmae); Myrtella beccarii (Myrtaceae) on Santa Isabel island only in the Solomons (but also recorded by van Royen 1960 in New Guinea); Pandanus lamprocephalus (Pandanaceae) on north-east San Cristobal island only; and an undescribed species of Xanthostemon (Myrtaceae). On the relatively dry island of Santa Isabel (and locally elsewhere) fires have caused extensive destruction of the forests on ultramafic soils and this has resulted in open heaths dominated by ferns (Gleichenia spp.), the clubmoss Lycopodium cernuum and Commersonia bartramia (Sterculiaceae).

Sabah

The most famous areas of ultramafic rock out-crops in Sabah are in the area of Gunung (Mount) Kinabalu. Fox & Tan (1971) described rainforests near the Ranau area of Kinabalu at 550–720 m altitude. The general forest canopy level varied from 18–55 m with emergent trees up to 52 m. There was a middle storey of trees 15–28 m in height and an understorey was locally dense. Climbers were present, though not large or frequent. In six 0.04 ha plots there were 132 trees above 9.7 cm dbh and 4 trees which exceeded 58 cm diameter. The largest tree in the general area was an individual of *Artocarpus melinoxylus* (Moraceae) which had a diameter of 121 cm. The recorded emergent trees were in the Dipterocarpaceae, Moraceae, Meliaceae, Burseraceae and Sapotaceae. In the general canopy there were Fagaceae, Sapotaceae, Leguminosae, and in the middle storey, Euphorbiaceae, Flacourtiaceae, Xanthophyllaceae, Sapotaceae, Lauraceae, Myrsinaceae, Leguminosae, Naucleaceae, Anacardiaceae, Dipterocarpaceae, Myrtaceae, Moraceae, Olacaceae and Meliaceae. Of the tree species, Fox & Tan (1971) claimed that only *Dipterocarpus orchraceous* was confined to the ultramafic soils: all the large trees were found elsewhere. (*D. orchraceous* is now known not to be confined to ultramafic soils and no member of its family is so restricted, Proctor *et al.* 1988).

There are various notes on the flora of the ultramafic soils of the Kinabalu area in Meijer (1965). He recorded several species of Dipterocarpaceae at about 610 m. Near the summit (about 1220 m) of the ultramafic hill, Bukit Hampuan, he recorded that mossy forest occurs at a much lower altitude than it does on nearby sandstone and quartzite ridges where it begins to occur at about 1830 m. He noted a local abundance of the thin climbing bamboo *Racemobambos*, 'a genus which tends to be common on ultrabasic but on higher altitude it may occur on sandstone or quartzite.' Some species of rattans (not named by Meijer) are described as being typical for this forest. *Borneodendron anaegmaticum* (Euphorbiaceae) which is endemic to Sabah and recorded only on ultramafic substrata is recorded from Bukit Hampuan (Meijer 1971). He mentions some Kinabalu species which occur also on ultramafic rocks in the Philippines: *Buxus rolfei* (Buxaceae) (which also occurs on the lowland ultramafic rocks on the island, Pulau Sakar, off the east coast of Sabah, near Gunung Silam); *Pittosporum pentandrum* (Pittosporaceae); *Euonymous glandulosus* (Celastraceae) and *Scaevola* spp.

Meijer (1965) discussed the abrupt contrast on the boundary between sandstone and ultramafic rocks at about 2450 m 'The soil is much barer than on the sandstone, *Dipteris* (Dipteridaceae) is completely lacking, *Leptospermum flavescens* (Myrtaceae) is replaced by *L. recurvum*, *Dacrydium gibbsiae* (Podocarpaceae) takes over from *D. beccarii*. The small leaved *Rhododendron ericoides* (Ericaceae) starts here. In the herb layer we notice *Hedyotis macrostegia* (Rubiaceae), *Elatostemma bulbothrix* (Urticaceae), *Machaerina falcata* (Cyperaceae), *Schoenus curvulus* (Cyperaceae), *Trachymene* (Hydrocotylaceae), *Stypelia suaveolens* (Epacridaceae), *Vaccinium coriaceum* (Ericaceae), *Euphrasia borneensis* (Scrophulariaceae) and the tiny fern *Schizaea fistulosa*. A species of *Scapania* is most common among the bryophytes. The most dominant features in the shrub-layer are the dark green *Dacrydium gibbsiae* and the greyish *Leptospermum recurvum*. This combination of species is very peculiar for the ultrabasic at this altitude. This means however that not all the species occurring here are restricted to ultrabasic soil. Except *Dacrydium gibbsiae* and *Schizaea fistulosa* they all occur on the granite above the ultrabasic. Right of the path to the east we find places where practically no vegetation occurs at all except here and there a tuft of *Euphrasia borneensis*, *Schoenus curvulus* and *Machaerina micrantha*.' Sato (1985) has given further floristic information on the ultramafic area of Gunung Kinabalu and shows diagramatically that it is of much lower stature than the forest at higher altitude on non-ultramafic rocks.

Near the summit of Mount Tambuyokon (2600 m) Meijer (1965) found 'a composite (possibly *Lactuca* spec., *Cyrtandra clarkei* (Gesneriaceae) and a dense vegetation of a hairy *Scaevola*.' Meijer (1971) refers to 'the famous Marai-Parai trail, an ultrabasic ridge full of botanical rarities and the best locality for the giant pitcher plant *Nepenthes rajah*. Botanists will be interested to know that the famous treelet *Scyphostega borneensis* (Scyphostegiaceae) still occurs here at its type locality . . . '

Meijer (1964) has recorded clear cut boundaries between forest on ultramafic rocks and that on sandstone at the base of the Mesasau Mountains. 'Many of the dipterocarps with large crowns, occurring on the sandstone are completely replaced by other species on the ultrabasic. The differences are especially striking in colored air photographs during flowering of the lowland dipterocarp forest on sandstone, with *Parashorea malaanonan* and *Dryobalanops lanceolata* in full bloom. These species strictly avoid the ultrabasic.'

The vegetation of the ultramafic mountain Gunung Silam (altitude 884 m) is the most intensively investigated on this substratum in Sabah. Proctor *et al.* (1988) studied the structure and floristics of ten plots ranging from 280 m to 870 m on a main ridge of the mountain. Over this narrow altitudinal range the vegetation changed from large stature lowland dipterocarp forest to stunted myrtaceous forest from which the Dipterocarpaceae were absent. Table 2 lists the contributions of each family to each plot. Proctor *et al.* (1988) could find no obvious edaphic causes of the rapid altitudinal changes and ascribed them to an unexplained '*Massenerhebung* effect' which refers to the general lowering of altitudinal limits of forest types on small mountains. They quantified many features of the forest and it was established that the lower forest had emergent mesophyllous trees with broad crowns. Leaf size tended to decrease with altitude and there was a rapid reduction of crown width above 330 m altitude. Proctor *et al.* (1988) pointed out that the distinctive features shown by forests on ultramafic soils may often involve the *Massenerhebung* effect since ultramafic rocks often occur on small mountains. Gunung Silam is very species-rich and the 0.24 ha plots at 610 m and 700 m both had 91 species of tree ($\geqslant$ 10 cm dbh) – probably the highest values recorded for contiguous plots of this size. Only two of these tree species (*Borneodendron anaegmaticum* and *Buchaniana arborescens*) on Gunung Silam are known to be restricted to ultramafic soils although it is possible that more such restrictions might be found if the taxonomy of large genera such as *Eugenia* were better

known. There is more evolutionary response amongst the palms and bamboos on Gunung Silam. Dr J. Dransfield recorded fifteen palm species from the mountain of which one (*Salacca* sp. nov.) has been recorded from nowhere else; a further six species or varieties occur elsewhere but only on ultramafic soils; and one is variable but suspected to have varieties restricted to soils on this substratum. Dr S. Dransfield (personal communication) recorded a bamboo on Gunung Silam which is confined to ultramafic soils.

There is a contrast between the vegetation on Gunung Silam and that described for other ultramafic rocks in the Western Tawau and Lahad Datu districts of Sabah. Wright (1975) gave the following description: 'These forests have a similar appearance on aerial photographs to Heath forests. The trees are low and dense and few exceed 120 cm in girth or 12 m in height. One forest was described by Wright (1975) as being 'rich in *Tristania grandifolia*' (which was recorded from Gunung Silam (Proctor *et al.* 1988) and is a species which is frequent in heath forests. Fox (1972) has recorded pure stands of *Casuarina nobilis* from low-lying sites on ultramafic rocks on islands in Darvel Bay, a few km from Gunung Silam.

Sulawesi

Van Balgooy & Tantra (1986) have briefly described some of the forests over ultramafic rocks in southern Sulawesi mainly around the Lake Matano area. (This part of Sulawesi has three large lakes and unusually the ultrabasic area is studded with small lakes or ponds.) This area has many endemic species.

Lake Matano has shores on ultramafic and limestone rocks. The ultramafic shore has forest, mostly less than 15 m tall, and completely dominated by one family, the Myrtaceae. The canopy is formed by *Metrosideros petiolata* and *Xanthostemon confertiflorum* (also often seen as a shrub). There is a remarkable *Kjellbergiodendron* which may be a new taxon and locally *Eugenia* and *Tristania* species. The Rubiaceae are well represented with small trees (up to 4 m) and shrubs including species of *Timonius*, *Canthium* and *Gardenia*. Amongst the taller trees they noted *Gymnostoma sumatrana* (Casuarinaceae),

Table 2. The percentage contribution of each family to tree ($\geqslant 10$ cm dbh) basal area in ten plots at a range of altitudes on Gunung Silam, Sabah

Family	Plot altitude (m)									
	280	330	420	480	540	610	700	770	790	870
Annonaceae	0.4	0.3	0.2	0.4	0.2	–	0.2	–	–	–
Anacardiaceae	21.5	13.1	22.7	14.7	4.2	12.4	4.8	–	0.7	–
Barringtoniaceae	1.9	1.9	–	–	–	–	–	–	–	–
Burseraceae	0.7	0.4	0.4	0.4	1.5	0.4	1.1	2.1	–	–
Casuarinaceae	–	–	–	–	–	–	–	11.2	2.5	–
Celastraceae	–	2.0	0.1	2.9	3.9	1.4	–	–	–	–
Chrysobalanaceae	1.1	1.4	0.5	1.0	2.5	1.7	2.9	–	1.4	–
Ctenolophonaceae	–	–	–	–	–	–	0.6	–	–	–
Dilleniaceae	3.9	–	–	–	–	–	–	–	–	–
Dipterocarpaceae	27.9	46.3	50.0	38.6	26.8	27.1	4.5	–	–	–
Ebenaceae	0.5	3.0	2.6	1.6	1.4	0.8	2.5	0.2	–	–
Elaeocarpaceae	–	–	–	–	0.1	0.1	–	–	–	2.5
Epacridaceae	–	–	–	–	–	–	–	–	–	1.0
Escalloniaceae	–	–	–	–	0.4	–	2.2	–	0.2	1.4
Euphorbiaceae	2.7	1.2	3.2	14.6	18.9	23.8	16.3	13.4	13.5	–
Fagaceae	–	1.0	2.1	0.4	2.4	1.8	7.4	1.2	6.9	–
Flacourtiaceae	–	–	0.1	–	0.7	–	0.5	–	–	–
Guttiferae	2.7	2.7	1.1	5.0	5.2	4.7	6.1	0.9	–	–
Icacinaceae	0.2	0.3	0.1	0.3	0.3	–	0.3	–	–	–
Ixonanthaceae	–	0.1	0.1	–	–	–	–	–	–	1.0
Lauraceae	3.9	2.4	0.6	0.3	1.4	0.1	2.9	–	0.7	2.5
Leguminosae	7.1	1.1	–	0.1	1.2	–	–	–	–	–
Magnoliaceae	–	–	–	0.1	–	0.4	–	–	–	–
Memecylaceae	–	–	0.5	–	0.3	–	1.5	1.1	–	–
Meliaceae	0.4	–	–	–	–	–	–	–	–	–
Moraceae	1.1	–	–	–	–	–	–	–	–	–
Myristicaceae	–	–	0.5	0.1	0.2	0.3	0.6	–	–	–
Myrsinaceae	0.1	–	0.2	0.4	0.6	1.2	0.1	–	–	–
Myrtaceae	6.9	10.0	3.7	5.6	10.7	10.1	19.9	37.6	30.7	60.7
Olacaceae	–	0.6	0.1	0.5	–	0.3	0.5	–	–	–
Oleaceae	0.1	0.2	0.5	0.9	0.3	0.2	0.1	–	–	–
Podocarpaceae	–	–	–	–	0.4	1.9	4.9	–	19.6	–
Proteaceae	–	–	–	–	–	–	0.2	–	–	–
Rhizophoraceae	–	–	0.4	0.2	–	0.4	0.9	–	1.2	–
Rubiaceae	0.2	0.3	1.6	5.8	3.1	4.5	3.3	9.1	2.8	–
Rutaceae	–	–	0.1	0.2	0.5	–	1.0	1.4	5.9	24.0
Sapindaceae	1.4	–	0.1	0.1	–	–	–	–	–	–
Sapotaceae	1.0	8.4	5.2	3.6	5.6	3.6	6.8	18.2	12.8	–
Staphyleaceae	–	–	–	–	–	–	0.2	–	–	–
Sterculiaceae	–	–	–	0.9	0.6	–	–	–	–	–
Symplocaceae	–	–	–	–	0.4	–	–	–	–	–
Theaceae	–	–	0.4	0.8	2.4	1.1	6.7	1.6	–	5.7
Tiliaceae	6.3	–	–	–	–	–	–	–	–	–
Ulmaceae	–	–	0.8	0.3	0.6	1.2	0.5	–	–	–
Verbenaceae	7.8	3.4	1.7	0.4	0.1	0.3	–	–	–	–
Violaceae	0.1	–	–	–	–	–	–	–	–	–
Xanthophyllaceae	–	–	0.1	–	0.8	–	0.1	–	–	1.1

Planchonella spp. (Sapotaceae), *Gluta papuana* (Anacardiaceae) and a local endemic *Terminalia supitiana*. Several other genera of trees and shrubs were noted including *Buchanania arbores-* *cens* (Anacardiaceae) which occurs on Gunung Silam and is restricted to ultramafic substrata in Sabah (Proctor *et al.* 1988). Other genera in common with Gunung Silam, Sabah are *Calophyllum*

(Guttiferae), *Dillenia* (Dilleniaceae), *Pandanus* (Pandanaceae) and *Memecylon* (Melastomataceae).

On ultramafic rocks near the village of Soroako on the shores of Lake Matano the forest is characterized by a regular canopy 25–40 m high. Formerly there were large emergent *Agathis* species which are now felled. The density of the trees was probably high, the trunks straight boled and generally less than 50 cm dbh. Few trees had buttresses or stilt roots. The bulk of the forest is made up of a few families. In the upper strata the Myrtaceae is represented by species of *Eugenia*, *Kjellbergiodendron* and *Metrosideros*. Myristicaceae are common in the lower strata with *Horsfieldia*, *Gymnacranthera* and *Knema* (including the endemic *Knema celebica*). Other common families are the Burseraceae, Fagaceae, Melastomataceae and Sapotaceae. Many genera occurring in this area have been recorded from Gunung Silam, Sabah by Proctor *et al.* (1988).

Van Balgooy & Tantra (1986) made an interesting comparison between the vegetation plots of 20 × 100 m on Gunung Konde an ultrabasic hill west of Soroako at 700 m altitude and those on Gunung Wawonseru, a limestone hill further to the west at 1000 m altitude. They reported that 'structurally the forest on the two mountains are obviously different, floristically too they have little in common. The Konde forest makes an orderly impression by its closed regular canopy at 30–35 m, its straight non-buttressed trunks, few of which exceed 60 cm diam. at breast height. To the tallest trees belong *Eugenia*, *Ficus*, *Kjellbergiodendron*, *Lithocarpus*, *Pouteria* and *Santiria*.' The density of the trees (≥10 cm dbh) averaged 1050 ha^{-1}. The foliage was small and coriaceous and dark coloured. They comment on the poverty of the avifauna. They showed that there was a dominance of individuals of certain families: Myrtaceae (*Eugenia*, *Kjellbergiodendron* and *Tristania*); Burseraceae (*Canarium* and *Santiria*); Myristicaceae (*Gymnacranthera* and *Myristica*); Sapotaceae (*Palaquium* and *Pouteria*); Fagaceae (*Castanopsis* and *Lithocarpus*); and Melastomataceae (*Memecylon*). Several other species and genera are listed including *Psychotria* (Rubiaceae) which are known from ultramafic outcrops elsewhere.

The forest at Batu Besi, on particular iron-rich parts of the ultramafic rocks, resembles some types of heath forest which have been described from Sarawak and Brunei by Brünig (1974) and from Kalimantan by Kartawinata (1980). The Batu Besi forest is low, 10–15 m tall, with straight trunks, rarely exceeding 20 cm dbh with few climbers, epiphytes, ferns and orchids. The soil consists of solid rock, pebbles, and stones of various sizes and is covered by a thick layer of bryophytes and no undergrowth apart from a few seedlings. The tallest trees include *Hopea celebica* (Dipterocarpaceae) and especially abundant *Austrobuxus nitida* (Euphorbiaceae). Other common smaller trees belong to the Myrtaceae (*Eugenia* and *Xanthostemon*) and Myristicaceae (*Horsfieldia*). The forest is very species poor, has small stature, and wide spacing of the trees.

Further notes on the flora of the same general ultramafic area have been made by Meijer (1984).

Talaud Islands

Lam (1927) described the vegetation of the ultramafic mountain, Gunung Piapi which is on the middle part of the eastern coast of the central island of the Talauds, Karakelong. The summit of the mountain was mapped at 690 m. In general the vegetation of Karakelong is rainforest on non-ultramafic rocks. The upper part of Gunung Piapi however has a vegetation which is perhaps the most remarkable on ultramafic soils in the Far East. This mountain has a fairly long, undulating north-west/south-east oriented ridge which is covered by a short yellowish-green vegetation of grasses, shrubs and small trees which makes a peculiar contrast to the dense dark green forest around it. There is a small remnant of slightly higher forest where the slope is less steep, at Panansaran.

The vegetation looks of an alpine type at an altitude as low as 150 m. The exposed slopes are covered with dense short clumps of the grass *Themeda gigantea* mixed with a few ferns, sedges and orchids. Locally there are small shrubs and occasionally taller shrubs and small trees. The pandan *Pandanus tectorius* is widespread. Lam (1927) claimed 'The impression of an alpine flora was so strong, that I thought myself time and again on a high mountain top close to the tree line. There were "Kruppelholz" trees with irregular, chaotic, dense twigs, often of a black to light-grey colour; and small hard leathery leaves, closely packed, especially at the branch ends; and several Myrtaceae and a species of *Vaccinium*.'

Lam (1927) described the soil as about 10–20 cm in depth with much emergent bedrock. He emphasized the importance of the low retention of water by the soil which, in spite of the humid climate, he felt was limiting the growth of trees. He listed fifty-eight species of vascular plants from the more open areas of Gunung Piapi. The fifty-eight species could be categorized into four groups based on their probable origin: I, plants from rainforest; II, plants from dry areas such as clearings; III, plants from the coastal flora; and IV, plants from an alpine flora. The Group II plants (23 species) seemed pre-adapted to this area but those of Group I (19 species) are in an unusually dry environment. Group III (6 species) includes those which are likely to have suffered drought from osmotic stress (and may be preadapted to the high Mg/Ca quotient of ultramafic soils since in sea water, in milliequivalents, this quotient is about 5). Group IV (10 species) is perhaps the least expected and raises the question of its origin. Gunung Piapi is the highest mountain in the Talaud Islands and these alpine species are related to more or less remote species. None of the named species is endemic to the ultramafic outcrop although Lam felt that some of the unnamed taxa may be new species. The orchid *Dendrobium hasseltii* which is part of the alpine element (Group IV) is recorded as an epiphyte in Java and Sumatra ranging from 1,400–2,300 m. On Gunung Piapi it occurs as a terrestrial white-flowered form.

Chemical analyses of naturally occurring species

Extensive foliar analyses for maquis and forest plants from ultramafic soils in New Caledonia have been reported by Jaffré (1980). These involved the analyses of nearly 5000 oven-dry leaf samples from more than 500 species. Less extensive oven-dry foliar analyses, of 149 species have been made by Proctor *et al.* (1989) and J. Proctor (unpublished) for larger trees on Gunung Silam, Sabah. Few foliar analyses have been published from other ultramafic sites in the Far East.

Macronutrients and sodium

In the New Caledonian maquis, the mean foliar nitrogen concentration was about 1.0%, in the forests it was about 1.4%. In the maquis, some species had very low values (the least was 0.34%) and nitrogen concentrations of more than 2% were rare. Many forest species had more than 2% foliar nitrogen however. The mean foliar phosphorus concentrations were 0.03% in the maquis and 0.04% in the forest: the least values were 0.008% (Cyperaceae) whilst the highest was 0.095%. Potassium was usually low in maquis species (0.59–0.81%) and was higher in the leaves of forest species. Sodium concentrations varied greatly and there was no clear difference between maquis and forest species: most species had between 0.01% and 0.5% sodium but one species had 2.57% of this element and several had concentrations exceeding 0.80%. The mean foliar calcium concentrations were about 1.0% for maquis species whilst for forests the value was more than 1.4%. Some ferns had as little as 0.15% foliar calcium, whilst the highest values recorded were about 5.6%. For magnesium there was no clear difference between maquis and forest species: the mean foliar concentrations were about 0.35%. Few species had more than 1% magnesium whilst the least values were in the Cyperaceae which were less than 0.1%. Most species had a Mg/Ca quotient of less than 1, the highest value for this quotient was 5.

There were some patterns discernible in macronutrient concentrations. Some species which had very low phosphorus had also less than 1% nitrogen, potassium and calcium. The Cyperaceae and Proteaceae had notably low concentrations of potassium and calcium. The Cunoniaceae had very low potassium concentrations. The families which generally had low macronutrient concentrations were: the Epacridaceae, Myrtaceae, Sapindaceae and Sapotaceae. Those which had high concentrations of macronutrients were the Apocynaceae, Rubiaceae and Rutaceae.

A comparison of species foliar concentrations between maquis types on different soils showed that nitrogen, potassium and magnesium had higher concentrations on the 'sols bruns eutrophes hypermagnésiens' and least on various types of oxisol. Even on the hypermagnesian soils, in the majority (68%) of species the Mg/Ca quotient was less than 1. For phosphorus, sodium and calcium there were no important differences in the overall means and ranges from one maquis type to another. In some cases comparisons were made between foliar concentrations within the same species for individuals on different soils. For these species there were no clear differences for nitrogen, phosphorus and potassium but calcium was least and magnesium highest on the 'sols bruns eutrophes hypermagnésiens.'

In general, when compared with foliar concentrations of species from non-ultramafic soils, the species on the New Caledonian ultramafic soils have low concentrations of nitrogen and calcium, very low concentrations of phosphorus and potassium, and relatively high concentrations of sodium and magnesium. The concentrations are relatively lower in the maquis than in the forest.

Proctor et al. (1988) have presented analytical data on foliar concentrations for trees on Gunung Silam, Sabah. Their data for some species are given in Table 3. There are many similarities with the results for New Caledonia but the Silam trees are distinctive in their very low potassium concentrations. Foliar calcium concentrations on Silam were generally less than those of Jaffré (1980) but as for New Caledonia, the foliar Mg/Ca quotient was usually less than unity.

Chromium, cobalt, manganese and nickel

In natural vegetation in New Caledonia, chromium concentrations were found to be low and half the species had less than 5 μg g^{-1} in their dry matter. Relatively high values were found in the Cunoniaceae and Violaceae (families which have nickel-accumulating species) but only two species had more than 20 μg g^{-1} of chromium. These were *Hybanthus austro-caledonicus* (Violaceae) for which the range of individuals analyzed had from 5–260 μg g^{-1} foliar chromium and *Symplocos rotundifolia* (Symplocaceae) which had from 24–140 μg g^{-1} of this element. It is noteworthy that the *Hybanthus* is a nickel accumulator with up to 25,000 μg g^{-1} of that element and the *Symplocos* accumulates up to 15,000 μg g^{-1} aluminium. Jaffré (1980) has observed some higher chromium concentrations (maximum 430 μg g^{-1} in the leaves of *Geissois pruinosa* (Cu-

noniaceae) in a few plants from chromium-rich mining areas in New Caledonia. The generally low chromium concentrations in plants on the New Caledonian ultramafics are slightly surprising since unusually, many of the soils (Table 1) have relatively high concentrations of exchangeable chromium. Very high chromium concentrations (up to 5,000 μg g^{-1}) have been reported by Lee et al. (1977a) in the moss *Aerobryopsis longissima* which occurred mainly on the trunks of *Homalium guillamii* (Flacourtiaceae), a species which accumulates nickel but has low chromium concentrations in the bark. This moss is widespread in the tropics and its chromium accumulation in this situation in New Caledonia merits further investigation. Proctor et al. (1989) reported that chromium concentrations were below 20 μg g^{-1} in all the species they analysed from Gunung Silam.

Cobalt concentrations in New Caledonian species were found to be usually low and 58% of the species analysed by Jaffré (1980) had less than 2.5 μg g^{-1} of this element. Values exceeding 25 μg g^{-1} occurred in 5.7% of the species and the highest value recorded was one of 450 μg g^{-1} in an individual of *Phyllanthus serpentinus* (Euphorbiaceae). Proctor et al. (1989) reported that cobalt concentrations were less than 20 μg g^{-1} in all the individuals they analyzed for Gunung Silam.

A feature of the New Caledonian flora is the wide range of manganese concentrations; 20% of the species had less than 50 μg g^{-1}, whilst 21% had more than 1,000 μg g^{-1} and nine species had more than 10,000 μg g^{-1}. The highest value recorded was 50,000 μg g^{-1} in *Macadamia neurophylla* (Proteaceae). The Proteaceae on ultramafic soils in New Caledonia often have high manganese concentrations: 50% of the species had more than 1,000 μg g^{-1} and no species had less than 1,000 μg g^{-1}. It is remarkable that several individuals showed a marked manganese accumulation even when they grew on basic or less acidic soils from which manganese should have a low availability. On non-ultramafic soils the Proteaceae show a tendency to accumulate aluminium. Manganese accumulation was recorded for eleven other families apart from the Proteaceae. The species *Eugenia clusioides* (Myrtaceae) and *Maytenus bureauiana* (Celastraceae) were similar to the Proteaceae species mentioned

Table 3. The means (with ranges in parentheses) of element concentrations (mg g^{-1} over dried (105°C) matter) in leaves of some species from four plots on Gunung Silam, Sabah

Species and family	Plot altitude (m)	Number of individuals	N	P	K	Na	Ca	Mg	Mn	Fe	Ni
Lophopetalum rigidum (Celastraceae)	330	3	18.5 (16.0–21.1)	0.50 (0.44–0.56)	2.1 (1.3–2.9)	6.2 (5.5–6.7)	10.1 (8.2–10.2)	12.8 (10.6–15.9)	0.10 (0.06–0.13)	0.088 (0.069–0.127)	0.036 (0.022–0.044)
Payena lucida (Sapotaceae)	330	3	14.2 (12.9–15.5)	0.53 (0.41–0.63)	5.4 (2.4–7.5)	2.0 (1.2–1.7)	10.9 (9.4–14.7)	3.1 (2.6–3.7)	0.02 (0.02–0.03)	0.067 (0.062–0.074)	0.003 (0.002–0.005)
Shorea multiflora (Dipterocarpaceae)	330	4	17.3 (13.9–18.8)	0.56 (0.45–0.63)	3.0 (1.8–4.2)	0.69 (0.36–1.2)	11.8 (8.8–15.7)	4.1 (2.8–5.3)	0.08 (0.07–0.10)	0.091 (0.051–0.183)	0.061 (0.054–0.070)
Borneodendron anaegneticum (Euphorbiaceae)	610	3	19.4 (17.7–22.6)	0.63 (0.54–0.80)	12.8 (11.6–13.6)	4.7 (2.8–6.6)	23.1 (21.3–24.3)	11.2 (9.5–12.3)	0.29 (0.15–0.44)	0.014 (0.008–0.018)	0.014 (0.009–0.023)
Eugenia sp. 'F' (Myrtaceae)	610	4	9.0 (7.5–9.9)	0.33 (0.32–0.35)	1.5 (1.3–1.8)	1.3 (0.9–1.6)	10.9 (8.3–12.7)	1.9 (1.6–2.2)	0.06 (0.04–0.08)	0.023 (0.012–0.027)	0.043 (0–0.014)
Shorea tenuiramulosa (Dipterocarpaceae)	610	3	15.3 (14.0–16.3)	0.53 (0.50–0.56)	3.9 (2.6–4.8)	1.4 (0.93–2.0)	9.4 (7.0–10.9)	3.7 (3.3–4.0)	0.11 (0.09–0.16)	0.053 (0.048–0.056)	0.650 (0.46–1.0)
Timonius flavescens (Rubiaceae)	610	3	10.3 (8.6–11.5)	0.47 (0.43–0.51)	5.3 (2.8–6.6)	2.9 (2.0–3.6)	7.5 (4.3–12.1)	3.0 (1.4–5.4)	0.08 (0.02–0.18)	0.062 (0.040–0.083)	0.038 (0.002–0.103)
Eugenia cerina (Myrtaceae)	790	2	9.0 (8.8–9.1)	0.42 (0.41–0.42)	2.7 (2.4–2.9)	3.9 (3.6–4.1)	13.0 (13.0–13.1)	2.8 (2.2–3.4)	0.04 (0.03–0.05)	0.026 (0.025–0.027)	0.005 (0.004–0.006)
Eugenia sp. 'A' (Myrtaceae)	790	1	25.4	0.59	3.5	3.4	7.1	1.9	13.7	0.050	0.020
Lithocarpus sp. 'D' (Fagaceae)	790	2	20.5 (18.2–22.8)	0.39 (0.39–0.39)	3.4 (3.3–3.6)	1.5 (1.5–1.5)	6.5 (6.3–6.8)	2.3 (2.3–2.4)	0.16 (0.16–0.17)	0.025 (0.023–0.026)	0.009 (0.007–0.011)
Tetractomia tetrandum (Rutaceae)	790	3	21.6 (17.1–24.6)	0.50 (0.47–0.52)	4.2 (3.0–5.0)	4.7 (4.6–4.8)	6.8 (5.2–7.8)	4.2 (3.9–4.4)	0.31 (0.06–0.46)	0.034 (0.023–0.043)	0.020 (0.016–0.026)
Eugenia subdessuata (Myrtaceae)	870	3	6.6 (4.6–8.5)	0.33 (0.26–0.38)	1.7 (1.2–2.2)	2.8 (2.5–3.5)	6.4 (4.8–7.4)	3.2 (2.8–3.4)	0.32 (0.26–0.41)	0.260 (0.065–0.466)	0.005 (0.002–0.007)
Syzygium sp. 'D' (Myrtaceae)	870	2	13.0 (12.3–13.7)	0.50 (0.49–0.51)	3.0 (2.6–3.3)	3.0 (2.6–3.3)	3.9 (3.6–4.1)	3.0 (2.6–3.3)	0.17 (0.16–0.17)	0.062 (0.061–0.063)	0.0 (0–0)
Tetractomia tetrandum (Rutaceae)	870	1	17.9	0.64	2.1	1.7	9.0	8.7	0.62	0.023	0.021

above, which could accumulate manganese from slightly acid soils. Some families always had low manganese concentrations: the Dilleniaceae (very low), the Escalloniaceae, Flacourtiaceae, Leguminosae, Rubiaceae and Rutaceae. Maquis species (29% of which had more than 1,000 μg g^{-1}) tended to have higher concentrations of this element than those of the forests (where 14% of the species had more than 1,000 μg g^{-1} manganese). On 'sols bruns eutrophes hypermagnésiens' and 'sols ferrallitiques érodés' most maquis species had less than 250 μg g^{-1} manganese and 7% and 9% respectively of their species had more than 1,000 μg g^{-1} of this element. On 'sols ferrallitiques colluviaux' 32% of the species had more than 1,000 μg g^{-1} whilst on 'sols ferrallitiques gravillonaires ou courassés' 46% of the species exceeded this value.

On Gunung Silam, Proctor et al. (1989) found that foliar manganese had much higher concentrations in the plots at 790 m and 870 m compared with those at 330 m and 610 m. The two higher plots included some individual trees with high manganese concentrations: at 790 m, an unidentified *Polyosma* species (Escalloniaceae) had 1,700 μg g^{-1}, *Polyosma integrifolia* had 2,200 μg g^{-1} and an unidentified *Eugenia* species (Myrtaceae) had 13,700 μg g^{-1}; at 870 m *Ixonanthes* cf. *crassifolia* (Ixonanthaceae) and *Cinnamomum* cf. *bintulensis* (Lauraceae), both had 1,500 μg g^{-1} whilst an unidentified *Lithocarpus* species (Fagaceae) had 1,700 μg g^{-1}. For the plots at 330 m and 610 m one species alone had a foliar manganese concentration in excess of 1,000 μg g^{-1}. This was a *Stemonurus grandiflorus* (Icacinaceae), from the plot at 330 m, with 1,700 μg g^{-1} manganese.

Of the species, analysed for foliar nickel, from New Caledonian ultramafic soils, Jaffré (1980) reported that 8% had less than 10 μg g^{-1}, 60% had 10–100 μg g^{-1}, 27% had 100–1,000 μg g^{-1} and 5% had more than 1,000 μg g^{-1}. Of the forty-eight species which have more than 1,000 μg g^{-1} nickel, nine had more than 10,000 μg g^{-1} foliar nickel (Table 4). High-nickel species are found in the soils which have higher concentrations of this element. The lowest nickel concentrations occurred in the leaves of shrubs on 'les sols ferrallitique cuirassés ou gravillonaires' where only 10% of the species had more than 50 μg g^{-1} nickel and

none exceeded 250 μg g^{-1}. In the maquis on 'les sols bruns eutrophes hypermagnésiens' and in the 'maquis lignoherbacés' on ferrallitic soils half the species had less than 50 μg g^{-1} nickel and 10% of the species had more than 250 μg g^{-1} nickel. In the maquis on 'sols ferralitique érodés' 40% of the species had less than 100 μg g^{-1} nickel, 25% had more than 250 μg g^{-1} and 10% had more than 1,000 μg g^{-1}. The New Caledonian forests have a high proportion of species with high concentrations of foliar nickel. Thus 41% of the forest species had more than 100 μg g^{-1} nickel and 9% of the forest species had more than 1,000 μg g^{-1}.

In the tree species on Gunung Silam, Sabah, Proctor et al. (1989) found nearly all had less than 50 μg g^{-1} foliar nickel and only two species had more than 250 μg g^{-1} nickel. The highest value recorded was 1,000 μg g^{-1} in one individual of *Shorea tenuiramulosa*, the first record of nickel accumulation in the Dipterocarpaceae. High nickel values have also been recorded in a few species from other sites in south-east Asia (Brooks & Wither, 1977 and Wither & Brooks, 1977).

New Caledonia has nine out of thirty-five species in the world which are known to accumulate more than 10,000 μg g^{-1} nickel. The others include twenty-one species of *Alyssum* (Cruciferae) from the Mediterranean region (Brooks et al. 1979), *Hybanthus floribundus* (Violaceae) of Australia (Severne & Brooks 1972), *Pearsonia metallifera* (Leguminosae) of Zimbabwe (Wild 1974), *Planchonella oxyhedra* (Sapotaceae) of Obi Island, Indonesia (Wither & Brooks 1977), and *Rinorea bengalensis* (Violaceae) of Indonesia (Brooks & Wither 1977).

Nickel accumulation is a marked feature of certain families in New Caledonia (Table 4). In the New Caledonian Flacourtiaceae ten of the eighteen species of *Homalium* and seven out of sixteen species of *Xylosma* have individuals with more than 1,000 μg g^{-1} nickel. By contrast *Psychotria douarrei* (Rubiaceae) is the sole nickel accumulator out of 210 species of the genus analysed by Baker, Brooks & Kersten (1985). Most of the species in Table 4 belong to primitive families of flowering plants (Sporne 1969) except for *Psychotria douarrei* and it seems that nickel accumulation may be an ancient character.

Table 4. The families and genera with New Caledonian species which have individuals with more than 1,000 μg g^{-1} foliar nickel and the species which have individuals with more than 10,000 μg g^{-1} foliar nickel (from Jaffré 1980)

Family	Genera (with number of species with >1,000 μg g^{-1} Ni in parentheses)	Species with >10,000 μg g^{-1} Ni
Cunoniaceae	*Geissois* (7), *Pancheria* (1)	*Geissois pruinosa,*[a] *G. intermedia*[bc]
Escalloniaceae	*Argophyllum* (2)	–
Euphorbiaceae	*Phyllanthus* (10), *Baloghia* (1), *Cleidion* (1)	*Phyllanthus serpentinus*[ab]
Flacourtiaceae	*Xylosma* (10), *Homalium* (7), *Casearia* (2)	*Homalium francii,*[bc] *H. guillaimii*[b]
Oncothaceae	*Oncotheca* (1)	–
Rubiaceae	*Psychotria* (1)	*Psychotria douarrei*[b]
Sapotaceae	*Sebertia* (1)	*Sebertia acuminata*[b]
Violaceae	*Agatea* (1), *Hybanthus* (3)	*Hybanthus austro-caledonicus.*[bc] *H. caledonicus* var. *linearifolia*[a]

[a]Maquis species.
[b]Forest species.
[c]Also occurring on non-ultramafic soils.

The highest concentrations of nickel for any part of any species is the 257,400 μg g^{-1} found in the oven-dried sap of *Sebertia acuminata* (Jaffré *et al.* 1976). Apart from this, the highest nickel concentrations are all reported for *Psychotria douarrei* (Jaffré & Schmid 1974, Jaffré *et al.* 1976 and Jaffré 1980); 47,000 μg g^{-1} in the leaves, 24,000 μg g^{-1} in the flowers, 28,000 μg g^{-1} in the fruits, 80,000 μg g^{-1} in the trunk bark and 2,300 μg g^{-1} in the trunk wood. The relatively low nickel values in the trunk wood are noteworthy and Jaffré (1980) has commented that in the nickel-hyperaccumulator species the concentrations of the element are usually highest in those parts where metabolism is active.

Where biochemical investigations have been made (Kelly *et al.* 1975, Lee *et al.* 1978), they have shown that the nickel is bound with citric acid except in the case of *Psychotria douarrei* which contains nickel complexed mainly (63%) as a negatively charged malate complex balanced by a cationic aquo-complex (Kersten *et al.* 1980). This species belongs to the more evolutionarily advanced Rubiaceae and it may be that the malate complex reflects this, although *Phyllanthus serpentinus* of the more primitive Euphorbiaceae also has substantial quantities of nickel malate complex as well as the preponderant citrate complex.

The metabolic and ecological significance of nickel accumulation remain unclear. There is no evidence that nickel is an essential element for the New Caledonian hyperaccumulators. Jaffré (1980) reported that *Geissois pruinosa, Hyb-*

anthus austro-caledonicus and *Psychotria douarrei* when transplanted to non-nickelferous soil grew well with no unusual symptoms despite a reduction of foliar nickel to concentrations of 90–220 μg g^{-1}. Jaffré (1980) reported nickel concentrations of less than 10 μg g^{-1} in individuals of *Hybanthus austro-caledonicus* growing naturally on non-nickeliferous soils. There are no great differences in major nutrient concentrations in accumulator species (Jaffré 1980) and no correlation between nickel concentrations and those of other elements (Lee *et al.* 1977b).

Jaffré (1980) discussed the suggestion that nickel accumulation may be part of an adaptation against drought but has concluded that this is unlikely since most of the New Caledonian accumulator species occur in forests (which are better supplied with water than maquis) and also in cultivation the hyperaccumulators are sensitive to dryness. A further possibility is that nickel accumulation may render leaves toxic to herbivores but Proctor *et al.* (1989) found no evidence of reduced herbivory in the nickel-accumulating dipterocarp species *Shorea tenuiramulosa*.

Plant resistance

Little is known about mechanisms of plants' resistance to ultramafic soils in the Far East.

Jaffré (1980) felt that low concentrations of macronutrients (except magnesium) are the most constant general cause of infertility in the ultramafic soils of New Caledonia. He cited the diver-

266

sity of nutrient concentrations over the range of species he analysed as showing that there are a range of methods by which native species have adapted to the low nutrient concentrations. There is apparently selective absorption of some nutrients by many species and an ability to grow at low internal nutrient concentrations in others. In the case of magnesium, the only macronutrient in excess, there was apparently a limited absorption in many species and a tolerance of high tissue concentrations in a few others. The analytical results of Proctor et al. (1989) for macronutrients in trees on Gunung Silam have many parallels with those of Jaffré (1980) and his suggestions about adaptations apply to their site also. A detoxification mechanism for nickel is suggested by the comparisons of Proctor et al. (1989) of metal concentrations in fresh leaves and leaf litterfall. For most elements, fresh leaf concentrations were higher than those for litterfall but for nickel the reverse was true. The effect was greatest in the plot at 330 m where there was a mean fresh-leaf nickel concentration of 32 μg g^{-1} and a litterfall nickel concentration of 420 μg g^{-1}. This may reflect nickel excretion into organs which are about to be shed from the plant.

Non-ultramafic soils in New Caledonia often have low nutrient concentrations too but have a different flora from the ultramafic soils. It would seem that the relatively high soil nickel concentrations are a major floristic determinant, particularly since so many New Caledonian species have developed the accumulator response to nickel.

Proctor et al. (1989) concluded that resistance to ultrabasic soils was achieved without excessive metabolic cost for the lowland rainforests of Gunung Silam. They found that the rates of leaf herbivory, tree mortality and diameter increment, and small litterfall were not unusually low on this mountain in spite of the high Mg/Ca quotients and nickel concentrations in the soil.

Experimental work

Very little experimental work has been done. Jaffré (1980) observed nickel toxicity in Avena sativa (Gramineae) growing in a 'peridotite soil' and found 160 μg g^{-1} nickel in the leaves. He observed nickel toxicity in Deplancheana sessiliflora (Bignoniaceae) growing in mining areas and found the same symptoms in experiments in which he added nickel to natural soils. As a result of this work Jaffré concluded that nickel-tolerant species cannot tolerate nickel indefinitely. They are selected for habitats in which the tolerance limits are not exceeded.

Experiments on the nickel-accumulating species Psychotria douarrei have been made by Baker, Brooks & Kersten (1985). They studied the tolerance of seedlings of this species to nickel by adding nickel nitrate at six 14-day intervals to a growth medium of perlite and peat to which ground dolomite and nutrients had been added. As a result of the additions the total nickel concentration of the substrate rose from 37 μg g^{-1} to 10,000 μg g^{-1}. (The last figure would include a high proportion of soluble nickel). During the experiment the nickel concentrations of the foliage increased from 2,800 μg g^{-1} to 9,500 μg g^{-1} after which the plants died (although much higher foliar concentrations of nickel have been found in Psychotria species collected from the field).

Concluding remarks

The vegetation of the ultramafic rocks of the Far East is very variable and the causes of the variation are not yet explained. It is still not clear why ultramafic soils should bear stunted vegetation in some areas and well developed forest in others. For example the large stature forest on Gunung Silam is on soils with a high Mg/Ca quotient and substantial nickel concentrations whilst the small stature forest on Gunung Kinabalu is on soils with low values of both these factors. As Proctor & Cole (this Volume) suggested for Zimbabwean ultramafic soils, stunting of vegetation may largely reflect hydrological rather than chemical causes and this hypothesis requires further testing. It is not obvious why ultramafic substrata should be associated with such a high degree of specificity to the substratum in New Caledonia (Morat, Veillon & McKee 1984, Morat et al. 1986) and much less elsewhere (Proctor et al. 1989). It is curious that manganese and nickel accumulation are such a feature of the New Caledonian ultramafic flora but apparently less developed on Gunung Silam.

There is an urgent need to survey the vegetation over ultramafic rocks in the Far East and to investigate it experimentally. Van Balgooy & Tantra (1986) have pointed out the need to conserve vegetation over ultramafic rocks since it remains little known and is a potential source of important chemicals for medicinal and other uses. Although cultivation of ultramafic soils is generally avoided by local farmers (Wall *et al*. 1979, Binnie *et al*. 1979) some ultramafic areas are coming under cultivation in reponse to increasing population pressure. There are large areas of ultramafic rocks in Indonesia, particularly in Sulawesi where there is an area of 8,500 km^2 which is the largest contiguous outcrop of these rocks in the world. Many of these areas are designated by the Government of Indonesia for agricultural development and a study of the chemistry and potential of their soils is of importance. Parry (1985) has cautioned against any attempt to cultivate ultramafic soils until more research has been made. Analyses of these soils (e.g. Proctor *et al*. 1988) have shown an increasingly unfavourable chemistry with depths and hence they may be very difficult to manage if their surface layers are eroded.

References

Acres, B. D., R. P. Bower, P. A. Burrough, C. J. Folland, M. S. Kalsi, P. Thomas & P. S. Wright. 1975. The soils of Sabah, Volume 5 references and appendices. Land Resources Division, Ministry of Overseas Development, Surbiton, U.K.

Baker, A. J. M., R. R. Brooks, & W. J. Kersten. 1985. Accumulation of nickel by *Psychotria* species from the Pacific Basin. Taxon 34: 89–95.

Binnie & Partners and Hunting Technical Services Ltd. 1979. Water resources and potentially irrigable land, central Sulawesi Province (main volume). Directorate Water Resources Development. Government of Indonesia, Jakarta.

Brooks, R. R., J. Lee & T. Jaffré. 1974. Some New Zealand and New Caledonian plant accumulators of nickel. Journal of Ecology 62: 493–499.

Brooks, R. R. & E. D. Wither. 1977. Nickel accumulation by *Rinorea bengalensis* (Wall) O.K.J. Journal of Geochemical Exploration 7: 295–300.

Brooks, R. R., R. S. Morrison, R. D. Reeves, T. R. Dudley & Y. Akman. 1979. Hyperaccumulation of nickel by *Alys-*

sum Linnaeus (Cruciferae). Proceedings of the Royal Society of London B 203: 387–403.

Brooks, R. R., J. M. Trow, J.-M. Veillon & T. Jaffré. 1981. Studies on manganese-accumulating *Alyxia* species from New Caledonia. Taxon 30: 420–423.

Brünig, E. F. 1974. Ecological studies in the kerangas forests of Sarawak and Brunei. Borneo Literature Bureau, Kuching.

Burnham, C. P. 1984. The forest environment: soils. In Whitmore, T. C., Tropical rain forests of the Far East (2nd edition), pp. 137–154. Oxford University Press, Oxford.

Fox, J. E. D. & T. H. Tan. 1971. Soils and forest on an ultrabasic hill north east of Ranau, Sabah. Journal of Tropical Geography 35: 38–48.

Fox, J. E. D. 1972. The natural vegetation of Sabah and natural regeneration of the dipterocarp forests. Ph.D. thesis, University of Wales, U.K.

Hamilton, W. 1979. Tectonics of the Indonesian region. Geological Survey Professional Paper 1078, United States Government Printing Office, Washington, U.S.A.

Hansell, J. R. F. & J. R. D. Wall. 1976. Land resources of the Solomon Islands, Volume 5: Santa Isabel, Land Resource Study 18. Land Resources Division, Ministry of Overseas Development, Surbiton, U.K.

Jaffré, T. 1974. La végétation et la flore d'un massif de roches ultrabasiques de Nouvelle-Calédonie: Le Koniambo. Candollea 29: 427–456.

Jaffré, T. 1976. Composition chimique et conditions de l'alimentation minérale des plants sur roches ultrabasiques. Cahiers ORSTOM Série Biologie 11: 53–63.

Jaffré, T. 1977. Accumulation du manganèse par des espèces associées aux terrains ultrabasiques de Nouvelle-Calédonie. Compte Rendu de l'Acadamie des Sciences Paris Série D 284: 1573–1575.

Jaffré, T. 1977. Composition chimique elementaire des tissus foliares des espèces vegetales colonisatrices des anciennes mines de nickel en Nouvelle Calédonie. Cahiers ORSTOM Série Biologie 12: 323–330.

Jaffré, T. 1980. Etude ecologique du peuplement végétal des sols dérivés de roches ultrabasiques en Nouvelle Calédonie. ORSTOM, Paris.

Jaffré, T. & M. Schmid. 1974. Accumulation du nickel par une Rubiacéae de Nouvell-Calédonie, *Psychotria douarrei* (G. Beauvisage) Däniker. Compte Rendu de l'Academie des Sciences Paris Série D 278: 1727–1730.

Jaffré, T. & M. Latham. 1974. Contribution a l'étude des relations sol-végétation sur un massif des roches ultrabasiques de la côte ouest de la Nouvelle-Calédonie: Le Boulinda. Adansonia 14: 311–336.

Jaffré, T., R. R. Brooks, J. Lee & R. D. Reeves 1976. *Sebertia acuminata*: a hyperaccumulator of nickel from New Caledonia. Science 193, 579–580.

Jaffré, T., M. Latham & M. Schmid. 1977. Aspects de l'influence de l'extraction du minerae de nickel sur la végétation et les sols en Nouvelle Calédonie. Cahiers ORSTOM Série Biologie 12: 307–321.

Jaffré, T., R. R. Brooks & J. M. Trow. 1979. Hyperaccumulation of nickel by *Geissois* species. Plant and Soil 51: 157–162.

Jaffré, T., W. Kersten, R. R. Brooks & R. D. Reeves. 1979.

Nickel uptake by Flacourtiaceae of New Caledonia. Proceedings of the Royal Society of London B 205: 385–394.

Kartawinata, K. 1980. A note on a kerangas (heath) forest at Sebula, East Kalimantan. Reinwardtia 9: 419–447.

Kelly, P. C., R. R. Brooks, S. Dilli & T. Jaffré. 1975. Preliminary observations on the ecology and plant chemistry of some nickel-accumulating plants from New Caledonia. Proceedings of the Royal Society of London B 189: 69–80.

Kersten, W. J., R. R. Brooks, R. D. Reeves & T. Jaffré. 1980. Nature of nickel complexes in *Psychotria douarrei* and other nickel-accumulating plants. Phytochemistry 19: 1963–1965.

Lam, H. J. 1927. Een plantengeografisch Dorado. Handelingen. IV Nederlandsch-Indisch natuurwetenschappelijk congres, Welteneden, pp. 386–397.

Latham, M. 1975a. Le sols d'un massif de roches ultrabasiques de la Côte Ouest de Nouvelle Calédonie Le Boulinda: 1er Partie Généralités-Répartition des sols dans le massif. Le sols à accumulation humifère. Cahiers ORSTOM Série Pédologie 13: 27–40.

Latham, M. 1975b. Les sols d'un massif de roches ultrabasique de la Côte ouest de Nouvelle Calédonie 2e Partie. Les sols à accumulation ferrugineuse relative. Cahiers ORSTOM Série Pédologie 13: 159–172.

Latham, M. 1975c. Geomorphologie d'un massif de roches ultrabasiques de la côte ouest de la Nouvelle-Calédonie, Le Boulinda. Cahiers ORSTOM Série Géologie 7: 17–37.

Leakey, R. A. & J. Proctor. 1987. Invertebrates in litter and soil at a range of altitudes on Gunung Silam, a small ultrabasic mountain in Sabah. Journal of Tropical Ecology 3: 119–129.

Lee, K. E. 1969. Some soils of the British Solomon Islands Protectorate. Philosophical Transactions of the Royal Society B 225, 211–258.

Lee, J., R. R. Brooks, R. D. Reeves & T. Jaffré. 1977a. Chromium accumulating bryophyte from New Caledonia. Bryologist 80: 203–205.

Lee, J., R. R. Brooks, R. D. Reeves, C. R. Boswell & T. Jaffré. 1977b. Plant-soil relationships in a New Caledonian serpentine flora. Plant and Soil 46: 675–680.

Lee, J., R. D. Reeves, R. R. Brooks & T. Jaffré. 1978. The relation between nickel and citric acid in some nickel accumulating plants. Phytochemistry 17: 1033–1035.

Meijer, W. 1964. Forest botany in North Borneo and its economic aspects. Economic Botany 18: 256–265.

Meijer, W. 1965. A botanical guide to the flora of Mount Kinabalu. Symposium on Ecological Research on Humid Tropical Vegetation, Kuching 1963: 325–366.

Meijer, W. 1971. Plant life in Kinabalu National Park. Malaysian Nature Journal 24: 184–189.

Meijer, W. 1984. Botanical explorations in Celebes and Bali. National Geographic Society Research Reports 17: 583–605.

Morat, Ph., J-M. Veillon & H. S. MacKee. 1984. Floristic relationships of New Caledonian rain forest phanerogams. In: Biogeography of the tropical Pacific (ed. by F. J. Radovsky, P. H. Raven & S. H. Sohmer), pp. 71–128. Association of Systematics Collections and Bernice P. Bishop Museum Special Publications No. 72, Honolulu.

Morat, Ph., T. Jaffré, J-M. Veillon & H. S. MacKee. 1986. Affinités floristiques et considérations sur l'origine des ma-

quis miniers de la Nouvelle-Calédonie. Adansonia 2: 133–182.

Paijmans, K. (ed.). 1976. New Guinea vegetation, Elsevier, Amsterdam.

Parry, D. E. 1985. Ultramafic soils in the humid tropics with particular reference to Indonesia. Unpublished report of Hunting Technical Services Ltd., U.K.

Podzorski, A. C. 1985. The Palawan botanical expedition final report. Hilleshög Forestry AB, Landskrona, Sweden.

Proctor, J., Y. F. Lee, A. M. Langley, W. R. C. Munro & T. Nelson. 1988. Ecological studies on Gunung Silam, a small ultrabasic mountain in Sabah, Malaysia. I. Journal of Ecology 76: 320–340.

Proctor, J., C. Phillipps, G. K. Duff, A. Heaney & F. N. Robertson. 1989. Ecological studies on Gunung Silam, a small ultrabasic mountain in Sabah, Malaysia. II. Journal of Ecology 77: 317–331.

Sato, T. 1985. A vegetational sketch of Mt. Kinabalu. Tukar-Menukar 4, 19–29.

Severne, B. C. & R. R. Brooks. 1972. A nickel-accumulating plant from Western Australia. Planta 103: 91–94.

Sporne, K. R. 1980. A reinvestigation of character correlations among dicotyledons. New Phytologist 80: 419–449.

Takhtajan, A. 1969. Flowering plants: origin and dispersal. Oliver and Boyd, Edinburgh.

Trescases, J. J. 1969a. Premiéres observations sur l'altération des péridotites de Nouvelle-Calédonie. Pédologie-Géochimie-Géomorphologie. Cahiers ORSTOM Série Géologie 1: 27–57.

Trescases, J. J. 1969b. Géochimie des eaux de surface et altérations dans le massif ultrabasique du sud de la Nouvelle-Calédonie. Bulletin du Service de la carte géologique d'Alsace et de Lorrainne 22: 329–354.

Trescases, J. J. 1973. Weathering and geochemical behaviour of the elements of ultramafic rocks in New Caledonia. Bureau of Mineral Resources Geology and Geophysics Department of Minerals and Energy Canberra Bulletin 141: 149–161.

Trescases, J. J. 1975. L'évolution géochimique supergène des roches ultrabasiques en zone tropicale. Mémoire ORSTOM 78, Paris.

Trescases, J. J. 1976. Geochemistry of waters as related to weathered ultramafic rock in New Caledonia in J. Cadek & T. Paces (eds.) Proceedings of the International Symposium on water-rock interaction pp. 109–117. Geological Survey, Prague.

Van Balgooy, M. M. J. & I. G. M. Tantra. 1986. The vegetation in two areas in Sulawesi, Indonesia. Forest Research Bulletin (special edition), Bogor, Indonesia.

Van Royen, P. 1960. The vegetation of some parts of Waigeo Island. Nova Guinea, Botany 5: 25–62.

Van Royen, P. 1963. The vegetation of the island of New Guinea. Department of Forests, Lae.

Virot, R. 1956. La végétation canaque. Mémoires du Muséum nationale d'histoire naturelle Paris Série B 8, 388.

Wall, J. R. D. & J. R. F. Hansell. 1976. Land resources of the Solomon Islands, Volume 6: Choiseul and the Shortland Islands, Land Resource Study 18. Land Resources Division, Ministry of Overseas Development, Surbiton, U.K.

Wall, J. R. D., J. R. F. Hansell, J. A. Catt, F. C. Ormerod,

J. A. Varley & I. S. Webb. 1979. The soils of the Solomon Islands, Volume I. Technical Bulletin 4, Ministry of Overseas Development, London.

Whitmore, T. C. 1969. The vegetation of the Solomon Islands. Philosophical Transactions of the Royal Society B 225: 549–566.

Whitmore, T. C. 1984. Tropical rain forests of the Far East 2nd edition. Oxford University Press, Oxford.

Wild, H. 1974. Indigenous plants and chromium in Rhodesia. Kirkia 9: 233–241.

Wither, E. D. & R. R. Brooks. 1977. Hyperaccumulation of nickel by some plants of south-east Asia. Journal of Geochemical Exploration 8: 579–583.

Wright, P. S. 1975. Western parts of Tawau and Lahad Datu districts. The soils of Sabah volume 3. Land Resource Study 20, Ministry of Overseas Development, Surbiton, England.

Author's Address:
J. Proctor
Dept. of Biological and Molecular Sciences
University of Stirling
Stirling
Scotland FK9 4LA
U.K.

XI

The distribution and extent of serpentinized

areas in Japan

N. MIZUNO & S. NOSAKA

Abstract. Japan contains a major part of the ultramafic rock belts in the Pacific Basin. The concentrations of nickel and chromium in the serpentine soils are high, but it is only nickel which is high in both its total and soluble forms, the latter being at a toxic level to plants. Chromium is not toxic to plants, because its solubility is low. Nickel toxicity in the plants causes a mineral imbalance between iron and nickel. Furthermore, associated with this, one can sometimes observe deficiencies in zinc and molybdenum.

Natural vegetation regions are recognized as the following four regions:
–*Camellietea japonicae* region
–*Fagetea crenatae* region
–*Vaccinio-Piceetea* region
–Alpine vegetation retgion.

In the *Camellietea japonicae* region and the *Fagetea crenatae* region, rather thin, evergreen needle-leaved forest vegetations occurs at serpentine vegetation. The main type of these forests is *Pinus densiflora* forest.

The *Fagetea crenatae* and the *Vaccinio-Piceetea* cover northern Japan, where, especially in Hokkaido, the serpentine vegetation consists of *Picea glehni* forest.

In the alpine vegetation region, the serpentine vegetation is recognized as a thin herbaceous community belonging to the order *Minuartetalia vernae japonicae*, which includes two alliances, *Drabo-Arenarion katoanae* and *Cerasteo-Minuartion vernae japonicae*. The former alliance includes four associations.

There are about 50 species of serpentine characteristic plants in Japan. 37 spp. of these are recognized as typical serpentinophytes, and the remainder are relics.

Introduction

Japan abounds in ultramafic rocks. The Japanese ultramafics are one of the main parts of the ring of ultramafic rock belts around the Pacific Basin, comparable with the belts of Sulawesi, New Caledonia, Tasman and Wakatipu of New Zealand, Curepto and Pichilemu of Chile, the Sierra Nevadan and Franciscan belt in California, the Skagit and Shuksan belt in Washington, and the belts of Canada and Alaska (Miyashiro and Kushiro 1977).

It is well known that ultramafic soils have characteristic floras with endemics, relict species and morphological races, or, 'serpentinomorphoses.' In Japan, ultramafic soils are not often cultivated because they are found mostly in remote mountainous areas, where they form steep slopes which are prone to landslips, and they have a tendency to form a layer of solidified clay which is impermeable to water. Alluvial and fluvial soil originating from serpentinite as well as peat soil which contains weathered serpentinite cause nickel toxicity in crops and lack a balance of mineral elements which results in deficiencies of zinc and molybdenum.

Geology and metamorphic belts in Japan

Hidaka metamorphic belt

The main characters of the Hidaka metamorphic belt were made clear by Hashimoto and collaborators (Hashimoto 1956, Hunahashi et al. 1956, Kizaki 1964, Shimizu 1965). The belt runs roughly in the north-south direction and occupies the main part of the central mountain range of Hokkaido. The temperature of metamorphosis was lower on both sides away from the central line of the long axis of the belt, but a detailed thermal structure has been studied only in the southwestern part. Ultramafic rocks that intrude along the western border are peridotites, the largest intrusion being the Horoman massif in the southwestern part.

Kamuikotan metamorphic belt

The Kamuikotan metamorphic belt runs in the north-south direction and occupies the western side of the central mountain range side by side with the Hidaka belt. The Kamuikotan belt consists of glaucophane schist, jadeite, lawsonite; and the belt is attended by many groups of serpentinite. The belt is divided into four main groups: Toikanbetsu-Nakagawa, Horokanai-Kamuikotan, Yubaridake, and Mitsuishi.

Kitakami mountains

The Kitakami mountains are composed chiefly of middle to late Paleozoic to late Mesozoic sediments intruded by granitic and ultramafic rocks. Along a curved dislocation line running from NW to SSE and convexing toward NW, there are several bodies of serpentinized peridotite and associated mafic rocks. Among these intrusions of igneous masses, the Miyamori and Hayachine masses have been studied in great detail (Hashimoto et al. 1970).

Hida metamorphic belt

An exposed part of the Hida metamorphic belt is observed mainly in the Hida terrain. A large part of the western extension of the belt is beneath the Sea of Japan (Miyashiro & Kushiro 1977). This belt is composed chiefly of quartzose-feld-spathic gneisses and amphibolites along with subordinate crystalline limestone. Kyanite, staurolite, sillimanite and andalusite occur in pelitic rocks in the northeastern part, while from near the north central margin sillimanite, alusite, and even cordierite are found (Isomi & Nozawa 1960, Kobayashi 1958, 1962, Sato 1968, Suwa 1969).

Sangun metamorphic belt

The Sangun metamorphic belt consists of many separate areas, which are distributed mostly in north and west Chugoku and north Kyushu. The eastern part of the Sangun belt constitutes a narrow arcuate zone surrounding the Hida belt; the zone is called the Hida marginal structural belt. High-grade metamorphic rocks as well as glaucophanitic rocks crop out in the Omi district in the zone. Furthermore, rocks resembling the Sangun schists are sporadically exposed in fairly small areas of the Joetsu and Sado districts. Turning to the western end, Sangun rocks are distributed in the northern half of Kyushu.

In the Sangun belt a large number of mafic and ultramafic igneous rocks intrude into schists. Ultramafic rocks are largely serpentinized dunites and peridotites, and are associated with gabbroic rocks. Jadeite, which does not coexist with quartz, but with albite, occurs in some serpentinites and metagabbros (Miyashiro & Kushiro 1977, Banno 1958, Chihara 1960, Hashimoto 1964, 1968a, 1968b, Hayama et al. 1969, Karakida et al. 1969, Nureki 1969).

Ryoke metamorphic belt

The Ryoke-Abukuma belt is of low-pressure facies, and is accompanied by abundant granitic intrusives. The granite gneiss complex is discontinuous, being distributed in several isolated terrains, such as Higo-Ryuhozan, Yanai, Kii peninsula, central Honshu, Tsukuba, Hitachi and central Abukuma. The areas between these terraines are covered by younger sediments or represented by granitic rocks and associated fornfelses. The southern border of the Ryoke belt is defined by the Median Line. Toward the north, Ryoke metamorphic rocks grade into virtually unmetamorphosed Upper Paleozoic rocks. Such a transition has been observed in some areas of central Honshu and in the Yanai and Yatsushiro areas. The age of Ryoke-Abukuma metamorphism is considered to be late Mesozoic.

The Ryoke belt consists chiefly of slates, mica schists and gneisses. Amphibolites are not common. Intrusion of abundant granitic rocks is noticed; most, if not all, of them are considered to postdate the major epoch of regional metamorphism. The Al_2SiO_5 polymorphs are found here as alusite and sillimanite. Cordierite is common, but garnet occurs only in areas where metamorphism took place at high temperature. In the highest temperature part of the Ryoke belt, sillimanite and K-feldspar are found as they were formed when muscovite reacted with quartz, and later by granitic intrusion, which again stabilized muscovite (Hashimoto et al. 1970). Figure 1 outlines the distribution of ultramafic rocks and belts in Japan.

Chemical Composition

Silicate, magnesium and calcium in serpentinite of Japan

As previously stated, the concentration of magnesium is markedly high and that of calcium is markedly low in serpentinite. Besides, the concentration of silicate is generally low.

Table 1 gives concentrations of SiO_2, MgO and CaO in serpentinite sampled from various areas in Japan. It shows that the concentration of SiO_2 ranges from 35% to 40% and that of MgO from 38% to 44%; however, most of the CaO concentrations are less than 1%. The value of the three vary very little with the region.

Distribution of heavy metals in serpentine soils

The concentrations of heavy metals in serpentine soils are often so high that crop toxicity at one time was considered to be caused by the presence of nickel and chromium (Crooke 1956, Soane & Saunder 1959). However, it is only nickel that is high in both total and soluble concentrations, the latter being at a toxic level to plants. Chromium is not toxic to plants, because its solubility is low despite that the total concentration of it is high (Masuda 1962, 1963). Why does solubility differ between nickel and chromium? Where in the minerals of the soils do they accumulate? These questions were answered as follows (Suzuki et al. 1971). The soil was fractionated to each mineral, as is seen in Table 2, the chromium concentration is high in ferro-magnetic minerals and non-magnetic heavy minerals; in particular, it is noted that the content is 8% in non-magnetic heavy minerals of fine sand. On the other hand, the nickel concentration is low in both minerals; particularly, it is the lowest in non-magnetic heavy minerals. But the nickel concentration is the highest in clay.

From the results of electron microprobe X-ray analysis, large quantities of chromium, manganese and iron were detected in the black hard mineral which was picked up from heavy minerals in fine sand. Thus, it was considered that the larger part of chromium in the soil was concen-

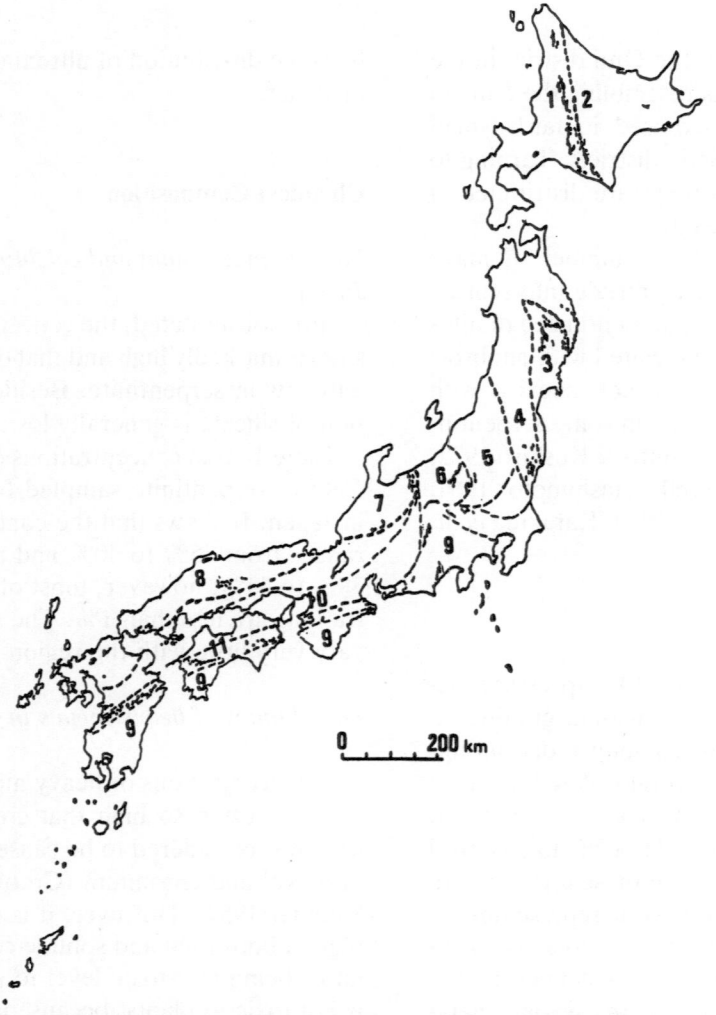

Fig. 1. Distribution map of ultramafic rocks in Japan. 1: Kamuikotan belt, 2: Hidaka belt, 3: Kitakami belt, 4: Abukuka belt, 5: Ashio belt, 6: Joetsu belt, 7: Hida belt, 8: Sangun belt, 9: Shimanto belt, 10: Ryoke belt, 11: Sambagawa belt.

trated in the black hard mineral which was regarded as chromite. When a soil is fused with anhydrous sodium carbonate for determination of various elements contained in it, we will realize how hard it is to determine them if the soil contains a larger amount of sandy chromite that is extremely stable.

The concentration of extractable nickel and chromium are shown in Table 3. It shows that the total concentration of chromium is about twice that of nickel. A comparison of water soluble or 1 N-ammonium acetate extractable chromium is very low.

Thus, chromite is extremely stable in nature and not easily weathered. So, it is not released from the mineral. It is considered that a slightly

Table 1. Concentrations of SiO_2, MgO and CaO in serpentinite of Japan (%)

Locality	SiO_2	MgO	CaO
Onishi, Gunma*	35.5	38.8	0.92
Ogose, Saitama*	39.3	39.7	0.43
Teshirogi, Fukushima*	39.5	39.5	0.18
Iwanebashi, Iwate*	37.2	37.9	1.42
Kotaki, Niigata*	37.9	37.9	0.28
Yamabe, Hokkaido*	35.2	38.6	0.87
Nakagawa, Hokkaido**	38.3	44.0	0.24

*Nakamura, 1959.
**Mizuno, 1979.

274

Table 2. Total element concentrations of separate fractions in serpentine soil (ppm).

Element	Original soil	Clay	Fine sand			Coarse sand		
				Nonmagnetic			Nonmagnetic	
			Ferr.	Heavy	Light	Ferr.	Heavy	Light
Cr	5,200	250	52,300	82,100	2,200	7,650	21,350	720
Ni	2,600	2,500	1,370	760	1,700	1.000	817	1,960
Weight ratio	100	33.7	0.20	1.28	11.3	0.60	0.17	6.72

Ferr.: ferromagnetic minerals.

soluble chromium may have originated in such a form of chromium that is attached to clay and other minerals.

Dichromate in an oxidation state (Cr^{6+}) has a very strong toxicity to many plants, but the toxicity of trivalent chromium (Cr^{3+}) is less (Kamada & Doki 1975, 1977). In the serpentine soils, dichromate is scarcely detectable; moreover, when dichromate is added to crops, their toxic symptoms are very different from those grown on serpentine soils.

Weathering

In the initial stage the weathering process of serpentinites tends to occur at pH 7 or above in reaction. The rainfall in the region, where the soil samples were taken is moderately high, so that the net movement of soil moisture is downward. The flora in this region consists principally of *Picea* (*L. picea*) and *Abies*, and there is a well developed raw humus horizon. The differentiation of soil horizons is clear and it is obvious that the A horizon is enriched by SiO_2 and the B horizon by Fe_2O_3 also, but the horizons have higher values of pH and base saturation than those of podsols on non-ultramafic parent materials in northern Hokkaido (Sasaki, Matsuno & Kondo 1968).

Table 3. Extraction of elements by different solvents ($\mu g\,g^{-1}$)

Element	Total	Solvent		
		H_2O	N-ammonium acetate	0.1-HCl
Cr	5,200	0	0.2	4.3
Ni	2,600	38	100	400

Engineering

The area of serpentinite presents a difficulty in construction of tunnels and roads because of the low coefficient of friction of the rocks. This is caused by the presence of steatite, as well as the property of serpentinite that the addition of a little water to it makes it quaggy easily because of its low water saturation. As is illustrated in Fig. 2a, in the case of montmorillonite, minerals swell by absorbing water, whereas in the case of serpentinite, minerals do not absorb water, but gaps between them are filled up with water instead (Noji 1983). Figures 2b and 2c illustrate tunnels constructed in serpentinized areas in the Kamuikoton mineral belt.

Grouping of serpentinites:

Serpentinites are classified into four groups according to their forms (Noji 1978).
1. *Massive serpentinite.* A massive rock with a highly developed joint system similar to igneous rock in general.
2. *Laminated serpentinite.* A rock with a characteristic of exfoliating easily and entirely lacking the joint system of the original rock.
3. *Psephitic serpentinite.* A psephitic or sandy rock resulting from granulation of the massive rock.
4. *Clayey serpentinite.* A rock which becomes clayey when water is added.

The clayey serpentinite often contains steatite (talc) (Table 4), but the massive serpentinite does not. The existence of talc (Table 4) has a marked effect on the property of serpentinite because it causes a decrease in the coefficient of friction in

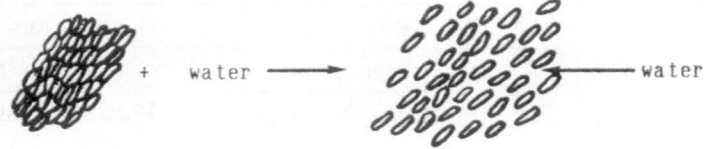

Serpentinite

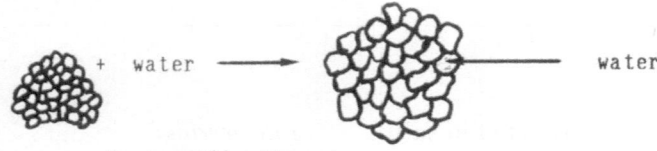

Montmorillonite

Fig. 2a. Expansion by water.

Fig. 2b,c. Tunnels which were made after difficult engineering to prevent swelling pressures and landslides in Kamuikotan, Hokkaido. (a) The road. (b) The railway.

276

Fig. 2d. A land slide preventer over a road on the serpentinite area in the Kamuikotan metamorphic belt.

the rocks, resulting in a sharp decrease in strength of resisting shear in a gap of joints.

Chemical composition of inland water and weathering

It is well known that water plays an important part in weathering of a rock whether it is a physical or a chemical process, and water is more important in the chemical process. Water which has passed through serpentine rocks has a specific chemical composition different from water which passed through other rocks; so, if we have information about it, we will have an effective way in coping with civil engineering work in areas of serpentine rocks.

Granite consists of feldspar, mica and quartz. Kaolin, montmorillonite and other clay minerals are known as weathered products originating from the granite. Playing a determining role in this weathering process is water into which carbon dioxide has been dissolved. Namely, such water causes Mg, K and SiO_2 to dissolve from biotite and form kaolin; it also causes K and SiO_2 to dissolve from orthoclase existing, in plagioclase and form kaolin. Their chemical formulas are given in equations 1–4.

$$2KMg_3AlSi_3O_{10}(OH)_2 + 14CO_2 + 7H_2O$$
$$\text{Biotite}$$
$$\rightarrow Al_2Si_2O_5(OH)_4 + 6Mg^{2+} + 2K^+ + 4SiO_2$$
$$\text{Kaolin}$$
$$+ 14HCO_3^- \quad (1)$$

$$2KAlSi_3O_8 + 2CO_2 + 3H_2O \rightarrow Al_2Si_2O_5(OH)_4$$
$$\text{Orthoclase} \qquad\qquad\qquad\qquad \text{Kaolin}$$
$$+ 2K^+ + 4SiO_2 + 2HCO_3^- \quad (2)$$

$$1.61Na_{0.62}Ca_{0.38}Al_{1.38}Si_{2.62}O_8 + 2.20CO_2$$
$$\text{Plagioclase}$$
$$+ 3.30H_2O \rightarrow 1.10Al_2Si_2O_5(OH)_4$$
$$+ 1.0Na^+ 0.61Ca^{2+} + 1.99SiO_2$$
$$+ 2.2OHCO_3^- \quad (3)$$

$$1.61Na_{0.62}Ca_{0.38}Al_{1.38}Si_{2.62}O_8 + 1.90CO_2$$
$$\text{Plagioclase}$$
$$+ 1.90H_2O \rightarrow 0.95Ca_{0.17}Al_{2.33}Si_{3.67}O_{10}(OH)_2$$
$$\text{Montmorillonite}$$
$$+ 1.0Na^+ + 0.45Ca^{2+} + 0.72SiO_2$$
$$+ 1.90HCO_3^- \text{(Kitano 1969)} \quad (4)$$

Consequently, if we know the amounts of elements which have dissolved into water from a rock, we can calculate the degree of weathering and the amount of the resultant clay. Namely,

Table 4. Minerals in serpentinite (Noji 1978)

Name of mineral	Chemical composition	Gravity	Hardness
Serpentine	$Mg_3Si_2O_5(OH)_4$	2.5–2.6	2.5–3.5
Steatite (talc)	$Mg_3Si_4O_{10}(OH)_2$	2.6–2.8	1
Brucite	$Mg(OH)_2$	2.39	2.5
Magnetite	Fe_3O_4	5.2	6
Olivine	$(Mg, Fe)_2SiO_4$	3.2–4.4	6.5–7
Chromine	$FeCr_2O_4$	4.5–4.8	5.5

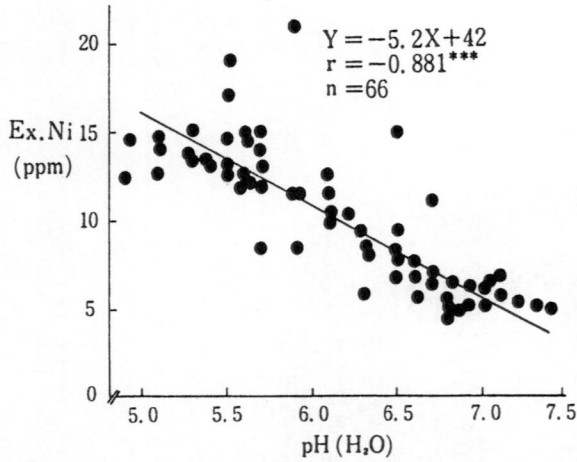

Fig. 3. Effect of soil pH on exchangeable nickel (Mizuno 1979).

$Y = -5.2X + 42$
$r = -0.881^{***}$
$n = 66$

from analyses of water we can tell the degree of weathering and forecast the occurrence of a landslide.

The toxicity of nickel to plants

Soil condition

The toxicity of nickel varies with the soil condition and plant species. Soils high in concentration of exchangeable nickel which is extracted with normal ammonium acetate are generally observed to be toxic, while the exchangeable nickel in the soils usually decreases with an increase in soil pH.

Generally, serpentine soils are very high in pH; so, many crops on any sedentary soils of serpentinite are not susceptible to the toxicity of nickel. Most of the nickel toxic symptoms of many plants occur on other soils such as alluvial soils, as well as peat soils to which weathered serpentinite is added with a large quantity of water or additional soil. And the total nickel concentration in these soils ranges from about 500 $\mu g\, g^{-1}$ to about 1000 $\mu g\, g^{-1}$, being less than 2000 $\mu g\, g^{-1}$, which is the total nickel in serpentinite. These soils are lower in pH than the sedentary soils of serpentinite because the Mg content has been decreasing by leaching. The relationship between pH and magnesium concentration in the soils is shown in

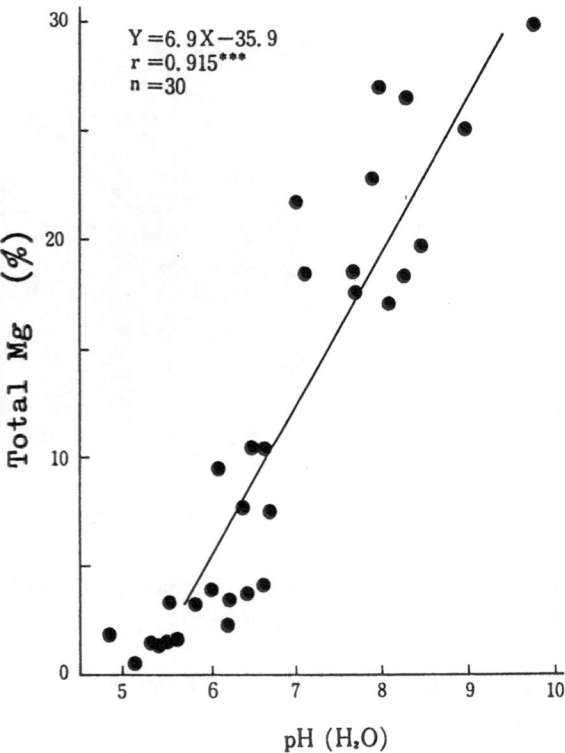

$Y = 6.9X - 35.9$
$r = 0.915^{***}$
$n = 30$

Fig. 4. Relationship between pH and total Mg concentration in soils and rocks (Mizuno 1979).

Fig. 4. It shows that both are highly correlated; i.e., the pH of serpentine soils is controlled by magnesium concentration.

It contrasts well with general soils, the pH of which is controlled by calcium concentration. Another characteristic relationship is observed between nickel and magnesium because the ratio of Mg/Ni decreases with increasing separation of serpentine soils from the parent rocks (Table 5), and the frequency of nickel toxic occurrence increases as a result.

Why does the leaching of weathered rocks bring about a difference in concentration between

Table 5. pH, total Ni, total Mg and Mg/Ni ratio of serpentinite and soils (Mizuno 1979)

Sample	Ni ppm	Mg %	Mg/Ni	pH
Serpentinite	2,200	18.2	83	8.3
Sedentary soil	800	4.0	50	6.6
Alluvial soil	500	3.6	72	6.4
	200	0.4	20	5.1
	800	1.9	24	4.8
Deluvial soil	1,100	1.2	11	5.4

278

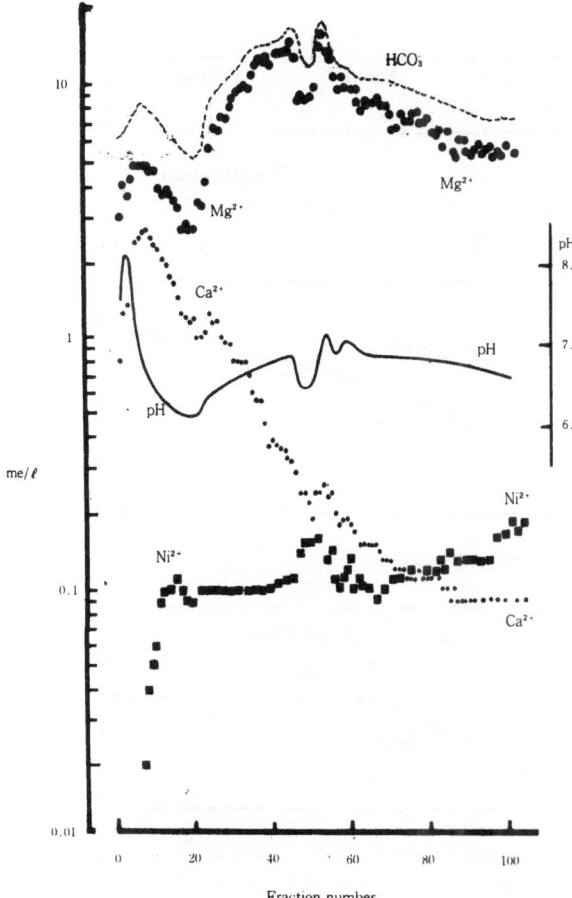

Fig. 5. Change in concentrations of Mg, Ca, Ni and pH when the elements are leached by a solution of carbon dioxide. (Mizuno 1979).

nickel and magnesium? It is considered that magnesium is leached away before nickel because magnesium has a weak bonding strength compared with nickel (electronegativity: Ni : 1.8, Mg : 1.2).

This phenomenon is confirmed by a simple chemical experiment; i.e., the saturated solution of carbon dioxide which is considered a main factor for weathering of many rocks is passed through the powder of serpentinite in a tube of a fraction collector. Then, the passed solution is subjected to determination of pH, and concentrations of magnesium, calcium, nickel and other elements, for example, Fig. 5 (Mizuno 1979).

As is evident from Fig. 5 with the leaching of magnesium and calcium, the solution's pH decreases from 8 to below 7, but soluble nickel makes a slow ascent in the solution, i.e., the

solubility of nickel begins to increase after the leaching of magnesium from serpentine soils.

It is observed that zinc deficiencies occur in Indian corn and onion plants particularly on serpentine soils whose pH is high. Other factors for abnormal growth of plants are calcium deficiency and molybdenum deficiency, but they are rarely observed on serpentine soils in Japan.

Cobalt which is highly concentrated in serpentinite is not a cause of metal toxicity to plants, because its toxicity is usually less than nickel toxicity (Chino & Kitagishi 1966); moreover, the cobalt concentration is markedly lower than the nickel concentrations (Mizuno 1968) in Japan's ultramafic soils.

The relationship between various elements and soil condition is summarized in Fig. 6. Magnesium concentration decreases in serpentine soils with the advance in weathering, and soluble nickel in the soils increases with increasing acidity of the soils. Calcium deficiency which brings about growth disorder occurs on a high pH soil before sufficient amounts of Mg is leached out, and zinc deficiency is due to high pH and high alkalinity, (Nelson *et al.* 1959). Molybdenum deficiency results mainly from the low concentration in the soil (Otsuka & Takahashi 1962).

As described above, the growth defects of plants on serpentine soils varies with the stage of weathering development from serpentinite, and with the kind of plant as well. Consequently, they are not caused only by soil conditions.

Nickel toxicity of plants

Disorders have long been observed in some plants grown on serpentine soils (Figure 6b, c) with evident symptoms of necrosis or chlorosis. Usually necrotic areas are found on the leaves of cabbage, white rape, radish, carrot, and watermelon. Meanwhile, chlorotic areas are found on the leaves of alfalfa, azuki beans, kidney beans and sugar beet, their symptoms changing to those of necrosis when the nickel toxicity is strong. The leaves of oats, wheat and grasses such as Italian rye grass, show the symptoms of longitudinal chlorosis. The leaves of oats show longitudinal stripes of chlorosis particularly well.

Studies in Japan on the nickel toxicity of serpentinite soils were conducted first by Masuda &

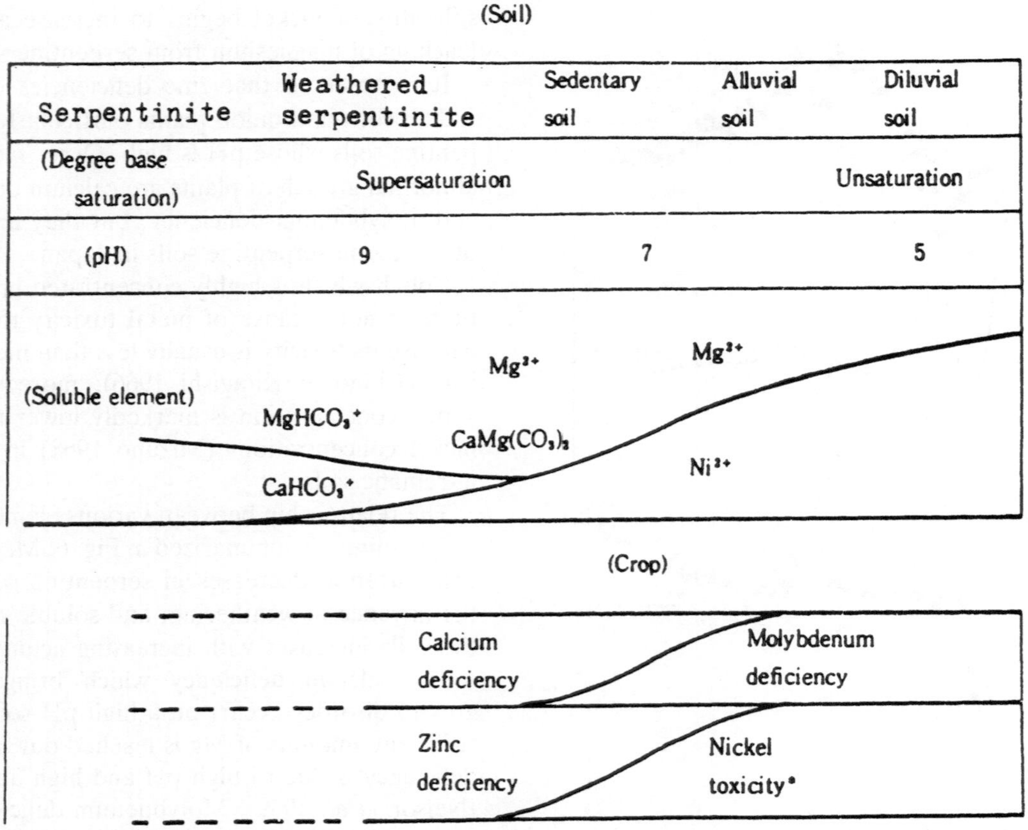

(Soil)

Serpentinite	Weathered serpentinite	Sedentary soil	Alluvial soil	Diluvial soil
(Degree base saturation)	Supersaturation		Unsaturation	
(pH)	9	7		5
(Soluble element)	$MgHCO_3^+$ $CaHCO_3^+$	Mg^{2+} $CaMg(CO_3)_2$	Mg^{2+} Ni^{2+}	

(Crop)

	Calcium deficiency	Molybdenum deficiency
	Zinc deficiency	Nickel toxicity*

Fig. 6a. Relationships of mineral deficiencies and toxicities of crops to conditions of soils: weathering of serpentinite, degree of base saturation, soil pH, and soluble elements. *Mostly accompanied by iron and/or copper deficiency.

Sato (1961). They found that disorders in oat plants on serpentine soils were usually caused by exchangeable nickel, but not by either chromium or the low Ca/Mg quotient (Masuda *et al.* 1961, 1963).

The degree of nickel toxicity symptoms was usually in agreement with the degree of nickel concentration in plants, but this concentration did not serve as a measure of the nickel toxicity of many kinds of plants because the nickel concentration varied greatly with the plant species (Mizuno 1967); besides, nickel tolerance varied with the species also. Usually rice and potato plants have a high tolerance for nickel, whereas oats, cabbage and alfalfa have a fairly low nickel tolerance.

Tolerance of a plant for nickel is distinguishable by the following two variables, the characteristics of the plant's uptake of nickel; and the Fe/Ni or Cu/Ni quotient in the plant.

Uptake characteristics of nickel:

Oats, alfalfa, azuki beans and cabbage, which have a low tolerance for nickel have a linear correlation between the uptake of nickel and the concentration of exchangeable nickel in soil. On the other hand, plants with high tolerance for nickel like potato and orchard grass do not have such a correlation (Fig. 7).

Fe/Ni and Cu/Ni quotients in various plants:

Some plants show no toxic symptoms despite the high concentration of exchangeable nickel in the soil. The result of analyses derived statistically, as shown in Fig. 8a indicate that most of the plants showing no toxicity symptoms or the plants tolerant of nickel have a high Fe/Ni quotient (>5) and a high Cu/Ni quotient (>1) (Mizuno 1968). It shows that the concentrations of copper and iron against that of nickel are consistently higher in those plants showing no toxic

Fig. 6b. Nickel toxicity in raspberry (*Rubus idaeus* L.) on serpentine soil in Horokanai, Hokkaido.

Fig. 6c. Nickel toxicity in Timothy (*Phleum pratense* L.) on serpentine soil in Horokanai, Hokkaido.

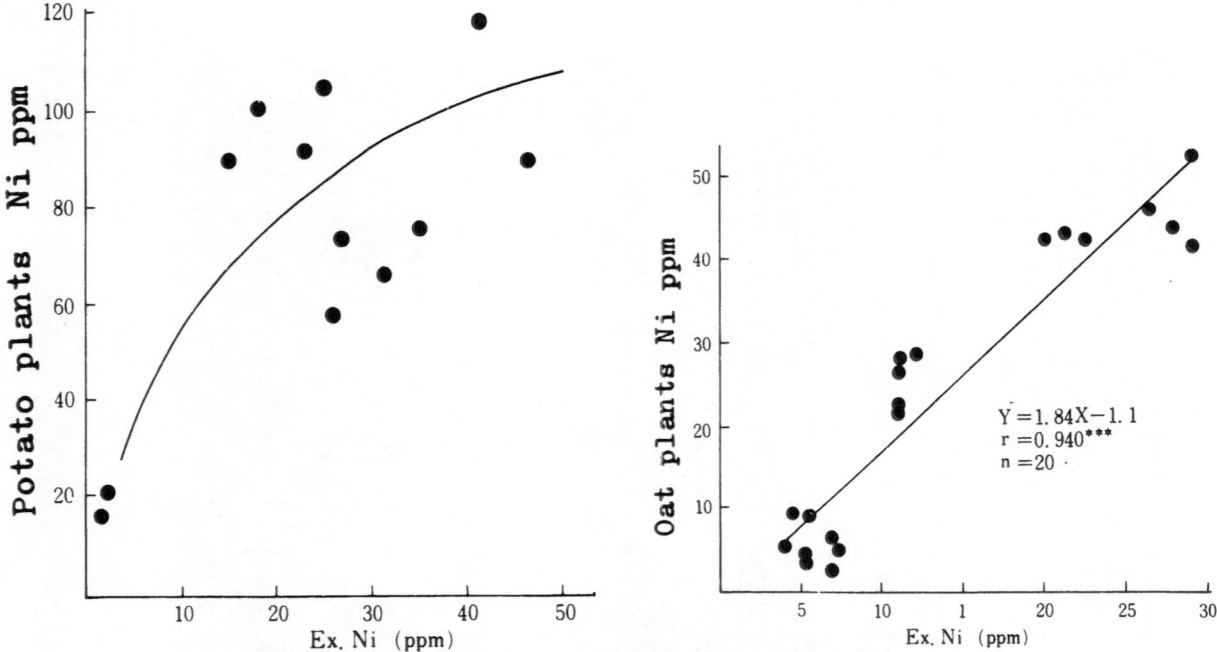

Fig. 7. Relation between exchangeable nickel and nickel concentration in potato and oats (Mizuno, 1979).

symptoms and no decrease in yield than in plants showing some toxic symptoms and a decrease in yield.

The crop yield of radish did not decrease in a soil with a high concentration of exchangeable nickel, but necrotic areas developed on the leaves of the plants. In this case the radish had a high Fe/Ni quotient and a low Cu/Ni quotient.

The result of the experiments indicates that (1) plants with a Cu/Ni quotient less than 1.0 had necrotic or chlorotic leaves, although the crop yield did not always decrease; (2) there was a decrease in crop yield when the Fe/Ni quotient in the plants was less than 5, although symptoms of necrosis or chlorosis did not always develop; (3) in most of the plants the crop yield decreased at the same time as toxic symptoms developed, but it was unusual for each of the two phenomena to occur independently; and (4) with decreases in soil pH and Fe/Ni and Cu/Ni quotients, the plants' tolerance for nickel was low.

Rice plants have a very high tolerance for nickel; so, symptoms due to nickel toxicity were not observed in the rice plants grown in serpentine soil areas despite that the Fe/Ni quotient was usually above 10 in the plants as a result of high level of iron concentration. Consequently, rice is the species recommended for cultivation on serpentine soils.

Interaction between nickel and other metals in plants

Studies were made on nickel toxicity symptoms in mulberry plants on serpentine soils (Takagishi *et al*. 1973); and the symptoms found resembled very much those which were caused by deficiencies in zinc and iron (Iizuka 1975) (e.g., Figure 8b).

The treatments which have been found effective in controlling the symptoms were: (1) application of iron and zinc to serpentine soils or by foliar spray to mulberry leaves; (2) supply of some calcareous materials to serpentine soil. It was recognized that the competitive interrelationship of nickel, iron and zinc contents in the leaves has a certain relationship to the development of nickel toxicity; the 'T value', which is expressed by Ni + 0.17 Fe + 0.38 Zn, was nearly constant except for both cases with heavy symptoms and with a healthy appearance, although the coef-

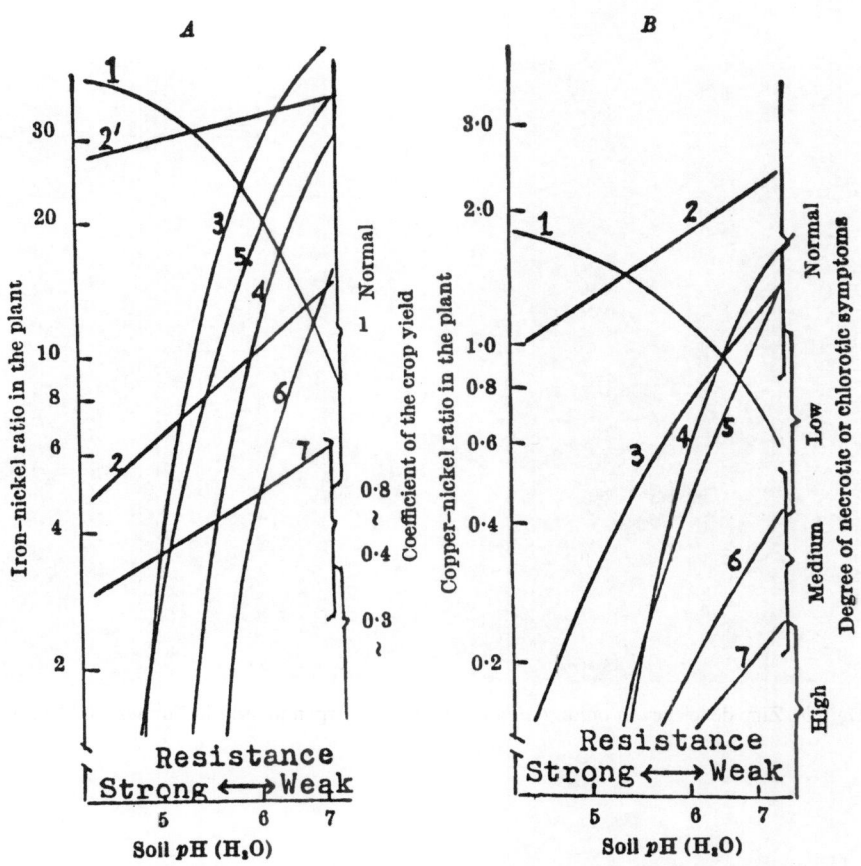

Fig. 8a. *A*, The relationship between the ratio of iron to nickel in the plant and the coefficient of the crop yield. (Normal yield = 1.0). *B*, The relationship between the ratio of copper to nickel in the plant and severity of the toxic symptoms in: (1) dent corn, (2) potato, (3) timothy, (4) oat, (5) red clover, (6) alfalfa, and (7) radish plants. (Mizuno 1968b).

ficients in the formula were obtained by trial and error.

The circulation of heavy metals through soils, plants and animals on serpentine soils

Mountainous areas in Hokkaido abound in serpentine soils, and a relatively large area of it is used as pasture. As is evident from the foregoing, the concentration of nickel is markedly high in the grasses and legumes there. It was very important, therefore, to look into the effect of nickel on domestic animals or the relationship between soils and plants and animals concerning the effect.

From an examination of the circulation of some heavy metals through plants and animals on the serpentine soils, Katayama *et al.* (1984) found that nickel is the only element that shows a high accumulation in the dried hay harvested. The concentration range from 10 μg g^{-1} to 40 μg g^{-1}, whereas the concentrations of cobalt and chromium are low or undetectable in the same hay (Table 6).

As shown in Table 6, there is no noticeable amount of nickel in the milk and hair of dairy cattle which are fed every day with the hay produced from a serpentine soil. Katayama *et al.* calculated the amount of nickel which circulates in soil, plants and cattle, with the result obtained shown in Fig. 9. It follows from the figure that most (about 90%) of the nickel taken by cattle is eliminated as the droppings; as a result, nickel is not accumulated in the body of the cattle. In Japan there are no reports about the harmful effects of nickel in hay and grass on cattle and other animals other than on serpentine soils.

Fig. 8b. Zinc deficiency in onion (*Allium cepa* L.) on serpentine soil in Furano, Hokkaido.

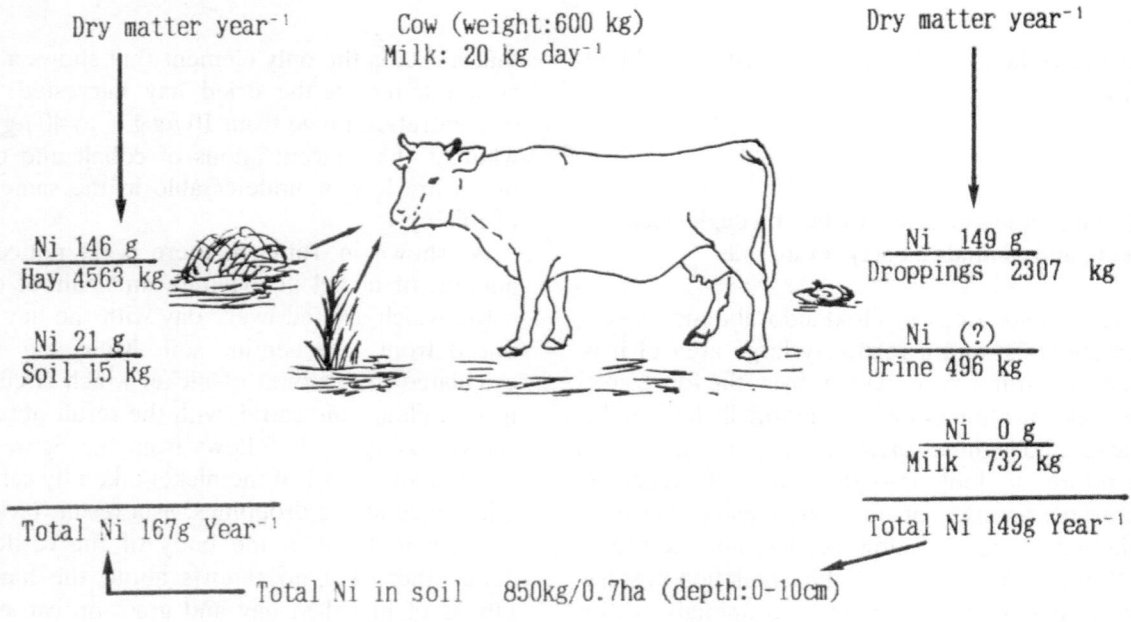

Dry matter year^{-1} Cow (weight:600 kg)   Dry matter year^{-1}
Milk: 20 kg day^{-1}

$\dfrac{\text{Ni } 146 \text{ g}}{\text{Hay } 4563 \text{ kg}}$

$\dfrac{\text{Ni } 21 \text{ g}}{\text{Soil } 15 \text{ kg}}$

$\dfrac{\text{Ni } 149 \text{ g}}{\text{Droppings } 2307 \text{ kg}}$

$\dfrac{\text{Ni } (?)}{\text{Urine } 496 \text{ kg}}$

$\dfrac{\text{Ni } 0 \text{ g}}{\text{Milk } 732 \text{ kg}}$

Total Ni 167g Year^{-1} Total Ni 149g Year^{-1}

Total Ni in soil   850kg/0.7ha (depth:0-10cm)

284

Utilisation of serpentinite

Use as fertilizers

Most of serpentinite is used in Japan for production of various fertilizers. It was formerly used as a colouring matter of cement, constituting about 3% of the mixture, but no such use is made of it now. The production of fused phosphatic fertilizer consumes about 180,000t of both serpentinite and peridotite. Moreover, about 50,000 t of the rocks are used annually to produce magnesian fertilizers. Hence, the total amount of consumption of the rocks in Japan reaches 230,000 t per year.

Areas deficient in MgO, MnO and B_2O_3 used to occupy about 520,000 ha in the farmland of Japan (Ministry of Agriculture and Forestry 1961), of which magnesium deficient areas reached about 240,000 ha or about 4% of the area under cultivation in Japan. When the amount of application is 20 kg/ha, then 100,000 t of MgO are sufficient to cover about 80% of the total area under cultivation. Today soils deficient in magnesium because of wide use of fused phosphatic fertilizer and other magnesian fertilizers are almost non-existent.

Use for hobbies

The Japanese are traditionally fond of stones and rocks. Since serpentinite is soft and easy to be worked on and has colours that bring about a subdued atmosphere, its rocks are made into an ornamental artwork after shaving and polishing. The rocks are also frequently used for fencing. These uses are marginal, however, if compared with the amount used for making fertilizer.

Platinum

It has been well known that platinum originates from serpentinite. Some of the rivers which run in or through serpentinite areas are named 'Shirakinsawa' or 'Hakkin-Kawa,' meaning a valley or a river of alluvial platinum mining, respectively. It shows that platinum used to be mined there. Platinum pieces are found at times in the silica residue when a finely ground serpentinite sample and anhydrous Na_2CO_3 are put together in a platinum crucible and heated in a muffle furnace at 900–950°C and then dissolved in hydrochloric acid. In Japan we can no longer find platinum miners, but the names of the places

Table 6. Concentrations of metals in soil, hay and others (in $\mu g\,g^{-1}$)

Pasturage No	Sample*	Fe	Mn	Cu	Zn	Co	Ni	Cr
1	Soil	46,000	747	25	66	80.3	735	213
	Ke(Ti)	140	204	7.8	29	0.3	38	nd
	Or	110	320	11.9	26	0.4	46	nd
2	Soil	72,000	230	35	96	190.0	1,390	275
	Ke(Ti)	170	113	5.3	20	0.4	27	nd
	Or	130	128	5.0	16	0.7	21	nd
3	Soil	41,000	530	23	61	65.0	350	190
	Ke(Ti)	150	55	6.3	19	0.2	12	nd
	Or	158	79	6.6	18	0.3	15	nd
	Lc	160	35	6.8	24	0.3	15	nd
4	Soil	42,000	873	26	83	90.0	1,429	300
	Ke(Ti)	270	58	3.7	51	0.4	31	nd
	Droppings	–	281	–	100	–	65	–
	Cow hair	–	nd	–	119	–	nd	–
	Milk	–	nd	–	33	–	nd	–

*Soil: total concentration with perchloric acid – nitric acid, Ke: Kentucky blue grass, Ti: timothy, Or: orchard grass, Lc: ladino clover, Droppings: cows droppings, nd: no determination.

Table 7. Climatic data of Japan

Locality	Temperature (mean value, °C)						Annual precipitation (P) (mm)	P/T
	Coldest month	Hottest month	Annual mean value (T)	Annual range	Warmth index	Coldness index		
1. Wakkanai	−5.7 (Jan.)	18.9 (Aug.)	6.3	24.6	53.8	−37.7	1187	188
2. Nemuro	−5.4 (Feb.)	16.9 (Aug.)	5.8	22.3	44.9	−35.1	1072	185
3. Sapporo	−4.9 (Jan.)	21.3 (Aug.)	8.0	26.2	68.1	−32.1	1158	145
4. Hakodate	−3.6 (Jan.)	21.2 (Aug.)	8.3	24.8	66.9	−27.1	1157	139
5. Aomori	−1.8 (Jan.)	22.5 (Aug.)	9.6	24.3	76.8	−21.4	1407	147
6. Akita	−0.5 (Jan.)	24.0 (Aug.)	11.0	24.5	87.8	−15.8	1787	162
7. Niigata	2.0 (Jan.)	25.7 (Aug.)	13.1	23.7	103.3	−6.0	1822	139
8. Toyama	2.1 (Jan.)	25.8 (Aug.)	13.5	23.7	106.7	−5.5	2346	174
9. Tottori	3.8 (Jan.)	26.4 (Aug.)	14.4	22.6	114.7	−2.0	2018	140
10. Matsumoto	−1.0 (Jan.)	23.9 (Aug.)	11.1	24.9	90.2	−15.6	1067	96
11. Miyako	0.0 (Jan.)	22.1 (Aug.)	10.5	22.1	79.6	−14.1	1278	122
12. Sendai	0.9 (Jan.)	23.9 (Aug.)	11.9	23.0	92.4	−9.9	1219	102
13. Mito	2.5 (Jan.)	24.8 (Aug.)	13.2	22.3	103.2	−4.5	1341	102
14. Tokyo	4.7 (Jan.)	26.7 (Aug.)	15.3	22.0	124.4	−0.3	1460	95
15. Shizuoka	6.0 (Jan.)	26.5 (Aug.)	16.0	20.5	132.0	0	2361	148
16. Nagoya	3.6 (Jan.)	26.8 (Aug.)	14.9	23.2	120.4	−2.1	1575	106
17. Kyôto	3.9 (Jan.)	27.5 (Aug.)	15.2	23.6	123.6	−1.5	1669	110
18. Hiroshima	4.3 (Jan.)	26.8 (Aug.)	15.0	22.5	120.2	−0.8	1603	107

above tell us that platinum is associated with the areas of serpentinite.

Serpentine vegetation of Japan

Natural vegetation of Japan

As climatic data show in Table 7 and Fig. 10, Japanese Archipelago, except the alpine and sand dune areas, ought to be covered by forest vegetation.

At present, approximately 65% of the area of Japan is covered with forest (primeval forest 19%, secondary forest 26%, plantation 20%). The area of farming (mainly paddy-field) is approximately 24%, that of natural grassland is 1%, that of substitutional grassland is 4.5% and that of town-site secondary is 4.5%.

Studies on the vegetation of Japan were firstly carried out by dendrologists (e.g., by Tanaka in 1887, by Honda in 1912 etc.) and subsequently have been carried out by many plant ecologists.

According to Miyawaki (1977), natural vegetation regions of Japan are divided into four regions, viz.; *Camellietea japonicae* region, *Fagetea crenatae* region, *Vaccinio-Piceetea* region and Alpine vegetation region.

Table 7. Continued

Locality	Temperature (mean value, °C)						Annual precipit-ation (P) (mm)	P/T
	Coldest month	Hottest month	Annual mean value (T)	Annual range	Warmth index	Coldness index		
19. Takamatsu	4.7 (Jan.)	26.8 (Aug.)	15.2	22.1	122.2	−0.3	1199	79
20. Kôchi	5.6 (Jan.)	27.1 (Aug.)	16.3	21.5	136.0	0	2666	164
21. Saga	5.0 (Jan.)	27.4 (Aug.)	16.0	22.4	131.6	0	1890	118
22. Oita	5.5 (Jan.)	26.6 (Aug.)	15.6	21.1	127.0	0	1708	109
23. Kagoshima	7.0 (Jan.)	27.7 (Aug.)	17.3	20.7	148.2	0	2375	137
24. Naze	14.2 (Jan.)	28.3 (Jul.)	21.3	14.1	195.7	0	3051	143
25. Naha	16.0 (Jan.)	28.1 (Jul.)	22.4	12.1	207.9	0	2128	95

−warmth index (WI) and coldness index (CI) were presented by Kira (1945).

warmth index (WI) $= \overset{n}{\Sigma}(t - 5)$.

coldness index (CI) $= -\overset{n'}{\Sigma}(5 - t)$

n) number of month mean monthly temperature of which is over 5°C.

n′) number of month mean monthly temperature of which is under 5°C.

Relation between WI and vegetation zone is as follows:

WI	Vegetation zone
0	Polar frost zone
0 ~ 15	Polar (tundra) zone
15 ~ 45	Subpolar zone
45 ~ 85	Cool temperate zone
85 ~ 180	Warm temperate zone*
180 ~ 240	Subtropical zone
over 240	Tropical zone

*In the district showing under −10 degree in CI, the evergreen broad-leaved forest is hardly formed. Therefore, in Akita district (87.8 in WI and −15.8 in CI) belongs to the deciduous forest zone. (Kira, 1945 & 1949).

(a) The *Camellietea japonicae* region covers the warmest district (southern half of Honshû, Shikoku, Kyûshû and Loochoo Isls.) showing over 85 degrees in warmth index. It is named the evergreen broad-leaved forest zone. The forest has been anthropogenically influenced for several thousand years, so the secondary vegetation (incl. plantations) occupies extensive area. At the southern end district of this region, Loochoo Isls. (Okinawa Pref.), shows over 180 degrees in warmth index, so the forest of the district includes some subtropical ligneous plants, e.g., *Ficus*, *Bruguiera*, *Kandelia*, *Pandanus*, *Cycas*, *Cyathea* and so on.

Characteristic plants of the forest of the *Camellietea japonicae* region are as follows:

—*in the T_1 layer*; evergreen broad-leaved trees – *Castanopsis* spp., *Machilus thunbergii*, *Cyclobalanopsis* spp. etc., evergreen needle-leaved trees – *Abies firma*, *A. homolepis*, *Tsuga sieboldii*, *Torreya nucifera* etc. (these needle-leaved trees occur on the mountain ridge or on the steep slope).

—*in the T_2 layer*; evergreen broad-leaved trees – *Camellia japonica*, *C. sasanqua*, *Ilex* spp., *Neolitsea* spp. etc.

—*in the Shrub layer*; evergreen broad-leaved shrubs – *Aucuba* spp., *Eurya* spp. etc.

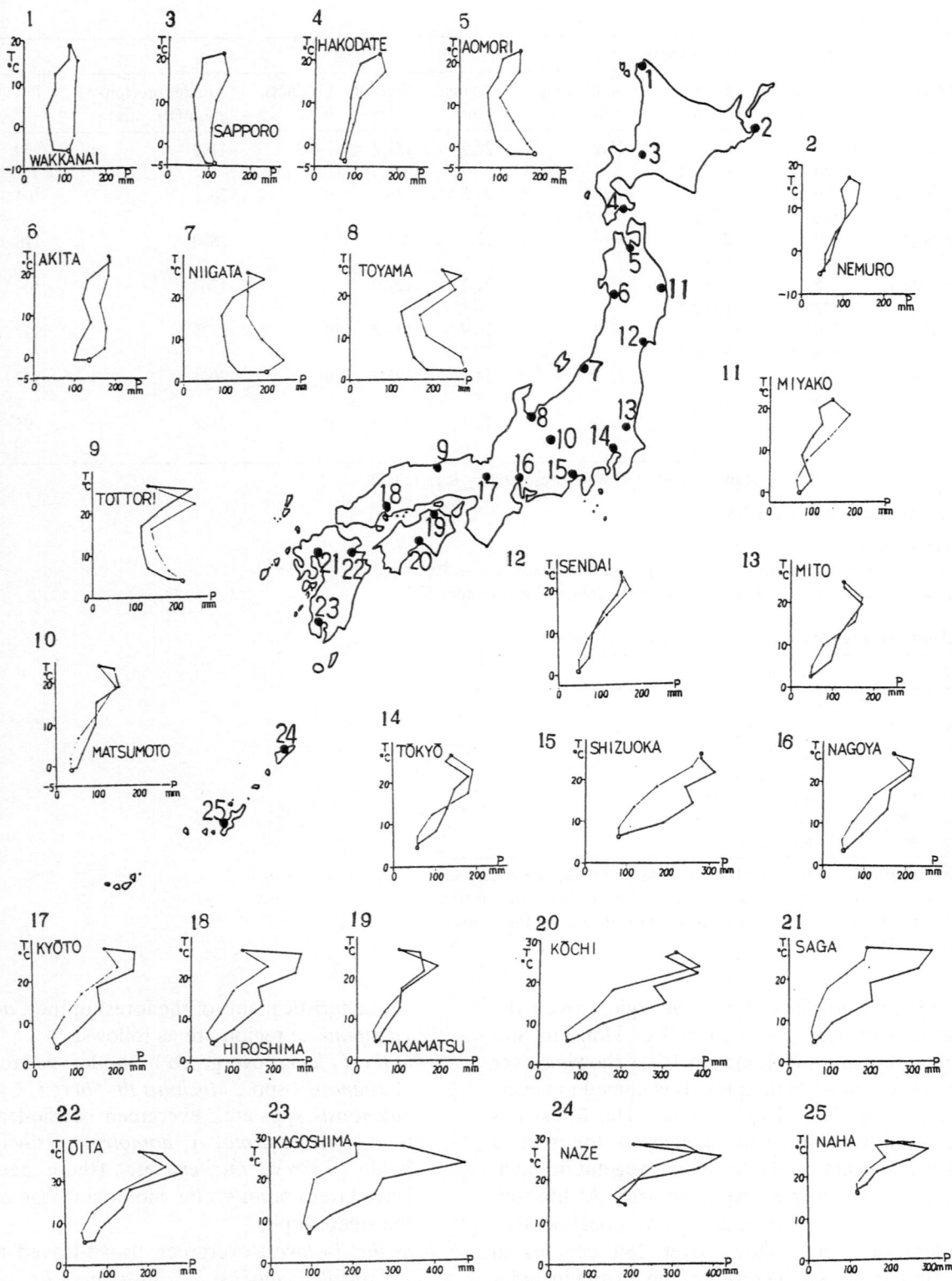

Fig. 10. Climographs of each locality in Table 7.

—*in the Herb layer*; evergreen broad-leaved shrublets – *Ardisia japonica*, herbs – *Cymbidium virescens*, *Liriope* spp. etc., and some evergreen ferns.

—*Woody climbers*; *Tracherospermum asiaticum*, *Kadsura japonica*, *Hedera rhombea* etc.

Main secondary forests in this region are the *Quercus serrata* – *Quercus acutissima* (Oak) forest and the *Pinus densiflora* (Japanese red pine) forest. The latter stands, originally on immature soil, formerly occupied rather limited area on the ridge, gravel etc., where they formed an open stands. Many years of deforestation have resulted in new gravel or scree by discharge of soil, and *Pinus densiflora*, one of the typical sun trees, has migrated to such barren areas as a pioneer species. It sometimes becomes the dominant of the edaphic climax community.

At present, the *Pinus densiflora* community occupies a considerably wide area, and from the view-point of plant-sociology, the community is recognized as *Rhododendro-Pinetum densiflorae*.

A synthetic table of the *Pinus densiflora* community is given in Table 8.

(b) The *Fagetea crenatae* region (deciduous broad-leaved forest zone) covers the northern half of Honshû and Hokkaidô, showing between 45 degrees and 85 degrees in warmth index. The montane areas of southern Honshû, Shikoku and Kyûshû are also covered with this vegetation. Communities such as *Saseto-Fagion crenatae*, *Thujopsis dolabrata* var. *hondai-Fagus crenata* community, *Lindero umbellatae-Fagetum crenatae*, *Sasamorpho-Fagion crenatae*, *Tilia maximowicziana-Quercus mongolica* var. *grosseserrata-association*, *Acer mono* var. *glabrum-Tilia japonica* community, *Ulmus davidiana* var. *japonica* community, *Alnus japonica* community, *Alnus pendula-Weigela hortensis* community, *Tsugion sieboldii* etc. are distinguished in this class. In the forest belonging to this class, such deciduous broad-leaved trees as *Fagus crenata* (Siebold's beech), *Quercus mongolica* var. *grosseserrata* (oak), *Alnus japonica* (Japanese alder), *Acer* spp., *Tilia maximowicziana* (Japanese linden), *Prunus* spp. etc. are dominant, and the forest floor is in most cases covered with *Sasa* or *Sasamorpha* plants. In the Pacific side district, *Sasmorpho-Fagion crenatae* is conspicuous, while in the Japan Sea side district, *Saseto-Fagion crenatae*

is common. The *Quercus mongolica* var. *grosseserrata* forest is confined to the arid habitats.

In Hokkaidô, the northern limit of distribution of *Fagus crenata* lies along the middle part of Oshima peninsula, Kuromatsunai Depression. So, *Fagus crenata* is lacking in the forest of Hokkaidô except the southern half of Oshima peninsula. The sea coast district and dunes are covered mainly with *Quercus dentata* (Daimyô oak) forest, while the inland district is covered with *Quercus mongolica* var. *grosseserata* forest. The humid lowland is covered with *Ulmus davidiana* var. *japonica-Fraxinus mandshurica* var. *japonica* forest. These deciduous forests are more or less mixed by the subarctic needle-leaved trees, and the needle-leaved forest dominated by *Abies sachalinensis*, *Picea jezoensis*, *Picea glehni* etc. which are more conspicuous in the eastern and northern end districts of Hokkaidô.

(c) The *Vaccinio-Picetea* region (subalpine and subarctic conifer forest zone) extends in the subalpine area of north of central Honshû, and in the montane, as well as the easternmost and the northernmost districts of Hokkaidô. In this class, *Piceion jezoenis* (Hokkaidô), *Abietion mariesii* (Honshû), *Betulo ermanii-Ranunculetea acris japonicae* and *Sasa* spp. -*Betula ermani* communities are distinguished.

In Honshû, *Abies mariesii* (Maries's fir), *A. veitchii* (Veitch's silver fir) and *Picea jezoensis* var. *hondoensis* (Hondo spruce) are generally dominant, and on rather xeric habitats, *Tsuga diversifolia* (Northern Japanese hemlock) is often dominant. In the subalpine zone of the Japan Sea side area of Tôhoku district of Honshû, those needle-leaved forests are less common because of heavy snowfall in winter. In place of the needle-leaved forest, the deciduous forest dominated by a small oak tree, *Quercus mongolica* v. *undulatifolia*, is characteristic.

Common species of the class in the undergrowth are *Cornus canadensis*, *Sterptopus* spp., *Listera* spp., *Pyrola* spp., *Rubus pedatus*, *R. pseudo-japonicus*, *R. ikenoensis* etc., and sometimes *Sasa* spp. are dominant. *Betula ermani*, *Acer ukurunduense*, *Sorbus* spp., *Vaccinium* spp., *Menziesia pentandra*, *Coptis trifolia* etc. occur as companions. In Hokkaidô, the canopy is occupied mainly by *Picea jezoensis* (Ezo spruce) and *Abies sachalinensis* (Sakhalin fir), sometimes on

Table 8. Synthetic table of the *Pinus densiflora* community (Miyawaki 1977)

	Locality				
	1	2	3	4	5
Carpinus laxiflora	II	+	.	.	.
Pleioblastus chino	II	2	.	.	.
Shirakia japonica	II	+	.	.	.
Rubus palmatus var. *coptophyllus*	I	+	.	.	.
Lysimachia clethroides	II	+	.	.	.
Carex siderosticta	IV	1	.	.	.
C. lanceolata	II	+	.	.	.
Pertya glabrescens	III	+	.	.	.
Leptogramma totta	I	+	.	.	.
Akebia trifoliata	II	+	.	.	.
Ilex pedunculosa	.	.	2	II	III
Pertya scandens	.	.	+	IV	III
Viburnum erosum forma *punctatum*	.	.	+	V	.
Rhododendron macrosepalum	.	.	.	II	IV
Miscanthus sinensis	.	.	.	V	.
Rhododendron reticulatum	.	.	1	.	IV
Pleioblastus distichus var. *nezasa*	.	.	2	.	II
Hydrangea hirta	IV	.	.	.	I
Fraxinus sieboldiana	IV	.	.	.	II
Rhus trichocarpa	IV	.	.	.	II
Pinus densiflora	IV	5	5	V	V
Castanea crenata'	III	1	+	.	I
Quercus serrata	III	1	2	IV	III
Clethra barbinervis	IV	+	1	.	I
Lyonia neziki	III	1	1	.	III
Rhododendron kaempferi	V	2	.	V	I
Smilax china	III	+	.	V	IV
Abies firma	II	1	.	.	I
Disporum smilacinum	V	+	.	.	I
Eurya japonica	.	.	1	V	V
Ilex crenata	.	+	+	.	III
Wisteria floribunda	III	+	.	III	.

Locality: 1 Utsunomiya, Tochigi Pref., (by Usui 1954), 2 Sendai, Miyagi Pref. (Yoshioka 1957), 3 Niimi, Okayama Pref. (Yoshioka 1957), 4 the Coast of the Inland Sea of Seto, Shikoku (Yamanaka 1957), 5 Mizuguchi, Shiga Pref. & Mt. Rokkô, Hyôgo Pref. (Suzuki & Kitagawa 1954).

the poorer soils *Picea glehni* (Sakhalin spruce) is a prominent dominant.

In most cases, the moss layer composed of mosses and lichens such as *Pleurozium*, *Ptilium*, *Cladonia* etc. develops on the forest floor in these conifer forests. Almost all species of those mosses and lichens are common to the members of the moss layer constituents of the *Pinus pumila* scrub in alpine zone.

The *Betula ermani* (Erman's birch) forest commonly spreads in the upper half of the subalpine zone, and its upper limit of zonal distribution indicates the timber line in Japan. In most cases, *Alnus crispa* ssp. *maximowiczii* is conspicuous in the upper part of the forest.

In the lowland district of Hokkaidô, the needle-leaved forest is more or less patchy with the deciduous broad-leaved forest, so the lowland forest of Hokkaidô is often treated as the mixed forest. Those districts covered with the needle-leaved forest are commonly treated as belonging to the subarctic region (e.g., Miyawaki's conception), while Tatewaki (1958) discussed that the lowland forest of Hokkaidô except that of the southern half of Oshima peninsula (south of the Kuromatsunai Depression) was not the subarctic conifer forest but the mixed forest because of the mosaic arrangements of the sub-arctic needle-leaved forest and the temperate deciduous forest.

In ultramafic rock areas of Hokkaidô, there

Fig. 11. Alpine vegetation on the serpentine plateau (ca. 1450 m alt.) in Mt. Yûbari. Dark parts are covered mainly with the *Pinus pumila* thickset, pale parts are mostly alpine grasslands and whitish parts are covered with the *Saussuretum chionophyllae*. The unstable serpentine scree slope stands on the right hand margin of the photograph. There very sparsely occurs the *Viola yubariana* community.

frequently occurs the prominent *Picea glehni* forest with *Sasa kurilensis* in the forest floor.

(d) The Alpine vegetation region contains as main communities *Vaccinio-Pinion pumilae*, dwarf scrub and heath, and windward grassland. The vegetation in the alpine zone of Japan is characterized by the *Pinus pumila* scrub with *Vaccinium vitis-idaea* var. *minus*, *Empetrum nigrum* var. *japonicum*, *Cornus canadensis* etc. in the undergrowth. This community belongs plant-sociologically to the alliance *Vaccinio-Pinion pumilae*.

Pinus pumila, north-eastern Asiatic dwarf pine (Fig. 11), occurs commonly in the alpine zone of Japan except in barren gravel, heavy windward slopes or snow bed areas. It spreads widely and the *Pinus pumila* scrub has distinct shrub, herb and moss layers. The scrub is treated plant-sociologically as the community belonging to the alliance *Vaccinio-Pinion pumilae* of north-eastern Asia. Characteristic species of the *Vaccinio-Pinion pumilae* are *Pinus pumila*, *Rhododendron aureum*, *Sorbus sambucifolia*, *Coptis trifolia*, *Streptopus streptopoides*, *Vaccinium smallii* etc.,

and some mosses and lichens such as *Hypnum plicatum*, *Lobaria linita* etc. The *Vaccinio-Pinion pumilae* is divided into the following four associations (Table 9):

(a) *Ledo-Pinetum pumilae* occurs in mesic or subhumid habitats in Hokkaidô and northern Honshû; dwarf shrubs such as *Ledum palustre* ssp. *diversipilosum*, *Vaccinium vitis-idaea* var. *minus*, *Empetrum nigrum* var. *japonicum* etc. occur commonly in the herb layer. Development of the herb and moss layer is poor.

(b) *Cetrario-Pinetum pumilae* is characterized by lichens such as *Cetraria ericetorum*, *Cladonia rangiferina* and *C. alpestris* in the moss layer. In the herb layer of this association, the coverage of ericaceous shrublets and *Empetrum* is over 80%, while in the moss layer, such xerophilous mosses as *Pleurozium schreberi*, *Drepanocladus uncinatus*, *Dicranum fuscescens* etc. and lichens are prominent. The community develops on the windward slopes and ridges in the vicinity of the summits of mountains in central Honshû and in Mt. Porojiri, Hokkaidô.

(c) *Rhododendro-Pinetum pumilae* is charac-

Table 9. Synthetic table of *Vaccinio-Pinion pumilae* (after Kobayashi 1973)
Associations: 1 *Ledo-Pinetum pumilae*, 2 *Cetrario-Pinetum pumilae*, 3 *Rhododendro-Pinetum pumilae*, 4 *Rubo-Pinetum pumilae*

Plant name	Association			
	1	2	3	4
Ledum palustre ssp. *diversipilosum*	V	·	·	r
Cetraria ericetonum	r	V	I	+
Cladonia alpestris	r	II	·	r
C. rangiferina	+	IV	r	II
Rhododendron fauriae	I	I	V	III
Rubus pedatus	I	·	I	V
Vaccinium ovalifolium	I	·	r	V
Tripetaleia bracteata	·	r	r	II
Gaultheria miqueliana	r	r	·	I
Pinus pumila	V	V	V	V
Hypnum plicatum	r	IV	III	III
Rhododendron aureum	I	III	I	III
Sorbus sambucifolia	r	I	II	II
Coptis trifolia	I	II	II	V
Empetrum nigrum var. *japonicum*	III	II	+	I
Vaccinium smallii	+	I	III	II
Streptopus streptopoides	+	r	+	III
Lobaria linita	·	+	+	+
Vaccinium vitis-idaea var. *minus*	IV	V	IV	IV
Dicranum fuscescens	II	IV	IV	V
D. majus	II	IV	IV	V
Ptilium crista-castrensis	I	I	III	IV
Hylocomium splendens	II	III	IV	IV
Cornus canadensis	II	+	I	IV
Peltigera aphthosa	r	+	r	I
Trientalis europaea	r	·	·	I
Bazzania trilobata	r	·	I	I
Barbilophozia lycopodioides	I	·	r	·
Vaccinium uliginosum	r	I	·	r
Rhytidiadelphus subpinnatus	·	·	r	I
Blepharostoma trichophyllum	·	·	r	r
Linnaea borealis	·	r	r	r

terized by *Rhododendron fauriae* in the shrub layer. In the Japan Sea side district, its shrub layer sometimes includes the components of the *Fagus* forest such as *Ilex*, *Acer*, *Hydrangea* and *Sasa* spp., while in the Pacific side district those shrubs are lacking. A few species of mosses such as *Hylocomium splendens*, *Pleurozium schreberi*, *Dicranum* spp. occur in the moss layer. This association is not common in Hokkaidô.

(d) *Rubo-Pinetum pumilae* is characterized by *Rubus pedatus*, *Vaccinium ovalifolium*, *Tripetaleia bracteata* and *Gaultheria miqueliana*, having rather rich floristic constituents in the herb and moss layers. *Coptis trifolia*, *Schizocodon soldanelloides*, *Solidago virgaaurea* ssp. *leiocarpa*, *Deschampsia flexuosa* etc. are conspicuous in the herb layer. Many species of liverworts and foliose

lichens are found in the moss layer. This association develops mainly in the lower half of the alpine zone.

Dwarf scrub and heath spread on the poorer habitat that is too poor for *Pinus pumila* scrub. The community is mainly on arid sites and is classified as *Arcterico-Loiseleurietum*. This association is characterized by dwarf shrubs such as *Arcterica nana*, *Loiseleuria procumbens*, *Arctous alpina*, *Diapensia lapponica* ssp. *obovata* etc. and lichens and mosses such as *Thamnolia vermicularis*, *Cetraria ericetonum*, *Rhacomitrium hypnoides* etc. It develops on the stable slopes or the area along the margin of the *Pinus pumila* scrub.

Alpine windward grassland is represented by the *Oxytropis-Kobresia* community (*Kobresio-Oxytropidetum japonicae*). It is characterized by

Table 10. Floristic composition of alpine communities (belonging to the *Dicentro-Violetum crassae*) in Taisetsu Mountains, Hokkaidô (after Ito & Nishikawa 1977)

Species	Subassociation			
	a	b	c	d
Dicentra peregrina var. *pusilla*	V	V	I	II
Viola crassa	I	I	V	IV
Saussurea yanagisawae	II	I	III	I
Pentstemon frutescens	·	V	I	I
Pleuropteropyrum ajanense	·	I	IV	I
Stellaria pterosperma	·	·	I	·
Cardamine nipponica	·	·	I	·
Lagotis yesoensis	·	·	·	IV
Artemisia trifurcata var. *pedunculosa*	·	·	·	II
Saxifraga laciniata	·	·	·	III
Oxytropis yezoensis	·	·	·	I
Bistorta vivipara	·	·	·	I
Pedicularis oederi	·	·	·	I
Saxifraga merkii	·	·	·	I
Potentilla matsumurae	·	·	I	I
P. miyabei	·	I	·	·
Carex pyrenaica	·	·	·	I
C. stenantha var. *taisetsuensis*	I	I	II	I
Festuca ovina var. *alpina*	·	·	I	II
Patrinia sibirica	·	I	II	I
Minuartia arctica	·	·	I	I
Deschampsia flexuosa	·	I	·	·
Luzula wahlenbergii	·	·	·	I

Subassociations:
a Typical subassociation
b *Pentstemonetum frutescentis*
c *Pleuropteropyretosum*
d *Lagotidetosum yesoensis*

such hemicryptophytic herbs and shrubs as *Oxytropis japonica*, *Minuartia* spp., *Pedicularis verticillata*, *P. apodochila*, *Kobresia myosuroides*, *Bryanthus gmelini* etc., on the severe windward habitats. Sometimes, *Dryas octopetala* var. *asiatica* occurs in this community.

On unstable soil the *Hedysarum vicioides* community is found and characterized by such hemicryptophytic plants as *Hedysarum vicioides*, *Sedum rosea*, *Tilingia tachiroei*, *Leontopodium fauriei*, *Saxifraga cherlerioides*, *Thymus quinquecostatus* etc.

Alpine frigorideserta in Japan are represented by a very sparse community consisting of *Dicentra peregrina*, *Viola crassa*, *Pleuroptero-pyrum* spp., *Lagotis* spp. etc. Floristic composition of the community is more or less different due to its locality, however, it is plant-sociologically regarded as the association *Dicentro-Violetum crassae*. Table 10 shows the floristic composition of the *Dicen-tro-Violetum crassae* in Taisetsu Mts. of central Hokkaidô (Ito & Nishikawa 1977).

In the gravel adjacent to snow beds, the *Anaphalio-Phyllodocetum aleuticae* is often encountered. It is characterized by *Phyllodoce aleutica*, *Anaphalis alpicola*, *Lycopodium sitchense* var. *nikoense*, *Sieversia pentapetala* etc. In more humid stands, the *Faurio-Caricetum blepharicarpae* is a typical snow bed community. It is characterized by *Fauria cristagalli*, *Carex blepharicarpa*, *Sieversia pentapetala*, *Primula* spp. (*P. cuneifolia*, *P. nipponica* etc.), etc., and in most cases, one of the former three species is dominant.

Serpentine vegetation in the alpine zone is very sparse and belongs to frigorideserta. The characteristic feature of its floristic composition is the occurrence of many species of serpentine plants. In most cases those serpentine plants are dominant or characteristic species of the communities.

(e) *Herbaceous vegetation*. Natural herba-

293

ceous vegetation is roughly classified into six classes, viz., *Oxycocco-Sphagnetea* in high moor, *Phragmitetea* in low moor, *Potamogetonetea* in lakes or ponds, *Asteretea tripolii* in salt marshes, *Glehnietea littoralis* on coastal dunes and *Circio-Campanulatum hondoensis* on volcanic areas. These associations cover limited areas in which trees hardly grow because of climatic, edaphic, or topographical conditions.

Sometimes, on the landslips in the ultramafic rock areas, there appear sparse herbaceous pioneer communities. As secondary herbaceous communities, the following are common; e.g., the *Sasa* community and the *Miscanthus-Arundinaria* community as windward grassland, the *Epilobium angustifolium* community and the *Erechtites hieracifolia* community in clear-cut and fire-ravaged areas. The *Poa pratensis* community and the *Zoisia* community as artificial meadows, the fieldweed community and the paddy-field community in farming area are also examples of pioneer communities some resembling the serpentine vegetation communities.

Serpentine vegetation in Japan

(a) The Pinus densiflora community in serpentine areas
Within the *Camellietea japonicae* region (the evergreen broad-leaved forest zone), the serpentine vegetation is mainly characterized by the *Pinus densiflora* community. This community is generally an open stand of *Pinus densiflora* trees. In the shrub layer of this community *Spiraea* spp., *Rhododendron* spp., *Abelia* spp., *Zabelia integrifolia*, *Corylopsis* spp. etc. are conspicuous, and in the herb layer, *Smilax china*, *Pertya glabrescens* etc. show high cover values. Sometimes, *Quercus phylliraeoides*, one of the character species of the sea-coast shrub, occurs in the shrub layer. Among the above mentioned plants, *Spiraea nipponica* var. *ogawae*, *S. blumei*, *Corylopsis spicata*, *Rhododendron sanctum* etc. are typical serpentine plants and they are hardly encountered in non-serpentine areas.

Table 11 shows a comprehensive overview of the floristic composition of the community published by Yamanaka (1979).

In comparison with the floristic composition of the community occurring in non-serpentine areas

shown in Table 8, the characteristics of the community of serpentine area seem to be quite clear.

In general, the *Pinus densiflora* community on the serpentine lacks deciduous trees such as *Castanea crenata* and *Quercus serrata*. On the other hand, such xerophilous shrubs as *Abelia* spp., *Zabelia integrifolia*, *Spiraea* spp., *Corylopsis* spp., etc. show considerably high presences in the shrub layer. Among those shrubs, *Spiraea nipponica* var. *ogawae* and *Corylopsis spicata* are typical serpentine plants and are rarely encountered in the non-serpentine areas.

(b) Other types of needle-leaved forests in the evergreen broad-leaved forest zone
In the ultramafic rock areas in Shikoku, there sometimes occurs needle-leaved forest dominated by *Chamaecyparis* (Hinoki-cypress), *Thuja* (Japanese arbor-vitae) and *Thujopsis* (Hiba arbor-vitae) accompanied by *Tsuga* (Japanese hemlock) and *Pinus* species. The shrub layers of these forests are often dominated by *Rhododendron* species. Typically these forests are found on Mt. Higashi-akaishi of Ehime Pref. and on Mt. Shirakami of Kôchi Pref. Similar communities occur on Mt. Shibutsu of Gunma Pref. and on Mt. Hayachine of Iwate Pref., Honshû.

The main features of these needle-leaved forests are described by Sasaki (1973) and include
a) The *Chamaecyparis obtusa-Thuja standishii-Pinus pentaphylla* community (Mt. Higashi-akaishi, Ehime Pref., Shikoku).
b) The *Chamaecyparis obtusa-Tsuga diversifolia-Pinus pentaphylla* community (Mt. Shirakami, Kôchi Pref., Shikoku)
c) The *Tsuga diversifolia-Thuja standishii-Pinus pentaphylla* community (Mt. Shibutsu, Gunma Pref. central Honshû)
d) The *Thujopsis dolabrata* var. *hondai-Tsuga diversifolia* community (Mt. Hayachine, Iwate Pref., northern Honshû)
They contain several typical serpentine plant species.

(c) The Picea glehni forest
In ultramafic rock areas of Hokkaidô, there sometimes occurs the forest dominated by *Picea glehni* (Sakhalin spruce) as the characteristic forest. In Japan, exclusive of Hokkaidô, the forest including *Picea glehni* develops only on Mt.

Table 11. Synthetic table of the *Pinus densiflora* community occurring in the ultramafic rock areas in Japan (after Yamanaka 1979)

Plant name	\[Locality\] 1	2	3	4	5	6	7	8	9	10	11	12
Pinus densiflora	V	V	4	V	3	V	V	III	V	V	V	V
P. thunbergii	(+)	.	.	.	.	.	II	IV	III	.	.	.
Fraxinus sieboldiana	.	.	4	V	3	IV	IV	V	V	V	V	V
Quercus phillyraeoides	.	.	.	.	.	.	IV	IV	I	.	.	.
Juniperus rigida	.	.	.	.	2	.	V	I	IV	.	V	I
Spiraea nipponica var. *ogawae*	.	.	.	.	.	.	V	.	.	.	.	.
S. nervosa (incl. var. *angustifolia*)	.	V	.	.	.	.	.	.	.	.	.	.
S. blumei (incl. var. *hayatae*)	.	.	.	.	3	V	.	.	.	.	.	.
S. japonica	.	.	.	.	3	.	.	.	IV	.	II	.
Rhododendron decandrum	V	.	.	.	.	.	.	.	.	.	.	.
R. mayebarae	.	.	.	V	.	V	.	V	.	V	V	V
R. reticulatum	.	.	.	.	.	.	.	.	II	.	.	.
R. sanctum	.	II	.	.	.	.	.	V	.	.	IV	II
R. weyrichii	.	.	3	.	.	(+)	I	III	(+)	.	.	.
R. macrosepalum	.	.	.	.	1	.	V	.	I	.	.	.
R. obtusum var. *kaempferi*	.	.	.	.	.	V	.	I	III	.	IV	.
Enkianthus perulatus var. *japonicus*	.	.	.	V	.	.	.	I	III	I	.	.
Abelia spathulata	.	V	4	V	3	.	V	V	IV	V	V	.
A. serrata	.	.	.	.	1	.	.	.	.	.	.	.
Zabelia integrifolia	.	.	.	.	1	.	.	.	.	.	.	.
Corylopsis pauciflora	.	.	.	.	.	.	.	.	.	.	V	V
C. gotoana	.	.	.	II	(+)	.	.	(+)	(+)	.	.	.
C. spicata	.	I	4	.	.	III	.	II	(+)	.	.	.
Buxus microphylla var. *japonica*	.	V	4	V	2	V	V	V	V	V	V	IV
Smilax china	.	V	3	V	3	V	V	V	V	V	V	.
Pertya glabrescens	.	V	3	V	2	V	V	.	V	V	V	III
Miscanthus sinensis	V	.	.	.	.	.	.	.	.	.	.	.

Locality: (1) Taura, Kumamoto Pref., Kyūshū; (2) Ichinomiya, Kōchi Pref., Shikoku; (3) Tosayama, Kōchi Pref., Shikoku; (4) Hidaka, Kōchi Pref., Shikoku; (5) Kamiyama, Tokushima Pref., Shikoku; (6) Kizawa, Tokushima Pref., Shikoku; (7) Kurosawayama, Wakayama Pref., Honshū; (8) Asamayama, Mie Pref., Honshū; (9) Shinshiro, Aichi Pref., Honshū; (10) Shimada, Shizuoka Pref., Honshū; (11) Sekunomiya, Hyōgo Pref., Honshū; (12) Ooeyama, Kyōto Pref., Honshū.

295

Table 12. Relation between Forest Types and Habitats of the Picea glehni forest in Hokkaidô (after Tatewaki 1943, simplified)

	Swamp or bog	**Serpentine**	Volcanic gravel	Sand dune	Gravel & scree	Fire-ravaged area
Sphagnum Type	+	−	−	−	−	−
Osmunda Type	+	−	−	−	−	−
Ledum Type	+	−	+	−	−	−
Moss Type	−	−	−	+	+	−
Carex Type	−	−	−	−	+	−
Menziesia Type	+	−	−	−	−	−
Rhododendron Type	−	−	−	−	+	−
Sasa Type	−	+	+	−	+	+

Hayachine, Iwate Pref., northern Honshû. Mt. Hayachine consists of serpentinite, and is famous for its endemic alpine plants.

Tatewaki (1943) studied the *Picea glehni* forests in Hokkaidô, and classified the forest into the following eight Types.
1. *Sphagnum* Type,
2. *Osmunda* Type,
3. *Ledum* Type,
4. Moss Type,
5. *Carex* Type,
6. *Menziesia* Type,
7. *Rhododendron* Type, and
8. *Sasa* Type.

These Forest Types correspond to the following Associations respectively:
(1) *Picea glehni-Sphagnum*-Association, (2) *Picea glehni-Osmunda cinnamomea*-Association, (3) *Picea glehni-Ledum palustre*-Association, (4) *Picea glehni-Moss*-Association, (5) *Picea glehni-Carex sachalinensis*-Association, (6) *Picea glehni-Menziesia pentandra*-Association, (7) *Picea glehni-Rhododendron fauriae*-Association and (8) *Picea glehni-Sasa*-Association.

The *Picea glehni* forest in Hokkaidô stands on the following six edaphically different habitats: swamp or bog, serpentine, volcanic gravel, sand dune, gravel and scree, and fire-ravaged areas. Relations between Forest Types and habitats are shown in Table 12.

In a recent paper Nakata and Kojima (1987) report on the *Picea glehni* forest type indicating that in comparison with other serpentine areas of Japan and elsewhere, northern Hokkaido is unique in that this forest type is represented by a relatively closed forest of fair-sized trees with a podzolic soil.

As shown in Table 12, the Forest Type of the *Picea* forest standing on serpentine soil is recognized as *Sasa* Type (the *Picea glehni-Sasa*-Association). Table 13 is a list of presence (P) of each plant occurring in the typical *Sasa* Type forest and Fig. 12 is a profile of the *Picea* forest at Teshio University Experiment Forest (serpentine area), reported by Tatewaki (1943).

(d) The open needle-leaved forest characterized by Pinus pentaphylla var. laevis on Mt. Apio, Southern Hidaka, Hokkaidô
Mt. Apoi (811 m. in alt.) consists of peridotite, and on the slopes between 400 m and 600 m, we find the open needle-leaved forest characterized by *Pinus pentaphylla* var. *laevis* with *Miscanthus* in the herb layer (Fig. 13). This is a unique community, being peculiar to Mt. Apoi. The *Miscanthus* herb layer includes an endemic species, *Cirsium apoense*. A plant-sociological analysis of the forest is given by Ohba (1968).

(e) Alpine vegetation
In general, the alpine vegetation of Japan is recognized as a relic vegetation of the last glacial period. Remnants of the vegetation of the past glacial period have been preserved to this date as isolated patches primarily on high mountains. The alpine vegetation of Japan has many species common to the arctic vegetation. Each geologically ancient mountain generally has several endemics and disjunctively distributed plants. The serpentine areas of alpine zones therefore had advantages to preserve relic plants both from a climatic and also a geochemical perspective. Some endemic plants are 'serpentinomorphosed' ones, while some plants showing limited occurrences are preserved disjunctively in the alpine serpentine areas. The distribution of the serpen-

Table 13. Presence (P) of each plant occurring in the *Picea glehni* forest at Teshio University Experiment Forest (after Tatewaki 1943)

Life form	P	Plant name
Tree	5	*Picea glehni*
	3	*Acanthopanax sciadophylloides, Sorbus commixta, Alnus hirsuta, Betula platyphylla* var. *japonica*
	2	*Quercus mongolica* var. *grosseserrata, Betula ermani, Abies sachalinensis*
	1	*Tilia japonica, A. mono* var. *glabrum, A. japonicum, Maackia amurensis* var. *buergeri, Magnolia obovata, Betula maximowicziana, Picea jezoensis, Taxus cuspidata*
Small tree & shrub	5	*Skimmia japonica* var. *intermedia* farma. *repens, Sasa kurilensis*
	3	*Viburnum furcatum, Menziesia pentandra, Vaccinium smallii, Vacc. ovalifolium* var. *coriaceum, Ilex sugeroki* var. *brevipedunculata, I. crenata* ssp. *radicans, Rhus trichocarpa*
	2	*Eubotryoides grayana, Vaccinium hirtum, Ilex rugosa, Daphniphyllum macropodum* ssp. *humile, Hydrangea paniculata*
	1	*Ledum palustre* ssp. *diversipilosum, Vaccinium praestans, Acer tschonoskii, Rosa acicularis, Corylus sieboldiana* var. *brevirostris*
Woody lianas	2	*Rhus ambigua*
	1	*Hydrangea petiolaris, Schizophragma hydrangeoides*
Herb	3	*Carex blepharicarpa, Maianthemum dilatatum*
	2	*Cirsium kamtschaticum, Tripterospermum japonicum, Cornus canadensis, Anemone yezoensis, Lycopodium serratum*
	1	*Solidago virgaaurea* ssp. *leiocarpa, Galium boreale, Pterygocalyx volubilis, Trientalis europaea, Angelica anomala, Conioselinum filicinum, Euphorbia sieboldiana* var. *montana, Thalictrum minus, Asiasarum heterotropoides, Ephippianthus schmidtti, Orchis aristata, Platanthera tipuloides, Hosta rectifolia, Heloniopsis orientalis, Lysichiton camtschatcense, Symplocarpus renifolius, Carex sachalinensis, Osmunda cinnamomea* var. *fokiensis, Pteridium aquilinum* var. *latiusculum, Lycopodium obscurum* farma *flabellatum* & forma *strictum*

tine areas of alpine zones in Japan is shown in Fig. 14.

Many floristic research reports were published earlier on the serpentine areas of Japan e.g., by Tatewaki (1928), Nakai (1930), Honda & Takenaka (1930), Kitamura (1952–1957), Kikuchi & Komizunai (1961), Toyokuni (1955–1960), Hara & Mizushima (1954), Nosaka (1960–1962, 1974) etc., and although scattered throughout the literature and floristic characteristics have been clarified to considerable extent. The vegetational ecology, especially plant-sociological classifications began more recently, and Ohba's publication titled 'Uber die Serpentine-Pflanzengesellschaften der alpinen Stufe Japans' was published in 1968.

Ohba (1968) treated the serpentine vegetation of alpine zone in Japan in three categories: alpine frigorideserta, alpine windward grassland and al-

pine-subalpine snow bed grassland. The vegetation of the first category was recognized as belonging to an order *Minuartetalia vernae japonicae*, that of the second category was recognized as belonging to an order *Caricetalia tenuiformi*, and the last, as the *Japonolirion osense* community.

(*a*) *Minuartetalia vernae japonicae*. Vegetations belonging to this order are most frequently recognized in Japanese mountains. Table 14 is the synthetic table of Minuartetalia vernae japonicae (Ohba 1968). Two alliances, *Drabo- Arenarion katoanae* and *Cerasteo-Minuartion vernae japonicae* are distinguished in this order.

The *Drabo-Arenarion katoanae* consists of four associations, those are *Violetum yubarianae* on Mt. Yûbari, *Saussuretum chionophyllae* on Mt. Yûbari and Mt. Tottabetsu, (Fig. 15), *Sanguisor-*

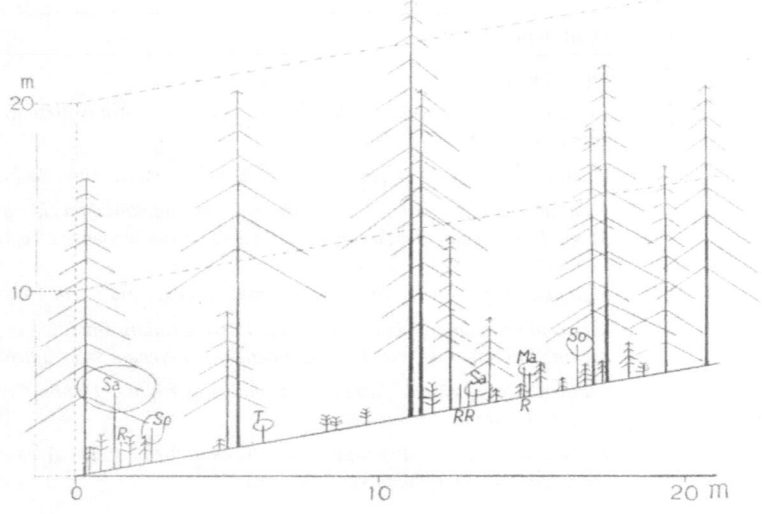

㣺:Picea glehni   ⋎:Abies sachalinensis
T:Taxus cuspidata   Sa:Sorbus alnifolia
So:Sorbus commixta   Ma:Magnolia obovata
R :Rhus trichocarpa

Fig. 12. A schematic profile of the *Picea glehni* forest at Teshio University Experiment Forest (serpentine area) (after Tatewaki 1943).

Fig. 13. The *Pinus pentaphylla* v. *laevis* forest on a peridotite slope on Mt. Apoi, Hokkaido.

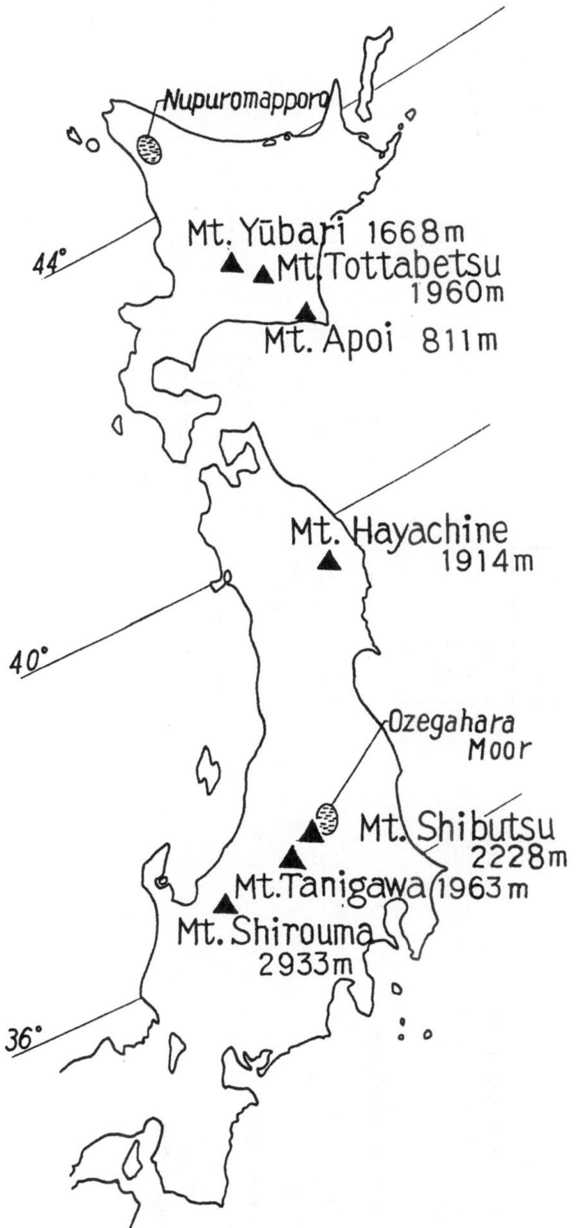

Fig. 14. Distribution of the serpentine areas of alpine zone in Japan.

bo-Minuartetum vernae japonicae on Mt. Hayachine and *Leontopodetum fauriei angustifolii* on Mt. Tanigawa and Mt. Shibutsu.

Violetum yubarianae occurs on the serpentine landslips of Mt. Yûbari, being characterized by an endemic and typical serpentine plant *Viola yubariana*.

Saussuretum chionophyllae was reported from Mt. Yûbari and Mt. Tottabetsu. It contains some 10 species, including one typical serpentine plant, *Saussurea chionophylla*, as the characteristic species, and covering 15% on average. Table 15 gives the results from the belt-transect in this community reported by Nosaka (1974).

On Mt. Tottabetsu, the vegetation is found mainly on the windward serpentine gravel slope of about 20 degrees. It consists of some 10 species, and there are no endemics. Total cover is about 10 percent.

On Mt. Apoi there is a similar community with some endemic taxa (Fig. 16a, b).

The *Sanguisorbo-Minuartetum vernae japonicae* is from Mt. Hayachine. This association contains 10–15 species, including endemics such as *Sanguisorba obtusa*, *Aruncus dioicus* v. *astilboides* and *Polygonum hayachinense*. Cover of the vegetation is 2–20%.

The *Leontopodetum fauriei angustifolii* was reported from steep slopes on Mt. Tanigawa and Mt. Shibutsu (Ohba, 1968). The association is characterized by two endemic herbs, *Leontopodium fauriei* var. *angustifolium* and *Erigeron thunbergi* var. *heterotrichus*.

The *Cerastio-Minuartion vernae japonicae* consists of one association, the *Cerastio-Minuartetum vernae japonicae* reported from Mt. Shirouma (Ohba 1968). It is found on the windward slopes at 5–40 degrees, from altitudes of 2000 to 2800 m.

(*b*) *Caricetalia tenuiformi in the windward grassland.* A special association belonging to the alliance *Bupleurumo-Patrinion sibiricae* was reported from the windward grassland of Mt. Apoi, on slopes in altitudes from 300 m to 800 m. This stable association shows considerably high cover values (>50%), and consists of some 30–40 species, many endemic to Mt. Apoi.

At lower altitudes the community includes a considerable number of species of lowland grassland shows rather high cover values. At higher altitudes, called 'Umanose', the unique serpentine relic plant *Callianthemum miyabeanum*, also an endemic to Mt. Apoi, occurs.

(*c*) *The Japonolirion osense community.* On Mt. Tanigawa and Mt. Shibutsu in central Honshû, as well as in the Nupromapporo, Teshio district

299

Table 14. Synthetic table of *Minuartetalia vernae japonicae* (Ohba 1968, 1972)

	A								B		
	a	b 1	b 2	c 1	c 2	d 1	d 2	e 1	e 2	1	2	3
Differential spp. of *Drabo-Arenarion katoanae*:												
Arenaria katoana	3	I	1	·	·	·	·	·	·	·	·	IV
Draba japonica	·	III	·	·	·	·	·	·	·	I	·	·
Differential sp. of *Violetum yubarianae*:												
Viola yubariana	3	·	·	·	·	·	·	·	·	·	·	·
Differential sp. of *Saussuretum chionophyllae*:												
Saussurea chionophylla	·	IV	4	·	·	·	·	·	·	·	·	·
Differential spp. of *Lagotis glauca* var. *takedana*-subassociation:												
Lagotis glauca var. *takedana*	·	V	·	·	·	·	·	·	·	·	·	·
Saxifraga laciniata	2	V	·	·	·	·	·	·	·	·	·	·
Differential spp. of *Campanula chamissonis*-subassociation:												
Stellaria nipponica var. *yezoensis*	·	·	3	IV	IV	·	·	·	·	·	·	·
Carex stenantha var. *taisetsuensis*	·	·	3	IV	III	·	·	·	·	·	·	·
Campanula chamissonis	·	·	3	IV	I	·	·	II	·	I	·	·
Festuca ovina var. *alpina*	·	·	3	II	V	·	·	II	·	·	I	IV
Differential spp. of *Sanguisorbo-Minuartetum vernae japonicae*:												
Sanguisorba obtusa	·	·	·	IV	IV	·	·	·	·	·	·	·
Aruncus dioicus var. *astilboides*	·	·	·	IV	III	·	·	·	·	·	·	·
Polygonum hayachinense	·	·	·	IV	I	·	·	·	·	·	·	·
Calamagrostis deschampsoides var. *hayachinensis*	·	·	·	IV	II	·	·	·	·	·	·	·
Differential spp. of *Leontopodium hayachinense*-subassociation:												
Leontopodium hayachinense	·	·	·	V	·	·	·	·	·	·	·	·
Primula macrocarpa	·	·	·	IV	·	·	·	·	·	·	·	·
Aletris foliata	·	·	·	III	·	·	·	·	·	·	·	·
Differential spp. of *Calamagrostis sachalinensis*-subassociation:												
Calamagrostis sachalinensis	·	·	·	·	V	·	·	·	·	·	·	·
Carex doenitzii	·	·	·	·	IV	·	·	·	·	·	·	·
Anaphalis margaritacea var. *angustior*	·	·	·	·	III	·	·	·	·	·	·	·
Differential spp. of *Leontopodeum fauriei angustifolii*:												
Leontopodium fauriei var. *angustifolium*	·	·	·	·	·	V	V	·	·	·	·	·
Erigeron thunbergii var. *heterotrichus*	·	·	·	·	·	·	I	·	·	·	·	·
Differential spp. of *Potentilla togashii*-subassociation:												
Adenophora nikoensis var. *stenophylla*	·	·	·	·	·	V	·	V	·	·	·	·
Potentilla togashii	·	·	·	·	·	·	·	I	·	·	·	·
Euphrasia insignis	·	·	·	·	·	IV	·	·	·	·	·	·
Ixeris dentata var. *kimurana*	·	·	·	·	·	V	·	·	·	·	·	·
Angelica actiloba var. *iwatensis*	·	·	·	·	·	IV	·	·	·	·	·	·
Seseli coreana	·	·	·	·	·	IV	·	·	·	·	·	·
Agrostis borealis	·	·	·	·	·	III	·	·	·	·	·	·
Thalictrum minus var. *stipellatum*	·	·	·	·	·	III	·	·	·	·	·	I

Table 14. (Continued)

	A					d		B		
	a	b 1	b 2	c 1	c 2	d 1	d 2	e 1	e 2	e 3
Differential spp. of Festuca rubra var. hondoensis-subassociation:										
Festuca rubra var. hondoensis	.	.	.	.	.	.	V	.	.	.
Pedicularis verticillata	.	.	.	.	.	.	III	.	.	I
Differential spp. of Cerasteo-Minuartetum vernae japonicae:										
Cerastium schizopetalum var. bifidum	.	.	.	.	.	.	.	V	IV	V
Arabis serrata var. grandiflora	.	.	.	.	.	.	.	V	IV	IV
Dianthus superbus var. amoenus	.	.	.	.	.	.	.	.	V	II
Festuca ovina var. tateyamensis	.	.	.	.	.	.	.	.	I	I
Minuartia hondoensis	.	.	.	.	.	.	.	V	III	III
Differential spp. of Erigeron thunbergii var. glabratus-subassociation:										
Erigeron thunbergii var. glabratus	.	.	.	.	.	.	.	.	V	r
Festuca rubra	.	.	.	.	.	.	.	.	V	r
Deschampsia caespitosa var. festucaefolia	.	.	.	.	.	.	.	.	IV	.
Differential spp. of Eritrichium nipponicum-subassociation:										
Eritrichium nipponicum	.	.	.	.	.	.	.	.	.	V
Tilingia tachiroei	.	.	.	.	.	.	.	.	.	V
Viola crassa	.	.	.	.	.	.	.	.	.	IV
Differential spp. of Minuaretalia vernae japonicae:										
Minuartia verna var. japonica	2	IV	4	V	IV	.	IV	V	IV	III
Carex capillaris	.	.	1	.	.	.	.	.	II	I
Companion spp.										
Dianthus superbus var. speciosus	.	.	1	II	III	II	.	.	IV	III
Thymus quinquecostatus	.	.	.	I	.	V	IV	.	.	IV
Parnassia palustris var. tenuis	.	.	.	II	.	.	III	.	I	r
Viola biflora	.	.	.	V	II	.	III	.	.	r

A *Drabo-Arenarion katoanae*
a) *Violetum yubarianae* (Mt. Yūbari)
b) *Saussuretum chionophyllae* (Mt. Yūbari & Mt. Tottabetsu)
  1 *Lagotis glauca* var. *takedana*-subassociation (Mt. Yūbari)
  2 *Campanula chamissonis*-subassociation (Mt. Tottabetsu)
c) *Sanguisorbo-Minuartetum vernae japonicae* (Mt. Hayachine)
  1 *Leontopodium hayachinense*-subassociation
  2 *Calamagrostis sachalinensis*-subassociation
d) *Leontopodium fauriei angustifolii* (Mt. Tanigawa & Mt. Shibutsu)
  1 *Potentilla togashii*-subassociation
  2 *Festuca rubra* var. *hondoensis*-subassociation

B *Cerasteo-Minuartion vernae japonicae*
e) *Cerasteo-Minuartetum vernae japonicae* (Mt. Shirouma)
  1 Typical subassociation
  2 *Erigeron thunbergii* subsp. *glabratus*-subassociation
  3 *Eritrichium nipponicum*-subassociation

Fig. 15a. Summit and ridge of Mt. Tottabetsu (built of peridotite), Hokkaido. The scree slope is covered with very sparse alpine herbaceous vegetation.

of northern Hokkaidô, there occurs an endemic herb *Japonolirion osense*, a member of the Liliaceae. *Japonolirion* is monotypic and it is suggested that the related genus is probably *Helonias* of Eastern North America. This plant characterizes serpentinic alpine-subalpine snowbed grassland on these mountains. The habitats of the community of Mt. Tanigawa and Mt. Shibutsu are mostly humid grassland especially near the snow beds. Many constituents of snow bed vegetation such as *Carex blepharicarpa*, *Heloniopsis orientalis*, *Anemone narcissiflora*, *Drosera rotundifolia*, *Viola biflora*, *Andromeda polifolia*, *Primula cuneifolia* v. *hakusanensis*, *Fauria crista-galli* etc. commonly occur. (Ohba 1968).

The community of Nupuromapporo, Teshio district of Hokkaidô, includes one more striking serpentinicolous relic plant. This is a white-flowered primrose, *Primula takedana*, endemic to the district.

Serpentine plants in Japan – a summary

Characteristic-serpentine (serpentinicolous) plants in Rune's sense (Rune 1953), or 'typical serpentinophytes' and 'serpentinicolous relics' in Japan are summarized in the following text. In Japan 37 'typical serpentinophytes' from 15 families are described, with the Compositae (9 species), the Rosaceae (5 species), and Violaceae (4 species) accounting for more than one-half the total species. The 37 species are listed 1–37 in the following paragraph.

Typical serpentinophytes

1. *Aconitum apoense* Nakai, closely related to *A. yuparense* in alpine region of Hokkaidô, having glabrous carpels. Endemic on Mt. Apoi and its neighborhood. Stenophyllism and glabrism. (Ranunculaceae)

2. *Aconitum ito-seiyanum* Miyabe et Tatewaki related to *A. sachalinense* in Hokkaidô and Sakhalin, having trisected and narrowly cleft leaves. Endemic on Nupuromapporo etc., N. Hokkaidô. Dissected stenophyllism. (Ranunculaceae)

3. *Thalictrum foetidum* L. var. *apoiense* Shimizu, a variety of Euro-Siberian *T. foetidum*, having smaller and glabrous habit. Endemic on Mt. Apoi and its neighborhood, S. Hokkaidô. Nanism and glabrism. (Ranunculaceae)

4. *Epimedium grandiflorum* Morren subsp. *koreanum* (Nakai) Kitamura var. *coelestre* (Nakai) Shimizu, a variety, with smaller size and entire leaves, of the subspecies in Korea and C. Honshû. Endemic on Mt. Tanigawa and Mt. Shibutsu, C. Honshû. Nanism and glabrism. (Berberidaceae)

5. *Corylopsis spicata* Siebold et Zuccarini, a deciduous shrub closely related to *C. pauciflora* of S. Honshû, Shikoku and Kyûshû, having rather thickened leaves. Endemic on serpentine area of Kôchi Pref., Shikoku. Crassism. (Hamamelidaceae).

6. *Aruncus dioicus* (Walter) Fernald:
6-1. *A. dioicus* (Walter) Fernald var. *astilboides*

Fig. 15b. *Minuartia verna* v. *japonica* on serpentine gravel on Mt. Yûbari, Hokkaido.

Table 15. Cover degree and frequency of the plants appearing in the belt-transect (1 × 12 m, 12 Q in directions from west to east) on the serpentine scree of Mt. Yûbari (Nosaka 1974).

Plant name	Quadrat number												Average Cover Degree	F (%)
	1	2	3	4	5	6	7	8	9	10	11	12		
*Saussurea chionophylla**	2	+	1′	1	+	1′	2	1	+	1	1′	1	0.59	100.0
Lagotis glauca var. *takedama**	+	1′	1′	1	1′	–	+	1	1	+	1′	+	0.33	91.7
Allium schoenoprasum var. *shibutsuense**	+	+	–	–	+	–	+	+	–	+	+	–	0.02	58.3
Potentilla matsumurae var. *yuparensis**	+	–	–	+	+	–	1	+	+	–	–	–	0.10	50.0
Pleuropteropyrum ajanense	+	+	+	–	+	+	–	+	–	–	–	–	0.02	50.0
Carex melanocarpa	–	–	+	+	+	+	–	–	+	–	+	–	0.02	50.0
Saxifraga laciniata	+	+	+	–	–	+	–	–	–	–	–	–	0.01	33.3
Minuartia verna var. *japonica*	–	1	+	–	–	–	+	–	+	–	–	–	0.09	33.3
Empetrum nigrum var. *japonicum*	–	–	+	–	2	1′	–	–	–	–	–	1	0.37	33.3
Carex capillaris	–	–	–	–	+	+	–	–	–	+	+	–	0.01	33.3
Patrinia sibirica	–	–	–	–	–	–	+	–	–	–	–	+	0.007	16.7
Erigeron thunbergii ssp. *glabratus*	–	–	–	–	–	–	–	–	–	–	–	+	0.003	8.3
Vaccinium vitis-idaea var. *minus*	–	–	–	–	–	–	–	–	–	–	–	+	0.003	8.3
Minuartia arctica	–	–	–	–	–	–	–	–	–	+	–	–	0.003	8.3
Stellaria nipponica var. *yezoensis*	+	–	–	–	–	–	–	–	–	–	–	–	0.003	8.3
Tofieldia coccinea var. *fusca*	–	–	–	–	–	+	–	–	–	–	–	–	0.003	8.3
Festuca ovina var. *alpina*	–	–	–	–	–	+	–	–	–	–	–	–	0.003	8.3
Pinus pumila	–	–	–	–	–	–	+	–	–	–	–	–	0.003	8.3

In the 7th Quadrat, a few individuals of *Saussurea chionophylla* f. *albiflora* were found.
*Marked plants are typical serpentine plants.
Alt. 1430 m, exposure – level.

303

Fig. 16a. Serpentine scree on Mt. Bôzu (300–500 m alt.), near Hobetsu Town, Hokkaido. Surroundings are covered with the deciduous *Quercus-Acer* forest. In the center of the scree a small patch of *Picea glehni*.

Fig. 16b. Serpentine scree on Mt. Bôzu (300–500 m alt.) near Hobetsu Town, Hokkaidô. On the scree slopes occur unstable sparse plant communities characterized by *Gymnocarpium jessoense*, *Arenaria katoana*, *Angelica stenoloba*, *Hypericum tatewakii* v. *nigropunctatum*, *Veronica schmidtiana* v. *yezoalpina*, *Carex humilis* v. *nana* etc.

(Maxim.) Hara, a variety of circumpolar *A. dioicus*, having 3–5 carpels and pointed fruits. Endemic variety on Mt. Hayachine, N. Honshû.

6-2. *A dioicus* (Walter) Fernald var. *subrotundatus* (Tatewaki) Hara, the other variety of *A. dioicus*, having thickened and orbicular leaves. Endemic on Mt. Apoi, S. Hokkaidô. (Rosaceae)

7. *Potentilla matsumurae* Wolf var. *yuparensis* Kudo, a variety of *P. matsumurae* in alpine zone of Japan, having narrowly and deeply dissected and somewhat lustrous, glabrous leaves. Endemic on Mt. Apoi and Mt. Yûbari etc., Hokkaidô. Dissected stenophyllism. (Rosaceae)

8. *Sanguisorba obtusa* Makino, closely related to *S. hakusanensis* of C. Honshû, but differs from it by having four stamens in a flower and brownish frizzly haired axis of leaves. Endemic on Mt. Hayachine, N. Honshû. (Rosaceae)

9. *Spiraea blumei* G. Don var. *pubescens* (Koidzumi) Ohwi, a pubescent variety of Asiatic *S. blumei*. Endemic on Mt. Higashi-akaishi, Shikoku. (Rosaceae)

10. *Spiraea nipponica* Maximowicz var. *ogawae* (Nakai) Yamanaka, a narrow-leaved variety of Japanese *S. nipponica*. This variety bears narrow oblong to oblong leaves. Endemic on serpentine areas of Wakayama Pref., S. Honshû. Stenophyllism. (Rosaceae)

11. *Hypericum tatewakii* S. Watanabe, closely related to *H. yamamotoi* of Hokkaidô, but having smaller petals. It is classified into two varieties, the typical variety and the black-dotted leaved variety, var. *nigro-punctatum* S. Watanabe. Endemic on serpentine areas of Mt. Shiratori, Kamuikotan etc., Hokkaidô. (Hypericaceae)

12. *Viola hidakana* Nakai, related to *V. brevistipulata*, differs from it by having smaller habit, somewhat glabrous, and thickened and lustrous leaves. Endemic on serpentine areas of Mt. Shiratori, Hidaka Mountain Range etc., Hokkaidô. Crassism, glabrism, nanism, lucency. (Violaceae)

13. *Viola sacchalinensis* H. Boiss. var. *alpina* Hara, an endemic variety of Hokkaidô, having a smaller habit, somewhat thickened leaves lustrous on upper surface and purplish on under surface. Nanism, crassism. (Violaceae)

14. *Viola yubariana* Nakai, related to *V. stipulata*, differs from it by having smaller habit, shorter stem and thickened leaves purplish on under surface. Endemic on Mt. Yûbari, C. Hokkaidô. Nanism and crassism. (Violaceae)

15. *Angelica stenoloba* Kitagawa, related to *A. iwatensis* (=*A. acutiloba* var. *iwatensis*), having strongly narrow and dissected leaves. Endemic on serpentine areas of alpine zone, Hokkaidô. Stenophyllism. (Umbelliferae)

16. *Bupleurum nipponicum* Koso-Poliansky var. *yesoense* (Nakai) Hara, a narrow-leaved variety of *B. nipponicum* of Japan. Endemic on Mt. Apoi, S. Hokkaidô. Stenophyllism. (Umbelliferae)

17. *Peucedanum multivittatum* Maximowicz var. *linearilobum* Tatewaki, a variety of Japanese *P. multivittatum*, having narrow and dissected leaves. Endemic on Mt. Apoi, S. Hokkaidô. Stenophyllism, dissected leaves. (Umbelliferae)

18. *Rhododendron sanctum* Nakai, related to *R. weyrichii* of S. Honshû. An endemic plant having thickened and lustrous leaves on serpentine areas of Mie, Aichi and Shizuoka Prefectures, Honshû. Crassism, lucency. (Ericaceae)

19. *Primula hidakana* Miyabe et Kudo, related to *P. jesoana* var. *pubescens* of Hokkaidô, differs from it by having smaller and more densely hairy habit, thickened and somewhat lustrous leaves. Endemic on serpentine areas of Hidaka Mountain range, Hokkaidô. Nanism and crassism. (Violaceae)

20. *Primula macrocarpa* Maximowicz, related to Euro-Asiatic *P. farinosa*, differs from it by having a smaller habit and white flowers. Endemic on Mt. Hayachine, N. Honshû. Nanism. (Primulaceae)

21. *Primula yuparensis* Takeda, closely related to Euro-Asiatic *P. farinosa* var. *denudata*, differs from it by having few-flowered inflorescence with larger flowers, leaves slightly white-farinose on under surface. Chromosome number is 36, a tetraploid plant. Endemic on serpentine area of Mt. Yûbari, C. Hokkaidô. (Primulaceae)

22. *Lagotis glauca* Gaertner var. *takedana* (Miyabe et Tatewaki) Kitamura, a variety of N. Pacific *Lagotis glauca*, having smaller habit bearing pale-colored flowers. Sometimes treated as a distinct species, *Lagotis takedana* Miyabe et Tatewaki. Endemic on serpentine scree of Mt. Yûbari, C. Hokkaidô, the characteristic plant of the community on serpentine scree of Mt. Yûbari. Nanism. (Scrophulariaceae) (Fig. 17):

23. *Lonicera mochidzukiana* Makino var. *filiformis* Koidzumi, a variety of *L. mochidzukiana*

Fig. 17. Lagotis glauca v. *takedana*, the character and differential plant of a subassociation of the *Saussuretum chionophyllae*, on serpentine gravel on Mt. Yûbari.

of Japan, having narrow bract. Endemic on Mt. Higashi-akaishi, Shikoku. (Caprifoliaceae)

24. *Adenophora pereskiaefolia* (Roem. et Schult.) Fischer var. *uryuensis* (Miyabe et Tatewaki) Toyokuni et Nosaka, a variety of NE Asiatic *A. pereskiaefolia*, with much patent and rather shorter corollae and sub-coriaceous leaves much reticulate on both sides. Endemic on serpentine areas of central and northern Hokkaidô. (Campanulaceae)

25. *Adenophora triphylla* (Thunberg) DC var. *puellaris* (Honda) Hara, a variety of Asiatic *A. triphylla*, with smaller habit and narrow leaves. Endemic on Mt. Higashi-akaishi, Shikoku. Nanism and stenophyllism. (Campanulaceae)

26. *Achillea ptarmica* L. subsp. *macrocephala* (Ruprecht) Heimerl var. *yezoensis* Kitamura, a narrow-leaved variety of *A. ptarmica* ssp. *macrocephala* of E. Asia. Endemic in N. Hokkaidô. Stenophyllism. (Compositae)

27. *Cacalia hastata* L. subsp. *orientalis* Kitamura var. *hayachinensis* Kitamura, a variety of NE Asiatic *C. hastata* ssp. *orientalis*, having smaller habit with deeply dissected leaves into five lobes. Endemic on Mt. Hayachine, N. Honshû. Nanism and dissected leaves. (Compositae)

28. *Cirsium apoense* Nakai, an endemic species of Mt. Apoi and its neighborhood. Closely re-

lated to *C. pectinellum* of N. Japan, distinguished from it by having shorter external involucral scales, leaves not so deeply dissected as *C. pectinellum*. (Compositae)

29. *Erigeron thunbergii* A. Gray ssp. *glabratus* (A. Gray) Hara:

29-1. *E. thunbergii* A. Gray ssp. glabratus (A. Gray) Hara var. *angustifolius* (Tatewaki) Hara, a narrow-leaved and pale-colored flowering and somewhat glabrous variety of N. Japanese *E. thunbergii* ssp. *glabratus*. Endemic on Mt. Apoi and its neighborings, Hokkaidô. Stenophyllism and glabrism. (Compositae)

29-2. *E. thunbergii* A. Gray ssp. *glabratus* (A. Gray) Hara var. *heterotrichus* (Hara) Hara, a narrow-leaved variety having smaller head. Endemic on Mt. Tanigawa and Mt. Shibutsu, C. Honshû. Nanism and stenophyllism. (Compositae)

30. *Heteropappus hispidus* (Thunberg) Lessing subsp. *leptocladus* (Makino) Kitamura, a narrow-leaved subspecies of Asiatic mother species, occurring on the lowland serpentine areas of Shikoku. Nanism. (Compositae)

31. *Leontopodium fauriei* (Beauv.) Hand.-Mazz. var. *angustifolium* Hara et Kitamura, a narrow-leaved variety of *L. fauriei* in C. Honshû.

Fig. 18a. Saussurea chionophylla, the character species of the *Saussuretum chionophyllae*, on serpentine gravel on Mt. Yûbari, Hokkaido.

Endemic on Mt. Tanigawa and Mt. Shibutsu, C. Honshû. Stenophyllism. (Compositae)

32. *Picris hieracioides* L. subsp. *jessoensis* (Tatewaki) Kitamura, a narrow-leaved subspecies of Euro-Asiatic *P. hieracioides*. Endemic in valleys of Nupuromapporo, Mt. Shiratori etc., N. Hokkaidô. Stenophyllism. (Compositae)

33. *Saussurea riederi* Herder subsp. *kudoana* (Tatewaki et Kitamura) Kitamura, a subspecies of NE Siberian *S. riederi*, having smaller and thicker cauline leaves. Endemic on Mt. Apoi, S. Hokkaidô. Crassism and nanism. (Compositae)

34. *Taraxacum yuparense* H. Koidzumi, somewhat related to *T. hondoense* of N. Japan, but differs from it by having pinnatisected leaves and smaller involucral scales. Endemic on serpentine area of Mt. Yûbari, C. Kokkaidô. (Compositae)

35. *Deschampsia caespitosa* (L.) Beauv. var. *levis* (Takeda) Ohwi, a variety having slender, nodding inflorescens. Endemic on serpentine area of Mt. Yûbari, C. Hokkaidô. (Poaceae)

36. *Hierochlöe pluriflora* Koidzumi, closely related to an amphi-arctic alpine *H. alpina*, but differs from it by having awnless spikelets and much shorter leafblades on the culm, and some-

what lustrous in general appearance. Endemic on serpentine scree of Mt. Yûbari, C. Hokkaidô. (Poaceae)

37. *Allium schoenoprasum* L.:
37-1. *A. schoenoprasum* L. var. *shibutsuense* Kitamura, a variety of circumpolar *A. schoenoprasum*, having smaller habit and rather small, thin umbel, shorter tepals as long as filaments. Distributed in Mt. Tanigawa and Mt. Shibutsu of central Honshû, and in Mt. Yûbari of central Hokkaidô. Nanism. (Liliaceae)
37-2. *A. schoenoprasum* L. var. *yezomonticola* Hara, the other variety of *A. schoenoprasum*, having smaller habit and rather small, thin umbel, shorter tepals and filaments shorter than tepals. Endemic on Mt. Apoi, S. Hokkaidô. Nanism. (Liliaceae)

Serpentinicolous relics

In Japan 13 serpentinicolous relics from 10 different families are recognized. The Compositae contains 4 species and all the remaining 9 families contain a single relic species. They are listed 1–13 as follows:

Fig. 18b. Colonies of *Saussurea chionophylla* on serpentine gravel on Mt. Yûbari, Hokkaido.

1. *Betula apoiensis* Nakai, described from Mt. Apoi together with *B. Miyoshii* and *B. niijimai* in 1930. These three *Betula* species were united into one species, *B. apoensis*, in 1934. Closely related to *B. fruticosa* of Siberia. Endemic on upper part of Mt. Apoi, S. Hokkaidô. A related species, *B. ovalifolia* of EN Asia, occurs in mossy bogs of Kamisarabetsu and Nishibetsu, eastern Hokkaidô. (Betulaceae)

2. *Polygonum hayachinense* Makino, described from Mt. Hayachine in 1903. This is related to the Chino-Himalayan *P. macrophyllum*, differs from it by having longer inflorescences and larger leaves. Endemic on Mt. Hayachine of N. Honshû, no related species occurs in Japan. (Polygonaceae)

3. *Arenaria katoana* Makino, described from Mt. Hayachine in 1905. Afterwards, its occurrences were reported from Mt. Tanigawa and Mt. Shibutsu of C. Honshû, Mt. Apoi, Mt. Tottabetsu, Mt. Yûbari, Mt. Furano-nishi and Mt. Kirigishi of Hokkaidô. These mountains are composed of serpentine except the last, of limestone. No related species appears in Japan. The plant of Mt. Apoi is recognized as a narrow-leaved variety, var. *lanceolata* Tatewaki. (Caryophyllaceae)

4. *Draba japonica* Maximowicz, described from Mt. Hayachine in 1876, afterwards, its occurrences were reported from Mt. Yûbari and Mt. Tottabetsu of Hokkaidô. These localities are all serpentine scree or gravel in alpine areas. This tiny yellow-flowered *Draba* plant is probably related to some arctic *Draba* species, the relation is uncertain. No related species occurs in Japan. (Cruciferae)

5. *Callianthemum miyabeanum* Tatewaki, described from Mt. Apoi in 1928, an endemic plant of Mt. Apoi and its neighborhood. This is closely related to *C. sachalinense* Miyabe et Tatewaki of S. Sakhalin, *C. hondoense* Nakai in Mt. Kitadake of C. Honshû and *C. insigne* Nakai of N Korea. These four species resemble *C. sajanense* of Siberia. (Ranunculaceae)

6. *Rhamnus ishidae* Miyabe et Kudo, related to *R. alnifolia* of N. America, but differs from it by having apiculate leaves with apparent lateral nerves and larger fruits. Described from Mt. Yûbari in 1924, known habitats today are Mt. Apoi, Mt. Tottabetsu, Mt. Yûbari and Mt. Kirigishi of

Hokkaidô. No related species occurs in Japan. (Rhamnaceae)

7. *Enkianthus perulatus* (Miquel) Schneider, firstly described by Miquel as *Andromeda perulata* in 1863. *Enkianthus* is a SE Asiatic genus, this species is related to *E. quinqueflorus* of SE Asia, but the present species is deciduous, while *E. quinqueflorus* is evergreen. Occurs in Mt. Nishiki-yama, Kôchi Pref. of Shikoku, and planted widely as a garden shrub. A broad-leaved form, form. *japonica* (Hooker f.) Kitamura occurs in Tôkai district of Honshû, Shikoku and Kyûshû, mainly in serpentine areas, occasionally in non-serpentine areas. (Ericaceae)

8. *Primula takedana* Tatewaki, described from the valley of Nupuromapporo, Teshio province of Hokkaidô, in 1928. Somewhat resembles to *P. jesoana* var. *pubescens* in general appearance, however, this amoenous white-flowered primrose has rhizomes, thickened leaves and narrower, smaller flowers with not so patent corolla-lobes. Inagaki & Toyokuni (1966) published a monotypic new section of genus *Primula*, *Primula* sect. Takedana, to which *P. takedana* belonged. Occurs in Nupuromapporo of Teshio prov. and Mt. Shirikoma of Kitami prov., Hokkaidô. The latter locality is non-serpentine. (Primulaceae)

9. *Crepis gymnopus* Koidzumi, described from Mt. Yûbari in 1917, related to *C. praemorsa* of Siberia. Endemic in serpentine areas of Hokkaidô, mainly in subalpine to alpine zone. No related species occurs in Japan. (Compositae)

10. *Hypochaeris crepidioides* Tatewaki et Kitamura, described from Mt. Apoi in 1934, related to *H. maculata* and *H. uniflora* of Euro-Siberian region. Endemic on Mt. Apoi, S. Hokkaidô, no related species appears in Japan. Only one native Japanese species belonging to the genus. (Compositae)

11. *Saussurea chionophylla* Takeda, described from Mt. Yûbari in 1915, somewhat related to *S. nuda* of N. Pacific region, however, differs from it, the present species has ovate-oblanceolate, cordate leaves with densely white tomentose undersurface of leaves. Only one species belonging to sect. *Depressae* of *Saussurea* in Japan. Endemic on alpine serpentine scree of Mt. Yûbari, Mt. Tottabetsu, Mt. Chiroro etc. of Hokkaidô (Compositae, Figs 19a,b).

12. *Leontopodium hayachinense* (Takeda) Hara et Kitamura, firstly reported from Mt. Hayachine in 1909, as *L. discolor* Beauv. of Isl. Rebun, described in 1911 as *L. alpinum* ssp. *campestre* v. *hayachinense* by Takeda, and in 1912, treated as *L. discolor* v. *hayachinense* Takeda et Beauverd. Afterwards, by having larger achenes twice as long as those of *L. discolor* and slightly amplexicaul leaves, it was regarded as a distinct species in 1935. Closely related to *L. kurilense* of S. Kuriles, probably an epibiotic plant of the last glacial period. Endemic on Mt. Hayachine, N. Honshû. A variety having more number of round-obtuse cauline leaves and hairy corolla-tubes occurs on Mt. Ohira and Mt. Kirigishi of Hokkaidō, both are built of limestone. (Compositae)

13. *Japonolirion osense* Nakai, described from Mt. Shibutsu and Ozegahara Moor of central Honshû in 1930. *Japonolirion* is a monotypic genus endemic to Japan, somewhat related to *Helonias* of eastern North America. Occurs on Mt. Shibutsu, Mt. Tanigawa and Ozegahara Moor of C. Honshû. A variety having larger habit, var. *saitoi* (Makino et Tatewaki) Ohwi, occurs in Nupuromapporo and Nobukanai of N. Hokkaidô. (Liliaceae)

References

Banno, S. 1958. Glaucophane schists and associated rocks in the Omi district, Japan. J. Geol. Geogr. 29: 29–44.

Chihara, K. 1960. Jadeite in the Omi and Kotaki districts, Niigata Prefecture. Rept. of Investigation of Cultural Properties in Niigata Prefecture. 6: 35–78.

Chino, M. & K. Kitagishi. 1966. Heavy metal toxicities in rice plants (Part 1). The growth of rice plants at high levels of copper, nickel, cobalt, zinc and manganese. J. Sci. Soil Manure, Japan. 37: 342–347.

Crooke, W. M. 1956. Effect of soil reaction on uptake of nickel from a serpentine soil. Soil Sci. 81: 269–276.

Hara, H. 1934–1939. Preliminary Report on the Flora of Southern Hidaka, Hokkaido (Yezo) I-XXXVI. Bot. Mag. Tokyo 48–53.

Hara, H. & M. Mizushima. 1954. List of Vascular Plants of the Ozegahara Moor and its surrounding District. Sci. Res. Ozegahara Moor 428–479. Nihon Gakujutsu Shinkokai, Tokyo.

Hashimoto, S. 1956. The Hidaka metamorphic zone. Jubilee Publication in the Commemoration of Prof. J. Suzuki, M. J. A., Sixtieth Birthday. 17–36.

Hashimoto, M. 1964. A review of the petrology of the Sangun metamorphic rocks, Japan. Bull. Natl. Sci. Mus. 7: 323–337.

Hashimoto, M. 1968a. Sangun metamorphic terrain of the

Asahi-cho area, Okayama Pref. J. Geol. Soc. Japan. 74: 433–437.

Hashimoto, M. 1968b. Glaucophanitic metamorphism of the Katsuyama district, Okayama Pref., Japan. J. Fac. Sci., Univ. Tokyo, II, 17: 99–162.

Hashimoto, M., S. Igi, Y. Seki, S. Banno, & G. Kojima, 1970. Geol. Survey of Japan. 6–7.

Hayama, Y., Y. Kizaki, K. Aoki, S. Kobayashi, K. Toya, & N. Yamashita. 1969. The Jōetsu metamorphic belt and its bearing on the geologic structure of Japanese Islands. Mem. Geol. Soc. Japan. 4: 61–82.

Hunahashi, M., S. Hashimoto, H. Asai, S. Igi, Y. Sotozaki, K. Kizaki, S. Hirota, & A. Kasugai. 1956. Gneisses and migmatites in the southern extreme region of the Hidaka metamorphic zone, Hokkaido. J. Geol. Soc. Japan. 62: 401–408; 464–471; 541–549.

Iizuka, T. 1975. Interaction among nickel, iron and zinc in mulberry tree grown on serpentine soil. Soil Sci. Plant Nutr. 21: 47–55.

Inagaki, K., H. Toyokuni, K. Matsunaga, S. Horochi, M. Chida, M. Yamaguchi, T. Kobayashi, M. Hanzawa, S. Kita, & T. Odazima. 1966–1969. On the flora of ultrabasic rock areas in central and northern Hokkaido, Japan (Part 1 – Part 5). Journ. Hokkaido Univ. Educ. (Sect. IIB) 16: 99–113; 18: 16–23; 19: 20–26, 79–81; 20: 16–19.

Isomi, H. & T. Nozawa. 1960. On the structure of the Hida metamorphic rocks. Chikyu Kagaku 48: 11–20.

Kamada, K. & K. Doki. 1975, 1977. Movement of Cr in submerged soil and growth of rice plant. J. Sci. Soil Manure, Japan. 46: 478–482; 48: 457–465.

Karakida, Y., H. Yamamoto, S. Miyachi, T. Oshima, & T. Inoue. 1969. Characteristics and geological situations of metamorphic rocks in Kyushu. Mem. Geol. Soc. Japan. 4: 3–21.

Katayama, M. 1984. Personal communications.

Kikuchi, M. & C. Komidzunai. 1961. Flora of Mt. Hayanchine and the adjacent Mountains, Prov. Rikuchù in North Hondo. Soc. Kaminishi Sci. Educ., Tono.

Kitamura, S. 1952a. Serpentine flora of Mt. Shibutsu, Prov. Kōzuke, Japan. Acta Phytotax. Geobot. 14: 174–176.

Kitamura, S. 1952b. Serpentine flora of Mt. Hayachine, Prov. Rikuchū, Japan. Acta Phytotax. Geobot. 14: 177–180.

Kitamura, S. 1956. Serpentine flora of Mt. Apoi, Prov. Hidaka, Hokkaido. Acta Phytotax. Geobot. 16: 143–148.

Kitamura, S. 1957. Serpentine flora of Nupromapporo Valley, a branch of Teshio River, Teshio, Hokkaido. Acta Phytotax. Geobot. 17: 41–45.

Kitamura, S., G. Murata, & K. Torii, 1953. Serpentine flora of Mikawa Prov. Acta Phytotax. Geobot. 15: 1–3.

Kitano, Y. 1969. Mizu no Kagaku. 89–90. NHK Books, Tokyo.

Kizaki, K. 1964. On migmatites of the Hidaka metamorphic belt. J. Fac. Sci., Hokkaido Univ. IV, 12: 111–169.

Kobayashi, H. 1958. The Hida metamorphic complex. Jubilee Publication in the Commemoration of Prof. J. Suzuki, M. J. A., Sixtieth Birthday. 141–161.

Kobayashi, H. 1962. The Hida gneisses. Geological Studies of the Hida plateau. 14–32.

Masuda, T. & R. Sato, 1961. The injury to growth of plants on serpentine soil. Bull. Hokkaido Prefect. Agric. Exp. Stn. 8: 37–48.

Masuda, T. & R. Sato, 1962. The injury to growth of plants on serpentine soil. J. Sci. Soil Manure, Japan. 33: 201–204.

Masuda, T. & R. Sato. 1963. The injury to growth of plants on serpentine soils. Bull. Hokkaido Prefect. Agric. Exp. Stn. 11: 52–58.

Miyashiro, A. & I. Kushiro. 1977. Gansekigaku. III. pp. 222–224. Kyoritsu Shuppan Co., Tokyo.

Miyawaki, A. (ed.) 1977. Vegetation of Japan compared with other region of world. Gakushū Kenkyūsha, Tokyo.

Mizuno, N. 1967–1968. Studies on chemical characteristics of serpentine soil in Hokkaido. Bull. Hokkaido Prefect. Agric. Exp. Stn. 15: 48–55; 16: 1–9; 17: 62–72.

Mizuno, N. 1968b. Interaction between iron and nickel and copper and nickel in various plants. Nature. 219: 1271–1272.

Mizuno, N., H. Kaneta, K. Kamada, T. Meguro, K. Doki, & K. Goto. 1977. Soil components of farming land in Hokkaido. Misc. Pub. Hokkaido Prefect. Agric. Exp. Stn. 8: 1–62.

Mizuno, N. 1979. Studies on chemical characteristics of serpentine soils and mineral deficiencies and toxicities of crops. Report Hokkaido Prefect. Agric. Exp. Stn. 29: 1–79.

Nakamura, T. 1959. Studies on the preparation of fertilizers containing magnesium, manganese, boron, or silicon, and on the improvement of their qualities. Bull. National Institute Agric. Sci. B9: 1–122.

Nakata, M. & S. Kojima. 1987. Effects of serpentine substrate on vegetation and soil development with special reference to *Picea glehnii* forest in Teshio District, Hokkaido, Japan. Forest Ecology and Management 20: 265–290.

Noji, M. 1978, 1981. Serpentinites and civil engineering. Monthly Report Civil Engineering Research Institute, Hokkaido Development Bureau. 303: 1–10; 342: 1–22.

Nosaka, S. 1974. The Phanerogam Flora of Mt. Yūpari, Prov. Ishikari, Hokkaido, Japan. Journ. Fac. Sci. Hokkaido Univ. Ser. V (Bot.) 9: 55–300.

Nureki, T. 1969. Geological relations of the Sangun metamorphic rocks to the "nonmetamorphic" Paleozoic formations in the Chūgoku province. Mem. Geol. Soc. Japan. 4: 23–39.

Ohba, T. 1968. Über die Serpentin Pflanzengesellschaften der alpinen Stufe Japans. Bull. Kanagawa Pref. Museum 1: 37–64.

Ohwi, J. 1965. Flora of Japan. Shibundo Co., Tokyo.

Rune, O. 1953. Plant life on serpentines and related rocks in the north of Sweden. Acta Phytogeogr. Suec. 31: 1–139.

Sasaki, S., T. Matsuno, & Y. Kondo. 1968. A podsol derived from serpentine rocks in Hokkaido, Japan. Soil Sci. Plant Nutrition, 14: 99–109.

Sasaki, Y. (ed.) 1973. Shokubutsu-shakaigaku. Kyōritsu Shuppan Co., Tokyo.

Sato, S. 1968. Precambrian-Variscan polymetamorphism in the Hida Massif, basement of the Japanese Islands. Sci. Rept. Tokyo Univ. Education, C. 10: 15–129.

Sato, T., H. Wada, & M. Kojima. 1977. Color of the soil derived from serpentine in Mikkabi, Shizuoka Prefecture. J. Sci. Soil Manure, Japan. 48: 339–340.

Shimizu, N. 1965. Metamorphic rocks in the southeastern part of the Hidaka metamorphic terrane, Hokkaido, Japan. M.Sc. thesis, University of Tokyo.

Shido, F. & Y. Seki. 1959. Jadeite and hornblende from the

Kamuikotan metamorphic belt. J. Geol. Soc. Japan. 65: 673–677.

Soane, B. D. & D. H. Saunder. 1959. Nickel and chromium toxicity of serpentine soils in Southern Rhodesia. Soil Sci., 88: 322–330.

Suzuki, S., N. Mizuno, & K. Kimura. 1971. Distribution of heavy metals in serpentine soil. Soil Sci. Plant Nutrition 17: 195–198.

Suzuki, T. 1954. Forest and bog vegetation within Ozegahara basin. Sci. Res. Ozegahara Moor. 205–268. Nihon Gakujutsu Shinkokai, Tokyo.

Suwa, K. 1969. Polymetamorphism in the Hida metamorphic complex. Mem. Geol. Soc. Japan 4: 113–116.

Takagishi, H., S. Higashino, & T. Iizuka 1973. Studies on abnormal features of mulberry plant growing on the soil derived from serpentine. Part 1. Chemical analysis of serpentine tineous soils and mulberry plants injured by nickel toxicity. J. Agricultural Sci. Japan. 42: 135–143.

Tatewaki, M. 1943. Aka-ezomatsu-rin no Gunrakugaku teki Kenkyū. Res. Bull. Coll. Exp. For. Hokkaido Imp. Univ. 13: 1–181.

Tatewaki, M. 1958. Forest ecology of the Island of North Pacific Ocean. Journ. Fac. Agr. Hokkaido Univ. 50: 371–472.

Tatewaki, M. 1963. Alpine plants in Hokkaido. Sci. Rep. Tohoku Univ. Ser. IV (Biol.) 29: 165–188.

Toyokuni, H. 1955–1960. On the Ultrabasicosaxicolous Flora of Hokkaido, Japan. Journ. Geobot. 4: 97–101; 5: 12–15, 81–84, 115–116; 6: 17–20, 63–67; 7: 37–38; 9: 10–13, 38–41.

Watanabe, S. 1971. Phytogeographical Studies of the Alpine Plants (Vascular Plants) on the Hidaka-Yūbari Ranges, Hokkaido. Mem, Natn. Sci. Mus. Tokyo. 4: 95–124.

Yamanaka, T. 1958. Serpentine flora of Mt. Nishi-Akaishi, Shikoku Acta Phytotax. Geobot. 17: 94–96.

Yamanaka, T. 1959. Serpentine flora of Mt. Higashi-Akaishi, Shikoku, Japan. Acta Phytotax. Geobot. 18: 80–96.

Yamanaka, T. 1972. Sociological Studies on the Serpentine Vegetation IX, X. Bull. Fac. Educ. Kochi Univ. 24 (Ser. 2): 29–33; 35–38.

Yamanaka, T. 1973. Sociological Studies on the Serpentine Vegetation XI-XIV. Bull. Fac. Educ. Kochi Univ. 25 (Ser. 2): 15–17; 19–21; 23–25; 27–29.

Yamanaka, T. 1974. Sociological Studies on the Serpentine Vegetation XVII. Resp. Rep. Kochi Univ. (Nat. Sci.) 23(4): 21–35.

Yamanaka, T. 1979. Nihon no Shinrin Shokusei. Tsukiji Shokan, Tokyo.

Authors' Addresses:
Naoharu Mizuno
Tokyo University of Agriculture
Onnenai 59–8
Abashiri City
Hokkaido 099–35
Japan

Shiro Nosaka
Department of Biology
Education University of Aichi
Kariya-shi
Aichi Prefecture, 448
Japan

XII

The ecology of ultramafic areas in Zimbabwe

J. PROCTOR & M. M. COLE

Abstract. Studies of the vegetation of ultramafic soils in Zimbabwe have concentrated on and around the Great Dyke, one of the world's largest outcrops of ultramafic rocks. This area occurs under a wet-dry tropical climate with annual rainfalls ranging from about 450 to 880 mm. The ultramafic vegetation studied occurs mainly at altitudes from about 1000–1500 m and some is subject to frost. The soils have fairly high nickel concentrations and rarely have exceptionally low calcium concentrations. The vegetation of these soils is generally grassland or open shrubland and is often in marked contrast with well-developed woodland on adjacent non-ultramafic soils. Along river banks on ultramafic soils there is a well-developed riverine forest. Since these forest soils show no substantial chemical differences from those under more open vegetation it is suggested that hydrological factors may be very important in determining physiognomy. The flora of the ultramafic soils is relatively species-poor but includes about twenty important endemics. There have been many chemical analyses of native plants and these have shown that although there are few nickel accumulators, one species *Pearsonia metallifera* (Leguminosae) had a recorded 14,100 μg g^{-1} nickel in its leaf dry matter. Experimental work on crop species has apparently demonstrated the existence of acute nickel toxicity on several Zimbabwean ultramafic soils. A very important study made on *Zea mays* has clarified many aspects of the interactions between nickel, magnesium and calcium supplied in water culture. Zimbabwean work on native species has shown tolerance of nickel in two species and in the case of the non-tolerant tree, *Brachystegia spiciformis* has given clues about the precise toxic effects of this element.

B. A. Roberts and J. Proctor (eds), The ecology of areas with serpentinized rocks. A world view, 313–331.
© *1992 Kluwer Academic Publishers.*

Introduction

The vegetation of ultramafic soils in Zimbabwe is the best known for any African country and its study has made an important contribution to our general understanding of the ecology of ultramafic soils. This is largely due to the enthusiasm of the late Professor H. Wild who stimulated many workers to research on this topic as well as writing valuable papers himself. The field studies have largely been on or near the Great Dyke, a massive intrusion which includes one of the largest outcrops of ultramafic rocks in the world. The Great Dyke runs practically across the Zimbabwean plateau (average altitude about 1000 m) and rises from a few metres to about 500 m above it.

Nearly all the areas of serpentinized ultramafic rock in Zimbabwe experience a very seasonal rainfall regime (Table 1). There is a rainy season with convective-type rainfall from the beginning of December to mid-March followed by a period of prolonged dryness. In the period from about mid-September to early November, pronounced desiccation occurs with high temperatures being combined with low relative humidities. The mean annual rainfall is greater in the north of the Great Dyke (876 mm at Umvukwes, Table 1) than in the south (455 mm at Tuli, 90 km west of the southern extremity of the Great Dyke). There are marked diurnal and annual ranges of temperature (Table 2). Frosts are frequent between mid-May and mid-August (Werger *et al.* 1978a).

Geology and geomorphology

The main areas of ultramafic rocks in Zimbabwe occur on the Great Dyke. These ultramafic and mafic rocks cut through the Archaean and early Precambrian rocks of Zimbabwe with a SWW–NNE orientation. The great Dyke has a strike length of some 530 km, a width of 5–11 km and a total area of about 4,400 km². About 70% of its outcrop is serpentinite and the remainder pyroxenite, norite and gabbro (Fig. 1). Between Norton (south of Darwendale) and the Ngesi river, a 100 km-long intrusion of norite narrows the outcrop of serpentinite and virtually divides it into two sections, a northern and southern one.

The relief of the Great Dyke varies. Northwest of the Empress mine, Cole (1971) has noted 'a pronounced east facing escarpment marks the edge of the Mafungabusi plateau'. This plateau, about 1,330 m high and developed over Karoo sediments and lavas, forms part of the Miocene African surface whose formation under hotter and wetter conditions than those prevailing today was accompanied by intensive leaching and laterization. Today, outliers of Karoo rocks south of Harare testify to the former extent of these rocks. Following continental uplift, the Miocene African surface has been progressively destroyed by erosion during the succeeding drier Victoria Falls and later planation cycles. These have removed the younger Karoo rocks and exposed Precambrian and Archaean Basement rocks including those of the Great Dyke. They have left their legacy in the central plateau of Zimbabwe regarded as Highveld, flanked on either side by lower dissected terrain with ranges of hills and clusters of inselberge forming the Middleveld that descends to the trough valleys of the Zambesi, Limpopo and Sabi, regarded as Lowveld. The Great Dyke cuts across these physiographic regions, throughout most of its length forming a prominent ridge feature rising above the plateau surfaces.

The impact of erosion on the different bedrock types along the Great Dyke has been influenced by its relationship to the planation surfaces and by regional differences in the rainfall. In the south near Selukwe the Great Dyke is associated with rugged hills rising above the lower planation surfaces (Figs 2, 3). Here the pyroxenite unit forms the crests of ridge features whose lower and flatter margins are of serpentinite. From the vicinity of the Bulawayo – Fort Victoria road to the Lundi river, the Dyke forms a flat-topped ridge over serpentinite. Near Lalapanzi, outcropping pyroxenite is responsible for irregular relief but further north this gives way to softer relief over serpentinite. North of Darwendale however differential weathering of alternating more resistant serpentinite has resulted in a series of pyroxenite scarps and serpentinite sills. Thus overall the Dyke is associated with prominent rugged hills standing above the lower planation surfaces in the north and south and more subdued ridge features rising above the Miocene plateau surface in the centre.

Table 1. Mean monthly rainfall (mm) at three stations in Zimbabwe (from Thompson 1965)

Station	Location	Altitude (m)	Years for data	Jan	Feb	Mar	Apr	May	Jun	Jul	Aug	Sept	Oct	Nov	Dec	Year
Umvukwes	17°00' S 30°53' E	1474	1926–1956	204	168	152	27	4.8	0.8	0.8	1.0	2.8	14	102	200	876
Gwelo	19°27' S 29°51' E	1433	1898–1963	150	139	87	19	7.4	2.5	0.8	1.3	5.1	22	97	152	682
Tuli	21°23' S 28°59' E	767	1933–1961	97	80	65	20	3.3	4.3	1.0	1.5	3.3	28	67	86	455

Table 2. Mean maximum and minimum temperatures °C for three months at the three stations described in Table 1

Station	Years for data	January max	min	July max	min	October max	min
Umvukwes	1962–1964	24.2	15.5	20.2	6.2	28.8	13.9
Gwelo	1951–1961	25.4	15.2	20.1	4.0	28.6	13.4
Tuli	1954–1961	30.2	17.9	24.3	5.5	32.1	16.2

Soils

The legacies of past climates and geomorphological processes are evident in the presence of lateritic materials along some sections of the Dyke, notable in the north near Gwebi and Banket where conglomeratic laterite occurs (Fig. 4). Elsewhere soils are closely related to bedrock. Over the hills developed over pyroxenite the soils are dark reddish brown loams relatively high in iron, magnesium and silica. The norite and gabbro, which usually are associated with lower level terrain, give rise to black clay loams or clay soils that have relatively high concentrations of calcium but may suffer from drainage impedance.

Thompson (1965) in his survey of the soils of Zimbabwe has described two types on the ultramafic rocks of the Great Dyke. There are 'Vertisols' which are generally dark coloured soils characterized by internal soil movement (self mulching), resulting from alternate seasonal wetting and drying of their expanding montmorillonitic clays. They occur on uplands of normal relief and also in broad depressions where rainfall is insufficient to give rise to soils that are primarily hydromorphic. (Truly hydromorphic ultramafic gley soils probably exist in some 'Vlei' areas judging from the brief descriptions in Ellis (1951)).

The other type of ultramafic soil described by Thompson (1965) occurs within the zone of 'Fersiallitic soils.' These are largely confined to uplands with good external drainage. It is odd that vertisols often occur with these fersiallitic soils in situations where relatively high rainfall and good drainage should favour the formation of the latter. Grim (1953) has stated that the presence of an appreciable amount of magnesium in basic or ultrabasic rocks favours the formation of montmorillonite.

There are other types of soil on or near the Great Dyke and over ultramafic rocks elsewhere in Zimbabwe including lithosols that sustain xero-

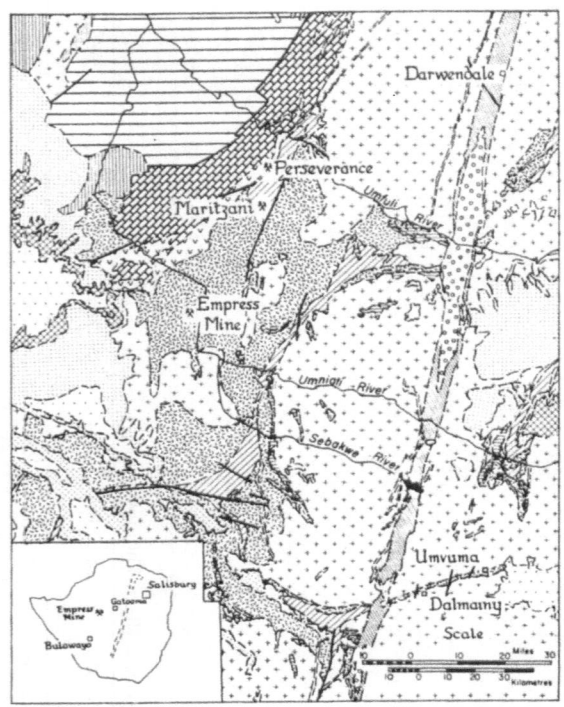

Geological Legend

		SYSTEM	ERA
	Aeolian sands	Kalahari	Tertiary
	Basalt		Jurassic
	Grits, sandstone etc.		Triassic
	Mudstones, grits, coal measures and sandstones	Koroo	Permion
	Slates, quartzite, dolomite		Late
	Metasediments and metavolcanics		Precambrian
	Phyllites, slates and minor quartzites		Precambrian
	Metasediments		Early
	Metavolcanics, interbedded sediments porphyries and felsites	Basement	Precambrian Various
	Dolorite and gabbro		Ages
	Norite and gabbro		Probably
	Serpentine and pyroxenite with chrome seams	Great Dyke	Post-Lomagunde
	Granite and allied rocks		Various
	Serpentine, talc-schist etc.		Ages

Fig. 1. Geology of the area around and including the Great Dyke, Zimbabwe.

316

Fig. 2. View eastwards from the Peak mine, Selukwe, cultivated Umtebekwe valley to the Great Dyke forming a pronounced ridge feature beyond which is granite terrain with a series of inselberge.

Fig. 3. View towards the Peak mine Selukwe showing grass covered shelf like feature of the Great Dyke with mine buildings below the rugged pyroxenite hills. The tree covered terrain in the foreground and middle distance is underlain by schist. Granite forms the line of low hills to the right of the pyroxenite hills of the Great Dyke.

Fig. 4. Conglomeratic laterite exposed at the foot of the Great Dyke near Banket.

Fig. 5. View to the grass covered prominent ridge formed by the Great Dyke between Gwebi and Banket. Mines working the chromite horizons for chrome ore occur on the middle slopes.

Fig. 6. The Great Dyke near Vanad Pass.

morphic species and deep humus-rich soils supporting riverine forest.

A large number of chemical analyses have been made for soils over Zimbabwean ultramafic rocks and a selection of these is given in Table 3. They show that the pH is slightly acid and although some high values have been recorded for Mg/Ca quotients, calcium concentrations are not usually very low. Nickel concentrations are always fairly high or high. The soil solutions have substantial quantities of calcium and relatively high nickel concentrations which are above the threshold known to be toxic to non-tolerant plants (Johnston & Proctor 1981). Johnston (1979) showed that the nickel in the soil solutions from the two Zimbabwean sites analyzed was largely in the form of Ni^{2+}.

Vegetation

On non-ultramafic soils *Brachystegia-Julbernardia* savanna woodlands occupy the Zimbabwean central plateau above 1300 m. Below this altitude there are *Burkea-Terminalia* woodlands over sandy soils, *Acacia* parklands on heavier soils,

and *Colophospermum mopane* woodland and *Acacia*-dominated thickets along the low-lying valleys (Cole 1986). More complex regional and local distributions result from vegetation- soil- relief catenas (Boughey 1961) that are related to stages in the dissection of the old planation surfaces which in some areas, including parts of the Great Dyke, have exposed bedrock. Over serpentinite, trees are usually sparse or absent and there are a number of types of grassland and communities rich in succulents. Werger *et al.* (1978a) have commented : 'As to contrast between serpentine and non-serpentine vegetation, both in structure and in floristic composition, the Rhodesian Great Dyke must certainly be among the most dramatic sites in the world. Where granite, norite or pyroxenite contact serpentine, the boundary between well-developed, dense miombo woodland and very open tree or bush savanna or grassland over considerable distances is very sharp.'

The most detailed descriptions of vegetation on ultramafic soils of Zimbabwe have been given by Werger *et al.* (1978a) for an area in the northern part of the Great Dyke. They discussed four commmunity types:

a) The *Setaria lindenbergiana – Euphorbia me-*

Table 3. The pH, percentage loss-on-ignition, exchangeable potassium, sodium, calcium, magnesium and nickel and total chromium, cobalt and nickel in a range of ultramafic soils from Zimbabwe (– means not determined)

Locality and soil type		Depth (cm)	pH	K	Na	Ca (meq 100 g⁻¹)	Mg	Mg/Ca	Ni	Co	Cr (μg g⁻¹)	Ni	Author
Great Dyke. Skeletal soils.		0–23	–	0.18	–	2.6	7.0	4.1	9	–	–	–	Hunter (1954)
Ngezi		0–12	6.3	0.14	0.004	0.9	5.4	6.0	15	270	480	4,600	Proctor et al. (1980)
Noro Chrome mine		–	6.6	0.26	0.006	2.0	10.7	5.4	41	250	110	6,600	
Kingston Hill		–	6.1	0.24	0.009	3.0	4.5	1.5	53	230	2,200	6,800	
Selukwe		–	6.4	0.22	0.023	3.5	9.9	2.8	0.4	73	1,200	430	
Tipperary Claims		–	6.2	0.28	0.007	2.5	8.0	3.2	56	240	2,400	7,100	
Umvuma		–	6.1	0.31	0.01	3.8	5.6	1.5	21	200	1,900	3,500	
Great Dyke, Mtoroshanga Pass. Soil under woodland.		0–15	5.8	0.24	0.02	2.10	6.7	3.2	14	300	1,760	3,630	Proctor & Craig (1978)
Soil under grassland near the woodland.		0–15	5.8	0.15	0.02	1.73	5.3	3.1	35	310	2,080	4,000	
Great Dyke, soil under riverine forest.													
Caesar's Pass A		0–15	6.1	0.27	0.07	3.4	17.7	5.2	17	180	1,700	2,300	Soane & Saunder (1959)
Caesar's Pass B		–	6.6	0.19	0.03	0.6	17.2	28.7	2.8	160	1,400	1,800	
Mtoroshanga Pass		–	6.3	0.79	0.04	2.7	16.2	6.0	31	250	950	4,800	
Great Dyke. (locations not specified)	A	0–15	6.1	0.14	–	2.6	6.3	2.4	70	–	39,000	4,000	
	B	–	6.4	0.12	–	0.8	13.2	16.5	15	–	20,000	1,400	
	C	–	6.4	0.20	–	2.7	12.7	4.7	24	–	8,000	1,100	
	D	–	5.6	0.16	–	0.8	1.2	1.5	11	–	8,500	1,000	
	E	–	6.1	0.13	–	0.7	1.9	2.7	40	–	46,000	3,600	
	F	–	6.1	0.11	–	0.9	2.4	2.7	39	–	41,000	3,500	
	G	–	6.3	0.31	–	0.7	12.9	18.4	23	–	–	3,900	
	H	–	6.0	0.33	–	1.5	2.0	1.3	8	–	–	700	
	I	–	5.6	0.09	–	0.6	0.1	1.7	4	–	–	–	
Great Dyke, near Selous. Vertisol under open grassland with scattered trees.		0–13	6.0	0.2	–	10.8	53.6	5.0	–	–	550	–	Thompson (1965)
		18–33	6.3	0.1	–	11.9	62.7	5.3	–	–	–	–	
		56–71	6.5	0.1	–	12.1	62.3	5.1	–	–	–	–	
Great Dyke, near Darwendale. Fersiallitic soil under scattered trees.		0–9	5.6	0.2	–	5.2	10.1	1.9	–	–	–	–	Thompson (1965)
		23–41	5.3	0.1	–	1.1	12.3	11.2	–	–	–	–	
Skeletal soil, Tipperary Claims.		–	6.1	0.16	–	25.0	8.7	0.35	55.2	–	2,410	–	Wiltshire (1974)

Fig. 7. The endemic *Euphorbia memoralis*, Vanad Pass.

moralis community is typical of exposed ridges in the higher part of the Great Dyke. The outcropping rock covers 80–95% of the area and the soil is restricted to rock fissures. The vegetation is an open shrubby community with both woody and succulent emergents up to 1.5 m high and a field layer up to about 1 m high. Prominent species include the grasses *Andropogon gayanus, Cymbopogon excavatus, Digitaria diagonalis, Loudetia simplex* and *Setaria lindenbergiana*; the succulents *Euphorbia memoralis* and *E. wildii*; and the dwarf trees or shrubs *Barleria aromatica, Protea welwitschii, Rhus wildii, Tapiphyllum velutinum* and *Xerophyta equisetoides*. Werger *et al.* (1978a) also described a savanna, which has floristic similarities with the *Setaria lindenbergiana – Euphorbia memoralis* community, with trees from 5–8 m high including *Croton gratissimus, Diplorhynchus condylocarpon, Olax obtusifolia* and *Ozoroa longepedunculata*).

b) The *Dicoma niccolifera – Loudetia simplex* community is an open grassland with a few shrubs or small trees. It usually occurs on slopes (>5°). Amongst the shrubs and small

trees are *Albizia antunesiana, Combretum molle, C. zeyheri, Diplorhynchus condylocarpon, Euphorbia wildii, Ozoroa longepedunculata, Protea welwitschii* and *Securidaca longepedunculata*. The field layer includes *Andropogon gayanus, Blepharis acuminata, Dicoma niccolifera, Digitaria diagonalis, Diheteropogon amplectens, Elephantorrhiza elephantina, Eragrostis racemosa, Scleria bulbifera* and *Tripogon minimus*.

c) The *Alloteropsis semialata – Loudetia simplex* community is a fairly dense grassland up to 0.8 m high. It occurs on badly drained soils at the base and tops of hills. In some stands a few emergents of the same species as in the previous community occur. Werger *et al.* (1978a) have listed several diffential species of this community although it has several species in common with the previous community. *Digitaria milanjiana*, though not restricted to this community, typically reaches high cover values.

d) Riverine forest community. Along the larger rivers and streams in the northern part of the Great Dyke a riverine forest can develop. It occurs on deep humic soils and the trees are

321

Fig. 8. The late Prof. H. Wild, with the endemic *Euphorbia wildii*, Vanad Pass.

12–20 m high. In the example studied by Werger *et al.* (1978a) there is a layer of lower trees, shrubs and saplings 1–4 m high. The tree layer was entirely made up of *Syzygium guineense* subsp. *guineense* with the epiphyte *Microcoelia exilis* and in the lower layers occurred *Acacia karroo*, *Bequaertiodendron magalismontanum*, *Ficus burkei*, *F. capensis*, *Ilex mitis*, *Maesa lanceolata*, *Myrica serrata*, *Nuxia oppositifolia*, *Rapanea melanophloeos* and *Rhus longipes*.

The grassland and savanna communities on the ultramafic soils of the Great Dyke provide a dramatic structural and floristic contrast with the woodlands which occur where granite, norite or pyroxenite contact serpentinites (although small patches of *Brachystegia* woodland can occur on serpentinites (Werger *et al.* 1978a and Proctor & Craig 1978).

One of us (M.M.C.) has observed grassland over serpentinite near Gwebi above which is small tree and shrub savanna with *Diplorhynchus condylocarpon*, *Faurea speciosa* and *Protea* spp. In contrast to these areas those over pyroxenite support *Brachystegia – Julbernardia* savanna woodland in which *Brachystegia tamarindoides* and *B. spiciformis* are locally prominent. Significantly savanna woodland of *Brachystegia boehmii*, *B. spiciformis*, *J. globiflora*, *Diplorhynchus condylocarpon*, *Protea* and *Faurea* species occurs over conglomeratic laterite near Gwebi and Banket. In these woodlands the same grasses *Loudetia simplex* and *Andropogon gayanus* that characterize the serpentinite soils, dominate the ground layer. The vegetation over norite bedrock contrasts with that over the pyroxenite because the typical black soils are self mulching and, in places, ill drained. Characteristically they support grassland which in the better drained areas near Banket is studded with small *Acacia gerrardii* trees. The riverine forest on ultramafic soils is similar in all respects to that occurring over other rocks.

There is a limited amount of information on the vegetation of ultramafic rocks elsewhere in Zimbabwe. Wild (1965, 1970, 1974) has given brief information for several parts of the Great Dyke and elsewhere. There is a marked difference between the drier, more southerly areas of the Great Dyke described by Wild (1965) and those in the north. Thus in the Lundi River – Bannockburn area in the south, the Dyke forms an elongated grassy plain between granite hills and thus forms an apparent trough in the landscape. The ultramafic area is more grazed than in the north and the grass *Themedra triandra* largely occupies the position which *Loudetia simplex* holds in the north. Woody vegetation is restricted to occasional stunted bushes and there is no well-developed riverine forest although the surrounding granite hills are covered by *Brachystegia spiciformis – Julbernardia globiflora* savanna woodland.

There has been a general assumption, e.g., by Werger *et al.* (1978b) that the more open vegetation on the Great Dyke occurs on chemically more toxic soils. This explanation must be questioned in the light of the work of Proctor & Craig (1978) who showed that well-developed riverine

Fig. 9. Vegetation of *Brachystegia tamarindoides* on top of a pyroxenite kopje on top of the Great Dyke, with *Diplorhynchus condylocarpon* and some *Cussonia* sp trees on the slopes below, between Gwebi and Banket.

Fig. 10. Open woodlands with the left to right *Uapaca* sp, *Brachystegia boehmii*, *Combretum apiculatum* and, framing picture, *Brachstegia spiciformis* over pyroxenite bedrock on top of the Great Dyke between Gwebi and Banket.

323

forest occured on soils with Mg/Ca quotients of up to 40.9 (on a milliequivalent basis) and nickel up to 31 μg g^{-1} (all elements extracted with ammonium acetate). These values for Mg/Ca quotients exceed those in all published analyses of ultramafic soils under grassland in the Great Dyke whilst the forest nickel values are within the range reported for the grasslands. This suggests that hydrological factors may be of great importance on the Great Dyke and there may be an interaction between soil water and chemical factors. Ernst (1972) claimed that water-soluble concentrations of toxic metals are greatly increased during dry conditions.

Flora

Wild (1965) has pointed out the floristic poverty of the ultramafic soils on the Great Dyke compared with that on other soils in Zimbabwe. He recorded 322 species from about 3,280 km^2 on the Great Dyke and the following numbers on non-ultramafic soils elsewhere in Zimbabwe: 859 species from less than 390 km^2 in the Chimanimani Mountains; 730 species on 27.4 km^2 near Marandellas; and 700–800 species for an area of about 521 km^2 around Victoria Falls. Wild (1965) commented: 'No further emphasis is required to point out the relative impoverishment of numbers of species on our serpentines'; this is most obvious in the case of tree species which are almost absent from the serpentines and numerous in the adjacent granites, but also applies to the numbers of species of grasses which although dominant on the serpentine, are represented by only 51 species, whereas on Grasslands Research Station, Marandellas (a really intensively collected area) in *Brachystegia spiciformis* – *Julbernardia globiflora* savanna woodland approximately 136 grass species have been recorded from 27.4 km^2. Werger *et al.* (1978a) have also commented on the floristic poverty of ultramafic soils and on the greater importance of monocotyledonous species there. 'In the woodland communities on granite and pyroxenite the amount of monocotyledonous species in the total flora is about a quarter to a third; where serpentine is somewhat mixed with pyroxenite they increase somewhat in relative importance up to 41% and on the serpentine they

usually make up between about 40 and 60 per cent of the total number of species.' It is curious that amongst the 20 species or subspecies distinguished by H. Wild as endemic to the Great Dyke only one (*Aloe ortholopha*) is a monocotyledon. Werger *et al.* (1978a) regard this rarity of endemism as reflecting a large 'niche hypervolume' of the monocotyledonous species. Ernst (1972) suggested that monocotyledonous species may be at an advantage in ultramafic soils because of the adventitous nature of their roots which they can replace when toxic metal accumulation in the xylem becomes too high.

Of the 322 species listed for the ultramafic soils of the Great Dyke by Wild (1965), 245 occur in the northern area and 125 in the southern area. Sixty-one of the species represented on the northern serpentinites (including seven endemics) do not occur on the southern ones and twenty (including four endemics) of those occurring on the latter do not occur on the northern serpentinites. Wild (1965) has suggested that the decrease in the numbers of species southward is related to the decreasing rainfall (Table 1) but that the interposition of the large area of outcropping norite and gabbro that separates the northern and southern serpentinites may have restricted migration of the species. (The southward decrease in species probably has a number of causes and is not confined to vegetation over serpentinite. For example the *Brachystegia-Julbernardia* woodlands of Zimbabwe have fewer species than those of Zambia and both of these genera are absent from the vegetation south of the Limpopo in South Africa).

Most of the endemic species have relatives which occur on non-ultramafic soils, considerable distances away. Wild & Bradshaw (1977) regard all the endemics on ultramafic soils as 'palaeoendemics,' ancient populations which have survived unchanged after being isolated (perhaps since the Tertiary, 75×10^6 years ago) by climatic changes from other populations. Several of these endemics have their closest relatives in South Africa and it is suggested that these are survivors of a previously more widespread extra-tropical Cape Flora. Only six widespread species are so far known to show morphological differences associated with ultramafic substrata (Wild & Bradshaw 1977). There seems to be no frequent infra-speci-

Table 4. The pH and concentrations (mg l^{-1}) of potassium, calcium, magnesium and nickel in soil solutions from two ultramafic soils in Zimbabwe (from Proctor *et al.* 1981)

Site of soil collection	K$^+$	Ca^{2+}	Mg^{2+}	Ni^{2+}
Kingston Hill	20	19	28	0.6
Tipperary Claims	19	23	38	0.7

fic differentiation resembling those which occur on ultramafic soils in temperate areas (e.g. da Silva, Proctor this volume). This low level of infra-specific variation suggests a shortage of neo-endemism. Since resistant races of widespread species exist (they have been demonstrated by Ernst (1972), Wiltshire (1974) and Craig (1977a,b) then as Wild & Bradshaw (1977) point out the Zimbabwean ultramafic soils are colonised by two extremes, the ancient palaeoendemics and the morphologically undifferentiated races of widespread species. One explanation of this is that there was such a large fluctuation in past climate and environment that all but the few palaeoendemics were eliminated. Such fluctuations would all be very effective in wiping out species highly specialized to ultramafic soils since they would lack genetic flexibility. It is noteworthy that endemics are only found where there are large areas of ultramafic soils, possibly because only large areas offer a sufficient variety of habitats for the endemics to find refuge during climatic changes. The rest of the Great Dyke flora is possibly of recent origin, perhaps as recent as the flora of similar soils in Europe. Most of the present flora is clearly derived from the surrounding flora and suggests that the ability to evolve resistance to ultramafic soils is a widespread attribute amongst angiosperm families. Wild & Bradshaw (1977) commented that the shortage of neoendemic species is 'not easy to reconcile with present ideas on the ease with which parapatric speciation is supposed to be able to occur in adjacent but dissimilar habitats'.

Chemical analyses of native species

Analyses of native species in Zimbabwe were first made by Soane & Saunder (1959) who found a good correlation ($r = 0.93$) between nickel con-

centrations in dried leaves of indigenous grasses (y μg g^{-1}) and soil nickel extracted with ammonium acetate (x μg g^{-1}). A regression equation of $y = 5.25 + 0.806\,x$ was obtained.

Detailed analytical investigations were made in the vicinity of the Empress nickel/copper mine by Cole (1971). At the site there are marked vegetational changes in areas of diorite and serpentinite bedrock and over the ore body. Wild 1974, 1975a published analyses of concentrations in the leaf ash of nickel, chromium, copper and cobalt of several species from ultramafic soils. Unfortunately there is now doubt over his results for chromium which seem much too high in many of the samples and Brooks & Yang (1984) have suggested that there may have been contamination with chromium-rich dust. Wild's analyses seem reliable for the other elements and his chromium work should be repeated.

Brooks & Yang (1984) have analysed leaves from twenty endemic Zimbabwean species and their results are summarized in Table 5. Proctor *et al.* (1980) analysed five species from a range of sites and their results are summarized in Table 6. The data in Table 5 confirm Wild (1974) in establishing *Pearsonia metallifera* as an outstanding nickel accumulator and also indicate high concentrations of this element in *Blepharis acuminata*, *Merremia xanthophylla* and *Rhus wildü*. J. Proctor (unpublished) has confirmed that *Dicoma niccolifera* is an accumulator (1,000 μg g^{-1} dry matter) of nickel since an individual from Kingston Hill had 1,500 μg g^{-1} foliar nickel.

A discrepancy occurs between the data in Tables 5 and 6 for *Dicoma niccolifera*. Brooks & Yang reported very high iron concentrations with a mean value of 7,920 μg g^{-1} dry matter whilst Proctor *et al.* found that mean iron concentrations did not exceed 700 μg g^{-1} dry matter. Some of the lack of correspondence between analyses of Zimbabwean species may result from different analytical techniques or from sampling at different times of the year. Only in Proctor *et al.* (1980) is the sample time specified. The high manganese concentrations in *Pearsonia metallifera* (Table 5) are noteworthy and approach some of the high values recorded for species in the Far East (Jaffré 1980).

Brooks & Yang (1984) established inter-element leaf correlations for the twenty species in

Table 5. The concentrations of several elements in dried leaves of twenty endemic species on the Great Dyke, Zimbabwe (from Brooks & Yang 1984)

Species	P (%)	K (%)	Ca (%)	Mg (%)	Mg/Ca	Co (μg g^{-1})	Cr (μg g^{-1})	Fe (μg g^{-1})	Mn (μg g^{-1})	Ni (μg g^{-1})
Aloe ortholopha	0.13	0.45	0.77	1.69	2.19	1	3	73	23	130
Lotononis serpentinicola	0.10	0.85	2.03	1.44	0.71	2	3	351	109	22
Pearsonia metallifera	0.06	0.03	1.14	0.44	0.39	115	12	2,020	1,910	14,100
Euphorbia nemoralis	0.07	0.44	1.38	1.03	0.75	3	1	78	27	244
E. wildii	0.33	1.94	1.70	0.56	0.33	4	1	642	136	500
Rhus wildii	0.10	0.29	1.47	1.25	0.85	2	4	522	70	1,311
Ozoroa longipetiolata	0.14	0.08	1.17	0.84	0.72	1	2	100	61	46
Gnidia capitata	0.11	0.61	1.60	0.51	0.32	2	12	505	125	34
Convolvulus ocellatus var. plicinervius	0.13	0.38	0.51	0.36	0.71	7	24	2,760	87	121
Merremia xanthophylla	0.26	2.24	0.93	0.57	0.61	5	24	1,710	495	879
Leucas aggerestris	0.06	0.21	0.47	1.41	3.00	1	4	179	59	331
L. hephaëstis	0.11	0.57	0.81	1.00	1.23	1	2	94	37	45
Sutera sp.	0.13	0.19	1.13	1.30	1.15	1	11	950	65	50
S. fodina	0.11	0.26	1.68	0.92	0.55	1	2	324	57	44
Walafrida sp.	0.07	0.27	2.12	1.27	0.60	1	4	376	47	22
Barleria molensis	0.12	0.15	2.72	1.06	0.39	3	23	1,500	85	154
Blepharis acuminata	0.06	0.41	3.49	1.21	0.35	2	4	257	59	895
Dicoma niccolifera	0.07	0.64	1.57	0.76	0.48	19	77	7,920	142	457
Helichrysum serpentinicola	0.06	1.41	0.68	0.87	1.28	7	8	1 230	95	82
Vernonia accommodata	0.09	0.64	3.17	2.79	0.88	1	8	142	49	116

Table 6. The concentrations of several elements in dried leaves of five species from a range of sites (for which soil analyses are given in Table 3) in Zimbabwe (from Proctor *et al.* 1980)

Species	Site	K (%)	Ca (%)	Mg (%)	Mg/Ca	Co (μg g^{-1})	Cr (μg g^{-1})	Fe (μg g^{-1})	Ni (μg g^{-1})
Andropogon gayanus	Noro	0.85	0.07	0.36	5.2	0	25	35	7.5
	Kingston Hill	1.6	0.13	0.32	2.5	0	18	43	33
	Tipperary Claims	1.6	0.07	0.40	5.6	0	68	80	18
	Selukwe	0.96	0.11	0.45	4.1	0	68	53	2.5
Combretum molle	Noro	0.49	0.87	0.55	0.63	0	10	290	33
	Kingston Hill	0.43	0.42	0.47	1.1	5.0	2.5	200	180
	Tipperary Claims	0.38	0.79	0.70	0.89	5.0	18	200	65
	Selukwe	0.45	0.74	0.53	0.72	2.5	25	110	5.0
Dicoma niccolifera	Noro	2.2	0.30	0.62	2.1	0	0	410	230
	Kingston Hill	2.8	0.55	0.77	1.4	2.5	15	230	880
	Tipperary Claims	1.8	0.32	1.0	3.1	0	45	700	440
	Umvuma	2.0	0.51	0.55	1.1	0	23	380	280
	Ngezi	1.7	0.26	0.85	3.3	0	5	450	120
Ledebouria revoluta	Noro	3.3	0.32	1.20	3.8	2.5	0	170	150
	Kingston Hill	3.0	0.91	0.94	1.0	2.5	0	160	350
	Tipperary Claims	1.3	0.36	1.5	4.2	0	61	500	340
	Umvuma	2.8	0.55	0.81	1.5	0	5.0	130	50
Xerophyta equisetoides	Noro	1.3	0.49	0.53	1.1	0	68	75	180
	Kingston Hill	1.7	0.57	0.47	0.82	0	68	38	200
	Tipperary Claims	0.91	0.34	0.68	2.0	2.5	40	140	280
	Selukwe	0.60	0.60	0.81	1.4	0	25	40	7.5

Table 5. They showed a strong negative correlation between leaf magnesium and seven other elements (aluminium, boron, cobalt, manganese, phosporus and sodium). However, Proctor *et al.* (1980) studied inter-element correlations within each of five species and found no general negative correlation between leaf magnesium and other elements except for potassium.

Ernst (1972) has provided some information on the distribution of chromium and nickel within a few plant species from ultramafic soils. For nickel, the leaves and the cortex of the roots and shoots had the highest concentrations. There were relatively low concentrations within the xylem indicating that the endodermis is a barrier to the movement of nickel. Leaf-cell sap analyses of four species revealed from 13.0 mg l^{-1} nickel in *Becium homblei* from Tipperary Claims to 72.5 mg l^{-1} nickel in *Dicoma niccolifera* from Kingston Hill. Chromium was chiefly restricted to the roots but not so in *Dicoma niccolifera*. The populations of this species at Tipperary Claims had a foliar dry weight concentration of 240 μg g^{-1} chromium, a root cortex concentration of 100 μg g^{-1} chromium and a root wood concentration of 24 μg g^{-1} chromium. Leaf-cell sap

analyses of four species (including *Dicoma niccolifera*) revealed no detectable chromium.

Animals

The influence on freshwater snails of the high Mg/Ca quotient in streams draining the Great Dyke has been investigated by Harrison *et al.* (1966). They found that snails were rare or absent in these streams and in experimental studies produced few eggs in unaltered stream water but more eggs in stream water to which calcium had been added.

Mr D. K. Blake (unpublished) reported that young crocodiles in a crocodile farm fed with stream water from the Great Dyke had poorly developed teeth. The likely cause of this was the high Mg/Ca quotient of about 25 in the stream water (Government Analyst's Laboratory, Harare, unpublished sample No 839/71).

Wild (1975b) investigated termites in ultramafic soils in Zimbabwe. He commented that 'a wide range of termite species occur in most soil types there but appear on the whole to be absent from ultramafic soils. Two species (*Odontotermes transvaalensis* and *Trinervitermes dispar*) were found in a small area of the Great Dyke ultramafics. *Odontotermes transvaalensis* is mound forming and is normally found in seasonal swamps. In the areas with ultramafic soils this species is found (surprisingly) in well-drained sites. The vegetation on the termite mounds is a palatable *Setaria lindenbergiana* (or *S. anceps*) – *Panicum novemneive* grass cover compared with the surrounding unpalatable *Loudetia simplex* – *Andropogon gayanus* grasslands. The mounds had a higher pH, total nitrogen and a lower Mg/Ca quotient than the surrounding soils but both had similarly high concentrations of total nickel and chromium.

Workers of both species of termites had high nickel concentrations (about 0.5% dry weight) and chromium (about 0.15% dry weight). Soldiers had much less (about 0.01% nickel and 0.03% chromium). Queens of *Odontotermes* had much lower concentrations (0.002% of both nickel and chromium). (No queens of *Trinervitermes* were analysed). Wild hypothesized that the differences in nickel concentration between castes were because the workers fed directly on plant material which contains both nickel and chromium whilst the soldiers and queens are fed by secretions from the workers, Mg/Ca quotients are less in the soldiers than the workers and much less still in the *Odontotermes* queens. The long-lived queen and the eggs she lays are thus protected from potential toxicities in the soils.

Thus these two species of termite have been shown to adapt to the ultramafic soil in this one locality. Because they are generally absent from these soils it seems as though this adaptation is an unusual one.

Experimental work

Crop species

Hunter (1954) investigated a Great Dyke ultramafic soil which 'had previously given an abnormal, low yielding maize crop, and which was a proposed site for tobacco seedling production.' Analyses of the soil at this site are given in Table 3. Hunter (1954) grew oats and tobacco in this soil and found that they developed characteristic nickel toxicity symptoms. This work was followed by Soane & Saunders (1959) who investigated infertility at four widely separated parts of the Great Dyke. They made plant analyses, soil analyses and growth experiments on maize, oats and tobacco. Their results provided good evidence for acute nickel toxicity in some of the soils. Their claim to have demonstrated acute chromium toxicity in their soils 'B' and 'I' is not convincing. In soil 'B' they did not eliminate the possibility of magnesium toxicity. Soil 'I' was a non-ultramafic soil with a suspicion that it was 'influenced by the toxic material from the neighbouring serpentine soils.' It had a moderate total chromium concentration of 550 $\mu g\,g^{-1}$ and a resin extractable chromium of 0.4 $\mu g\,g^{-1}$ but tobacco grown on this soil had an exceptionally high foliar chromium concentration of 440 $\mu g\,g^{-1}$. Some doubt must be cast on this latter analysis since tobacco grown in sand culture with a solution concentration of 10 mg l^{-1} chromium had only 34 $\mu g\,g^{-1}$ chromium in the leaf dry matter. There was however a resemblance of symptoms for tobacco in soil 'I' and the same species grown in the high-chromium sand culture. Further work is needed on this soil and tobacco because if confirmed it would be the best example of acute chromium toxicity in a soil which is at least partly of ultramafic origin. In experiments in which fertilizers were added to ultramafic soils Soane & Saunder found (p. 325) that 'oats receiving calcium carbonate at all levels showed only slight symptoms of nickel toxicity. Small white chlorotic areas developed in early leaves but were slight or absent on later leaves. There was only very slight general chlorosis.' This statement has a different emphasis from their often quoted summary (p. 329) that 'On one soil toxicity was so intense that raising the pH to 8.2 with calcium carbonate did not eliminate nickel uptake or toxicity symptoms in oats.' An interesting effect of a micronutrient was observed for lucerne, which on their soil 'B' showed poor growth in general but 'a marked response to boron was found, both leaves and growth being normal.'

Further demonstrations of nickel toxicity to crop plants on Zimbabwean ultramafic soils have been made by Cooper (1978) for maize (in a large-scale experiment with about 100 soils), Proctor *et al.* (1980) for oats, and Wiltshire (1972) for a range of crops and weedy species. Wiltshire (1972) examined the effect of nitrogen source (i.e. ammonium or nitrate) on nickel toxicity. He grew non-resistant crops and weeds in soils, all from Kingston Hill and Tipperary Claims, with low or high nickel concentrations. The soil treatments were: unfertilized; and fertilized with complete nutrients with nitrogen supplied entirely as ammonium or nitrate. It was found that fertilization always increased yields and nickel uptake. Nickel uptake was less with ammonium than nitrate fertilization in low-nickel soils but higher with ammonium fertilization in high nickel soils. Nickel was always much higher in the roots than in the shoots but ammonium fertilization caused more to move to the shoots.

A very important study of the interactions of calcium, magnesium and nickel ions has been made by Robertson (1985). He grew maize in water cultures at a range of concentrations of these three elements and expressed his results using the probit method advocated by Craig (1977). He found that root growth was reduced to half that of the controls (RG_{50}) by as little as $0.1 \, mg \, l^{-1} \, Ni^{2+}$. This poisoning was prevented by $4 \, mg \, l^{-1} \, Mg^{2+}$ and by $32 \, mg \, l^{-1} \, Ca^{2+}$. RG_{50} for Mg^{2+} alone was $64 \, mg \, l^{-1}$ and for Ca^{2+}, more than $100 \, mg \, l^{-1}$. Root growth could be stopped within 8 h of exposure to $1 \, mg \, l^{-1} \, Ni^{2+}$. In the absence of Ni^{2+}, Ca^{2+} protected against Mg^{2+} toxicity but Mg^{2+} did not reduce the effects when $CaCl_2$ created osmotic loss at very high concentrations. In low concentrations, a ratio of 4 : 1 excess of Mg/Ca was necessary for Mg^{2+} to begin to reduce growth. At higher concentrations Mg^{2+} exerted its own toxicity despite the presence of Ca^{2+}. When Ni^{2+} would otherwise have poisoned growth, Ca^{2+} added only to the relief of poisoning by Mg^{2+} if the Ca^{2+} was in excess by more than 4 : 1 over Mg^{2+}. Thus Mg^{2+} appeared to be more fitted to prevent Ni^{2+} poisoning, yet at higher concentrations it was toxic itself, while Ca^{2+} could at still higher concentrations protect against both Ni^{2+} and Mg^{2+}. Ni^{2+} poisoned growth by stiffening the expansion zone tissues and by destroying the integrity of the meristems. This latter was the first and most important toxic effect, and was achieved by arresting mitosis. This work of Robertson's goes far to explain the interactive effects of ions in ultramafic soils and should be repeated on a range of species.

Native species

Ernst (1972), using the method of comparative protoplasmatology tested the plasmatic resistance of various populations of *Indigofera* species against graduated solutions of nickel nitrate. He found that the populations of *Indigofera setiflora* from the high-nickel soils at Tipperary Claims had leaf cells which were able to tolerate up to as much as one hundred times more nickel than races from non-ultramafic soils. The tolerance was specific for nickel and the race was not tolerant of copper and zinc. Craig (1977) showed that a clone of the grass *Loudetia simplex* from Tipperary Claims was more nickel tolerant (judged by root growth in water culture) than a clone of the same species from a non-ultramafic area near Salisbury. A feature of Craig's work was his pioneering the use of probit analysis in measuring metal tolerance. Wiltshire (1974) tested a wide range of species and found that populations from ultramafic soils usually grew better on these soils in the glasshouse than populations from non-ultramafic soils. However the differences were statistically significant in only three instances. Wiltshire showed that the fraction of nickel retained in the roots was larger in the more tolerant populations. The problem of the precise effect of toxic concentrations of nickel has been approached by Robertson & Meakin (1980). They studied the effect of single-salt solutions of nickel sulphate over the nickel concentration range of $0-0.25 \, mg \, l^{-1}$, on cell division and growth of roots of seedlings of the tree *Brachystegia spiciformis*. This tree is usually absent from ultramafic soils although it has been recorded by Proctor & Craig (1978) in one area with moderate concentrations (but lower than those in the adjacent grassland) of exchangeable nickel. The soil analyses for this site are summarized in Table 3. It seemed that *Brachystegia spiciformis* may be sensitive to nickel toxicity and Robertson & Meakin set out to discover what damage nickel does and at what concentration, to the growing root cells of the seedlings. They found that $0.016 \, mg \, l^{-1}$ nickel had little effect, that $0.03 \, mg \, l^{-1}$ and $0.06 \, mg^{-1}$ both permit considerable growth for one day and some further growth in the next two days; and that 0.125 and $0.25 \, mg \, l^{-1}$ nickel stop growth within one day. They concluded that the GD50 value (that concentration at which growth is reduced by 50% in three days) is $0.04\,mg\,l^{-1}$ but commented that this value depends on experimental conditions since for example, a change in oxygenation will alter it. Their anatomical investigation showed that nickel has two effects. It prevents cell division by interfering with entry into prophase and the low nickel concentration needed to bring about this effect suggests that the mechanism is highly specific. There is another effect in preventing the expansion of the protoplast in growing cells, many of which appear partially plasmolysed in nickel solutions. As well as showing the likely intracellular influence of nickel the work confirms the sensitivity of *Brachystegia spiciformis* to nickel toxicity. It would be most useful to extend the work to include simulated soil solutions to assess the likely influences of other ions on the nickel toxicity (cf. Johnston & Proctor 1981, Robertson 1985) and also to investigate if the population of *Brachystegia spiciformis* described by Proctor & Craig (1978) is nickel tolerant.

Concluding remarks

The knowledge of the ecology of ultramafic areas of Zimbabwe is still incomplete even though they are the best known in Africa. All analytical studies and experimental work indicates a major effect of nickel in the soils and yet it is difficult to explain vegetation differences within the ultramafic areas in terms of differing amounts of nickel. There is no simple correlation between vegetation type and soil nickel concentrations. For example riverine forest (the most luxuriant vegetation in many parts of Zimbabwe) may occur on highly nickeliferous soils. Our suggestion that soil hydrology plays an important role (perhaps interactively with chemical toxicity) in determining vegetation types on the ultramafic soils needs testing critically.

There are many differences between the vegetation of ultramafic areas in Zimbabwe and those for the best investigated tropical ultramafic area, which is New Caledonia in the Far East (Proctor, this volume). In Zimbabwe vegetation on ultramafic soils is usually of much smaller stature and more open and less species-rich than that of adjacent non-ultramafic soils. This does not

apply in New Caledonia. In Zimbabwe there are relatively few endemic species and apparently relatively few nickel-accumulating species on the ultramafic soils in contrast to the situation in New Caledonia. Much more work is required to provide satisfactory explanations for these differences.

References

Boughey, A. S. 1961. The vegetation types of Southern Rhodesia – a reassessment. Proceedings and Transactions of the Rhodesian Scientific Association 49: 54–98.

Brooks, R. R. & X.-H. Yang. 1984. Elemental levels and relationships in the endemic serpentine flora of the Great Dyke, Zimbabwe and their significance as controlling factors for the flora. Taxon 33: 392–399.

Cole, M. M. 1971. Biogeographical/geobotanical and biogeochemical investigations connected with exploration for nickel-copper ores in the hot wet summer/dry winter savanna woodland environment. Journal of the South African Institute of Mines and Metallurgy 71: 199–209.

Cole, M. M. 1986. *The Savannas – Biogeography and Geobotany*. Academic Press, London.

Cooper, G. R. 1978. Greenhouse experiments with maize grown in ultramafic soils. Proceedings of the 8th National Congress of the Soil Science Society of South Africa. Technical Communication 165: 152–157.

Craig, G. C. 1977a. A method of measuring heavy metal tolerance in grasses. Transactions of the Rhodesian Scientific Association 58: 9–16.

Craig, G. C. 1977b. Adaptation to nickel toxicity as a factor affecting colonisation of serpentine soils in Rhodesia. Unpublished Ph.D thesis, University of Zimbabwe, Zimbabwe.

Ellis, B. S. 1951. The soils of Rhodesia. The Rhodesia Agricultural Journal 48: 182–212.

Ernst, W. 1972. Ecophysiological studies on heavy metal plants in south central Africa. Kirkia 8: 125–145.

Grim, R. E. 1953. Clay mineralogy, McGraw Hill, London.

Harrison, A. D., W. Nduku, & A. S. C. Hopper. 1966. The effects of a high magnesium-to-calcium ratio on the egg-laying rate of an aquatic planorbid snail, Biomphalaria pfeifferi. Annals of Tropical Medicine & Parasitology 60: 212–214.

Hunter, J. G. 1954. Nickel toxicity in a southern Rhodesian soil. South African Journal of Science 133–135.

Jaffré, T. 1980. Etude écologique du peuplement végétal de sols dérivés de roches ultrabasiques en Nouvelle Calédonie. Travaux et Documents de l'O.R.S.T.O.M., 124.

Johnston, W. R. 1979. Biochemical and physiological studies on races of Festuca rubra L. from a serpentine and non-serpentine soil. Unpublished Ph.D. thesis, University of Stirling, UK.

Johnston, W. R. & J. Proctor. 1981. Growth of serpentine

and non-serpentine races of *Festuca rubra* in solutions simulating the chemical conditions in a toxic serpentine soil. Journal of Ecology 69: 855–869.

Proctor, J., J. Burrow, & G. C. Craig. 1980. Plant and soil chemical analyses from a range of Zimbabwean serpentine sites. Kirkia 12: 127–139.

Proctor, J. & G. C. Craig. 1978. The occurrence of woodland and riverine forest on the serpentine of the Great Dyke. Kirkia 11: 129–132.

Proctor, J., W. R. Johnston, D. A. Cottam, & A. B. Wilson. 1981. Field-capacity water extracts from serpentine soils. Nature, London 294: 245–246.

Robertson, A. I. 1985. The poisoning of roots of *Zea mays* by nickel ions, and the protection afforded by magnesium and calcium. New Phytologist 100: 173–89.

Robertson, A. I. & M. E. R. Meakin: 1980. The effect of nickel on cell division and growth of *Brachystegia spiciformis* seedlings. Kirkia 12: 115–126.

Soane, B. D. & D. M. Saunder. 1959. Nickel and chromium toxicity of serpentine soils in Southern Rhodesia. Soil Science 88: 322–330.

Thompson, J. G. 1965. The soils of Rhodesia and their classification. Salisbury, Rhodesia.

Werger, M. J. A., H. Wild, & B. R. Drummond. 1978a. Vegetation structure and substrate of the northern part of the Great Dyke, Rhodesia. Environment and plant communities. Vegetatio 37: 79–89.

Werger, M. J. A., H. Wild, & B. R. Drummond, 1978b. Vegetation structure and substrate of the northern part of the Great Dyke, Rhodesia: gradient analysis and dominance-diversity relationships. Vegetatio 37: 151–161.

Wild, H. 1964. The endemic species of the Chimanimani mountains and their significance. Kirkia 4: 125–157.

Wild, H. 1965. The flora of the Great Dyke of southern Rhodesia with special reference to the serpentine soils. Kirkia 5: 49–86.

Wild, H. 1970. Geobotanical anomalies in Rhodesia 3. The vegetation of nickel bearing soils. Kirkia 7: suppl. 1–62.

Wild, H. 1971. The taxonomy, ecology and possible method of evolution of a new metalliferous species of *Dicoma* Cass. Mitt. Bot. Staatssaml. München 10: 266–274.

Wild, H. 1974a. Variations in the serpentine floras of Rhodesia. Kirkia 9: 209–232.

Wild, H. 1974b. Indigenous plants and chromium in Rhodesia. Kirkia 9: 233–241.

Wild, H. 1975a. The uptake of heavy metals by some succulent species of Rhodesian serpentines. Excelsa 5, 17–22.

Wild, H. 1975b. Termites and the serpentines of the Great Dyke of Rhodesia. Transactions of the Rhodesia Scientific Association 57: 1–11.

Wild, H. & A. D. Bradshaw. 1977. The evolutionary effects of metalliferous and other anomalous soils in south central Africa. Evolution 31: 282–293.

Wild, H. 1978. The vegetation of heavy metal and other toxic soils. In Werger, M. J. A. & Van Bruggen (eds). *Biogeography and Ecology of Southern Africa.* pp. 1301–1332. W. Junk, The Hague.

Wiltshire, G. H. 1972. Effect of nitrogen source on translocation of nickel in some crop plants and weeds. Kirkia 8: 103–123.

Wiltshire, G. H. 1974. Growth of plants on soils from two metalliferous sites in Rhodesia. Journal of Ecology 62: 501–525.

Authors' Addresses:
J. Proctor
Dept. of Biological and Molecular Sciences
University of Stirling
STIRLING
Scotland FK9 4LA
U.K.

M. M. Cole
Dept. of Geography
Royal Holloway and Bedford New College
University of London
Egham Hill
EGHAM
Surrey TW20 0EX
U.K.

XIII

The vegetation over mafic and ultramafic rocks in

the Transvaal Lowveld, South Africa

M. M. COLE

Abstract. Following a review of the physical environment and characteristic vegetation of the Transvaal Lowveld the distributions of the distinctive shrub communities over serpentinite and pyroxenite bedrock within the Palabora Complex and in the Mashishimale area are evaluated with reference to levels of major, minor and trace elements in the soils. The distributions are shown to be associated with low calcium and phosphorus levels, high magnesium levels and low calcium/magnesium ratios. The plants however are found to derive adequate calcium and phosphorus from the small amounts present in the soils.

B. A. Roberts and J. Proctor (eds), The ecology of areas with serpentinized rocks. A world view, 333–342.
© *1992 Kluwer Academic Publishers.*

Introduction

Distinctive shrub communities that contrast with the prevailing dry savanna woodland and savanna parkland of the eastern Transvaal Lowveld of South Africa occur over serpentinite and pyroxenite bedrock near Phalaborwa. Several of these have been examined in the field and detailed geobotanical, biogeochemical investigations have been carried out over a pyroxenite/serpentinite dyke in the Mashishimale Bantu area and over similar rock types within the Palabora Complex.

Physical environment

The greater part of the western section of the Transvaal Lowveld is underlain by granite and granite gneiss of the Archaean Basement Complex (Fig. 1). More ancient metamorphic rocks of the Primitive System form the Murchison Range that extends WSW to ENE from south of Tzaneen to north of Phalaborwa and also the Barberton Mountain Land some 200 km farther south. In the eastern part of the Transvaal Lowveld the ancient rocks are covered with Karoo sediments and lavas. Near Phalaborwa the Old Granite has been intruded by a pipe-like body known as the Palabora Complex. This Complex is a multiple replacement that is expressed in concentric outcrops of different rock types (Fig. 2). Peripheral outcrops of syenite that give rise to a series of castle kopjes or tors are succeeded first by felspathic pyroxenite and then pyroxenite which occupies most of the centre of the Complex. The northern part of the pyroxenite body contains a pegmatoid consisting of pyroxene, vermiculite and olivine that has yielded large quantities of vermiculite. The centre is formed mainly of serpentine and vermiculite while the southern part contains small irregular pegmatoids of diopside, vermiculite, apatite and minor amounts of serpentinized olivine. A carbonatite, of Proterozoic age, that hosts an important copper orebody and a surrounding girdle of phoscorite that is mined for phosphate, intrudes the centre of the Complex. Near Phalaborwa, pyroxenite and serpentinite bedrock units occur both within the Palabora Complex and the Archaean System.

The relief, which is the product of the interplay of processes of pediplanation and drainage superimposition from a Karoo surface onto older underlying rocks (Cole 1987) is generally subdued in the vicinity of Phalaborwa. For the most part the terrain is level, but inselberge surmounted by tors mark the outcrops of resistant rocks like the syenite that rings the Palabora Complex and whalebacks of bare rock occur where the Old Granite outcrops.

The climate in the vicinity of Phalaborwa is characterized by hot rainy summers and warm dry winters with great extremes of temperature and humidity. The rainfall increases from 400 mm in the east to 750 mm near Letaba and the regional differences exert some influence on soil-forming processes.

The soils reflect the interplay of variations of rainfall and relief on weathering processes and pedogenesis over different types of bedrock. Over the higher ground and where the rainfall exceeds 750 mm the Old Granite has weathered to highly leached grey sandy soils whereas on level ground and under more arid conditions it has weathered to red sandy or sandy clay soils. By contrast dark brown loams and clay loams with calcite nodules of variable size and in varying concentrations occur over basaltic lavas and pyroxenite. Calcrete is widespread within the region while in areas of impeded drainage sodic soils are found. Over the Palabora Igneous Complex the soils vary with rock type, grey and red sandy soils occuring over syenite, dark brown loams over pyroxenite and black clay loams over carbonatite.

Vegetation

The characteristic vegetation over dark brown loams over pyroxenite bedrock in the Phalaborwa area is an open dry type of savanna woodland dominated either by *Acacia nigrescens* trees or by *Colophospermum mopane* trees associated in different parts of the area with *Ormocarpum trichocarpum*, *Dalbergia melanoxylon*, *Vangueira infausta*, *Lannea discolor* and *Securinega virosa* trees and *Grewia bicolor*, *G. flavescens*, *Euclea undulata* and *Ozorea paniculosa* shrubs and *Enneapogon cenchroides* and *Schmidtia pappophorides* grasses. The vegetation, however, is sensitive

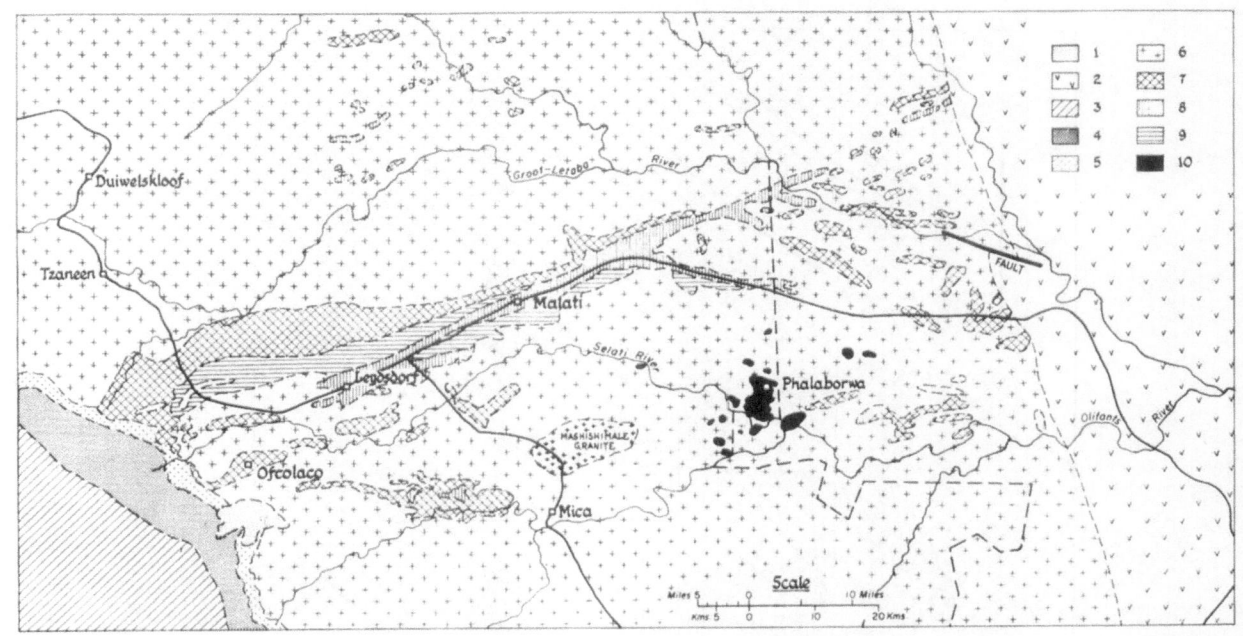

Fig. 1. Geology of part of the Transvaal Lowveld including the Phalaborwa and Mashishimale area.

to changes of soil mineral status associated with bedrock type and the presence of mineral deposits. The trees are absent from dolerite dykes which are marked by grassland of one species – *Schmidtia pappophoroides*. *Acacia nigrescens* trees favour pyroxenite with high concentrations of vermiculite whereas stunted *Colophospermum mopane* trees occur over high-apatite pyroxenite and phoscorite and at the periphery of areas of serpentinite. Prior to mining for copper, a dense and distinctive vegetation of *Acacia nigrescens*, *Dichrostachys cinerea*, *Ormocarpum trichocarpum*, *Zizyphus zeyheriana* and *Lannea stuhlmannii* trees and tall *Grewia hexamita* and *G. bicolor* shrubs covered the carbonatite at Loolekop and the small shrubs and herbs *Barleria affinis*, *B. galpinii*, *B. senensis*, *Ecbolium revolutum* and *E. lugardae*, the last two being copper indicator

plants, *Peristrophe cernua* and *Acalypha indica* were conspicuous over the cupriferous ground. At the periphery of the Complex *Combretum apiculatum* trees characterized the syenite outcrops beyond which a savanna woodland of *Colophospermum mopane*, *Combretum* and *Acacia* trees with an occasional *Sclerocarya birrea* occupied granite terrain.

With these variations in the vegetation over the prevailing rock types the occurrence of impoverished shrub communities over pyroxenite/serpentinite bedrock units near Phalaborwa is not surprising.

On the farm 'Laaste,' north of Phalaborwa township, the *Colophospermum mopane* woodland becomes dwarfed and then gives way to a shrub community of *Grewia bicolor* and *Euclea undulata* over serpentinite bedrock covered by

335

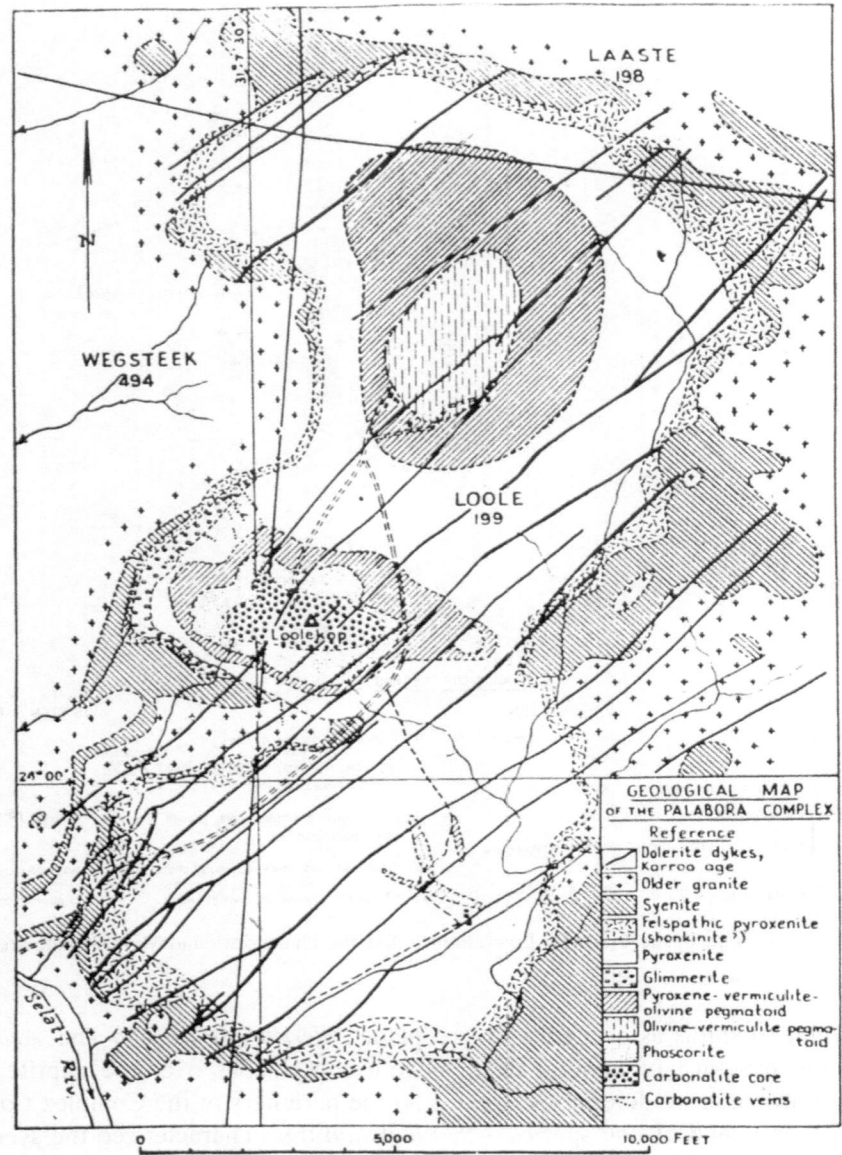

Fig. 2. Geology of the Palabora Complex (after Hoffman and Hanekan in Hanekan, van Staden, Smit and Pike. The Geology of the Palabora Igneous Complex. Memoir 54 Geological Survey of Republic of South Africa, 1965).

dark brown clay loam soils with a calcrete horizon in the profile (Fig. 3). Similar vegetation occurs over serpentinite south of the Selati river and also over pyroxenite/serpentinite bedrock in the Mashishimale Bantu area. Geobotanical, biogeochemical and geochemical investigations made along a series of transect lines on the last named

area produced some insight into the vegetation/environmental relationships.

The Mashishimale Bantu area lies about 15 km west of Phalaborwa on the Gravelotte – Tzaneen road. About 4 km south of the road a pyroxenite dyke with a WNW–ESE orientation cuts through the Old Granite, a highly weathered pink fel-

Fig. 3. Shrub vegetation dominated by *Grewia bicolor* and *Euclea undulata* over serpentinite bedrock on the farm Laaste, north of Phalaborwa. *Colophospermum mopane* trees at the periphery. Serpentinite hills at the periphery of the Palabora Complex in the background.

spathic granite that underlies most of the area. Open grassland with scattered *Sclerocarya birrea* and *Colophospermum mopane* trees cover most of the granite country much of which, however, is used for cultivation following clearance of savanna woodland. Owing to the unfavourable nature of the area for agriculture the vegetation over the pyroxenite dyke, however, remains undisturbed. It comprises a low shrub community dominated by *Euclea undulata* and *Glossochilus parviflorus* with a few taller shrubs of *Grewia hexamita*, *G. bicolor* and *G. caffra* at the periphery (Fig. 4). The grass cover consisting mainly of *Enneapogon cenchroides* with *Setaria woodii*, *Brachiaria deflexa* and *Schmidtia pappophoroides* is sparse.

Geobotanical, biogeochemical and geochemical data from these transects provide information on plant/soil relationships over the Mashishimale pyroxenite/serpentinite body but the prevalence of cultivation precluded the acquisition of comparable data over the adjacent granite areas. The geochemical data (Figs 5–8) show relatively high concentrations of iron, moderate concentrtions

of magnesium and very low concentrations of calcium and phosphorus over the pyroxenite/serpentinite body. Nickel and chromium concentrations are above average but are far below those over serpentinite bedrock along the Great Dyke and near the Empress nickel/copper mine in Zimbabwe or over serpentinite and other ultramafic rocks in Western Australia. The soil geochemistry of the Mashishimale pyroxenite/serpentinite therefore suggests that high magnesium/calcium quotients and very low concentrations of phosphorus exert important influences over the form and composition of the vegetation in that area.

Only one species, *Euclea undulata* was sampled along the transects. The analytical results (Table 1) show relatively high calcium concentrations and low magnesium/calcium quotients in both the leaves and first-year stems of this species as well as high concentrations of potassium. The species appears able to obtain the calcium and potassium that it requires from the small amounts available in the soils. Phosphorus concentrations however, are very low and may be a limiting factor for plant growth.

337

Fig. 4. Shrub vegetation dominated by *Euclea undulata* with some taller *Grewia* spp over serpentinite (foreground) contrasting with open woodland of *Colophospermum mopane* and *Sclerocarya birrea* trees over granite (background). Mashishimale Bantu area, west of Phalaborwa.

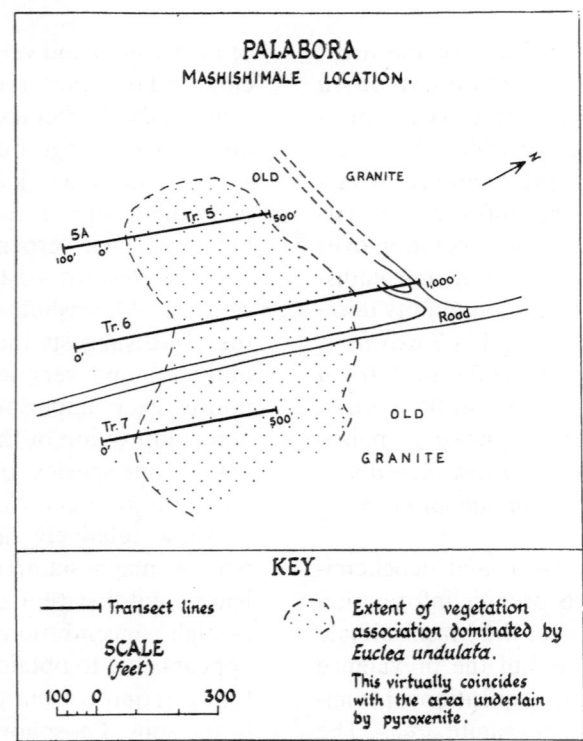

Fig. 5. Location of geobotanical, biogeochemical, geochemical transects across the Mashishimale pyroxenite/serpentinite outcrop.

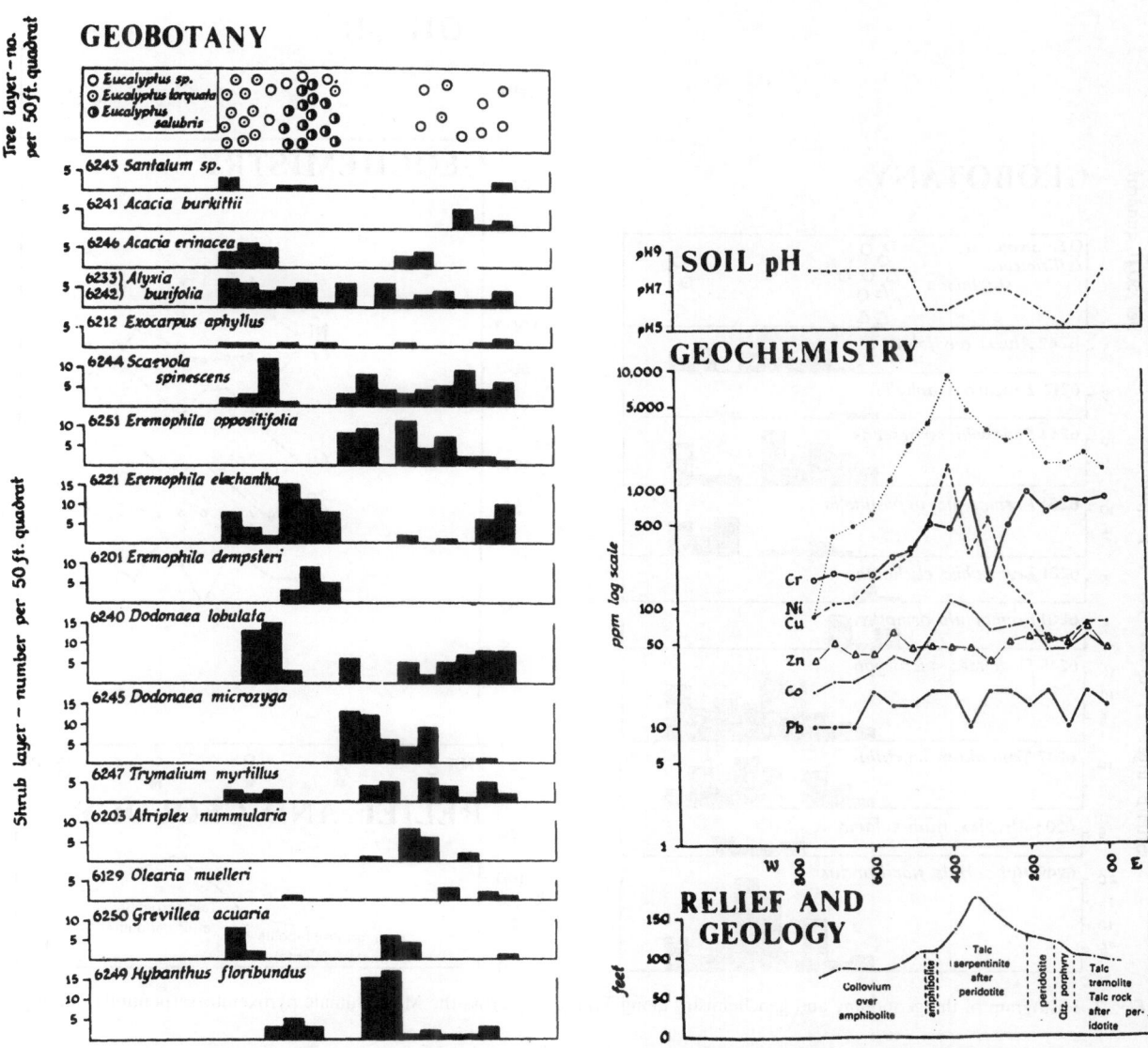

Fig. 6. Histogram of the geobotany, and geochemistry along Transect 5 across the Mashishimale pyroxenite/serpentinite outcrop.

While the geochemistry of the soils over the Mashishimale pyroxenite/serpentinite body cannot be compared with that of soils over the adjacent granite it can be compared with that of soils with thick calcrete layers developed over pyroxenite bedrock containing large amounts of apatite and mica in the northern part of the Palabora Complex. There the levels of total calcium, phosphorus and potassium average 100,000 μg g^{-1}, 25,000 μg g^{-1} and 4,000 μg g^{-1} respectively and exceed 300,000 μg g^{-1}, 50,000 μg g^{-1} and 5,000 μg g^{-1} at some sites. The character and composition of the open savanna woodland in that area, dominated either by *Colophospermum*

mopane or *Acacia nigrescens* trees associated with *Ormocarpum trichocarpum*, *Dalbergia melanoxylon*, *Vangueria infausta*, *Lannea discolor* and *Securinega viroza* trees, *Grewia* spp. shrubs and *Schmidtia pappophoroides* or *Enneapogon cenchroides* grass appear to be related to the favourable base status of the soils.

Thus the data indicate that the characteristic low shrub communities that occupy serpentinite and pyroxenite bedrock in the eastern Transvaal Lowveld are associated with low concentrations of phosphorus and calcium and relatively high concentrations of magnesium in the soils and further that the characteristic species are able to

339

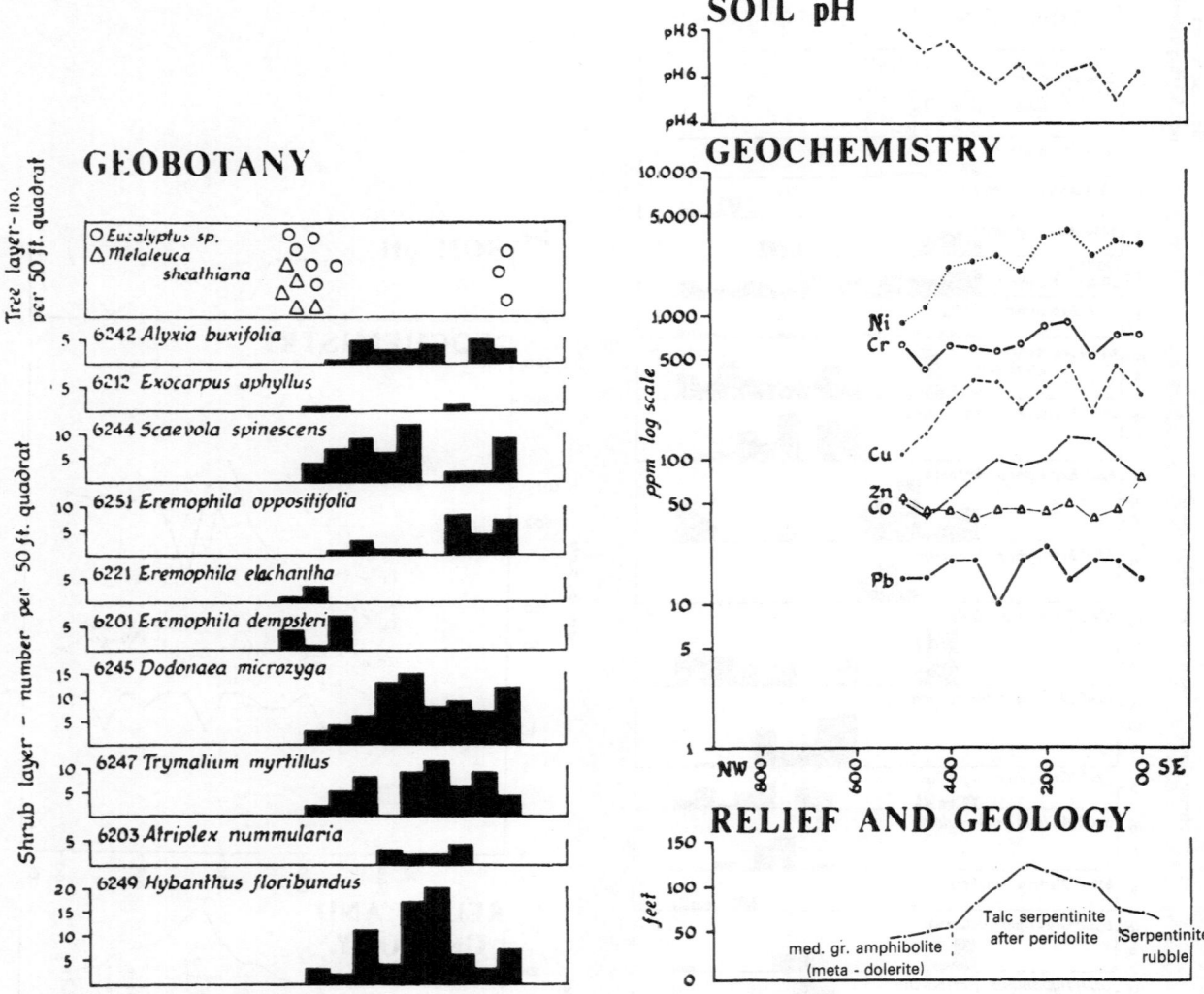

Fig. 7. Histogram of the geobotany and geochemistry along Transect 6 across the Mashishimale pyroxenite/serpentinite outcrop.

obtain their requirements of calcium and potassium from the small amounts present in the soil.

Appendix I: Species names with their authorities

ACANTHACEAE

Barleria affinis C. B. Clarke
B. galpinii C. B. Clarke
B. senenis Klotzsch
Ecbolium lugardae N. E. Br.
Ecbolium revolutum C. B. Clarke
Glossochilus parviflorus Nees

ANACARDIACEAE

Lannea discolor (Sond.) Engl.
Lannea stuhlmannii (Engl.) Engl.
Ozoroa paniculosa (Song.) R. and A. Fernandes
Sclerocarya birrea (A. Rich.) Hochst
S. caffra Sond

COMBRETACEAE

Combretum apiculatum Sand.

EBENACEAE

Euclea undulata Thunb.

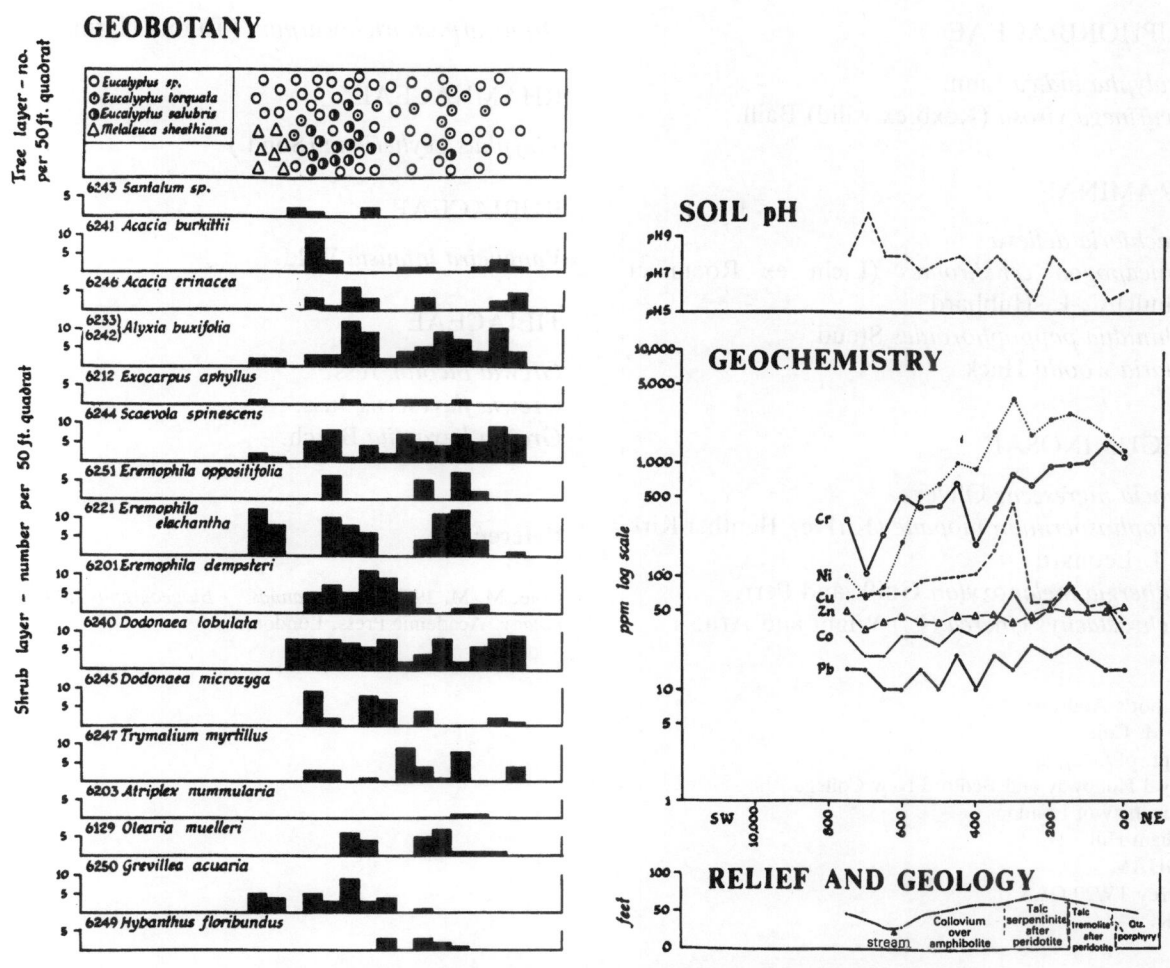

Fig. 8. Histogram of the geobotany and geochemistry along Transect 7 across the Mashishimale pyroxenite/serpentinite outcrop.

Table 1. The percentage ash and concentrations (μg g^{-1} dry matter) of phosphorus, potassium, sodium, calcium, magnesium, chromium, nickel and iron in (a) leaves and (b) first year stems of *Euclea undulata* from the Mashishimale pyroxenite/serpentinite area

Site	Part of plant	Ash (%)	P	K	Na	Ca	Mg	Mg/Ca quotient	Cr	Fe	Ni
Tr 6 300N	a	6.02	357	13300	287	8010	2530	0.316	0.9	90	4.7
	b	6.29	340	7960	439	11300	1180	0.104	1.7	133	6.9
400N	a	6.53	191	9330	505	11900	2260	0.190	1.0	100	5.2
	b	4.93	126	6380	537	13900	921	0.066	2.1	199	5.6
700N	a	8.03	45	6160	125	9080	1960	0.216	1.1	48	4.5
	b	5.80	151	9790	392	11700	1150	0.098	1.5	118	8.2
800N	a	8.52	89	5510	236	24400	1700	0.070	1.2	159	11.5
	b	7.31	54	3490	216	21100	846	0.040	2.1	161	9.7
Tr 7 100N	a	6.19	166	8790	88	11800	2370	0.201	1.2	63	4.9
	b	4.74	142	5440	201	11800	818	0.069	0.9	112	5.0
300N	a	6.81	172	7170	349	13100	2460	0.188	–	58	6.0
	b	5.39	134	6580	348	15600	1010	0.065	–	139	6.5
400N	a	7.29	169	10920	155	13900	1850	0.133	0.7	74	5.8
	b	5.92	150	6630	272	17300	7670	0.443	1.4	117	5.4
500N	a	6.31	177	8230	171	12500	2030	0.162	0.8	68	6.3
	b	5.45	83	5180	352	15800	731	0.046	1.3	167	5.2

341

EUPHORBIACEAE

Acalypha indica Linn.
Securinega virosa (Roxb ex willd) Baill.

GRAMINAE

Brachiaria deflexa
Enneapogon cenchroides (Licht ex Roem et Schult) C. E. Hubbard
Schmidtia pappophoroides Steud.
Setaria woodii Hack

LEGUMINOSAE

Acacia nigrescens Oliver
Colophospermum mopane (Kirk ex Benth.) Kirk ex J. Leonard
Dalbergia melanoxylon Guill. and Perr.
Dichrostachys cinerea (L.) Wight and Arn.

Ormocarpum trichocarpum (Taub.) Harms.

RHAMNACEAE

Zizyphus zeyheriana (Sond.)

RUBIACEAE

Vangueira infausta Wild.

TILIACEAE

Grewia bicolor Juss.
Grewia flavescens Juss.
Grewia hexamita Burch.

Reference

Cole, M. M. 1987. *The Savannas — Biogeography and Geobotany*. Academic Press, London.

Author's Address:
M. M. Cole
Dept. of Geography
Royal Holloway and Bedford New College
University of London
Egham Hill
EGHAM
Surrey TW20 OEX
U.K.

XIV

The vegetation of the greenstone belts

of Western Australia

M. M. COLE

Abstract. The vegetation of the Eastern Goldfields area of Western Australia which embraces extensive narrow belts of ultramafic and mafic rocks, comprises numerous sclerophyllous woodland and shrub communities exhibiting complex distributions that are related to the interplay of climatic and edaphic factors, the latter in turn influenced by geomorphology and geology. Following reviews of these relationships, the distributions of vegetation communities and individual species over ultramafic soils in a number of selected areas are examined with reference to levels of major, minor and trace elements, notably calcium, magnesium, iron, nickel, copper, cobalt and chromium. The role of *Hybanthus floribundus* (Lindl.) F. Muell as a nickel indicator and accumulator plant is evaluated.

B. A. Roberts and J. Proctor (eds), The ecology of areas with serpentinized rocks. A world view, 343–373.
© 1992 *Kluwer Academic Publishers.*

Introduction

In Western Australia ultramafic and mafic rocks occur within sequences of Archaean folded meta-volcanic and meta-sedimentary rocks, locally known as greenstones, that outcrop or suboutcrop between 34° and 25°S from the coast south of Norseman through Kalgoorlie, Menzies and Leonora to the vicinities of Mount Magnet, Meekatharra, Wiluna and Barwidgee in the north (Fig. 1). Because of the historic importance of gold mining this area is known as the Eastern Goldfields of Western Australia.

The vegetation of the Eastern Goldfields embraces a variety of sclerophyllous woodland and shrub communities exhibiting distributions that are related to the interplay of climatic and edaphic factors with the latter in turn related to geomorphology and geology. Information on the vegetation is available from Carnahan (1978), Cole (1973), Elkington (1969), Groves (1981) and Speck (1958).

The elucidation of the factors influencing the plant communities over ultramafic rocks in Western Australia is complicated by climatic variations related to the latitudinal extent over which these rocks outcrop or suboutcrop, by physiographic features, by the extent, variable thickness and variable nature of superficial cover and by the varied nature of the ultramafic rocks themselves. Recent studies of Landsat imagery, however, have provided information on the superficial and bedrock geology, relief and drainage features which assists the elucidation of vegetation distributions (Cole 1985).

Climate

The climate of the Eastern Goldfields is characterized by hot, dry summers and cool winters when an average 250–300 mm of rainfall is received. In January, the hottest month, the mean maximum temperatures range from 33°C in the south to 36°C in the north; the mean minima increase from 16.5°C to 21.5°C. July is the coldest month with average temperatures of only 11°C to 14°C. Temperatures rise in September and October to respective averages of 15.5°C and 21°C.

Generally speaking temperature extremes increase with distance from the coasts. The rainfall occurs mainly in the May–July period. About 12.5 mm of rainfall may be received in August and negligible amounts in September. The summer is normally rainless but occasionally cyclones penetrating inland from the northwest coast may bring torrential falls and cause widespread flooding and a temporary rapid lowering of temperature. The period when there is adequate moisture and the temperatures are sufficiently high for active plant growth is virtually limited to September and October. The favourable period shortens towards the interior as decreasing rainfall and increasing temperatures accentuate aridity.

Geology

The older Archaean greenstones occupy narrow belts enclosed within the younger Archaean granite or granite gneiss that makes up the greater part of the Yilgarn block of the Western Australian Precambrian Shield (Fig. 1). Near Norseman a major east-west trending norite dyke cuts through the greenstone sequence while in the Kambalda, Widgiemooltha and Dordie Rocks areas stocks and semi-stratigraphic dykes of felsic and quartz porphyry composition intrude all rock formations.

Throughout the Yilgarn block, outcrops are rare and are mainly confined to hills and ridges of more resistant rocks. Superficial deposits cover most of the area, concealing contacts between greenstone and granite, between sedimentary and volcanic, mafic and ultramafic rocks within the greenstone sequence (Fig. 2).

The ultramafic rocks are of several types. At Kambalda serpentinite outcrops in low eminences and also forms the lower slopes of more prominent meta-basalt hills. Over the lower ground it is masked by superficial deposits and soil cover. At Widgiemooltha talc serpentinite after peridotite forms a prominent hill feature whereas amphibolite and talc tremolite-talc rock after peridotite forms the adjacent lower ground. To the south the Mount Thirsty area is underlain by a succession of volcanic and sedimentary rocks

344

Fig. 1. The Eastern Goldfields of Western Australia – geology.

which have been intruded by a complex mafic-ultramafic sill exhibiting macro-layering from altered pyroxenite and/or peridotite to hypersthenite, gabbro and diorite (Yates 1966/67). Farther south, hypersthenite forms prominent ridges near Mission Ridge. At Kurnalpi, northeast of Kalgoorlie, lateritized serpentinite forms of low plateau that rises about 5 to 10 m above adjacent areas where colluvium masks serpentinite and amphibolite. North of Kalgoorlie lateritized serpentinite occurs again near Broad Arrow and on the Kurrajong property in the Mount Ida areas where it forms prominent hill features, near Agnew and near Mount Magnet. Nickel-ore bod-

345

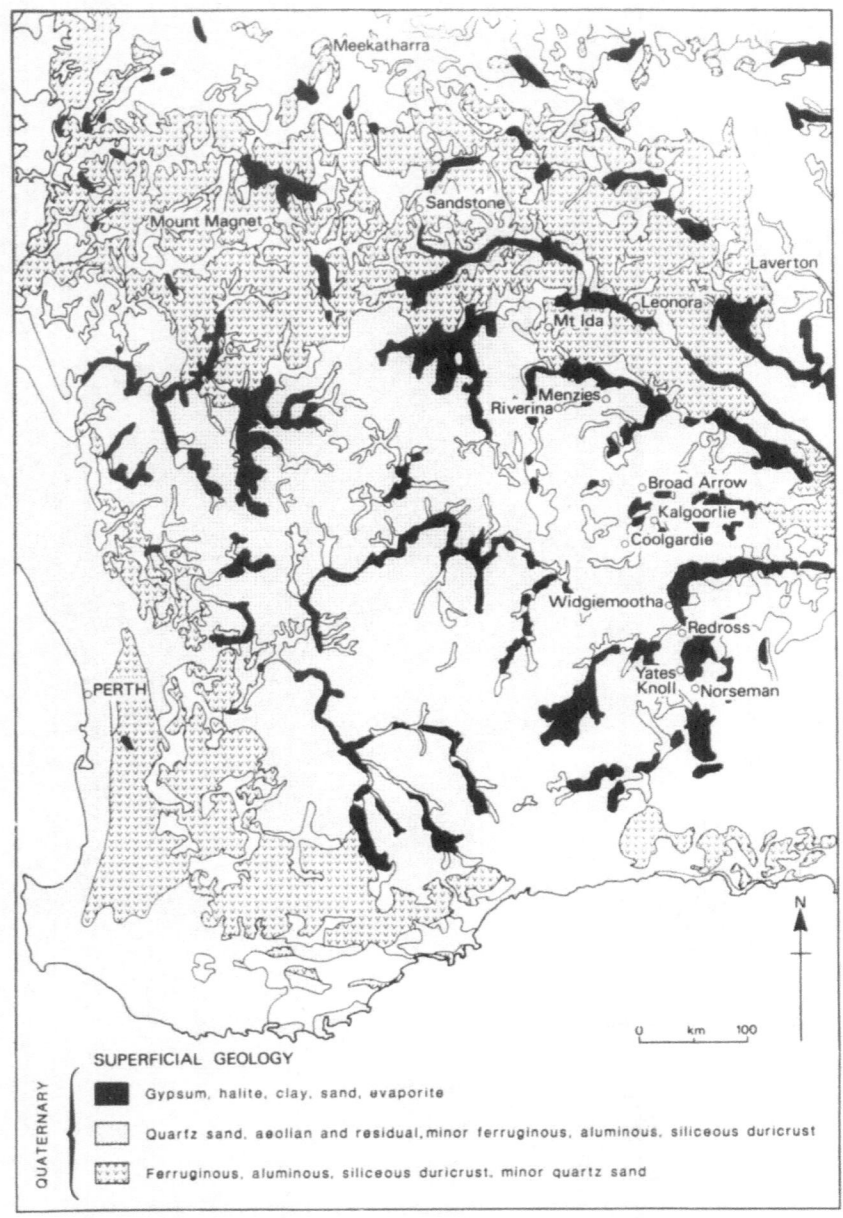

SUPERFICIAL GEOLOGY

▐ Gypsum, halite, clay, sand, evaporite

☐ Quartz sand, aeolian and residual, minor ferruginous, aluminous, siliceous duricrust

▨ Ferruginous, aluminous, siliceous duricrust, minor quartz sand

Fig. 2. The Eastern Goldfields of Western Australia – superficial geology.

ies occur in the serpentinite rocks at Kambalda, Widgiemooltha, Dordie Rocks and Agnew and high concentrations of chromium or nickel or both are found in the soils associated with the other named areas of serpentinite bedrock. Generally speaking serpentinite weathers more readily than other rock types. It usually forms low ground, normally covered by superficial deposits. Where it forms higher ground it is either affected by mineralization or capped by relict lateritic material.

Physiography and geomorphology

The physiography (Jutson 1934) is the legacy of the interplay of geomorphological processes act-

346

ing over the bedrock geology. In broad outline the area comprises a plateau exhibiting two distinct planation surfaces cut across the Archaean rocks. Whereas the trend of the greenstone belts is north-north-west-south-south-east, that of the drainage patterns, which are characterized by strings of salt lakes is generally east-west. The older and higher of the planation surfaces has an elevation of 330 to 450 m. The lower and younger one which cuts into it, has a height of 270 to 330 m; it is characterized by shallow depressions occupied by the salt lakes which now form local base levels for drainage and erosion. The higher surface dates from the Miocene. Originally capped by lateritic soils, its dissection following epeirogenic uplift has produced a sequence of erosional and depositional surfaces ranging in age from Tertiary to the present. Breakaways mark the limits of the old plateau. Sometimes they form cliff features over 7 m high, as at Sandstone, where the mottled and pallid kaolinitic zones are displayed beneath the laterite horizon. Elsewhere, as at the margin of the salt lakes near Goongarrie, where mottled and pallid zones are exposed, they form only minor ledges less than 1 m high. Where dissection is most advanced, all traces of lateritic material has been removed to reveal underlying bedrock but over most of the area relict lateritic material is widespread. In places, the eroded upper layers of the old lateritic soils have been deposited as a mantle of sand to produce the so-called sandplains. Additionally deposition of alluvium and of calcrete along trunk drainage lines took place during the Quaternary. Uranium deposits are associated with the trunk valley calcretes at Yeelirrie and southwest of Lake Way (Cole 1985). Loess blown from the limestone area of the Nullabor is believed to form a mantle of variable thickness in the more southerly parts of the area while gypsum and halite occur along the salinas (Fig. 2). Gypsum dunes of recent origin occur on the eastern and southeastern lake shores and are still growing by accretion of material from the lakes. Lunettes composed of sand, silt or clay with sometimes a band of fossilized shells of *Coxiella* species border the outer margins of the lakes. Thus, a variety of superficial material mantles much of the area and frequently influences the composition of the soils even over outcropping and near-surface bedrock.

Overburden and soils

The nature and thickness of the overburden is related to the relief and geomorphology and is highly variable. Over ridges, hills and low eminences where bedrock may outcrop, the residual and colluvial cover may be only a few cm thick. By contrast, over the lower lying ground, considerable depths of superficial deposits occur.

Relict lateritic soils occur over the peneplained or pediplained plateau surfaces whereas skeletal soils are characteristic of areas with near-surface bedrock. Solonized brown soils prevail over most of the lower level terrain but arid red earths are found over sandy parent material.

Depending on the parent material the solonized brown soils normally have either a yellowish red (5YR 4/6) to dark reddish brown (5YR 3/4) or dark brown (7.5YR 4/4) silty clay loam or silty loam A horizon, 15 to 30 cm deep, overlying a heavy textured B horizon containing calcareous nodules which extend to a depth of over 1 m in some profiles in the Kambalda area (Elkington 1969). The soils over the greenstones are generally alkaline. Over ultramafic parent material the pH values frequently exceed 8.0 or 9.0 in the surface horizons and increase towards the weathered bedrock zone. In the Kambalda, Widgiemooltha, Mount Thirsty and Kurnalpi areas the calcareous nodules in soils derived from serpentinite as well as from basic rocks are believed to be a legacy of loess that was blown from the limestone area of the Nullabor to form a mantle of variable thickness. Calcareous nodules, however, occur in soils derived from greenstone north of Menzies and calcrete occurs along trunk valleys at Yeelirrie and southwest of Lake Way. There is evidence of weathering of greenstones to depths of 70 m at the Sons of Gwalia mine near Leonora and the widespread occurrence of calcareous nodules in the soil profiles over the greenstones may be a legacy of the deep weathering processes in the Tertiary that produced lateritic profiles on the planation surfaces and led to the precipitation of the leached bases at lower levels. Whatever the origin the presence of calcium in many soil profiles distinguishes the greenstones of Western Australia from those of other countries.

The soils over near-surface granite or granite gneiss and over the sandy types of superficial deposits are frequently of the red earth type. They contrast with the solonized brown soils. Depending on the parent material they may be stony or clayey sands exhibiting texture contrast or hardpan development in the B horizon. Near lateritic residuals they may contain iron or manganese concretions.

Overall, because of the interplay of climatic changes and geomorphological processes over different rock types the soils of the greenstone belts of Western Australia vary considerably. These variations profoundly influence the vegetation.

The background vegetation

Within the area extending from Norseman through Kalgoorlie to Leonora and Mount Magnet the distribution of vegetation formations and associations is broadly related to climate and physiography/geomorphology. Within this broad framework there are regional and local relationships between the vegetation and bedrock geology. Distinctive vegetation associations occupy sites characterized by particular combinations of relief, soil and drainage conditions that are related to stages in the dissection of laterite-covered plateaux and the dismemberment of old drainage lines to saline tracts. Within the so-called greenstone belts differences in the form and composition of the vegetation distinguish areas underlain respectively by mafic and ultramafic rocks. Anomalous plant communities which include the nickel-accumulator plant *Hybanthus floribundus* occupy nickeliferous environments including those associated with the presence of nickel-copper ore bodies in ultramafic rocks.

The characteristic vegetation in the Norseman-Kalgoorlie area is open woodland dominated by *Eucalyptus* species with an understorey of both tall and low growing shrubs (Cole 1973). Where there are skeletal soils over near-surface bedrock *Eucalyptus* trees seldom grow. Depending on the type of bedrock these soils are characterized by shrub forms of *Acacia* or *Casuarina* species growing in associations with a variety of smaller shrubs. Saline tracts along existing and former water courses are likewise treeless, being occupied by halophytic shrubs, mostly of the genera *Arthrocnemum*, *Atriplex*, and *Cratystylis*.

North of Kalgoorlie broadly similar relationships between vegetation and environmental conditions occur but with increasing aridity, *Acacia aneura* woodland replaces the open *Eucalyptus* woodland over the deeper soils. Over outcropping and near-surface bedrock the shrub cover is sparser and its species composition varies with rock type. Associations of *Eucalytpus* spp exhibiting the mallee habit and *Triodia* spp grasses forming widely spaced hummocks occur over remnant lateritic plateaux while a great variety of shrubs, particularly *Grevillea* and *Hakea* species and *Calythrix*, *Thryptomene*, *Verticordia* and *Wehlia* species occupy the sandplains.

The division between the *Eucalyptus* woodlands in the south and the *Acacia aneura* woodlands and the shrublands in the north has been recognised by Carnahan (in Jeans 1978) and Groves (1981) who focussed attention on floristics and climatic influences. In a study of the vegetation of the Wiluna-Meekatharra area, which straddles the northern part of the greenstone belts, Speck (1958) however, recognised seventy communities whose distributions are related to habitat conditions that are controlled by geomorphology and geology, particularly the nature of the superficial deposits. The communities embrace woodlands, shrublands and grasslands (*Triodia* spp.). The upper layers of most woody communities are dominated by *Acacia aneura* and the individual communities are discriminated by the shrub cover and especially by *Eremophila* spp which are particularly sensitive to habitat. The large number of communities recognised in this area is indicative of the variety and complexity of both vegetation and habitat.

The vegetation of areas of ultramafic rocks

Consequent on variations in the climatic, edaphic and relief conditions, regional differences in the form and composition of the vegetation over areas of ultramafic rocks may parallel those over other rock types. Additionally some species display recurrent patterns of distribution that are related to the ultramafic environment.

Fig. 3. Eucalyptus salmonophloia woodland with understorey of tall *Eremophila interstans* shrubs and low *Atriplex paludosa* and *Olearia muelleri* shrubs over basalt, Norseman area.

Areas of ultramafic rocks south of Kalgoorlie

South of Kalgoorlie there are marked contrasts in the form and composition of the vegetation between areas of serpentinite and related rock types and those of meta-basalt, amphibolite and meta-sediments (Figs 3, 4). These changes are partly related to differing relief features and differing soils over the separate rock units.

The Kambalda area

At Kambalda meta-basalts form the prominent hills whereas serpentinite forms the lower slopes and underlies some of the low ground that has superficial cover. The characteristic vegetation of the areas of meta-basalt is an open woodland dominated by *Eucalyptus lesoueffi* which is associated with *E. salmonophloia*, *Santalum acuminatum* and *S. spicatum* on the deeper soils. This woodland gives way to communities of tall shrubs, *Acacia quadrimarginea*, *A. tetragonophylla*, *Dodonaea lobulata*, *Eremophila duttoni*, and *Scaevola spinescens* accompanied by the small shrub *Ptilotus obovatus* on skeletal soils on

the hill tops. *Eucalyptus foecunda* and *E. torquata* are the most common eucalyptus in the woodlands occupying the deeper soils over serpentinite. These woodlands yield to shrub communities of *Acacia tetragonophylla*, *Alyxia buxifolia*, *Cassia artemisioides*, *Dodonaea lobulata*, *Eremophila duttoni*, *Hybanthus floribundus*, *Ricinocarpus stylosus*, *Scaevola spinescens* and *Trymalium myrtillus* on skeletal soils over strong serpentinite outcrops (Fig. 5). *Hybanthus floribundus*, *Ricinocarpus stylosus* and *Trymalium myrtillus* occur only over serpentinite (Elkington 1969). Outcrops of quartz porphyry are marked by the occurrence of *Brachychiton gregorii* and *Acacia acuminata* but shrub communities of the last named associated with *A. quadrimarginea*, *Casuarina campestris*, *Dampiera trigona* var. *latealata*, *Dodoneae lobulata*, *Prosthanthera aspalathoides* and *Scaevola spinescens* occur where the outcrop is mantled by coarse scree from meta-basalt. Where the scree is finer and the soil deeper the vegetation changes to a woodland of *Eucalyptus lesoueffii* and *E. oleosa*. The distinctive vegetation over the outcropping serpentinite is considered to be related to high chromium and

Fig. 4. Eucalyptus foecunda trees, *Acacia graffiana* (right) and *Eremophila interstans* (left) shrubs dominating the vegetation over ultramafic rocks between Coolgardie and Norseman.

Fig. 5. Hybanthus floribundus shrub community over ultramafic rocks (foreground); at contact with meta-sediments, behind, taller *Eremophila* shrubs with *Eucalyptus torquata* at the periphery of the shrub vegetation: Mount Thirsty area.

350

nickel concentrations in the soils – respectively up to 970 μg g^{-1} and 1230 μg g^{-1} – rather than to the high Mg/Ca quotients associated with serpentinite bedrock. Many species are absent from soils with high nickel concentrations. However where, on lower ground, calcium carbonate nodules thought to be derived from windblown loess, are present in the soil a woodland of *Melaleuca sheathiana* occurs even where nickel concentrations are high. Here adequate supplies of calcium apparently ameliorate the effect of nickel. The relative importance of the influences of high Mg/Ca quotients, chromium, copper and particularly nickel on soils derived from ultramafic rocks, including those with mineral deposits, are considered later.

The Widgiemooltha-Dordie rocks area
Somewhat different relationships between vegetation, soils, relief and geology occur in the Widgiemooltha and Dordie Rocks areas to the south of Kambalda.

At Widgiemooltha the talc serpentinite after peridotite with which is associated a nickel-copper deposit at the contact with the basic amphibolite bedrock, forms a prominent hill feature whereas talc tremolite-talc rock after peridotite and amphibolite form the adjacent low ground mantled by colluvium and other superficial deposits. Elsewhere in this area ultramafic rocks which include talc rock, talc-carbonate rock, tremolite talc rock, talc-tremolite-carbonate rock, talc-schist, tremolite-talc schist, talc-carbonate schist, tremolite-talc-carbonite schist and tremolite-actinolite rock underlie generally level terrain largely covered by superficial material. Here outcrops of semi-stratigraphic dykes of felsic and quartz porphyry form minor relief features.

A woodland of relatively tall *Eucalyptus griffithsii*, *E. oleosa* and *E. salmonopholoia* trees associated with the tall shrubs *Alyxia buxifolia*, *Acacia graffiana*, *Eremophila glabra*, *E. interstans*, *E. ionantha*, *E. scoparia* and *Scaevola spinescens* and the low shrubs *Acacia colletioides*, *Kochia georgei* and *Olearia muelleri* occupies the level terrain where deep cover of soil and colluvium masks bedrock in the Widgiemooltha area (Cole 1973). The vegetation changes in both form and composition over elevated ground with near-surface bedrock. Over mafic rocks a woodland of

Eucalyptus celastroides with a variety of tall shrubs notably *Acacia burkittii*, *Alyxia buxifolia*, *Cassia nemophylla*, *Eremophila dempsteri*, *E. pachyphylla* and *Scaevola spinescens* and the low-growing *Olearia muelleri* prevails whereas one of *Eucalytpus campaspe* and *E. lesoueffii* with relatively fewer shrub species notably *Pomaderris forrestiana* and *Trymalium myrtillus* and the low-growing *Westringia rigida* occurs over ultramafic rocks. Over the nickel-copper deposit within the talc serpentinite after peridotite at the contact of the ultramafic and mafic rocks, however, the vegetation comprises a shrub community composed mainly of *Hybanthus floribundus* associated with *Dodonaea microzyga*, *Grevillea acuaria* and *Trymalium myrtillus* (Fig. 6). The *Hybanthus* and *Trymallium* are confined to skeletal soils with high concentrations of chromium and nickel over outcropping serpentinite as at Kambalda.

Investigations over the hill of talc serpentinite after peridotite at Widgiemooltha and over adjacent areas of covered ground have shown that the distribution of the community of *Dodonaea microzyga*, *Grevillea acuaria*, *Hybanthus floribundus* and *Trymalium myrtillus* is related to high concentrations of chromium, cobalt, copper and nickel in the surface soil (Fig. 7) and that *Hybanthus floribundus* accumulates nickel especially in its leaves (Cole 1973). Maximum amounts in the leaf ash of over 21% of nickel (which is equivalent to 1.2% expressed on a dry weight basis) were found in one sample collected over the nickel-copper deposit. Another plant collected along a drainage line over covered ground near the contact of ultramafic and mafic rocks contained ten times the amount of total nickel present in the underlying soil. The investigations also showed that the *Eucalyptus* species characteristic of the woodland vegetation were absent from the area with high toxic metal values in the soil and that only *E. torquata* and *E. salubris* were present at its periphery. The former appears resistant to quite high concentrations of nickel (of around 200 μg g^{-1}) in the soil and as at Kambalda marks the boundary between ultramafic and mafic rocks, even in areas of covered ground, while the latter appears resistant to concentrations of chromium of around 1000 μg g^{-1} in the soil.

Prominent relief features do not occur in the Dordie Rocks area where a woodland dominated

Fig. 6. Shrub community of *Hybanthus floribundus* over outcropping peridotite, Widgiemooltha hill. *Acacia tetragonophylla* (top right).

by *Eucalyptus griffithsii* and *E. oleosa* trees, associated with *Acacia acuminata*, *Atriplex paludosum* and *Olearia muelleri* occurs over mafic rocks and a shrub community of *Dodonaea microzyga*, *D. stenozyga*, *Eremophila pachyphylla*, *E. scoparia*, *Hybanthus floribundus*, *Phebalium filifolium*, *Pimelea microcephala* and *Trymalium myrtillus* occurs over the ultramafic rocks. *Eucalyptus torquata* again favours the contact between the two rock types. As at Widgiemooltha the distribution of *Hybanthus floribundus*, *Trymalium myrtillus* and *Dodonaea microzyga* appears to be related to high concentrations of nickel in the soil, with *Hybanthus floribundus* again accumulating large quantities of the metal in its tissues – the ash of one leaf sample containing nearly 23% (equivalent to over 1.4% on a dry weight basis).

Mapping of the distribution of *Hybanthus floribundus* in both the Widgiemooltha and Dordie rocks areas has shown that it is not confined to the vicinities of nickel-copper deposits but that it extends over other areas of serpentinite and is concentrated at the surface contact of this rock type with quartz feldspar porphyry dykes. This

distribution has prompted the suggestion that the species, which is distinguished from the sclerophyllous species characteristic of this winter rainfall area by the development of delicate deciduous leaves favours sites where meteoric moisture may be held temporarily near the surface in the joints of rocks (Tuite 1969). This however has not been substantiated.

The Mount Thirsty area
South of Widgiemooltha the occurrence of shrub communities within the prevailing *Eucalyptus* woodlands of the Mount Thirsty-Mission Ridge area is associated with relatively high concentrations of chromium or nickel or both in the soils.

The Mount Thirsty area is underlain by a succession of volcanic and sedimentary rocks that have been intruded by a complex mafic-ultramafic sill exhibiting macro-layering from altered pyroxenite and/or peridotite or both to hypersthenite, gabbro and diorite (Yates 1966/67). The rocks of the area were originally folded into a concentric anticline with steep limbs and then refolded to produce at least two domes within the major

352

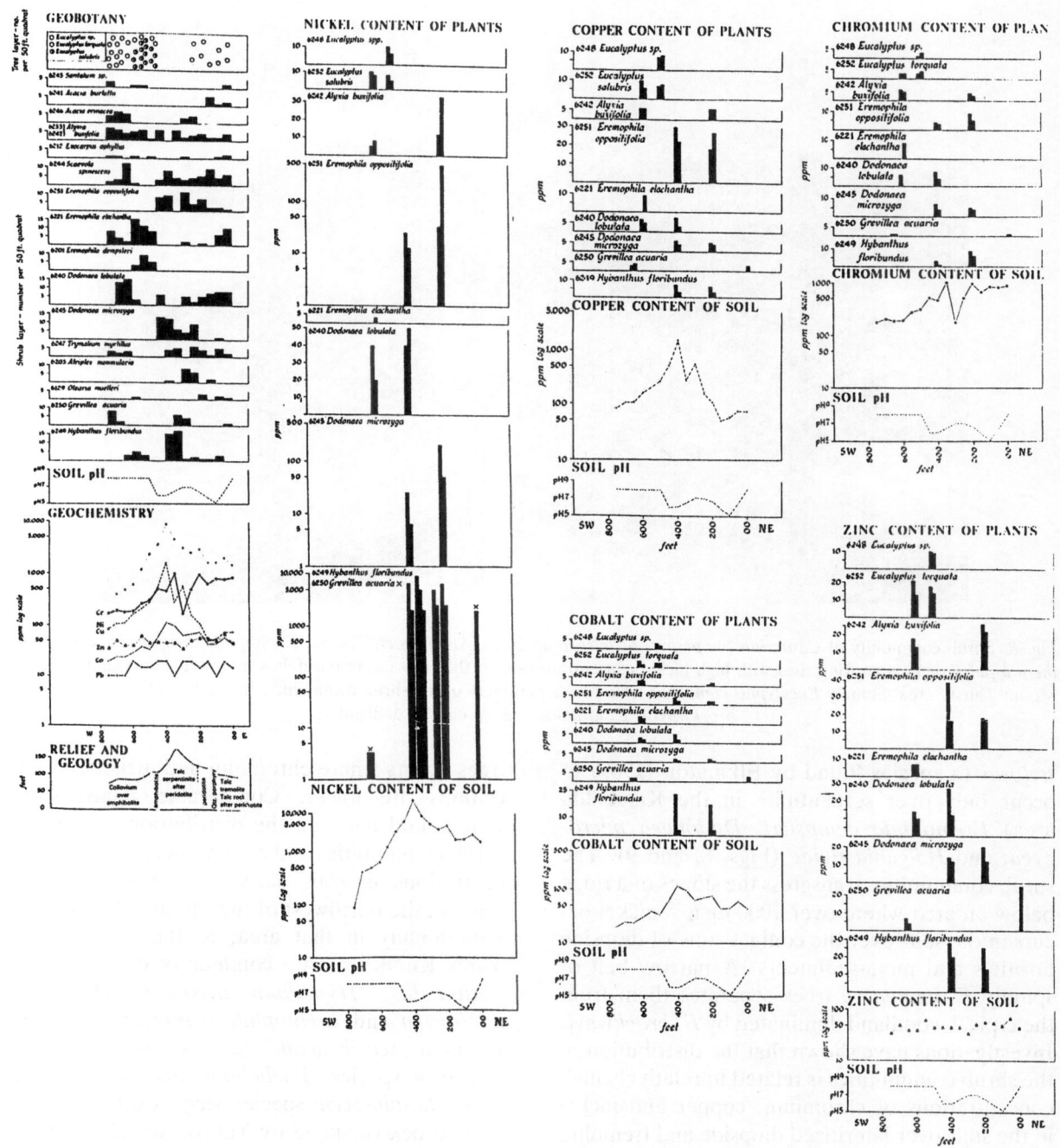

Fig. 7. Histogram of transect 66/3 run across Widgiemooltha hill showing the relationships between the distribution of plant species and the chromium, cobalt, copper, lead, nickel and zinc concentrations in the − 80 mesh fraction of the soil sampled at 10–15 cm. Recording and sampling by P. J. Nicholson in May 1967. (Reproduced by courtesy of the Journal of Applied Ecology.)

structure. The whole forms a prominent ridge feature with microrelief related to bedrock type (Cole 1973).

Two distinctive shrub communities that contrast sharply with the prevailing *Eucalyptus* wood-land dominated by *E. brockwayi*, associated with *E. lesoueffii* and *E. torquata*, occur in the southern part of the area. One is characterized by *Casuarina campestris*, *Acacia acuminata* and *Dodonaea microzyga* and the other by *Ricinocarpus*

Fig. 8. Shrub community of *Casuarina campestris*, *Melaleuca uncinata*, *Acacia cochliocarpa*, *Eremophila dempsteri* and *Prostanthera aspalathoides* occupying sites with high nickel concentrations over the calc-silicate hornfels – meta-sediments contact zone, Mount Thirsty area. Spindly *Eucalyptus campaspe* trees at the periphery of the shrub community. In the background *Eucalyptus brockwayi* – *E. torquata* – *E. lesoueffi* woodland.

stylosus (a species found by Elkington (1969) to occur only over serpentinite in the Kambalda area) *Eremophila dempsteri*, *Dodonaea microzyga*, and *Halgania rigida* (Figs. 8, and 9). The shrub communities transgress the slopes of a ridge below an area where over 1000 μg g^{-1} nickel occurs in the soils over the contact zone of diopside hornfels and meta-sediments. A narrow belt of spindly *E. campaspe* trees separates them from the typical woodland dominated by *E. brockwayi*. Investigations have shown that the distribution of the shrub communities is related to relatively high concentrations of chromium, copper and nickel in the soils over lateritized diopside and tremolite hornfels caused by the presence of sulphides. The community of *Acacia acuminata*, *Casuarina campestris* and *Dodonaea microzyga* appears to be associated with chromium concentrations of 2000 to 5000 μg g^{-1} and nickel concentrations of 1000 to 2000 μg g^{-1} and that of *Ricinocarpus stylosus*, *Dodonaea microzyga*, *Eremophila dempsteri*, and *Halgania rigida* with chromium and nickel concentrations below 2000 μg g^{-1} and 1000 μg g^{-1} respectively. The belt of spindly *E. campaspe*

trees occurs where chromium and nickel concentrations are lower. Comparable relationships were found between the distributions of similar shrub communities and chromium and nickel concentrations in soils derived from similar rock types in the northwest of the Mount Thirsty area. Additionally in that area, to the northeast of Yates Knoll, a shrub community of *Hybanthus floribundus*, *Trymalium myrtillus*, *Dodonaea microzyga* and *Eremophila dempsteri*, with less frequent *Acacia acuminata*, *Exocarpus aphylla*, *Grevillea* species, *Phebalium tuberculosum* and *Scaevola spinescens* species occurs over serpentinite bedrock (mapped by Yates 1966/67 as pyroxenite) covered by residual lateritic material (Fig. 10). Samples of *Hybanthus floribundus* collected from this area also contained large quantities of nickel suggesting again the important influence of this metal over plant distributions. A community of *Ricinocarpus stylosus* and *Melaleuca cymbifolia* intervened between the *Hybanthus floribundus*-dominated community and *Eucalyptus* woodland dominated by *Eucalyptus celastroides* over adjacent covered ground. These vegetation distri-

Fig. 9. Shrub community dominated by *Eremophila dempsteri* and *Ricinocarpus stylosus*; in the background *Eucalyptus campaspe* trees over altered pyroxenite, Mount Thirsty area.

Fig. 10. Shrub community of *Hybanthus floribundus* (foreground) with *Trymalium myrtillus* (flowering tall shrub) and *Dodonaea microzyga* (not visible) over covered ground underlain by ultramafic rocks northeast of Yates Knoll, Mount Thirsty area. *Eremophila dempsteri* (right and behind). In the background *Eucalyptus celastroides* woodland.

butions further demonstrate the relationships between species occurrence and edaphic conditions.

Within the *Eucalyptus* woodlands south of Mount Thirsty, shrub communities that contrast with those on serpentinite occur over hypersthenite. These communities are composed of *Casuarina campestris*, *Cryptandra glabriflora*, *Dodonaea boroniaefolia*, *Melaleuca coccinea*, *Prostanthera aspalathoides* and *Stackhausia huegelii*. *Boronia fabianoides* and *Westringia rigida* occur around outcrops and spindly *Eucalyptus calycogona* and *E. campaspe* at the periphery of the area underlain by hypersthenite and near the contact with gabbro.

Overall within the area south of Kalgoorlie, while there are variations in the composition of the prevailing *Eucalyptus* woodlands, a few species recur again and again in shrub communities over areas of ultramafic rock, notably serpentinite. These species are *Hybanthus floribundus*, *Trymalium myrtillus* and *Dodonaea microzyga*. In different areas they are variously associated with other shrub species including *Acacia acuminata*, *Casuarina campestris* and *Ricinocarpus stylosus*.

Areas of ultramafic rocks around and north and east of Kalgoorlie

In response to increasingly arid climatic conditions marked changes occur in the characteristic vegetation northwards and eastwards from the vicinity of Kalgoorlie.

The Kurnalpi area
The vegetation around Kurnalpi, which lies to the east of Kalgoorlie, comprises either very open woodlands of small *Casuarina* or *Eucalyptus* trees or open shrub associations dominated by *Acacia*, *Eremophila* or *Kochia* species. The generally level terrain is underlain by amphibolite and serpentinite with a variable cover of colluvium, alluvium or, in places, of lateritic material. The last is responsible for a flat topped hill feature where it caps serpentinite.

Where shallow soils prevail over a thin cover of colluvium or alluvium an open woodland of small *Eucalyptus lesoueffii* trees with a variable shrub understorey occurs over both mafic and ultramafic rocks. The composition of the shrub layer, however, changes from one of *Acacia spp.*, *Dodonaea lobulata*, *Eremophila spp* and *Olearia muelleri* over serpentinite to one of *Atriplex nummularia* and *Kochia sedifolia* where deeper soils have formed over alluvium over amphibolite. Here the change is related to both soil depth and bedrock type. Sharper changes of vegetation distinguish the skeletal soils over the laterite-capped serpentinite hill. Here the *Eucalyptus lesoueffii* woodland gives way to one of small *Casuarina obesa* trees. In parallel the characteristic *Eremphila scoparia* and *E. strongylophylla* shrubs of the former yield to *Dodonaea lobulata*, *E. oldfieldii* and *Santalum spicatum* over the latter where *Scaevola spinescens* is also abundant. Over the western summit and western slope of the hill further marked changes of vegetation occur where removal of the laterite cap has exposed serpentinite bedrock, and a gossan suggests that this is mineralized. Here the presence of *Hybanthus floribundus*, *Eremophila weldii*, *Grevillea acuaria* and *Acacia erinacea* and the reduced frequency or absence of other species coincides with high concentrations of chromium and nickel in the soils – respectively 2,000–5,000 μg g^{-1} and 1,000–4,000 μg g^{-1}.

Notwithstanding the differences in the characteristic vegetation occasioned by more arid climatic conditions and despite the prevalence of superficial cover, changes in the form and composition of the vegetation are evident over serpentinite in the Kurnalpi area. Species composition differs from that in the Kambalda, Widgiemooltha and Dordie Rocks areas south of Kalgoorlie but the presence again of *Hybanthus floribundus* and *Grevillea acuaria* and the appearance of *Acacia erinacea* and *Eremophila weldii* appears to be related to chromium and nickel concentrations in the soils.

Mount Hunt
The characteristic vegetation in the Mount Hunt area is an open woodland of *Eucalyptus foecunda* associated with either *Acacia leptoneura* and *Triodia pungens* or with *Atriplex paludosa*, *Eremophila interstans*, *E. lesoueffii* and other shrubs. A number of distinctive associations, however, occur over a relatively small area.

Mount Hunt, which attains a height of 450 m, is one of a few hills in the more or less flat or undulating terrain in the Boulder area. Smaller eminences are formed by outcropping serpentinite and porphyry, notably where the former underlies the surface west of Hannan's lake (which is dry for most of the year).

Elkington (1969) found that relief exerted little influence in this area. Different associations succeeded one another over the sides of the hill where slope and aspect were similar while the same associations that occupied the top and upper slopes of Mount Hunt also occurred on less exposed lower ground. She found that the distribution of the vegetation associations was controlled mainly by geology. This relationship was due largely to the skeletal nature of the soils but it persisted even where bedrock was not exposed.

The typical vegetation over ultramafic rocks comprised an open woodland of *Eucalyptus foecunda* trees with *Acacia graffiana*, *A. leptoneura* and *Cassia eremophila* shrubs and the spinifex grass *Triodia irritans*, but over carbonated serpentinite the associations changed to one of *Eucalyptus foecunda*, *E. gracilis*, *Eremophila interstans* and *E. weldii*, the last named being one of the species that occurred in soils with high chromium and nickel values at Kurnalpi. This vegetation contrasted with associations of *Acacia quadrimarginea*, *Dodonaea lobulata*, *Eremophila duttoni* and *Ptilotus obovatus* over meta-dolerite and of the first named together with *Dodonaea boroniaefolia* and an unnamed *Eremophila* sp, over meta-basalt, with the additional presence of *Helipterum adpressum*, *Prostanthera obovatus* and *P. wilkieana* over variolitic lava (Fig. 11). The large serpentinite outcrop on the western shore of Hannan's lake carried a vegetation of *Acacia leptoneura*, *Dodonaea stenozyga*, *Eucalyptus foecunda* and *Triodia irritans* but over carbonated serpentinite gave way to sparse cover of *Melaleuca lateriflora*, with halophytic species, notably *Arthrocnemum halocnemoides*, *Atriplex paludosa*, *Disphyma australe* and *Frankenia interioris* on the lower ground where the soils are saline and clay pans occur. Here it is interesting to note that these vegetation distributions are comparable with those found on the eastern side of the peninsular area that juts northwards into the southern part of Lake Cowan, between Widgiemooltha and Norseman. There in a comparable saline environment, a shrub community of *Melaleuca*

lateriflora and *Cassia nemophylla* occurs over serpentinized and talc carbonate rocks, whereas a woodland dominated by *Eucalyptus stricklandi* with a shrub understorey of *Atriplex* and *Arthrocnemum* species covers the adjacent areas where colluvium masks meta-sediments and porphyry rocks.

While there appeared to be close relationships between vegetation associations and bedrock geology, Elkington (1969) did not find close relationships between plant distributions and concentrations of either chromium and nickel or copper, lead and zinc. This may have been due to the fact that concentrations of chromium and nickel were much lower than in the areas already considered to the south and east of Kalgoorlie. A maximum value of 1200 μg g^{-1} chromium occurred at one site over carbonated serpentinite but elsewhere both chromium and nickel values were in the range of 200–600 μg g^{-1}. Perhaps significantly the species that occurred over high chromium and nickel values over serpentinite bedrock to the south and east of Kalgoorlie were absent from the Mount Hunt area. There the strong relationships between vegetation and geology may be occasioned either by differences in the texture and/or in the levels of other elements in the soils over the different bedrock types. Clearly near Hannan's lake, as near Lake Cowan to the south of Kalgoorlie, salinity exerts an important influence on vegetation distributions with the same species, *Melaleuca lateriflora* occurring over the carbonated serpentinite and talc carbonate rocks.

The Broad Arrow, Menzies, Riverina-Mount Ida, Leonora-Kurrajong and Barwidgee areas

North of Kalgoorlie, information on the vegetation over areas of ultramafic rocks is based only on reconnaissance studies.

In areas extending from Kalgoorlie to Broad Arrow and Menzies, residual, colluvial or lateritic material masks the underlying granite and greenstones which seldom outcrop. A low woodland characterized by *Eucalyptus salubris* and smaller *Santalum spicatum* trees with a variety of shrubs including *Atriplex* species, *Cratystylis* species, *Cassia eremophila*, *Eremophila interstans* and *E. oppositifolia* prevails where the greenstones are

Fig. 11. Community of small *Helipterum adpressum* shrubs over basaltic lavas, Mount Hunt. In the background, line of low serpentinite hills by Hannans Lake, near Kalgoorlie.

mantled by residual and colluvial cover but gives way to communities dominated by *Acacia* species, *Casuarina cristata* and *Eremophila miniata* over gravelly lateritic material and by *Acacia aneura*, *Grevillea* spp., *Thryptomene* and numerous other shrubs of the *Myrtaceae* family, over sands. Within this area, however, a distinctive association characterized by *Acacia acuminata, A. tetragonophylla, Cassia eremophila, Dodonaea lobulata, Ptilotus obovatus* and *Scaevola spinescens* distinguishes the continuation of the Golden Mile dolerite (with which the gold won on the Kalgoorlie goldfield is associated), while a hill of basic volcanics near Bardoc is marked by the presence of *Acacia quadrimarginea, Casuarina cristata, Eremophila serrulata, Helipterum adpressum* (which occurred over similar rocks at Mount Hunt) and *H. polygalifolium*. Nearby the vegetation over an area of altered peridotite consists mainly of *Eucalyptus griffithsii*, and a *Triodia* sp., whereas *Eucalyptus celastroides*, with *Eremophila oldfieldii* and *Ptilotus obovatus* predominates over amphibolite and a belt of small *Casuarina helmsii* trees and a *Dodonaea* sp, *Exocarpus aphyllus* and

Grevillea acuaria shrubs follows a zone of copper-nickel mineralization at the contact of the two rock types. No really distinctive vegetation occupies the ultramafic rocks in this area, doubtless because of the presence of lateritic cover but the recurrence of *Exocarpus alphylla* and *Grevillea acuaria* which respectively occurred within communities characterized by *Hybanthus floribundus* over nickeliferous serpentinite near Yates Knoll in the Mount Thirsty area and in the Widgiemooltha and Dordie Rocks areas is noteworthy.

In the greenstone belt to the west of Menzies and Broad Arrow, greater contrasts occur between the vegetation over mafic and ultramafic rocks in the Mount Ida, Riverina and Dunnsville areas. Near Snake Hill in the Mount Ida area, whereas a low open woodland of *Acacia aneura* trees and *Atriplex paludosa* and *Kochia sedifolia* shrubs covers areas underlain by meta-sediments, a very sparse vegetation in which *Exocarpus aphylla* recurs with *Dodonaea lobulata, Eremophila scoparia, Hakea preissii, Lepidium leptopetalus* and *Scaevola spinescens* dots the areas underlain by meta-pyroxenite and peridotite (Figs 12,

Fig. 12. Sparse cover of *Exocarpus aphylla* (foreground left and right) and *Dodonaea lobulata* over meta-pyroxenite and peridotite; hills beyond salt lake are of amphibolite. Snake Hill area.

13) and *Calytrix micromyrtaceae* shrubs occupy the crevices within the outcrops of the amphibolite.

North of Leonora only two species, *Rhagodia preissii* and *Grevillea acuaria* occur over siliceous ironstone over periodotite on Kurrajong station, whereas a woodland of *Eucalyptus oleosa*, exhibiting the mallee habit, occupies the adjacent covered ground (Fig. 14). Farther north the first named occurs again over serpentinite on Barwidgee station where the presence of *Eremophila fraseri* is indicative of underlying granite. In this area, however, striking relationships between vegetation and underlying bedrock are generally lacking, a reflection of widespread cover and climatic aridity.

Within the greenstone belts north of Kalgoorlie the most significant plant distributions are the occurrences of *Hybanthus floribundus* over serpentinite bedrock on Riverina station south of Mount Ida and west of Menzies, and on Kurrajong station north of Leonora. On both stations the occurrences are associated with high nickel concentrations in the soils (Cole 1973).

The factors influencing the distribution of vegetation associations and plant species over ultramafic rocks in the Eastern Goldfields region

Studies over a number of discrete areas within the greenstone belt of the Eastern Goldfields of Western Australia show that the vegetation varies greatly in both form and composition and that the relationships between the distributions of vegetation associations and of plant species and environmental factors are complex. The distributions represent the expressions of response to the interplay of climatic and edaphic conditions, in turn related to geomorphology and geology, that have changed over time. In some areas soil depth and texture are the most important whereas elsewhere soil chemistry exerts most influence.

The detailed investigations made during the 1964–7 period were orientated towards exploration for nickel orebodies and therefore focussed on relationships between bedrock geology and mineralization. Many plant and soil samples were analysed for nickel, chromium, cobalt, copper, lead, zinc and sulphur (Cole 1973, Elkington

Fig. 13. Vegetational change from shrub community of *Hakea preissii* and *Lepidium leptopetalum* (foreground) over meta-pyroxenite/peridotite to one with *Acacia aneura* trees over amphibolite with interbedded shales. Hills of amphibolite in background. Snake Hill-Mount Ida area.

Fig. 14. Shrub community of *Rhagodia preissii* and dwarf *Grevillea acuaria* (middle distance) with *Ptilotus obovatus* (foreground) over siliceous ironstone over peridotite. *Eucalyptus oleosa* woodland over basic rocks in the background, Kurrajong area.

1969). Calcium and magnesium determinations were made on a relatively small number of samples. Since then, however, many of the original samples, as well as new ones, have been analysed for phosphorus, potassium, sodium, calcium, magnesium and iron. Detailed chemical investigations have also been made on the nickel-accumulator plant, *Hybanthus floribundus* to ascertain the pathways of nickel uptake, the loci of nickel accumulation, the compounds in which it is held and its role in plant physiology.

The overall studies indicate that there are no clearly defined serpentinite floras but that there are plant species and assemblages of plant species that are characteristic of different types of outcropping or near-surface ultramafic rocks and of areas where these rocks are mantled by material of lateritic or loessic origin and of areas where conditions are saline. Individual species, notably *Hybanthus floribundus* occur where there are high concentrations of nickel, especially over sulphide nickel deposits. Because of the effects of lateritic, loessic or evaporitic cover, relationships between vegetation and environmental factors are often more complex than those over serpentinites in other countries where either the magnesium/calcium quotient may be regarded as the dominant influence over the distribution, physiognomy and composition of the vegetation (Kruckeberg 1954, 1964) or the toxic effects of nickel and chromium (Soane & Saunder 1959) and/or the lack of soil development may be the controlling influence (Spence 1957).

Determinations of the calcium and magnesium concentrations of soil and plant samples from the Widgiemooltha, Dordie Rocks, Mount Thirsty and Kurnalpi areas show that while, in some cases, calcium may be low in soils with high magnesium concentrations over outcropping ultramafic bedrock, in most instances plant tissues contain appreciably more calcium than magnesium (Tables 1–16).

Some variations in the magnesium/calcium quotients in plant tissues, however, do occur and may have some significance in connection with species distribution. The highest quotients occur in *Hybanthus floribundus* growing over the most extreme sites, with notably high concentrations of nickel, chromium, cobalt and copper (Fig. 7, Tables 1, 2) (Cole 1973, Farago *et al.* 1977) in the Widgiemooltha and Dordie Rocks areas. The quotient is high also in *Westringia rigida* which is common over ultramafic rocks, and higher in *Eucalyptus torquata* which favours the ultramafic/ mafic rock contact, than in the *Eucalyptus gracilis*, *E. griffithsii* and *E. salubris* (Tables 3, 4) growing over mafic rocks or covered ground.

In both the Widgiemooltha and Dordie Rocks areas the concentrations of phosphorus and iron are very low in all the plant samples whereas those of potassium and sodium are variable, being dependent on the superficial cover (Table 1).

In the Mount Thirsty area where the plant communities contrast with those in the Widgiemooltha and Dordie Rocks areas the magnesium/calcium quotients are marginally lower in plants growing over altered pyroxenite than over metasediments and diopside and tremolite hornfels but overall do not differ appreciably from those in plants growing over amphibolite in the Widgiemooltha area (Tables 1, 3, 10–13). In the Mount Thirsty area the distribution of the shrub communities that occur within the *Eucalyptus brockwayi-Eucalyptus lesoueffii* woodlands appear to be related primarily to concentrations of chromium and nickel in the soils with communities of *Casuarina campestris*, *Acacia acuminata* and *Dodonaea microzyga* occurring over the more toxic ground with chromium concentrations exceeding 5,000 μg g^{-1} and nickel concentrations of about 1,000 μg g^{-1} in places and being succeeded in turn by taller shrub communities of *Eremophila dempsteri* and *Ricinocarpus stylosus* and by stunted *Eucalyptus campaspe* trees over the soils with lower concentrations (Cole 1973).

Northeast of Yates Knoll within the Mount Thirsty area, however, analyses of soil samples collected along a transect run across a *Dodonaea microzyga-Hybanthus floribundus-Trymalium myrtillus-Dodonaea microzyga* community that occurs where lateritic material mantles ultramafic and mafic rocks provide interesting information on the influences of calcium, magnesium and iron on plant distributions (Fig. 15). Here the distribution of the *Hybanthus floribundus*-dominated communities is related to relatively high concentrations of nickel in the soil which are accompanied by low concentrations of both calcium and magnesium and high concentrations of iron caused by lateritization processes. Where the iron

Table 1. The percentage ash and the concentrations (μg g^{-1} oven dry matter) of phosphorus, potassium, sodium, calcium, magnesium (and the Mg/Ca quotient) and iron in (a) leaves and (b) stems of plants collected along transect line 66/3 Widgiemooltha

Species	Site	Rock type	Part of plant	Ash*	Phosphorus	Potassium	Sodium	Calcium	Magnesium	Mg/Ca quotient	Iron
Hybanthus floribundus	400 W	talc serpentinite after peridotite	a	5.04 (4.73)	389	890	120	4,820	4,670	0.97	158
			b	1.57 (1.43)	109	300	50	4,610	2,710	0.59	126
	200 W	peridotite	a	4.93	615	520	60	12,900	9,270	0.72	210
			b	2.60	287	810	140	11,300	3,380	0.30	228
Dodonaea microzyga	400 W	talc serpentinite after peridotite	a	3.19	296	720	60	11,100	1,200	0.11	87
			b	1.79	260	340	40	8,060	3,350	0.42	133
	200 W	peridotite	a	3.94	421	740	60	14,100	1,290	0.09	108
			b	2.09	111	100	20	4,050	1,000	0.25	65
Eremophila oppositifolia	400 W	talc serpentinite after peridotite	a	4.96 (5.12)	240	1,000	700	31,200	3,490	0.11	100
			b	2.76	311	815	390	6,730	2,370	0.35	241
	200 W	peridotite	a	4.52	329	650	750	12,700	3,570	0.28	125
			b	2.60	263	630	370	7,260	2,140	0.29	153
Alyxia buxifolia	200 W	peridotite	a	4.89 (5.03)	259	760	40	29,900	4,140	0.14	91
			b	3.49	247	830	40	12,900	8,040	0.62	151
	600 W	colluvium over amphibolite	a	5.14	240	1,050	60	28,800	6,180	0.21	73
			b	12.45	–	645	60	12,600	2,760	0.22	1,150
Grevillea acuaria	600 W	colluvium over amphibolite	a	1.94	–	400	150	13,500	3,860	0.29	1,226
			b	1.10	–	285	100	14,900	2,090	0.14	1,992
Eremophila elschantha	600 W	colluvium over amphibolite	a	5.35	–	420	970	16,000	4,430	0.28	1,217
			b	3.82	–	635	510	17,600	1,720	0.10	2,258
Eucalyptus salubris	600 W	colluvium over amphibolite	a	4.81	–	400	720	12,600	3,380	0.27	1,105
			b	4.20	–	560	400	25,200	4,500	0.18	1,022
	500 W	colluvium over amphibolite	a	3.99	264	1,000	790	13,000	3,310	0.25	85
			b	4.57	231	675	420	22,300	4,100	0.18	81

*In some cases calcium and magnesium were analysed on separate subsamples and for these the ash contents are given in parentheses.

Table 2. The percentage ash and the concentrations (μg g^{-1} oven dry matter) of phosphorus, potassium, sodium, calcium, magnesium (and the Mg/Ca quotient), iron and nickel in leaves of *Hybanthus floribundus*

Sample No	Site	Rock type	Ash*	Phosphorus	Potassium	Sodium	Calcium	Magnesium	Mg/Ca quotient	Iron	Nickel
7461	Widgiemooltha	peridotite	4.73	125	705	67	4,820	4,670	0.97	148	12,400
7445	Dordie Rocks	serpentinite	6.05	–	–	–	4,380	4,090	0.93	–	13,900
7517	Kurnalpi	serpentinite	3.92	–	–	–	4,700	5,010	1.07	–	758
7494	Kurrajong	carbonate	5.53	–	–	–	5,150	2,600	0.50	–	7,390
7529	Kurrajong	amphibolite	5.00 (4.17)	23	79	5	3,410	1,070	0.31	83	375
7528	Kurrajong	amphibolite	6.54 (5.71)	204	1,925	195	6,960	1,460	0.21	42	914
7532	Riverina	serpentinite	4.34 (4.24)	156	840	70	4,400	2,500	0.57	194	3,140

*In some cases nickel was analysed on separate subsamples and for these the ash concentrations are given in parentheses.

Table 3. The percentage ash and the concentrations (μg g^{-1}) oven dry matter of phosphorus, potassium, sodium, calcium, magnesium (and the Mg/Ca quotient) and iron in (a) the leaves or (b) the whole plant of four species from the Widgiemooltha area

Sample No	Species	Rock type	Part of Plant	Ash	Phosphorus	Potassium	Sodium	Calcium	Magnesium	Mg/Ca quotient	Iron
7128	*Eucalyptus torquata*	near mafic/ ultramafic contact	(a)	3.49	1,930	2,850	4,900	6,490	2,590	0.40	62
7275	*Eucalyptus torquata*	mafic/ ultramafic contact	(a)	3.57	2,070	3,800	6,250	5,380	2,870	0.53	75
7309	*Eucalyptus gracilis*	mafic	(a)	4.16	2,240	4,500	10,150	6,200	1,380	0.22	59
7280	*Westringia rigida*	ultramafic	(b)	1.66	910	3,800	3,700	1,820	830	0.46	149
7302	*Eremophila caerulea*	mafic	(b)	4.32	2,040	4,700	10,250	6,660	727	0.11	75
7306	*Eremophila caerulea*	mafic	(b)	3.74	1,500	5,950	4,800	6,210	563	0.09	111

364

Table 4. The percentage ash and the concentrations (μg g^{-1} oven dry matter) in (a) the leaves and (b) the stems of phosphorus, potassium, sodium, calcium, magnesium (and the Mg/Ca quotients), and iron in leaves of *Eucalyptus griffithsii* from the northeastern extension of transect 66/2, Widgiemooltha

Sample No	Site	Part of plant	Ash	Phosphorus	Potassium	Sodium	Calcium	Magnesium	Mg/Ca quotient	Iron
7231	230', covered	a	4.24	1,980	2,850	4,500	7,130	2,640	0.37	49
	ground	b	3.65	1,880	3,300	8,750	15,400	2,630	0.17	55
7234	320', covered	a	4.39	1,260	2,850	8,250	10,800	1,050	0.10	38
	ground	b	5.98	1,560	3,800	8,150	15,600	1,020	0.07	46
7238	570', covered	a	7.15	2,210	3,800	9,650	13,500	1,810	0.13	67
	ground	b	6.44	2,180	4,500	10,300	14,800	1,890	0.13	49

Table 5. The percentage ash and the concentrations (μg g^{-1} oven dry matter) of phosphorus potassium, sodium, calcium, magnesium (and the Mg/Ca quotient) and iron in (a) the leaves and (b) twigs of *Trymalium myrtillus* collected from the Dordie Rocks area

Sample No	Site field grid reference	Part of plant	Ash*	Phosphorus	Potassium	Sodium	Calcium	Magnesium	Mg/Ca quotient	Iron
7739	48100Y/9050X	(a)	4.30	4	10,450	650	14,200	3,310	0.23	136
		(b)	1.05	7	3,950	1,350	3,640	311	0.09	59
7755	48300Y/8950X	(a)	4.76	5	5,200	950	42,300	4,640	0.11	107
		(b)	2.71	23	3,850	850	11,100	1,820	0.16	44
7758	48300Y/8900X	(a)	5.58 (1.83)	18	2,300	1,200	11,100	1,470	0.13	80
		(b)	1.97 (2.05)	12	5,400	2,100	7,540	424	0.06	87
7770	48100Y/9000X	(a)	3.76 (4.65)	2	5,130	1,510	41,400	4,000	0.10	54
		(b)	1.10 (6.20)	10	2,100	700	17,500	1,950	0.11	56

*For samples 7758 potassium and sodium analyses and for sample 7770 phosphorus analyses were made on separate subsamples with a different percentage ash which is given in parentheses.

Table 6. The percentage ash and the concentrations (μg g^{-1} oven dry matter) of phosphorus, potassium, sodium, calcium, magnesium (and the Mg/Ca quotient) and iron in (a) leaves and (b) twigs of *Dodonaea lobulata* collected from the Dordie Rocks area

Sample No	Collection site field reference	Part of plant	Ash*	Phosphorus	Potassium	Sodium	Calcium	Magnesium	Mg/Ca quotient	Iron
7723	48300Y/9050X	(a)	6.12 (6.49)	25	7,100	1,200	14,700	2,080	0.14	152
		(b)	4.02 (4.12)	7	3,250	900	9,300	368	0.04	105
7730	48300Y/9100X	(a)	4.10 (3.02)	23	5,750	850	10,400	2,020	0.19	100
		(b)	2.66 (2.57)	12	3,000	900	12,400	3,690	0.30	104
7736	48300Y/9150X	(a)	3.98	11	7,700	750	4,450	968	0.22	94
		(b)	2.55	6	3,350	850	3,210	1,500	0.47	82
7746	48100Y/9125X	(a)	3.11	18	6,150	550	10,900	2,520	0.23	87
		(b)	5.62	18	11,050	750	4,960	1,000	0.20	56
7747	48100Y/9150X	(a)	5.03	10	18,400	5,200	7,440	2,600	0.35	70
		(b)	2.87	4	4,440	3,900	6,560	672	0.10	82
7754	48300Y/9000X	(a)	5.41	11	6,200	2,800	14,410	4,240	0.29	49
		(b)	2.50	11	4,850	2,400	5,990	659	0.11	28
7759	48300Y/8900X	(a)	5.34	15	9,100	1,050	9,980	2,880	0.29	51
		(b)	2.85	–	3,350	2,000	16,500	4,230	0.26	77
7761	48300Y/9200X	(a)	4.69	17	24,000	3,900	7,500	1,750	0.23	63
		(b)	2.58	–	2,300	–	6,970	312	0.05	38

*For sample 7723 potassium and sodium analyses, for sample 7730 (a) iron, and for sample 7730 (b) iron and sodium analyses, were made on separate subsamples with a different percentage ash which is given in parentheses.

Table 7. The percentage ash and the concentrations (μg g^{-1} oven dry matter) of phosphorus, potassium, sodium, calcium, magnesium (and the Mg/Ca quotient) and iron in the leaves of *Alyxia buxifolia* from the Dordie Rocks Area

Sample No	Site field grid reference	Ash*	Phosphorus	Potassium	Sodium	Calcium	Magnesium	Mg/Ca quotient	Iron
7727	48300Y/9075X	3.49	14	18,400	3,200	5,280	2,470	0.47	50
		3.20	42	28,000	3,600	4,600	1,750	0.38	80
7741	48100Y/9075X	3.99	7	6,550	500	8,640	3,600	0.42	52
		2.52	7	4,050	1,600	6,610	550	0.08	60
7744	48100Y/9100X	2.17	17	5,950	750	14,700	3,470	0.24	110
		6.43	4	2,600	700	1,940	223	0.11	110
7745	48100Y/9125X	3.13	10	4,800	400	11,500	2,820	0.25	65
		3.20	9	3,650	600	5,570	503	0.09	93
7750	48100Y/9200X	4.22 (3.90)	15	6,200	2,238	6,920	1,910	0.28	39
		2.24 (2.60)	9	4,800	800	2,170	483	0.22	88

*For sample 7750 potassium and sodium analyses were made on a separate subsample with a different percentage ash which is given in parentheses.

Table 8. The percentage ash and the concentrations (μg g^{-1} oven dry matter) of phosphorus, potassium, sodium, calcium, magnesium (and the Mg/Ca quotients) and iron in (a) leaves and (b) stems of *Olearia muellerii* collected from the Dordie Rocks Area

Sample No	Collection Site field grid reference	Part of plant	Ash	Phosphorus	Potassium	Sodium	Calcium	Magnesium	Mg/Ca quotient	Iron
7725	48300Y/9050X	(a)	5.92	4	3,800	5,850	6,040	1,200	0.20	166
		(b)	3.13	9	2,950	3,300	8,310	259	0.03	119
7728	48300Y/9075X	(a)	4.41	–	16,200	3,760	4,000	459	0.11	90
		(b)	6.72	49	5,480	6,000	7,240	1,980	0.27	62
7731	48300Y/9100X	(a)	6.41	15	21,100	1,650	7,170	824	0.11	153
		(b)	3.20	20	8,450	21,100	5,470	284	0.05	189
7742	48100Y/9075X	(a)	4.21	15	20,500	3,000	4,740	1,830	0.39	61
		(b)	3.29	12	3,700	2,100	9,850	575	0.06	110
7764	48300Y/9250X	(a)	6.59	12	1,560	1,000	6,970	2,680	0.38	114
		(b)	2.08	12	6,900	600	4,340	451	0.10	164

Table 9. The percentage ash and the concentrations (μg g^{-1} oven dry matter) of phosphorus, potassium, sodium, calcium, magnesium (and the Mg/Ca quotients) and iron in (a) leaves and (b) the twigs of *Eremophila glabra* collected from the Dordie Rocks area

Sample No	Collection site field grid reference	Part of plant	Ash	Phosphorus	Potassium	Sodium	Calcium	Magnesium	Mg/Ca quotient	Iron
7724	48300Y/9050X	(a)	8.10	10	7,600	5,000	12,500	1,250	1.00	109
		(b)	3.73	–	1,150	500	4,200	555	0.13	95
7729	48300Y/9100X	(a)	7.55	12	7,280	461	21,600	3,480	0.16	13
		(b)	4.63	15	7,000	590	8,180	832	0.10	6
7751	48100Y/9100X	(a)	8.52	26	4,500	5,900	2,570	577	0.22	94
		(b)	5.63	18	1,200	1,070	1,000	324	0.32	62
7760	48300Y/8900X	(a)	7.46	25	2,200	9,150	11,700	806	0.07	83
		(b)	4.62	9	2,440	6,000	4,010	509	0.13	109
7753	48300Y/9000X	(a)	2.98	8	4,400	680	5,760	1,690	0.29	108
		(b)	4.43	8	1,100	670	4,190	581	0.14	80

Table 10. The percentage ash and the concentrations (μg g^{-1} oven dry matter) of phosphorus, potassium, sodium, calcium (and the Mg/Ca quotient) and iron in (a) leaves and (b) twigs of *Ricinocarpus stylosus* collected in the Mount Thirsty area

Site	Rock type	Part of plant	% Ash	Phosphorus	Potassium	Sodium	Calcium	Magnesium	Mg/Ca quotient	Iron
Transect 64/6	altered	(a)	6.65	257	82,000	10,900	20,500	923	0.05	94
1000 W	pyroxene	(b)	2.77	189	42,700	5,600	17,900	1,150	0.06	65
Transect 64/5	tremolite and	(a)	5.08	144	77,100	5,290	20,500	923	0.05	70
200 W	diopside epidote hornfels	(b)	3.34	86	45,700	2,790	43,200	1,810	0.04	132
Transect 64/2	tremolite and	(a)	6.03	225	92,700	5,840	22,700	2,350	0.10	72
800 W	diopside epidote hornfels	(b)	2.04	84	32,900	544	8,210	640	0.08	53
Transect 64/1	lateritized	(a)	8.80	609	101,000	12,900	7,720	3,240	0.42	99
200	diopside and tremolite hornfels	(b)	4.89	307	65,500	12,700	26,800	2,570	0.10	107
Transect 64/1	lateritized	(a)	5.97	137	78,400	4,670	11,600	2,240	0.19	80
400 W	diopside and tremolite hornfels	(b)	4.27	158	66,500	2,860	5,530	1,140	0.21	102

Table 11. The percentage ash and the concentrations (μg g^{-1} oven dry matter) of phosphorus, potassium, sodium, calcium, magnesium (and the Mg/Ca quotients) and iron in the leaves of *Eucalyptus campestre* collected in the Mount Thirsty area

Site	Rock type	Ash	Phosphorus	Potassium	Sodium	Calcium	Magnesium	Mg/Ca quotient	Iron
Transect 64/6 1,000 W	altered pyroxenite	5.15	1,180	480	570	24,400	5,540	0.23	1,150
Transect 64/6 600 W		6.28	1,190	590	560	15,100	6,790	0.45	838
Transect 64/4 300 W		4.31	1,970	800	640	17,900	7,500	0.42	1,030
Transect 64/2 400 W	tremolite and diopside epidote hornfels	8.15	2,440	430	210	18,800	4,240	0.23	1,200
Transect 64/1 200 E	lateritized diopside and tremolite hornfels	3.94	2,620	414	490	10,200	2,830	0.28	3,260
Transect 64/3 1,000 W	pyroxenite	4.90	3,610	490	340	13,500	8,750	0.65	1,060
Transect 64/4 900 W		6.27	3,060	650	750	10,600	5,000	0.47	1,180

concentrations of the soil peaks and magnesium concentrations rise the community is replaced by one of *Melaleuca cymbifolia* which in turn gives way to *Eucalyptus salmonophloia* woodland where the concentrations of magnesium and iron decrease and those of calcium increase. Throughout this area the soil contains low concentrations of potassium and very low concentrations of phosphorus and sodium. Here the legacy of lateritization influences the levels of major elements in the soils and in turn the distribution of plant communities.

In the Kurnalpi area where the *Eucalyptus lesoueffii* woodland, that covers ultramafic and mafic rocks alike, gives way to an association of small *Casuarina obesa* trees over a laterite-capped

Table 12. The percentage ash and the concentrations (μg g^{-1} oven dry matter) of phosphorus, potassium, sodium, calcium, magnesium (and the Mg/Ca quotients) and iron in (a) leaves and (b) twigs of *Melaleuca streathana* collected in the Mount Thirsty Area

Site	Rock type	Part of plant	Ash	Phosphorus	Potassium	Sodium	Calcium	Magnesium	Mg/Ca quotient	Iron
Transect 64/2	tremolite	(a)	5.53	166	40,700	8,470	18,700	4,640	0.25	79
800 W	and diopside	(b)	7.90	122	44,400	19,500	24,200	2,730	0.11	135
Transect 64/2	epidote	(a)	3.71	332	64,400	424	4,380	4,730	1.08	189
306 W	hornfels	(b)	1.68	143	41,500	194	7,050	830	0.11	101

Table 13. The percentage ash and the concentrations (μg g^{-1} oven dry matter) of phosphorus, potassium, sodium, calcium, magnesium (and the Mg/Ca quotients) and iron in (a) the leaves and (b) the twigs of *Dodonaea stenozyga* collected in the Mount Thirsty Area

Site	Rock type	Part of plant	% Ash	Phosphorus	Potassium	Sodium	Calcium	Magnesium	Mg/Ca quotient	Iron
Transect	altered	(a)	3.56	280	67,800	1,920	14,400	3,020	0.21	61
400 E	pyroxenite	(b)	2.33	289	96,400	47,500	14,900	3,050	0.20	48
Transect 64/1	lateritized	(a)	4.94	249	78,100	36,000	22,500	2,030	0.09	122
400 W	diopside and tremolite hornfels	(b)	3.05	139	57,300	300	16,200	1,170	0.07	114

hill, analyses of the leaves and twigs of these species for phosphorus, potassium, sodium, calcium, magnesium and iron suggest that differences in the concentrations of the major elements present in the different rock types have some influence on species distribution (Tables 14, 15). With one or two exceptions the Mg/Ca quotients in the plant tissues are broadly similar over different bedrock units. The concentrations of phosphorus and iron are markedly higher in *Eucalyptus lesoueffii* than in *Casuarina obesa* over all rock types and in the latter, iron concentrations are higher over lateritized than over unweathered bedrock. The change from *Eucalyptus lesoueffii* woodland to the *Casuarina obesa* association thus appears to be related to the differing concentrations of available major elements associated with the legacies of lateritization. The change to shrub communities and notably that composed of *Hybanthus floribundus*, *Acacia erinacea*, *Enchylaena tomentosa*, *Eremophila weldii*, *Grevillea acuaria* and *Westringia rigida*, however, occurs where high nickel and chromium concentrations occur in the soil, with the distribution of *Hybanthus floribundus* being related to the highest nickel concentrations.

In the areas of detailed investigations cited above the major element concentrations of soils are available only for Mount Thirsty, the area near Yates Knoll and Halls Knob Island in the Dordie Rocks area. There calcium concentrations are low and well below those of magnesium, while those of iron, a legacy of lateritization, are high (Fig. 15 and Table 16). In the Kambalda area, Elkington (1969) found 2,000 μg g^{-1} calcium, 9,000 μg^{-1} magnesium, and 1230 μg g^{-1} nickel over serpentinite outcrops occupied by *Hybanthus floribundus* whereas 16,000 μg g^{-1} calcium, 12,600 μgg^{-1} magnesium and only 120 μgg^{-1} nickel were found over meta-basalt in the same area.

Despite the excess of magnesium over calcium in the soils over serpentinite all the plant species that have been sampled contain more calcium than magnesium. As indicated earlier the Mg/Ca quotient is highest in *Hybanthus floribundus*, being near unity in samples collected at Widgiemooltha, Dordie Rocks, Mount Thirsty, Kurnalpie and Kurrajong (Farago *et al.* 1977). This species accumulates large amounts of nickel which is held in the cell wall and the phloem. Some of the nickel occurs in pectinates but since the amount held in the plant tissue appears to increase in parallel with an increase in calcium it is unlikely

Table 14. The percentage ash and the concentrations (μg g^{-1} oven dry matter) of phosphorus, potassium, sodium, calcium, magnesium (and the Mg/Ca quotient) and iron in (a) the leaves and (b) the stems of *Casuarina obesa* collected along transect lines 4,213 and 4,212, Kurnalpi

Location	Site	Rock type	Part of plant	Ash	Phosphorus	Potassium	Sodium	Calcium	Magnesium	Mg/Ca quotient	Iron
Transect 42/3	400 E	Alluvium over amphibolite	(a)	3.92	380	9,460	9,170	26,000	5,530	0.21	72
	600 E	Amphibolite	(a)	2.64	261	4,650	46	18,900	7,540	0.40	67
			(b)	4.71	515	8,240	303	45,300	6,680	0.15	105
	800 E	Amphibolite	(a)	3.88	412	5,630	60	9,930	2,240	0.23	119
	1,000 E	Quartz porphyry	(a)	4.23	574	16,600	5,660	25,000	6,580	0.26	90
	1,200 E	Serpentinite	(a)	4.11	395	8,930	344	29,600	3,880	0.13	73
			(b)	5.81	175	5,630	354	62,700	2,450	0.04	62
	1,400 E	Lateritized serpentinite	(a)	6.20	2,150	422	120	18,800	2,890	0.15	1,320
			(b)	4.46	326	4,450	236	83,100	6,580	0.08	90
	1,600 E	Serpentinite	(a)	3.80	339	9,590	644	44,800	4,600	0.10	117
	1,800 E	Serpentinite	(a)	5.08	383	4,380	180	59,000	6,450	0.11	135
			(b)	5.51	271	2,290	387	29,500	1,500	0.05	71
	2,000 E	Lateritized serpentinite	(a)	5.74	454	4,900	160	55,200	11,000	0.20	307
			(b)	5.12	2,640	200	40	32,800	5,640	0.17	932
	2,400 E	Lateritized serpentinite	(a)	4.67	2,530	530	190	22,600	3,970	0.18	1,020
			(b)	4.67	476	8,130	233	41,400	1,020	0.02	60
Transect 42/2	2,300 W	Coarse-grained amphibolite	(a)	2.52	1,960	498	230	13,300	5,900	0.44	1,000
			(b)	6.21	2,040	210	140	28,900	5,150	0.18	804
	500 E	Lateritized and	(a)	2.83	1,210	445	125	16,900	4,050	0.24	1,090
			(b)	3.32	1,720	432	50	24,400	7,760	0.32	1,190
	550 E	weathered	(a)	2.39	1,740	480	160	16,500	4,940	0.30	1,210
	800 E	serpentinite	(a)	3.42	1,740	472	220	12,600	8,840	0.70	1,070

Table 15. The percentage ash and the concentrations (μg g^{-1} oven dry matter) of phosphorus, potassium, sodium, calcium, magnesium (and the Mg/Ca quotients) and iron in the leaves of *Eucalyptus lesoueffii* collected along transect 42/2, Kurnalpi

Site	Rock type	% Ash	Phosphorus	Potassium	Sodium	Calcium	Magnesium	Mg/Ca quotient	Iron
2,000 W	meta-sediments	4.69	3,770	860	685	17,900	5,290	0.30	1,770
1,700 W		4.64	4,960	865	675	17,700	8,000	0.45	2,020
1,600 W	coarse-grained	4.48	3,210	769	600	18,700	5,630	0.30	1,550
1,500 W	amphibolite	4.20	4,040	529	550	15,600	8,630	0.55	1,700
600 W	serpentinite	4.94	4,900	825	880	21,900	9,360	0.43	2,530
250 W		4.94	4,480	651	640	20,500	5,700	0.28	1,280
150 W		3.91	3,120	742	564	15,200	6,420	0.42	2,070
350 E	serpentinite	3.77	2,590	700	435	15,800	7,390	0.47	1,690
450 E		4.50	2,660	720	645	21,800	5,500	0.25	1,940
700 E		3.61	1,540	560	420	14,500	5,860	0.40	1,420

that it takes the place of calcium as initially thought possible. As yet many questions concerning the relationships between the occurrences of *Hybanthus floribundus*, particular soils and other environmental factors are unanswered.

Conclusions

Within the greenstone belts in the Eastern Goldfields area of Western Australia variations in the physiognomy and composition of plant communities reflect the interplay of many environmental factors, whose relative importance varies spatially. Different plant communities occupy the distinctive types of ultramafic rocks in different regions. In some areas the high Mg/Ca quotients in soils over serpentinites probably exert an important influence, but usually this quotient is low enough for normal plant growth with calcium deriving, in some cases, from superficial cover. Nowhere is there a distinctive serpentinite flora. Plant assemblages characterized by *Hybanthus floribundus* indicate the presence of high concentrations of nickel, emanating from sulphide ore deposits in some areas, notably Widgiemooltha and Dordie Rocks, while elsewhere the replacement of woodland by shrub vegetation reflects fairly high concentrations of chromium and nickel in the substrate. Overall the vegetation/environment relationships are very complex.

List of species

AMARANTACEAE

Ptilotus obovatus F. Muell.

APOCYNACEAE

Alyxia buxifolia (End.) Druce

BORAGINACEAE

Halgania rigida S. Moore

CASUARINACEAE

Casuarina campestris Diels
Casuarina cristata Miq.
Casuarina helmsii Ewart et Gordon
Casuarina obesa Miq.

CHENOPODIACEAE

Arthrocnemum halocnemoides Nees
Atriplex nummularia Lindl.
Atriplex paludosa R. Br.
Kochia georgei Diels
Kochia sedifolia F. Muell.
Rhagodia preissii Moq.

COMPOSITAE

Cratystylis spp.
Helipterum adpressum W.V. Fitzg.
Helipterum polygalifolium DC.
Olearia muelleri Benth.

Table 16. The concentrations of (μg g^{-1} oven-dried soil) of phosphorus, potassium, sodium, calcium, magnesium (and the Mg/Ca quotient), iron and nickel in soils from south of Mount Thirsty, from Halls Knob Island, and from the Dordie Rocks Area. The analyses have been made on soil sieved through a 2 mm mesh

Sample No	Site	Rock type	Phosphorus	Potassium	Sodium	Calcium	Magnesium	Mg/Ca quotient	Iron	Nickel
7,000	South of Mount Thirsty	hypersthenite	906	720	74	20	4,430	222	65,300	63
7,001		gabbro	715	2,930	400	418	8,430	20.2	88,300	31
7,002		calc-silicate hornfils	838	432	260	2,028	5,180	2.6	86,400	28
7,003		tremolite	285	432	313	78	4,140	53.0	65,800	28
7,123	Halls Knob Island	talc carbonate	985	3,550	2,680	148,000	138,000	0.93	62,500	2,900
7,124			1,050	1,490	400	144,000	103,000	0.72	54,800	2,130
7,125			395	432	1,040	261,000	157,000	0.60	57,400	4,150
7,126		peridotite	452	1,630	74	1,210	73,800	0.61	50,600	600
7,015	Northeast of Yates Knoll, Mount Thirsty Area	gabbro	642	6,790	1,520	51,200	46,300	0.90	105,000	519
7,009		serpentinite	778	2,450	3,400	20,800	65,300	3.14	177,000	800

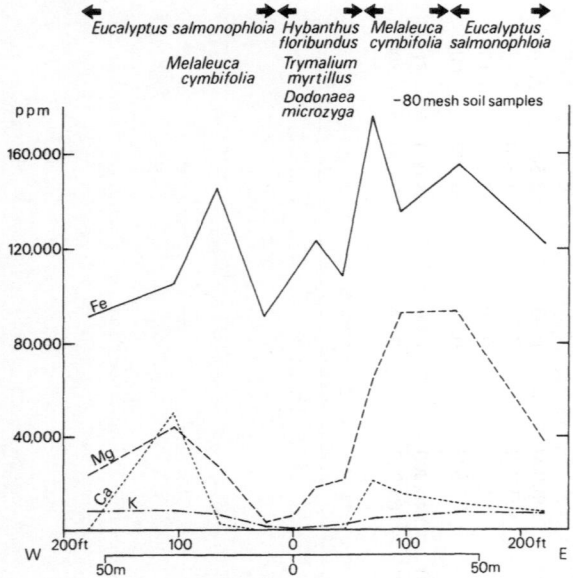

Fig. 15. Relationship between species distributions and concentrations of potassium, calcium, magnesium and iron in the soil where lateritic material mantles ultramafic and mafic rocks north-east of Yates Knoll, Mount Thirsty area.

CRUCIFERAE

Lepidium leptopetalum F. Muell.

EUPHORBIACEAE

Ricinocarpus stylosus Diels.

FICOIDACEAE

Disphyma australe Sol.

FRANKENIACEAE

Frankenia interioris Ostf.

GOODENIACEAE

Dampiera trigona var latealata De. Vt.
Scaevola spinescens R. Br.

GRAMINAE

Triodia irritans R. Br.
Tridoia pungens R. Br.

LABIATAE

Prostanthera aspalathoides A. Cunn.
Prostanthera wilkieana F. Muell.
Westringia rigida R. Br.

LEGUMINOSAE

Acacia acuminata Benth.
Acacia aneura F. Muell.
Acacia burkittii F. Muell.
Acacia colletioides A. Cunn.
Acacia erinacea Benth.
Acacia graffiana F. Muell.
Acacia leptoneura Benth.
Acacia quadrimarginea F. Muell.
Acacia tetragonophylla F. Muell.
Cassia artemisioidea Gaud.
Cassia eremophila A. Cunn.
Cassia nemophylla F. Muell.

MYOPORACEAE

Eremophila dempsteri F. Muell.
Eremophila duttoni F. Muell.
Eremophila fraseri F. Muell.
Eremophila glabra (R. Br.) Ostf.
Eremophila interstans (S. Moor) Diels.
Eremophila ionantha Diels.
Eremophila miniata C. A. Gardn.
Eremophila oldfieldi F. Muell.
Eremophila oppositifolia R. Br.
Eremophila pachyphylla Diels.
Eremophila scoparia F. Muell.
Eremophila serrulata (A. Cunn.) Druce
Eremophila stronglyophylla F. Muell.
Eremophila weldii F. Muell.

MYRTACEAE

Calythrix micromyctaceae
Eucalyptus brockwayi C. A. Gardner
Eucalyptus calycogona Turez
Eucalyptus campaspe S. Moore
Eucalyptus celastroides Turez
Eucalyptus foecunda Schauer
Eucalyptus gracilis F. Muell.
Eucalyptus lesoueffi Maiden
Eucalyptus oleosa F. Muell.
Eucalyptus salmonophloia F. Muell.
Eucalyptus salubris F. Muell.
Eucalyptus stricklandi Maiden
Eucalyptus torquata Luch.
Melaleuca coccinea A. S. George
Melaleuca cymbifolia Benth.
Melaleuca lateriflora Benth.
Melaleuca sheathiana W. V. Fitzg.
Thryptomene sp

Verticordia sp
Wehlia spp.

PROTEACEAE

Grevillea acuaria F. Muell.
Hakea preissii Meissn.

RHAMNACEAE

Cryptandra glabriflora Benth.
Pomaderris forrestiana F. Muell.
Trymalium myrtillus S. Moore

RUTACEAE

Boronia fabianoides (Diels) P.G. Wals.
Phebalium filifolium Turcz.
Phebalium tuberculosum F. Muell. Benth.

SANTALACEAE

Exocarpus aphylla R. Br.
Santalum acuminatum (R. Br.) D.C.
Santalum spicatum D.C.
Dodonoea boroniaefolia G. Don.
Dodonoea lobulata F. Muell.
Dodonoea microzyga F. Muell.
Dodonoea stenozyga F. Muell.

STACKHOUSIACEAE

Stackhousia heugelii Endl.

STERCULIACEAE

Brachychiton gregorii F. Muell

THYMELANEACEAE

Pimelea microcephala R. Br.

VIOLACEAE

Hybanthus floribundus (Lindl.) F. Muell.

References

Carnahan, J. A. 1978. Vegetation in D. N. Jeans (edit) Australia – A Geography pp. 175–195, Routledge and Kegan Paul, London.

Cole, M. M. 1973. Geobotanical and biogeochemical investigations in the sclerophyllous woodland and shrub associations of the Eastern Goldfields area of Western Australia with particular reference to the role of *Hybanthus floribundus* (Lindl). F. Muell. as a nickel indicator and accumulator plant. Journal of Applied Ecology 10: 269–320.

Cole, M. M. 1985. Simple remote sensing in prospecting for gold, uranium and base metals in desert areas in Australia and Africa; some case studies. Prospecting in areas of desert terrain. Institution of Mining and Metallurgy, London.

Elkington, J. E. 1969, Vegetation studies in the Eastern Goldfields of Western Australia with particular reference to their role in geological reconnaissance and mineral exploration. Ph.D. thesis, University of London.

Farago, M. E., A. J. Clarke & M. J. Pitt. 1977. Plants which accumulate metals. Part 1. The metal content of three Australian plants growing over mineralized sites. *Inorganica Chimica* 24: 53–56.

Farago, M. E. & Mahmoud, I. E. D. A. W. 1983. Plants that accumulate metals (Part VI): Further studies of an Australian nickel accumulating plant. Minerals and the environment 5: 113–121.

Groves, R. H. 1981. Australian vegetation. Cambridge University Press, Cambridge.

Jutson, J. T. 1934. The physiography of Western Australia. Bulletin of the Geological Survey of Western Australia, 11.

Kruckeberg, A. R. 1954. The ecology of serpentine soils III Plant species in relation to serpentine soils. *Ecology* 35, 267–274.

Kruckeberg, A. R. 1964. Plant life on serpentines and other ultramafic rocks in north western North America, Proceedings of the International Botanical Congress of Edinburgh, 13.

Speck, N. H. 1958. Vegetation of the Wiluna-Meekatharra area in General Report on Lands in the Wiluna-Meekatharra area, Western Australia. CSIRO, Land Research Series No. 7, pp. 143–161.

Spence, D. H. N. 1957. Studies on the vegetation of Shetland I. The Serpentine debris vegetation in Unst. Journal of Ecology 45: 917–945.

Walker, R. B. 1954. The Ecology of serpentine soils II. Factors affecting plant growth on serpentine soils. Ecology 35: 259–266.

Walker, R. B., H. M. Walker, & P. R. Ashworth. 1955. Calcium – Magnesium nutrition with special reference to serpentine soils. Plant Physiology 30: 214–221.

Yates, K. R. 1966/67. Geology of the Mt. Thirsty – Hayes Hill Area. Temporary Reserve 3508, Norseman, Western Australia. Unpublished report of Anaconda Australia Incorporated.

Author's Address:
M. M. Cole
Dept. of Geography
Royal Holloway and Bedford New College
University of London
Egham Hill
EGHAM
Surrey TW20 0EX
U.K.

XV

New Zealand Ultramafics

W. G. LEE

Abstract. Ultramafic rocks comprise less than 0.1% of the New Zealand land surface but their occurrence in areas of contrasting climate and glacial history has produced a wide range of plant habitats. Compared with other New Zealand soils, ultramafic soils, ranging from deeply weathered laterites to youthful skeletal regoliths, have low concentrations of most major nutrients, a wide Mg/Ca quotient, high concentrations of nickel, chromium and cobalt, and a high pH. Ultramafic vegetation, while often strikingly different from that on adjoining rock types, has close floristic and physiognomic similarities with communities on other edaphically extreme soils. Forest on ultramafic soils is restricted to lowland and montane sites in areas of higher rainfall, while in drier areas an open mixed shrubland with isolated stunted trees appears to be the maximum vegetation development. With increasing elevation and declining soil conditions, the vegetation becomes more open, stunted and sparse with a greater proportion of low-growing, microphyllous shrubs and several small grasses. Most of the major plant families and genera in New Zealand have representatives on ultramafic soil. At present 34 taxa are considered to be ultramafic endemics with distributions limited to outcrops which escaped direct effects of Pleistocene glaciations. Studies of ecotypic differentiation, element accumulation and species response to soil amendment suggest that plant species adopt diverse strategies to achieve ultramafic tolerance.

B. A. Roberts and J. Proctor (eds), The ecology of areas with serpentinized rocks. A world view, 375–418.
© *1992 Kluwer Academic Publishers.*

Photo 1. Red Mountain, north-west Otago. (Photograph by Lloyd Homer.)

Introduction

New Zealand has numerous outcrops of ultramafic rocks which support an impressive variety of habitats. Their frequency arises because of New Zealand's position astride the margins of two crustal plates along which sections of upper mantle and oceanic crust were emplaced during the Permian. The habitat diversity reflects the broad latitudinal spread of sites, from near subtropical conditions in the north (34°24′S) to cool temperate in the south (45°50′S), the contrasting climates of coastal and inland areas, and the wide altitudinal range of the larger sites from lowland through to alpine zones.

Apart from a small area at Surville Cliffs near North Cape in the far north of the North Island, the majority of New Zealand's ultramafic outcrops are clustered in the north-east and south-west regions of the South Island, where they form 0.35% of the total land area. The largest ultramafic areas are found along the Dun Mountain Ophiolite Belt (Coombs *et al.* 1976), a narrow group of rocks that originally formed a continuous structural unit but now comprises two seg-

ments offset 450 km due to movement along the Alpine Fault (Fig. 1).

The earliest observations, albeit unawares, by a botanist, on the effect of ultramafic rocks on vegetation were provided by Joseph Banks who visited New Zealand with Captain James Cook in 1770. While sailing north along the south-west coast of the South Island, Banks noted 'much snow on the ridges on the high hills, two were, however seen on which was little or none. What ever the cause of it I could not guess. They were quite bare of trees and vegetables and seemed to consist of bricks of bright red ochre' (Banks 1770). Banks was here aptly describing the ultramafic outcrops in the middle Cascade Valley, South Westland (Photo 1).

Over 100 years later surveyor-geologists exploring New Zealand also noticed the absence of vegetation on ultramafic rocks (Park 1887, Bell *et al.* 1911). Cockayne (1910, 1922, 1928) provided the first descriptions of the vegetation of the accessible sites with ultramafic soils in the Nelson–Marlborough region. Since then several accounts of the vegetation on ultramafic outcrops have appeared, the most detailed being Oliver

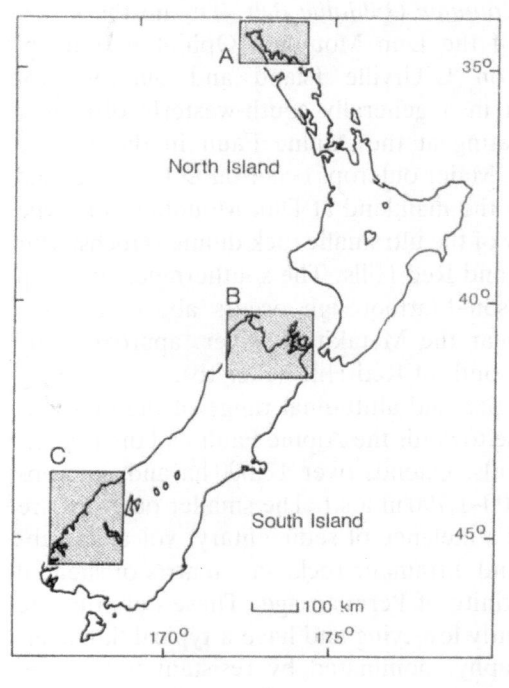

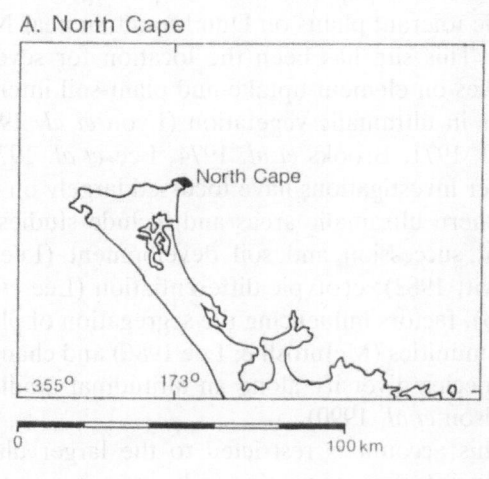

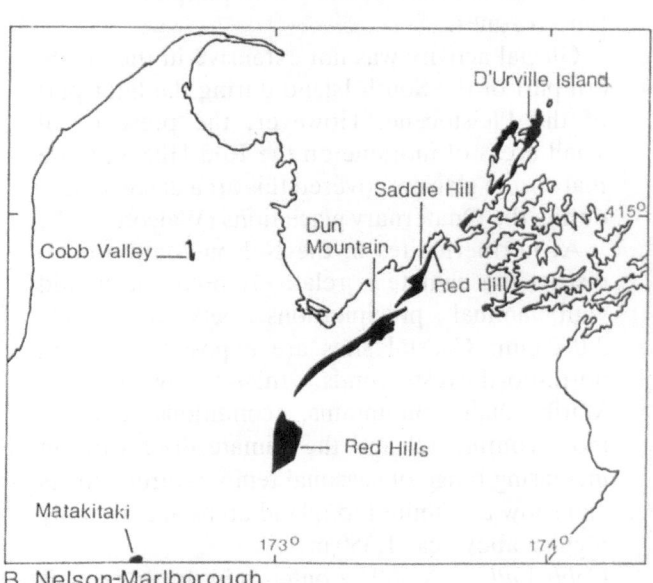

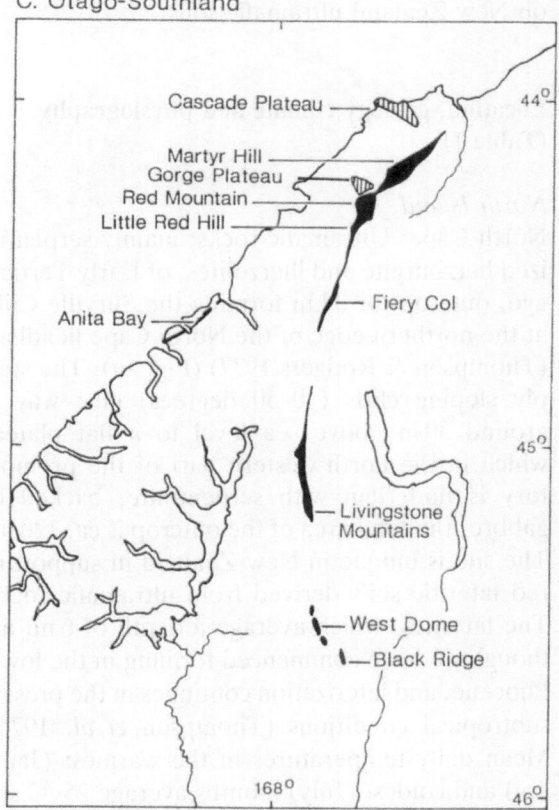

Fig. 1. Location of major ultramafic outcrops in New Zealand.

(1944), Wardle *et al.* (1971), Mark (1977), Druce *et al.* (1979), Lee (1980) and Walls & Laffan (1986).

The first ecological investigations of ultramafic plants were published by Betts (1918, 1919, 1920) in a series of papers on the morphology of ultramafic-tolerant plants on Dun Mountain near Nelson. This site has been the location for several studies on element uptake and plant-soil interactions in ultramafic vegetation (Lyon *et al.* 1968, 1970, 1971, Brooks *et al.* 1974, Lee *et al.* 1975). Other investigations have focussed largely on the southern ultramafic areas and include studies of plant succession and soil development (Lee & Hewitt 1982), ecotypic differentiation (Lee *et al.* 1983), factors influencing the segregation of plant communities (McIntosh & Lee 1986) and changes in species diversity along an altitudinal gradient (Wilson *et al.* 1990).

This account is restricted to the larger ultramafic outcrops supporting indigenous vegetation and refers only to vascular plant species, there being little information on bryophytes and lichens on New Zealand ultramafic soils.

Location, geology climate and physiography (Table 1)

North Island

North Cape: Ultramafic rocks, mainly serpentinized harzburgite and lherzolites, of Early Tertiary age, outcrop for 3 km forming the Surville Cliffs at the northern edge of the North Cape headland (Thompson & Rodgers 1977) (Fig. 1a). The steeply sloping cliffs (40–50 degrees) give way at around 30 m above sea level to a flat plateau which in the north-western part of the promontory is underlain with serpentinite, backed by gabbro, the total area of the outcrop is ca. 120 ha. The site is unique in New Zealand in supporting red lateritic soils derived from ultramafic rocks. The laterites, which average a depth of 6 m, are thought to have commenced forming in the lower Pliocene, and laterization continues in the present subtropical conditions (Thompson *et al.* 1974). Mean daily temperatures in the warmest (January) and coldest (July) months average 25 °C and 18 °C respectively while the annual precipitation is around 1500 mm.

South Island
Nelson–Marlborough

Dun Mountain Ophiolite Belt. The northern segment of the Dun Mountain Ophiolite Belt appears on D'Urville Island and outcrops for 150 km in a generally south-westerly direction, terminating at the Alpine Fault in the Wairau Valley. Major outcrops occur on D'Urville Island and on the mainland at Dun Mountain, the type locality of the ultramafic rock dunite (Hochstetter 1864), and Red Hills. The southernmost outcrop in Nelson–Marlborough occurs above the tree limit near the Matakitaki River, approximately 50 km south of Red Hills (Fig. 1b).

The size and altitudinal range of the outcrops increase towards the Alpine Fault and the largest, Red Hills, extends over 12,600 ha and outcrops from 700–1,790 m a.s.l. The smaller outcrops are part of a melange of sedimentary, volcanic, gabbroic and ultramafic rocks in a matrix of sheared serpentinite of Permian age. These outcrops are frequently low-lying and have a typical 'knocker' topography, dominated by resistant tors of included blocks rising above the more easily eroded sheared and crushed serpentinite matrix which forms moderate slopes and gently rolling terrain. At Red Hills and Dun Mountain parental ultramafics predominate and form high summit peaks and ridge massifs, typically with marginal serpentinized zones.

Glacial activity was not extensive in the northern part of the South Island during the later part of the Pleistocene. However, the presence of small areas of moraine on the Red Hills indicate that cirque glaciers covered this area at some time during the Quaternary glaciations (Walcott 1969).

At northern sites in the Nelson–Marlborough district the climate is relatively mild and humid with annual precipitations between 1,200–2,000 mm. Coastal sites are exposed to strong north-north-west winds. Inland towards the Marlborough mountains, conditions become more continental and the climate drier with an increasing range of seasonal temperatures. Frosts and snow are limited to inland areas and are only regular above ca. 1,350 m.

Cobb Valley. A 400 ha outcrop in north-western Nelson extends between Cobb Valley and the Takaka River catchment, and is considered to be

Table 1. Location, altitudinal range, size, climate and vegetation accounts for major New Zealand ultramafic outcrops

	Location NZMS 1 Map series	Altitudinal range (m)	Approx. area (ha)	Climatic types after Coulter (1975)	Accounts of vegetation
NORTH ISLAND					
North Cape	N1/2 495 554	0–30	120	A	Thompson *et al.* (1974), Druce *et al.* (1979)
SOUTH ISLAND					
Nelson–Marlborough					
D'Urville Island	S10 144 747	0–580	250	D2	Oliver (1944), Lee (1980)
Red Hills	S15 890 460	300–400	50	D2	Cockayne (1928)
Saddle Hill	S20 771 289	1,180–1,210	130	B	Lee (1980)
Dun Mountain	S20 717 208	400–1,128	1,500	B	Cockayne (1910), (1922), Bell *et al.* (1911), Betts (1918), Lee (1980)
Red Hills	S27 419 868	770–1,790	13,000	Co	Cockayne (1910), Lee (1980)
Matakitaki	S39 828 378	1,200–1,500	300	M	
Cobb Valley	S13 110 520	600–1,100	400	M	
Otago-Southland					
Martyr Hill	S97 403 734	120–1,030	1,400	M	Lee (1980)
Red Mountain– Little Red Hill	S105 247 556	450–1,700	4,550	M	Ogden (1970), Lee (1980), Lee & Hewitt (1982)
Fiery Peak	S113 403 284	880–1,690	720	M	Mark (1977), Lee (1980)
Livingstone Mountains	S141 077 496	1,200–1,400	100	F_0	Lee (1980)
West Dome–Black Ridge	S150 253 019	300–950	600	F_0	Lee (1980), McIntosh & Lee (1987)
Anita Bay ultramafites	S112 723 161	10,780	250	D	Wardle *et al.* (1971), Lee (1980)
Cascade Plateau	S97 360 910	30–550	2,500	D	Lee (1980)
Gorge Plateau	S105 210 670	300–1,000	5,000	D	Wardle *et al.* (1986)

older (Cambrian) and of separate origin to the Dun Mountain Ophiolite Belt (Fig. 1b). The ultramafics are completely serpentinized (Coleman 1966) and form gentle slopes.

Otago-Southland

Dun Mountain Ophiolite Belt. The southern segment begins south-east of the Alpine Fault in South Westland and continues southwards for 170 km with large outcrops of ultramafic rock at Martyr Hill, Red Mountain–Little Red Hill, Fiery Col and Livingstone Mountains, terminating at West Dome-Black Ridge in northern Southland (Fig. 1c). Geologically, the ultramafics follow a similar pattern to that described for the northern segment: extensive massifs of dunite and peridotite near the Alpine Fault with a decrease in outcrop size and increasing serpentinization in distal

parts of the belt. Red Mountain, the largest outcrop in the south, extends from 450–1,700 m a.s.l. over an area of 4,500 ha.

Outcrops along and west of the Southern Alps have a topography that reflects the recent glacial history of the region, with the presence of broad U-shaped valleys, hanging valleys, cirques, aretes, moraines and tarns. Subsequent modification by fluvial and periglacial processes has altered and topography considerably. The present ultramafic landscape is characterised by broad glacially scoured basins and benches (above 500 m), deeply incised streams, and talus slides now buttressing over-steepened slopes. Moraine heaps and possible landslide debris of ultramafic origin up to 100 m high are found in the Cascade River Valley, while at the mouth of the Cascade River there is a series of low, well-preserved lateral moraines, mostly of peridotite, trending north-west–south-east, which form a triangular-shaped tableland, the Cascade Plateau, of ca. 5200 ha. This material has been glacially translocated from the Martyr Hill–Mt Raddle section of the ultramafic belt, over a distance of 14 km. A veneer of ultramafic moraine also covers Gorge Plateau, 5 km west of Red Mountain.

Outcrops east of the Southern Alps in northern Southland are usually associated with broad spurs and have typical 'knocker' topography. These areas, particularly on West Dome and Black Ridge, may have escaped the direct effects of glaciation, at least in the later stages of the Pleistocene.

West of the Main Divide the climate is cool and wet, with rainfall exceeding 3,200 mm on the coast and increasing with altitude to over 6,000 mm a few km inland. Temperatures near the coast are mild (mean January temperature 14.2 °C–mean July temperature 5.2 °C), but decrease to the east. Snow is present for only a few days at a time in winter at altitudes below c. 1000 m.

East of the Main Divide precipitation declines rapidly and northern Southland receives less than 1,000 mm annually. In general, summers are relatively warm and dry and winters cold. Climatic conditions in the alpine zone of the Livingstone Mountains are less severe than those described by Bliss & Mark (1974) for mountain ranges of similar altitude further east in Central Otago where mean monthly air temperatures approach 0 °C and rise only to 5–6 °C in the warmest month. Frosts are frequent throughout the year and continuous snow usually persists for five months.
Anita Bay. Another band of ultramafic rocks, the Anita Bay Ultramafics (Wood 1972) (Fig. 1c) occurs in north-west Fiordland, where it extends for 23 km parallel to and 3 km from the coast. Rarely more than 1 km wide, it attains its maximum width and altitude (700 m) near Lake Ronald (Photo 2). These rocks are mainly harzburgite with pockets of dunite and are thought to have a separate origin from and to be much older (Cambrian or Carboniferous) than the Dun Mountain Ophiolite Belt.

The climate and topography is similar to that described for the lowland ultramafic areas 65 km north-west of the Southern Alps.

Disturbance

Much of the disturbance on accessible ultramafic areas was associated with the exploration and exploitation of their mineral potential. The Dun Mountain Belt is locally known as the 'Mineral Belt' and considerable quantities of copper and chromium were taken intermittently from the ultramafites near Dun Mountain (Photo 3) between 1855 and 1911. However, the main economic significance of ultramafic areas has been as a source of serpentinite, which until recently was quarried at North Cape, Cobb Valley and Black Ridge for use in agriculture. Mixed with phosphatic fertilizer, the serpentinite was at first introduced to conserve phosphate supplies but was retained because it improved the storage and handling properties of the fertilizer mixture and supplied magnesium where there was a deficiency (Metson 1974).

During the period of European settlement fire has severely modified the plant cover on the plateau at North Cape, the lower slopes of Dun Mountain, large areas of the Red Hills, and parts of the Cascade Plateau, West Dome and Black Ridge outcrops. Ultramafic vegetation contains a high proportion of resinous species making it particularly flammable.

On the Dun Mountain belt only D'Urville Island and Red Hills have outcrops grazed by sheep and cattle. However, all mainland sites support feral populations of introduced ungulates, primarily red deer and goats together with smaller herbivorous animals such as hares, possums and rabbits. Bell et al. (1911) attributed the disappearance of the scree plant *Stellaria roughii** from Dun Mountain to selective browsing by goats and deer.

In the south-west of the South Island only red deer are present. Herbivores are unlikely to have greatly affected the composition of the vegetation as ultramafic habitats are unattractive owing to their generally unpalatable flora, the sparse, relatively open vegetation cover, the sharp olivine crystals that protrude from harzburgite rocks, and the generally unstable terrain.

* Nomenclature follows Allan (1961), Moore and Edgar (1970), Webb et al. (1988) and name changes in the indigenous New Zealand flora updated in Connor & Edgar (1987).

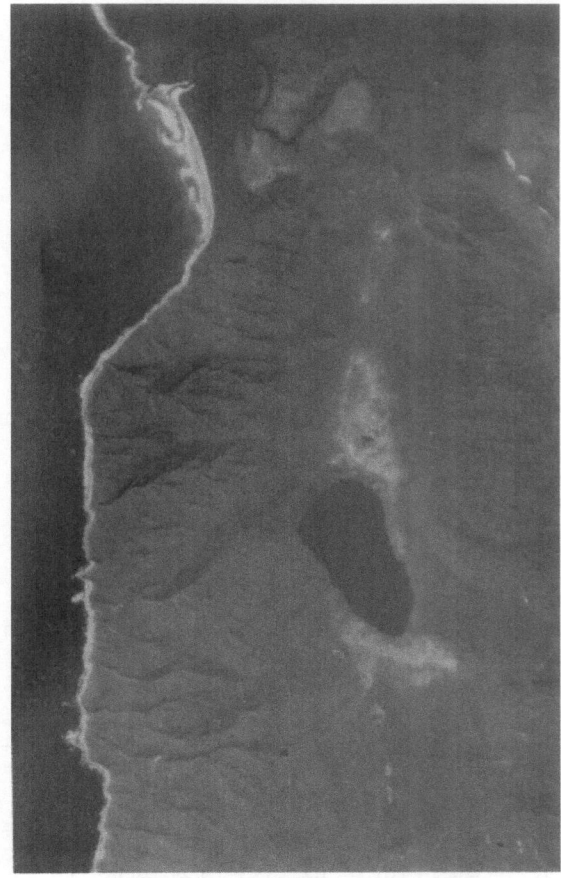

Photo 2. Ultramafic outcrops with a sparse vegetation cover occur north and south of Lake Ronald (central picture) on the Anita Bay ultramafites, Fiordland, contrasting with the tall forest on adjoining rock types. (Photograph by the Departments of Lands and Survey.)

Soils

Ultramafic soils are classified in the intrazonal brown granular loams and clays and currently three soil sets are recognized – Dun Steepland Soils, Cascade Soils and Black Ridge Hill Soils (New Zealand Soil Bureau 1968a). According to Soil Taxonomy (1975), New Zealand ultramafic soils are mostly dystrochrept, eutrochrepts or, at higher elevations, cryochrepts (Lee & Hewitt 1982, McIntosh & Lee 1986). However, criteria currently used in Soil Taxonomy fail to differentiate ultramafic soils in relation to the plant communities they support (McIntosh & Lee 1986).

Summary descriptions of ultramafic soils usually associated with maximum vegetation development, at representative ultramafic localities, are given in Table 2. There are no formal soil descriptions for the serpentine-derived laterites at North Cape, although Thompson & Rodgers (1977) illustrate different horizons and element concentrations in the intact laterites. However, the least-modified plant communities occur on cliffs on weakly weathered serpentinite with minor amounts of lateritic colluvium. On South Island ultramafites skeletal soils are widespread, particularly in the south. They are typically shallow, friable and stony with weakly developed horizons. However, considerable morphological and structural development can occur on ultramafites in depressions or on terraces and stable slopes associated with greater vegetation cover. Typically these soils are free-draining, friable, permeable stony or silt loams, with granular or crumb structure. Those derived from serpentinite are usually grey to dark brown in colour, in contrast to the yellow or reddish-brown soils derived from dunite or peridotite.

In comparison with other New Zealand soils, ultramafic soils have high pH values, medium N concentrations, high C/N quotients, low P, Ca and K concentrations, a medium-high cation exchange capacity, and very high Mg concentrations. High Ni, Cr and Co concentrations initially discovered in soils on Dun Mountain (Lyon *et al.* 1968), and a high Mg/Ca quotient, are typical of soils on ultramafic outcrops in New Zealand (Table 3). The combination of low fertility, high pH and moderate textures is rare in New Zealand.

Floristics

Approximately 800 vascular plant taxa are recorded from ultramafic soils in New Zealand; however it would be misleading to suggest that all these taxa exhibit ultramafic tolerance on a large scale. The North Cape ultramafic vegetation supports 140 taxa, representing the highest number of species per unit area on ultramafic soil in New Zealand, while the northern and southern regions of the South Island support 408 and 353 taxa, respectively (Table 4) and show little differ-

Photo 3. Dun Mountain, Nelson, showing the abrupt transition from tall forest on sedimentary rocks to open shrubland on ultramafites. However, much of the ultramafic vegetation has been burnt.

Table 2. Representative soil profiles from New Zealand ultramafic soils

Location	Horizon	Depth (cm)	Colour	pH	C%	N%	C/N	CEC*	TEB*	Ca*	Mg*	K*	Na*	P 0.5 M H₂SO₄
French Pass	A	0–8	10YR 3/2	6.4	5.2	0.29	18	26.2	22.6	4.0	17.9	0.42	0.27	3
	Bw	8–17	10YR 5/4	6.2	1.9	0.12	16	13.2	11.1	1.3	9.4	0.24	0.18	1
	Bg	17–45	5YR 5/8	6.3	2.0	0.10	20	27.2	24.1	1.7	21.9	0.24	0.26	1
Dun Mountain	A1	0–8	10YR 2/2	6.5	4.3	0.31	14	22.6	17.5	3.4	13.8	0.20	0.05	5
	A3	8–18	10YR 3/3	6.6	3.1	0.25	12	21.0	15.1	2.1	12.9	0.12	0.02	2
	B2	23–40	10YR 3/4	6.8	2.1	0.18	12	37.4	28.9	1.4	27.4	0.10	0.00	1
	C1	43–55	7.5YR 4/4	6.6	1.2	0.11	11	50.2	40.1	0.8	39.2	0.10	0.01	0
Red Mountain	01+02	0–25	5YR 3/2	4.3	54.0	1.61	34	25.1	49.0	1.2	49.3	0.91	0.69	9
	A	25–31	10YR 4/4	6.1	6.7	0.38	18	18.3	16.9	0.3	22.2	0.06	0.18	6
	B1	31–63	10YR 4/5	6.6	2.8	0.20	14	10.7	12.5	0.4	10.5	0.03	0.12	4
	B2L	63–69	7.5YR 2.5	6.8	6.6	0.41	16	22.2	23.5	0.7	27.6	0.06	0.14	6
		69+	10YR 4/3	6.8	1.1	0.07	16	5.9	7.5	0.2	6.1	0.02	0.04	3
Cascade Plateau	Al	0–18	10YR 4/2	6.4	19.7	0.62	32	50.0	33.4	6.7	25.8	0.44	0.46	4
	A2	18–30	10YR 5/2	6.0	7.6	0.37	21	23.3	11.4	2.0	8.9	0.22	0.26	3
	Bc	30–50	7.5YR 5/7	6.1	2.5	0.10	25	11.9	7.5	0.8	6.5	0.05	0.16	3
Black Ridge	A	0–13	10YR 3/3	6.2	nd	nd	nd	30.3	25.7	6.1	15.6	0.12	0.13	nd
	BW	13–23	10YR 3/3	6.0	nd	nd	nd	22.5	15.4	1.3	18.8	0.62	0.22	nd
	BW2	23–34	10YR 4/4	6.1	nd	nd	nd	21.5	14.0	0.5	13.8	0.15	0.14	nd

*Meq %.

ence in the relative contributions of the different plant groups.

A list of vascular species common or dominant on vegetation of ultramafic soils is presented in Table 5. The flora of these soils is mainly evergreen and perennial and life-forms such as deciduous trees, summer-green and winter-green herbs and annuals are rare or absent. Although

Table 3. Comparison of calcium, magnesium and metal ions (total concentrations) in representative New Zealand ultramafic soils. Samples collected from 0–10 cm depth on sparsely vegetated sites and analysed using standard techniques of atomic absorption spectrophotometry. Values expressed on an ash weight basis. On Dun Mountain two sets of samples were collected, one from mixed shrub-grassland (Dun Mountain) and another from mine spoil (Mine Tailings)

	Ca mg/g	Mg mg/g	Mg/Ca ratio	Ni μg/g	Cu μg/g	Co μg/g	Cr μg/g
North Cape	3.8	3.5	0.92	1,400	75	255	3,250
Dun Mountain	12.3	49.8	4.04	1,650	75	190	1,750
Mining Tailings	2.1	175.0	83.33	2,100	50	160	950
Red Hills	3.9	88.0	22.56	1,950	29	165	850
Red Mountain	2.6	185.0	71.15	3,800	28	255	900
Livingstone Mountains	6.9	179.0	25.94	1,950	30	127	1,100
Black Ridge	11.2	55.6	4.96	950	22	147	1,000
Anita Bay	7.1	47.2	6.64	1,550	23	770	1,300
John Innes P.C.	16.5	7.0	0.42	<10	37	30	<10

Table 4. Number of vascular plant species on New Zealand ultramafic soils

	North Cape No.	% of total	Nelson–Marlborough No.	% of total	Otago-Southland	% of total
Pteridophytes	20	14	59	14	45	13
Gymnosperms	1	1	12	3	13	4
Angiosperms						
Dicotyledons						
Woody	51	36	117	29	89	25
Herbaceous	27	19	100	25	102	28
Monocotyledons	42	30	120	29	104	30
Total	140		408		353	

this reflects the frequency of these lifeforms in the New Zealand flora generally, other common life forms such as lianes, epiphytes, tree ferns and divaricating shrubs, are under-represented on ultramafic soils. Nearly all the largest families and genera in New Zealand are represented in the ultramafic flora, a feature also found in the floras of other specialised habitats such as screes (Fisher 1952) and lake-edge turf communities.

About 70% of New Zealand's indigenous plant families have representatives on ultramafic soil. On South Island outcrops, the Podocarpaceae, Epacridaceae, Rubiaceae, Orchidaceae, Caryophyllaceae, Thymelaeaceae, Blechnaceae and Hymenophyllaceae have at least 50% of their species growing on ultramafic soils, but it is only the predominantly woody genera that consistently maintain this proportion. Families such as Asteraceae, Scrophulariaceae, Rosaceae and Caryophyllaceae have few ultramafic-tolerant species, although some families are necessarily excluded because of their general distribution (e.g., Fabaceae). However it is surprising that the Gentianaceae, Brassicaceae, Boraginaceae and Onagraceae, families noted for their ecological diversity, have few representatives on ultramafic soils in New Zealand, although their general role as ecological pioneers would expose them to extreme ultramafic conditions.

At the generic level, Coprosma and Dracophyllum have greatest ultramafic tolerance with just over 40% of their species important members of the ultramafic vegetation.

Naturalised species are rarely successful on ultramafic soils. They comprise less than 5% of the total flora, a pattern that reflects their general inability to tolerate nutrient-poor soils. They include three Pinus species, one shrub, eleven dicotyledonous herbs, five grasses and one rush. Most of the non-woody species are associated with human disturbance or sites of possible soil enrichment.

Table 5. Vascular flora of New Zealand ulramafic areas, excluding uncommon species. For the North Cape, species recorded as common by Druce *et al.* (1979) are noted. South Island ultramafic aras are grouped regionally. A relative importance value is given to species using the following code: 1 – occasional, 2 – common but not abundant, 3 – abundant but subdominant, 4 – dominant or codominant; 5 – locally abundant or dominant

	North Cape	Nelson–Marlborough	Otago-Southland
PTERIDOPHYTES			
ADIANTACEAE			
Adiantumhispidulum	+		
BLECHNACEAE			
Blechnum capense		2	2
Blechnum procerum		2	3
GLEICHENIACEAE			
Gleichenia dicarpa	+		5
GRAMMITIDACEAE			
Grammitis billardieri			1
HYMENOPHYLLACEAE			
Hymenophyllum armstrongii			1
Hymenophyllum multifidum		5	2
Hymenophyllum villosum		1	
LINDSAEACEAE			
Lindsaea linearis	+	5	5
LYCOPODIACEAE			
Lycopodium deuterodensum	+		
Lycopodium fastigiatum			1
Lycopodium ramulosum			5
Lycopodium scariosum		1	1
Lycopodium varium		1	
Lycopodium volubile		1	
POLYPODIACEAE			
Pyrrosia serpens	+		
PTERIDACEAE			
Pteridium esculentum	+	1	5
GYMNOSPERMS			
PODOCARPACEAE			
Halocarpus bidwillii		2	
Halocarpus biformis			4
Lagarostrobos colensoi			5
Dacrydium cupressinum			5
Lepidothamnus intermedius			4
Lepidothamnus laxifolius		2	2
Phyllocladus alpinus		4	4
Phyllocladus trichomanoides var.	+		
Podocarpus hallii		1	
Podocarpus nivalis			5
ANGIOSPERMS – DICOTYLEDONS			
AIZOACEAE			
Disphyma australe	+		
Tetragonia trigyna	+		
APIACEAE			
Anisotome aromatica		2	5
Anisotome filifolia		1	
Apium prostratum	+		
Centella uniflora	+		5

Table 5. Continued

	North Cape	Nelson–Marlborough	Otago-Southland
ARALIACEAE			
Pseudopanax arboreus		5	
Pseudopanax crassifolius		5	1
Pseudopanax lessonii	+		
Pseudopanax linearis			5
Pseudopanax simplex			5
ASTERACEAE			
Brachyglottis bellidioides			2
Brachyglottis lagopus			5
Brachyglottis repanda	+		
Cassinia leptophylla	+	5	
Cassinia vauvilliersii		4	5
Celmisia glandulosa			5
Celmisia gracilenta		3	2
Celmisia spectabilis var. *spectabilis*		5	
Celmisia spedenii			5
Craspedia sp.		1	1
Gnaphalium audax	+		
Gnaphalium sphaericum	+		
Helichrysum bellidioides		2	2
Hypochoeris radicata			5
Lagenifera pumila	+		
Leptinella pyrethrifolia var. *linearifolia*		5	
Olearia sp.		2	
Raoulia glabra		1	
Senecio lautus	+		
BORAGINACEAE			
Myosotis lyallii			5
Myosotis monroi		5	
BRASSICACEAE			
Notothlaspi australe		2	
CAMPANULACEAE			
Wahlenbergia albomarginata		2	2
CARYOPHYLLACEAE			
Colobanthus strictus			5
Colobanthus wallii ssp.			2
CONVOLVULACEAE			
Dichondra repens	+		
CORNACEAE			
Corokia cotoneaster	+	1	
Griselinia littoralis			5
CUNONIACEAE			
Weinmannia racemosa		5	2
DONATIACEAE			
Donatia novae-zelandiae			5
DROSERACEAE			
Drosera spathulata			1
ELAEOCARPACEAE			
Aristotelia fruticosa		2	

Table 5. Continued

	North Cape	Nelson–Marlborough	Otago-Southland
EPACRIDACEAE			
Cyathodes empetrifolia		2	1
Cyathodes juniperina	+	4	3
Cyathodes parviflora	+		
Dracophyllum longifolium		4	3
Dracophyllum pronum		5	
Dracophyllum prostratum			5
Dracophyllum uniflorum			4
Leucopogon fasciculatus	+		
Leucopogon fraseri	+	2	2
FAGACEAE			
Nothofagus solandri var. *cliffortioides*		3	4
Nothofagus truncata		5	
GENTIANACEAE			
Gentiana sp.		2	
HALORAGACEAE			
Gonocarpus aggregatus		5	
Gonocarpus incanus	+	5	
Haloragis erecta ssp. *cartilaginea*	+		
LOBELIACEAE			
Lobelia anceps	+	5	5
LOGANIACEAE			
Geniostoma rupestre	+		
MYRSINACEAE			
Myrsine australis	+		
Myrsine divaricata		3	2
Myrsine nummularia			1
MYRTACEAE			
Kunzea ericoides	+	1	
Leptospermum scoparium	+	4	
Metrosideros excelsa	+		
Metrosideros umbellata		2	4
Neomyrtus pedunculata		5	5
ONAGRACEAE			
Epilobium atriplicifolium		5	5
Epilobium pubens		5	
Epilobium tasmanicum			5
PAPILIONACEAE			
Ulex europaeus		5	
PIPERACEAE			
Macropiper excelsum var. *excelsum*	+		
PITTOSPORACEAE			
Pittosporum anomalum		5	
Pittosporum divaricatum		5	1
PORTULACACEAE			
Neopaxia australasica		5	5
POLYGONACEAE			
Muehlenbeckia axillaris		5	1
Muehlenbeckia complexa	+		
Rumex acetosella		1	
RANUNCULACEAE			
Clematis forsteri		5	

Table 5. Continued

	North Cape	Nelson–Marlborough	Otago-Southland
RHAMNACEAE			
Pomaderris kumeraho	+		
Pomaderris sp.	+		
Pomaderris edgerleyi	+		
RUBIACEAE			
Coprosma cheesemanii		5	
Coprosma colensoi			2
Coprosma foetidissima		2	1
Coprosma macrocarpa ssp.	+		
Coprosma obconica ssp.	+		
Coprosma sp.		1	5
Coprosma propinqua		3	5
Coprosma pseudocuneata s.s.			1
Coprosma repens	+		
Nertera depressa			5
SANTALACEAE			
Exocarpus bidwillii		1	
SCROPHULARIACEAE			
Euphrasia cuneata		5	
Hebe ligustrifolia	+		
Hebe odora		5	2
Hebe urvilleana		5	
Hebe venustula		2	
THYMELAEACEAE			
Pimelea gnidia			5
Pimelea oreophila			5
Pimelea prostrata	+		
Pimelea suteri var. *suteri*		5	
VIOLACEAE			
Melicytus alpinus		2	1
Melicytus micranthus	+		
Viola cunninghamii		5	1
Wahlenbergia sp.	+	1	1
MONOCOTYLEDONS			
ASPHODELACEAE			
Arthropodium cirratum	+		
Astelia banksii	+		
Astelia nervosa		2	
CYPERACEAE			
Baumea juncea	+		
Baumea rubiginosa			5
Baumea teretifolia			5
Carex berggrenii			5
Carex breviculmus		2	5
Carex devia		3	
Carex spinirostris	+		
Carpha alpina		1	2
Cyperus ustulatus	+		
Gagnia lacera	+		
Gahnia pauciflora		2	
Gahnia procera			3
Gahnia rigida			5
Gahnia setifolia		1	

387

Table 5. Continued

	North Cape	Nelson–Marlborough	Otago-Southland
CYPERACEAE (Continued)			
Isolepis aucklandica			1
Isolepis nodosa	+		
Lepidosperma australe		2	2
Lepidosperma filiforme	+		
Lepidosperma laterale	+		
Morelotia affinis	+		
Oreobolus impar			5
Oreobolus pectinatus		1	2
Schoenus brevifolius	+		
Schoenus pauciflorus		3	3
Uncinia divaricata		5	
Uncinia fuscovaginata		1	
Uncinia rubra			5
Uncinia uncinata			1
IRIDACEAE			
Libertia ixioides		5	
Libertia peregrinans		5	
Libertia pulchella			1
JUNCACEAE			
Juncus gregiflorus	+	1	
Juncus novae-zelandiae		1	
Luzula sp.		5	
Luzula crinita var. *petriana*			5
Luzula pumila			5
Luzula rufa			5
ORCHIDACEAE			
Cyrtostylis reniformis	+		
Prasophyllum colensoi		1	
Thelymitra longifolia		1	
PHORMIACEAE			
Dianella nigra	+	1	
Phormium cookianum		3	2
Phormium tenax	+		
POACEAE			
Agrostis dyera		5	
Chionochloa crassiuscula			5
Chionochloa macra			5
Chionochloa defracta		4	
Chionochloa cf. *rigida*			4
Cortaderia splendens	+		
Deyeuxia aucklandica		1	
Deyeuxia billardieri	+		
Dichelachne crinita	+		
Dichelachne micrantha	+		
Festuca sp.		5	
Lachnagrostis sp.		5	
Lachnagrostis richardii			5
Microlaena stipoides	+		
Oplismenus aemulus	+		
Poa acicularifolia spp.		5	
Poa colensoi		3	2

388

Table 5. Continued

	North Cape	Nelson–Marlborough	Otago-Southland
POACEAE (Continued)			
Rytidosperma biannulare	+		
Rytidosperma gracile		1	
Rytidosperma nigricans			1
Rytidosperma setifolium		2	1
Stipa stipoides	+		
Trisetum spicatum		5	
RESTIONACEAE			
Empodisma minus		5	5

Endemism

A total of thirty four taxa are thought to be restricted to New Zealand ultramafic outcrops although the majority have yet to be described formally (Table 6). Fifteen are considered distinct species while the remainder are recognised at the varietal or subspecific level.

These are segregated regionally with the majority of taxa exclusive to either the North Cape (fifteen taxa) or Nelson–Marlborough (eighteen taxa); and only one species is restricted to Otago-Southland. All but one of the taxa recorded by Druce *et al.* (1979) as being limited to the North Cape serpentinite area are woody whereas only three species endemic to the Nelson–Marlborough ultramafics are woody, the remainder comprising nine dicotyledonous herbs, three grasses, two sedges and one rush.

Habitat age is the primary factor responsible for regional differences in endemism on ultramafic soil. Probably most outcrops have been terrestrial since the late Miocene but only North Cape and the Nelson–Marlborough outcrops could have supported plant life throughout this period. Ultramafic habitats in Otago-Southland along and west of the Main Divide have become available again over the last 12,000 years, a period apparently insufficient for the differentiation of endemic taxa. The only ultramafic endemic in the southern areas is on the extreme eastern ultramafic outcrops, sites that lay beyond the major ice advances of the Pleistocene. Where the habitat age is similar, the number of ultramafic endemics is closely related to the area of the outcrop (Fig. 2), but size is clearly subordinate to age, as the distribution of ultramafic endemic taxa in New Zealand indicates.

With the exception of *Chionochloa defracta* in Nelson–Marlborough, ultramafic endemic species in the South Island are rarely dominant or abundant in ultramafic communities, characteristically being found in open, marginal habitats on younger mineral soils which represent extreme ultramafic conditions. On Dun Mountain, the ultramafic endemics *Pimelea suteri* and *Myosotis monroi* grow on soils with the highest Mg concentrations, in contrast to the non-endemic species *Leptospermum scoparium*, *Hebe odora* and *Cassinia vauvilliersii*. The endemic taxa are clearly distinguished by their response to K and Ca concentrations in the soil – *P. suteri* favouring high K/low Ca, *M. monroi* favouring low K/high Ca (Lee *et al.* 1975). In one analysis of species distribution in a mixed shrub-grassland community on Dun Mountain, Lee (1980) found that most of the endemic species were growing on level sites with shallow soils low in available Ca, K and P, but rich in Mg and with a high pH, compared with the more widespread ultramafic-tolerant species. *Celmisia spedenii*, the ultramafic endemic in Otago-Southland, also favours extreme sites where soil pH and Mg concentrations are highest (McIntosh & Lee 1986) (Fig. 4). Although they have broadly similar habitat preferences, ultramafic endemics show marked differences between the species in their elemental composition, even when growing on similar soils (Lyon *et al.* 1971). For example, *Pimelea suteri* had a high Cr, Ni and Co content, but a high Mg/Ca quotient. *Myosotis monroi*, in contrast, had lower concentrations of Ni, Co, Cu, Cr and Mg, the latter resulting in the lowest Mg/Ca quotient. *Celmisia spedenii* has high concentrations of Ni, Cr and Mg in plants on ultramafic soil and is more efficient at maintaining high Ca and K concentrations under these con-

Table 6. Taxa considered to be endemic to New Zealand ultramafic soils, compiled from Druce et al. (1979), Druce (unpubl. obs.) and Lee and Given (1984)

	Family	North Island North Cape	South Island Nelson–Marlborough	South Island Otago-Southland
TREES				
Phyllocladus trichomanoides var.	Podocaparceae	×		
SHRUBS				
Cassinia leptophylla var.	Asteraceae	×		
Coprosma spathulata ssp.	Rubiaceae	×		
Coprosma obconica ssp.	Rubiaceae	×		
Coprosma rhamnoides ssp.	Rubiaceae	×		
Corokia cotoneaster var.	Cornaceae	×		
Cyathodes parviflora var.	Epacridaceae	×		
Geniostoma rupestre var. *crassum*	Loganiaceae	×		
Hebe ligustifolia var.	Scrophulariaceae	×		
Hebe urvilleana	Scrophulariaceae		×	
Hebe sp.	Scrophulariaceae	×		
Helichrysum aggregatum var.	Asteraceae	×		
Pimelea suteri var. *suteri*	Thymelaeceae		×	
Pittosporum crassifolium var.	Pittosporaceae	×		
Pittosporum pimeleoides var. *maius*	Pittosporaceae	×		
LIANES				
Parsonsia sp.	Apocynaceae	×		
Dracophyllum longifolium var.	Epacridaceae		×	
DICOT HERBS				
Anisotome flexuosa var.	Apiaceae		×	
Cardamine sp.	Brassicaceae		×	
Celmisia spedenii	Asteraceae			×
Colobanthus masoniae ssp.	Caryophyllaceae		×	
Colobanthus sp.	Caryophyllaceae		×	
Craspedia sp.	Asteraceae		×	
Haloragis erecta ssp. *cartilaginea*	Haloragaceae		×	
Myosotis laeta	Boraginaceae		×	
Myosotis monroi	Boraginaceae		×	
Notothlaspi sp.	Brassicaceae		×	
Leptinella pyrethrifolia var. *linearifolia*	Asteraceae		×	
GRASSES				
Chionochloa sp.	Poaceae		×	
Poa acicularifolia ssp. *ophitalis*	Poaceae		×	
Trisetum sp.	Poaceae		×	
SEDGES				
Carex devia	Cyperaceae		×	
Carex traversii	Cyperaceae		×	
RUSHES				
Luzula sp.	Juncaceae		×	

ditions than the nontolerant species *Celmisia markii* (Lee & Reeves 1989).

There is no evidence that any New Zealand taxa are obligatory ultramafic and several have been successfully grown in garden loam (A. P. Druce personal communication, Lee & Given 1984). However, *Celmisia spedenii* has shown greater vigour and regenerated better from seed on ultramafic soil (Lee & Given 1984). It is debatable whether these endemics are specialists or

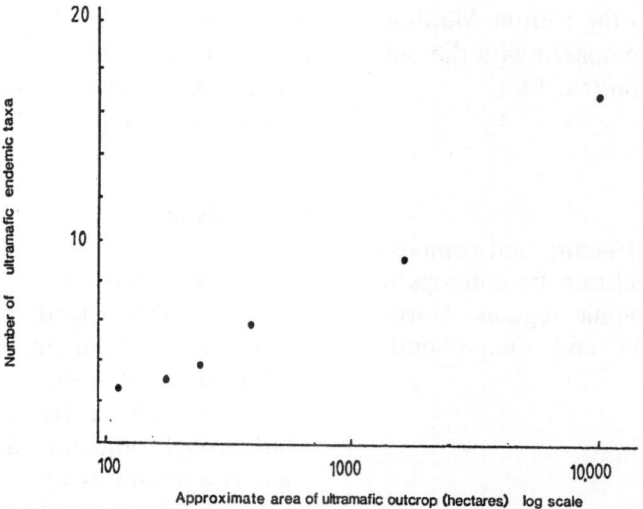

Fig. 2. Relation between the number of ultramafic endemic taxa and ultramafic outcrop area in the South Island, New Zealand. Outcrops in Nelson–Marlborough (●) and Otago-Southland (○) are treated separately.

Photo 4. Celmisea spedenii, the only ultramafic endemic known on the Otago-Southland ultramafites.

relic species that have become restricted to the open, sparse vegetation on ultramafic soils through their elimination on other substrates. Azonal soils generally buffer plants from changes in climate that allow tall, continuous vegetation to predominate on zonal soils. Thus the long-term availability of relatively open habitats, may account for the disproportionately high concentration of endemic taxa on the Nelson–Marlborough ultramafites (15%) compared with the outcrop area in the same region (ca. 1%).

Vegetation description

Brief descriptions of the structure and composition of the vegetation of ultramafic outcrops in the three major biogeographic regions, Northland, Nelson–Marlborough and Otago-Southland, are provided.

North Island

Surville Cliffs

The best-developed ultramafic communities are found on the Surville Cliffs (Druce *et al.* 1979). Eroding gullies and mobile debris have created fragmented habitats with the most stable sites supporting 2–10 m tall forest of *Metrosideros excelsa*, *Macropiper excelsum* var. *excelsum*, *Myoporum laetum* var. *laetum*, *Melicytus ramiflorus* ssp. *ramiflorus* and *Pseudopanax lessonii*. Shrubland, often found in gullies, contains a diverse woody flora of species with a distinctive semi-lianoid habit, including varieties of *Pseudopanax lessonii*, *Phyllocladus trichomanoides*, *Pittosporum crassifolium*, *Coprosma rhamnoides* and *Geniostoma rupestre* (Photo 5). Other species with a normal erect habit are *Myrsine australis*, *Melicope simplex*, and *Melicytus micranthus*. The most widespread community consists of scattered low-growing shrubs of both forest and shrubland species, namely *Kunzea ericoides*, species of *Coprosma* and *Hebe*, *Corokia cotoneaster*, *Pomaderris* sp., *Pomaderris edgerleyi*, *Pimelea prostrata* and *Cyathodes parviflora*. The giant tussocks *Astelia banksii*, *Phormium tenax* and *Cortaderia splendens* are conspicuous members of most communities, but generally herbs are of minor importance.

A species-rich low mixed shrubland of *Leptospermum scoparium*, *Kunzea ericoides*, *Cyathodes juniperina*, *Cyathodes parviflora*, *Hebe* sp., *Pomaderris edgerleyi*, *Cassinia amoena* and *Cassinia retorta* is reported from serpentine-derived

laterite on the North Cape plateau (Thompson *et al.* 1974). Its extension beyond the serpentinite contact on to gabbro shows that nonedaphic factors may be responsible for its occurrence.

South Island

Nelson–Marlborough
Sites: D'Urville Island, Red Hill, Whangamoa Saddle, Dun Mountain, Red Hills, Matakitaki Valley and Cobb Valley.

The non-ultramafic hillslopes of Nelson–Marlborough adjoining the ultramafic outcrops support a natural cover of *Nothofagus* tall forest up to an elevation of around 1,400 m. Below 500 m a.s.l. *N. truncata* is the major canopy species with emergent *Dacrydium cupressinum*. Between 500–700 m *N. fusca* and *Weinmannia racemosa* predominate, while at higher altitudes *N. menziesii* and, on drier sites, *N. solandri* var. *cliffortioides* are important (Walls & Laffan 1986).

Forest on ultramafic outcrops from D'Urville Island, at the north-western margin of Marlborough Sounds, through to the Red Hills in the Wairau Valley, is very limited at lower altitudes due to widespread burning of ultramafic vegetation (Photo 6). The natural climax on ultramafic rocks here is low forest of *Nothofagus truncata*, *Metrosideros umbellata*, *Weinmannia racemosa*, *Leptospermum scoparium* and the podocarps *Dacrydium cupressinum* and *Podocarpus hallii*. The dicotyledonous trees are often multi-stemmed with copious additional branching which creates a dense understorey in which the small-leaved shrubs *Neomyrtus pedunculata*, *Coprosma foetidissima* and *Myrsine divaricata* are numerous. The ground cover is poorly developed but amongst weathered serpentinite fragments the ferns *Blechnum minus* and the tussock herbs *Gahnia xanthocarpa*, *Dianella nigra* and *Uncinia uncinata* are consistently present.

Walls & Laffan (1986) also recognise at higher altitudes on the ultramafics, a low forest of *Nothofagus solandri* var. *cliffortiodes*, *Phyllocladus alpinus* and *Libocedrus bidwillii*, and this is present on most outcrops, particularly along the outcrop margins (Photo 7).

Photo 5. Tall trees of *Melrosideras excelsa* amongst low shrubs of *Pseudopanax lessonii*, *Phyllocladus Kichomanoides* var., *Geniostoma rupestre* and *Myrsine auskalis*, with frequent tussocks of *Astelia banksii* and *Phormium tenax* on ultramafic rocks at Sarville Cliffs, North Cape. (Photograph by G. C. Kelly.)

The most widespread community on the ultramafics is low scrub or heathland of *Leptospermum scoparium*, *Dracophyllum longifolium*, *Cassinia vauvilliersii* and stunted, isolated forest trees. Leptospermum scoparium 1–3 m tall occupies stable sites, in gullies or on margins transitional to forest forest. Although induced by fire in many places, the scrub may be a relatively stable community in some areas on the ultramafites and not seral to forest as it is elsewhere in the region. The diverse shrub flora contains one ultramafic endemic, *Pimelea suteri*. The most frequent and

Photo 6. Whangamoa Saddle, with mixed *Nothofagus* forest giving way to shrubland and grassland on the ultramafic rocks in the foreground. Dead tree trunks indicate previous burning of vegetation on the ultramafic outcrop.

largest herbs are *Chionochloa defracta* and *Phormium cookianum*, but much smaller herbs such as *Gentiana* sp., *Helichrysum bellidioides* and *Senecio lagopus* grow on a thin humus layer amongst rocks (Photo 8).

On ridge summits where the combined effects of soil conditions and exposure produce impenetrably matted shrublands less than 1 m tall, with areas of exposed humus, mineral soil and, less often, bare bedrock (Photo 9). Low sprawling shrubs of *Corokia cotoneaster*, *Myrsine divaricata*, *Leucopogon fasciculatus* and *Coprosma*

parviflora enter the community at these sites. The interstitial species are characteristic of bog or wetland habitats and include *Lepidothamnus laxifolius*, *Carpha alpina*, *Drosera auriculata*, *Oreobolus pectinatus*, *Lindsaea linearis* and *Lepidosperma australe*.

Chionochloa defracta tussock grassland is widespread on the ultramafites, particularly on dunite and peridotite, and increases in importance with altitude (Photo 10). The tussock cover approaches 70%, with rare low shrubs of *Cassinia vauvilliersii*, *Dracophyllum pronum*, *D. filifolium*

394

Photo 7. Red Hills, Nelson, showing low mixed forest of *Leptospermum scoparium*, *Nothofagus solandri* var. *cliffortioides* and emergent *Libocedrus bidwillii*. (Photograph by P. Wardle.)

and *Leptospermum scoparium* breaking through the tussock canopy at 0.75 m. The woody species are taller and more abundant amongst rock piles or rocky slopes on the upper slopes of Dun Mountain. Interstitial herbs are few, chiefly *Wahlenbergia albomarginata*, *Brachyglottis lagopus*, *Anisotome aromatica* and *Helichrysum bellidioides*. On serpentinite rock, erosion frequently produces a series of anastomosing channels which dissect the grassland into small, slightly pedestalled islands.

Mine tailings, erosion channels and stone pavements, usually on serpentinite, contain a sparse cover of a few species, several of which are endemic to ultramafic outcrops (e.g. *Carex devia*, *C. traversii*, *Colobanthus* sp. and *Myosotis monroi*), while others (e.g. *Notothlaspi australe* var. *stellatum* and *Montia australasica*) are common in similar scree habitats on surrounding mountains. Depressions, flushes and other permanently moist

sites have species typical of bogs – *Schoenus pauciflorus*, *Isolepis aucklandia* and *Celmisia gracilenta*. Species from both mine tailings and wet habitats on Dun Mountain colonise eroded mineral soil, typically in association with young plants of *Chionochloa defracta* and appear to be forerunners of the grassland community. The only ultramafic outcrops extending above treeline (1,400 m) are Red Hills and Matakitaki. These support 40% cover of *Chionochloa* grassland interspersed amongst bare ground of angular rock fragments, mineral soil or small rock outcrop (Photo 9). The dwarf shrub *Dracophyllum pronum* and the small tussock grasses *Poa colensoi*, *Poa acicularifolia* and *Rytidosperma setifolium* are associated with less-exposed sites, while weakly weathered ultramafic rock usually carries *Montia australasica*, *Notothlaspi australe* and the endemic herb species found in the tall tussock

395

Photo 8. Open mixed shrub-grassland on ultramafic soil, Dun Mountain, Nelson. (Photograph by P. Wardle.)

Photo 9. Red Hills, Marlborough. Alpine rockfields with patches of *Chionochloa defracta* and prostrate shrubs of *Melicytus alpinus* and *Dracophyllum pronum*. (Photograph by P. Wardle.)

grassland at lower elevation. Alpine flushes hold typical bog species, notably *Schoenus pauciflorus* and *Oreobolus pectinatus*. Several larger shrub species – *Hymenanthera alpina, Lepidothamnus laxifolius, Cyathodes empetrifolia* and *Pittosporum anomalum*, also reach the low-alpine zone, forming compact low-growing bushes.

Otago-Southland

North-west Otago
Sites: Cascade Plateau, Martyr Hill, Red Mountain, Gorge Plateau, Little Red Hill and Fiery Peak.

The tall forest that grows on non-ultramafic rocks in the area is largely *Nothofagus menziesii*. Emergent podocarps (*Dacrydium cupressinum* and *Prumnopitys ferruginea*) are present below 500 m, and other dicotyledonous trees (*Weinmannia racemosa* and *Metrosideros umbellata*) are associated with *Nothofagus* up to 700 m. Above this the latter tends to form pure stands until treelimit is reached at 1,100 m.

The forest on ultramafic parent material is stunted and patchy, and is described in relation to landform.

Ultramafic moraine. The most widespread forest community on ultramafic rocks is on till, although on the Cascade (Photo 11) and Gorge plateaux its extent has been greatly reduced by fire. It comprises a variable mixture of the podocarps *Podocarpus hallii, Dacrydium cupressinum, Lepidothamnus intermedius, Halocarpus biformis, Lagarostrobos colensoi,* the hardwoods *Weinmannia racemosa* and *Metrosideros umbellata,* and *Nothofagus solandri* var. *cliffortioides.* The canopy height (3–15 m) is influenced by local soil conditions although none of the trees attain the stature of the same species off the ultramafites. The shrub species are *Coprosma parviflora, C. colensoi, Neomyrtus pedunculata, Leptospermum scoparium, Myrsine divaricata* and less commonly *Cyathodes juniperina.* Bryophytes to 10 cm thick cover the forest floor, with a sparse cover of the ferns *Blechnum minus, Hymenophyllum multifidum* and *Gleichenia dicarpa* amongst large tussocks of *Gahnia procera.*

Leptospermum scoparium is common where soil conditions prevent forest establishment. *Halocarpus biformis, Weinmannia racemosa* and *Me-*

trosideros umbellata rarely exceed the height of the *Leptospermum scoparium* shrubs (2 m). Interspersed throughout the community on generally bare soil are plants of *Lepidothamnus intermedius, Dracophyllum longifolium, Pimelea prostrata* and minor small herbs such as *Poa colensoi* and *Drosera spathulata.* A sparse cover of *Chionochloa* cf. *rigida* and *Chionochloa acicularis* is also associated with this stunted shrubland.

Ultramafic alluvium. Terraces of ultramafic alluvium support a stunted forest, 6–8 m tall, dominated by *Nothofagus solandri* var. *cliffortioides, Lepidothamnus intermedius, D. cupressinum* and *Metrosideros umbellata.* Less common but regular components of the canopy include *Lagarostrobus colensoi, Halocarpus biformis, Weinmannia racemosa, Podocarpus hallii* and *Libocedrus bidwillii.* These overtop a dense layer of largely young trees, but including the shrubs *Phyllocladus aspleniifolius* var. *alpinus, Coprosma parviflora, Neomyrtus pedunculata, Cyathodes juniperina* and *Leptospermum scoparium.* Ground cover plants are sparse, with only *Blechnum minus* and *Gleichenia dicarpa* locally common. On sites with impeded drainage, the broadleaved trees are absent leaving an open canopy of *L. scoparium* and several small gymnosperms. Negligible change is seen amongst the shrubs, while *Carex gaudichaudiana, Eleocharis gracilis* and *Lepidosperma australe* on the forest floor indicate wetter soil conditions.

Ultramafic outcrops. On lower slopes, between 350 m and 720 m (Photo 12), the tallest vegetation on ultramafic rocks (typically talus scree soils) contains in decreasing importance *Nothofagus solandri* var. *cliffortioides, Lepidothamnus intermedius, Halocarpus biformis, Metrosideros umbellata, Leptospermum scoparium, Phyllocladus aspleniifolius* var. *alpinus* and *Weinmannia racemosa,* 3–8 m tall and forming an irregular diffuse canopy. *Dracophyllum uniflorum, Cyathodes juniperina, Coprosma colensoi, C. foetidissima* and regenerating tree species provide a dense understorey, beneath which are few herbs. Where red, unweathered rock breaks the ground surface, it is occasionally covered with a thin bryophyte layer. The larger tree species and *Coprosma* shrubs decline in abundance above 600 m and are replaced in the canopy mainly by *Leptospermum scoparium.*

Photo 10. Red Hills, Marlborough. Extensive areas of *Chionochloa defracta* cover the gentle topography of New Zealand's largest ultramafic outcrop.

Glacially-sculptured spurs and benches above c. 650 m, usually formed on peridotite rock, hold an open, low woodland community which continues to c. 1000 m elevation (Photo 13). Approximately 20% of the ground is bare rock or soil. The podocarps *Phyllocladus aspleniifolius* var. *alpinus* and *Lepidothamnus intermedius* are noticeably less abundant in the shrubland than in the forest but *Halocarpus biformis* and *Lepidothamnus laxifolius* increase in dominance and, combined with *Leptospermum scoparium*, cover 25%. *Nothofagus solandri* var. *cliffortioides* shows a similar decline in the shrubland. The trees present are not much taller than shrubs, although typically erect in form. The epacrid *Dracophyllum uniflorum* is the most common shrub, but several microphyllous species such as *Cyathodes juniperina* and *Pimelea oreophila* are usually present, as is the liane *Muehlenbeckia axillaris*. A thin cover of less than 5% is provided by several subalpine and alpine tussock grasses, notably *Chionochloa* cf. *rigida*, *Poa colensoi* and *Rytidosperma setifolium*, between stands of woody vegetation. The wetland cyperad *Schoenus pauciflorus* is important on stony soils and

amongst the tussock grasses. However, there are very few species of dicotyledonous herbs: only *Celmisia gracilenta*, *Wahlenbergia albomarginata* and *Brachyglottis bellidioides* are consistent representatives, all growing on bare ground.

Above 900 m, rocky bluffs and screes prevail (Photo 14) and vegetation usually occurs only where soil has accumulated in cirque basins or on gentle slopes. Prostrate shrubs of *Halocarpus biformis* and *Phyllocladus aspleniifolius* var. *alpinus* reach 1050 m elevation, above which *Melicytus alpinus*, *Myrsine nummularia*, *Hebe odora*, *Muehlenbeckia axillaris*, *Dracophyllum uniflorum* and *Cyathodes juniperina*, 0.10–0.20 m tall, form pockets of vegetation amongst tussock grasses and the three herbs found in the shrubland.

Northern Southland
Sites: Livingstone Range, West Dome and Black Ridge.
Montane. Black Ridge and West Dome are surrounded by developed pasture on sites that would have supported either *Nothofagus menziesii* forest or *Chionochloa rubra* and *C. rigida* tall tussock grassland. The lower slopes of the ultramafic

398

Photo 11. Cascade Plateau, a tableland capped with lateral morains of ultramafic material supporting low mixed forest and shrubland. (Photograph by D. L. Homer.)

outcrops carry an open shrubland of *Leptospermum scoparium* up to 3 m tall, scattered amongst smaller shrubs of *Cyathodes juniperina*, *Dracophyllum uniflorum*, *Pimelea oreophila* and *Melicytus alpinus*, interspersed with a sparse turf of *Carex berggrenii*, *Schoenus pauciflorus*, *Poa colensoi*, *Helichrysum bellidioides*, *Wahlenbergia albomarginata*, *Gentiana patula* and *Celmisia gracilenta* (Photo 15). *L. scoparium* thickens to form an almost closed canopy near the benign influence of the knockers, and throughout grows on shallow stony soil, 20 cm above bedrock. Small stands of *Halocarpus bidwillii* and *Podocarpus nivalis* grow on similar soil at ca. 700 m in the open woodland on Black Ridge. Where the *L. scoparium* is

dense, fire has removed previous forest cover but such areas are very localised and the more open typical woodland covers ca. 80% of undisturbed parts of both West Dome and Black Ridge.

Small seepage areas occur throughout the woodland, characterised by *Empodisma minus*, *Oreobolus pectinatus*, *Carpha alpina*, *Lepidosperma australe* and *Juncus articulatus*. Intrusive rocks support vegetation distinct from that on serpentinite, with small trees up to 4 m tall of *Nothofagus solandri* var. *cliffortioides* which generally are confined to these sites and, less frequently, *Podocarpus hallii*. Understorey composition depends on the degree of sheep or cattle usage. On unbrowsed sites *Corokia cotoneaster*,

Photo 12. Low mixed forest at 500 m on Red Mountain, north-west Otago, surrounded by successional shrublands that have established on eroded sites.

Photo 13. Scattered shrubs of *Hebe odora* and *Halocarpus biformis* (foreground) and occasional trees of *Nothofagus solandri* var. *cliffortiodes*, grow amongst ultramafic boulders and a sparse cover of tussock grasses at 950 m on Red Mountain, north-west Otago.

Photo 14. The barren bluffs and screes on Red Mountain present a striking contrast with the subalpine *Nothofagus menrusii* forest and alpine fall fussock grassland on adjoining schist rock at Simonin Pass (1,030 m), north-west Otago. (Photograph by D. L. Homer.)

Dracophyllum longifolium, Myrsine divaricata, Coprosma parviflora and *C. pseudocuneata* abound, with herbs. *Phormium cookianum* and *Celmisia durietzii* and several fern species, particularly *Blechnum minus, B. penna-marina* and *Pteridium esculentum.*

Heaps of raw crushed serpentinite spoil at Black Ridge cannot be aged accurately, but most have been undisturbed for at least 5 years. They support a sparse cover of *Carex breviculmis, C. berggrenii, Poa colensoi,* and mat-forming *Raoulia subsericea* and *R. glabra,* the main native components of the slag flora, together with the versatile naturalised grasses *Anthoxanthum odoratum, Agrostis capillaris, Dactylis glomerata* and *Holcus lanatus.*

Alpine. The areas with serpentinite soils in the Livingstone Mountains are above treeline (1,050 m) (Photo 16). In contrast to the near continuous tall tussock grassland on adjoining areas, the sparsely vegetated serpentinite outcrop is covered with 20–30 cm of regolith. The most frequent species are *Schoenus pauciflorus, Myosotis lyallii, Isolepis aucklandia, Anisotome aromatica, Wahlenbergia albomarginata, Luzula crinita* var. *petriana, Cardamine debilis, Poa colensoi, Craspedia uniflora* and *Epilobium alsinoides.* Depauperate tussock grassland of *Chionochloa macra* occurs as scattered plants or tussocks grouped precariously on soil pedestals where stable rock nears the surface.

Both seepage areas and protruding knockers harbour species not found on ultramafic skeletal soils. Wet, boggy sites, occasionally underlain by 0.10–0.20 m of peat, resting on serpentinite, hold a thin cover mainly of *Forstera tenella, Dolichoglottis lyallii, Hebe pauciramosa, Celmisia laricifolia* and *Lycopodium australianum,* in addition to the more widespread sedges. Species on intrusive rock vary, perhaps with rock type, but *Haastia sinclairii, Raoulia grandiflora, Geum parviflorum, Blechnum penna-marina, Helichrysum bellidioides, Leucogenes grandiceps* and *Lycopodium australianum* are usual members.

Fiordland

The tall forest of *Nothofagus menziesii* and *Weinmannia racemosa* that grows on adjoining terrain, is replaced along the margins of the Anita Bay ultramafics by a low forest (<8 m) of *Weinmannia racemosa, Nothofagus solandri* var. *cliffortioides, Metrosideros umbellata, Lepidothamnus intermedius* and *Phyllocladus aspleniifolius* var. *alpinus. Nothofagus menziesii* is present as a minor species.

On the ultramafites in sheltered, stable sites of moderate slope there is a dense low forest floristically and structurally similar to that on the lower slopes of ultramafic outcrops in north-west Otago (Photo 17), viz. an uneven canopy of *Lepidothamnus intermedius, Nothofagus solandri* var. *cliffortioides, Metrosideros umbellata, Weinmannia racemosa* and *Halocarpus biformis.* Maturing dominants and shrubs of *Cyathodes juniperina* and *Coprosma colensoi* form a dense understorey, usually with *Gahnia procera* and *Phormium cookianum.*

On shallow, wet soils, freshly-eroded ultramafic skeletal soils, exposed slopes and ridge summits, trees and shrubs are gnarled and stunted. *Weinmannia racemosa, Phyllocladus aspleniifolius* var. *alpinus* and *Coprosma foetidissima* are unable to grow on these harsher sites but the majority of the ultramafic forest species are present, in addition to *Melicytus alpinus, Dracophyllum uniflorum* and *Hebe odora. Lepidothamnus intermedius* and *Leptospermum scoparium* with a combined cover of 50% are the dominant woody species occurring as uniformly low-growing bushy shrubs.

The most extreme ultramafic sites, frequently located on raised mounds or crests, support a mixture of stunted shrubs and a thin cover of grasses, chiefly *Chionochloa* cf. *rigida, Poa colen-*

Photo 15. An open shrubland of *Leptospermum scoporium* on serpentinite (foreground) with *Nothofagus solandri* var. *cliffortioides* on intrusives (background), at 500 m on Black Ridge, Southland.

soi and *Rytidosperma setifolium.* Forty percent of these areas are bare rock or soil and others have a sparse cover of bog ferns *Gleichenia dicarpa* and *Schizaea fistulosa,* and the juncad *Lepidosperma australe.*

Comparisons of ultramafic and adjoining non-ultramafic vegetation

The presence of ultramafites is marked by abrupt changes in vegetation form, stature and composi-

Photo 16. Tussock grassland (foreground) gives way to barren fell field on the alpine serpentinite outcrops in the Livingstone Mountains, Southland.

tion. On ultramafic soil below tree line, the tall podocarps, southern beech species, and broad-leaved trees that form the major forest cover in New Zealand are either absent or infrequent. Instead, tall forest gives way to a mosaic of low forest, woodland or shrubland communities on ultramafic soil in which a number of low-statured gymnosperms, a diversity of microphyllous shrubs and grasses and herbs are dominant. At higher altitudes, the subalpine shrubland and alpine snow-tussock grassland communities characteristic of non-ultramafic soils are replaced by discontinuous dwarf grassland or shrubland with intervening outcropping rocks, screes and boulder slopes.

An altitudinal comparison of vegetation on schist and ultramafic-derived soils at Red Mountain–Little Red Hill in north-west Otago high-

lights the major features of ultramafic vegetation (Tables 7, 8).

The total vascular flora of sampled sites from schist habitats numbers 191 of which about half are excluded from the ultramafites. Species diversity is consistently higher in schist communities, although to a lesser extent at lower elevations. Alpine and subalpine communities on schist are floristically richer than the forest communities at lower elevation, but this trend is reversed on ultramafic soils where species diversity declines with increasing elevation, largely because of the paucity of small herbaceous species on the ultramafites (Wilson *et al.* 1990). The area of bare ground increases markedly with elevation, but more so on ultramafites where, in the alpine zone, 55% of the ground area is devoid of plant life. Gymnosperms, chiefly small trees of the Podocarpaceae, are more important, both numerically and in abundance, on ultramafic soil, while dicotyledonous trees are heavily selected against. Of the latter, only *Weinmannia racemosa*, *Nothofagus solandri* var. *cliffortioides* and *Metrosideros umbellata* make a major contribution to the ultramafic forest, particularly at lower elevations. Other growth forms are also severely depleted on ultramafic soil. Pteridophyte cover in lowland and montane forest on schist ranges from 5–10%, but drops to 1% in the mixed forest on ultramafic soil. Most herbaceous species are also less successful on ultramafic soil, but shrubs are of similar importance on both substrates.

Tall forest on schist soils declines progressively in height from 16 m to ca. 7 m at its abrupt upper limit (1,060 m). Mean canopy height in forest on ultramafic soil is only a third of that normally found in the region, and the upper tree limit is gradual and attenuated.

Several species have wider altitudinal ranges on ultramafic soils in the absence of tall forest trees and continuous vegetation cover. For example *Dracophyllum uniflorum* and *Lepidothamnus laxifolius*, two low growing shrubs, are restricted to low alpine shrublands on schist parent material but enter lowland vegetation on the ultramafics.

Similar changes in growth form, dominance and physiognomy are evident in vegetation on the Anita Bay ultramafites in Fiordland (Tables 9, 10). Here different communities are associated

Photo 17. Dwarf shrubland at 300 m on the Anita Bay ultramafics, Lake Ronald, Fiordland. Dominant species include *Leptospermum scoparium*, *Metrosideros umbellata*, *Nothofagus solandri* var. *cliffortioides*, and *Lepidothamnus intermedius*.

with increasing ultramafic infertility and are not segregated altitudinally.

Of the important dicotyledonous trees and shrubs in the nonultramafic tall forest, only *Weinmannia racemosa*, *Metrosideros umbellata* and *Coprosma colensoi* extend beyond the transition forest onto the ultramafic outcrop, where they are associated with a range of small gymnosperms and several new dicotyledonous shrubs such as *Leptospermum scoparium*, *Dracophyllum longifolium* and *Cyathodes juniperina*. The major herbaceous ground cover shifts from the ferns in the tall forest to a small group of grasses and sedges on the ultramafics.

Species diversity is highest in the transition forest due to co-occurrence of several groups of species on the mixed parent material, *viz.* the extension of species out of the adjoining tall forest that are tolerant of mild ultramafic influence (e.g. *Pseudopanax simplex*, *Coprosma astonii*, *Blechnum discolor*); species that are common on the ultramafics but cannot compete with, or tolerate shade from, the tall trees on fertile soils (e.g. *Halocarpus biformis*, *Lepidothamnus intermedius*, *Coprosma pseudocuneata*, *Gahnia procera*,

Phormium cookianum); and two small groups of species, one locally restricted to this community (e.g. *Archeria traversii*, *Podocarpus hallii*, *Coprosma pseudocuneata*) and the other including species that grow in most communities (e.g. *Coprosma colensoi*, *Metrosideros umbellata*).

On the extreme ultramafic sites, sclerophyllous shrubs and graminoids replace the gymnosperms and broadleaved trees that form a low forest on well developed ultramafic soil. The cover of bare soils and rock increases as the mean canopy height declines to 0.3 m.

Regional differences between ultramafic habitats

Because of the limited outcrops elsewhere, the comparison of interest is between ultramafic areas in Nelson–Marlborough and those in Otago-Southland. While most of the ultramafites share a common geological origin, they have experienced quite different glacial and climatic conditions during the Quaternary, and it is these that account for nearly all the floristic and vegetational differences.

Table 7. Floristic comparison of vegetation on ultramafic and adjoining non-ultramafic soils along an altitudinal gradient (450–1,650 m) at Red Mountain, north-west Otago. Data from 20×1 m plots (106 on ultramafic soil and 59 on schist substrate) include cover (based on 40 height intercepts per plot), and species presence. Community types were obtained from a cluster analysis, using flexible sorting strategy (beta – −0.25) and Euclidean Distance (squared) Measure of the quadrant scores on the first axis of a Reciprocal Averaging ordination based on species cover values in each plot. Species with a mean percentage cover value of greater than 0.3% are listed. Major community types: – Nonultramafic communities on schist rock: (i) Lowland mixed beach podocarp forest, (ii) Montane-subalpine mixed Nothofagus menziesii forest, (iii) Low-alpine tall snow tussock grassland, (iv) High-alpine short tussock grassland; – Ultramafic communities: (i) Lowland-montane heath forest, (ii) Montane mixed grass shrubland, (iii) Subalpine tall tussock grassland – fellfield

	Schist parent material				Ultramafic parent material		
	(i)	(ii)	(iii)	(iv)	(i)	(ii)	(iii)
Gymnosperms							
Podocarpus cunninghamii	13.60				1.83		
Phyllocladus aspleniifolius var. *alpinus*	2.60				3.96	0.47	0.25
Lepidothamnus laxifolius			0.91			6.00	3.44
L. intermedius					11.98	1.47	
Halocarpus biformis					5.08	7.54	0.38
Dicotyledons Trees							
Pseudopanax crassifolius	4.00						
Griselinia littoralis	0.90						
Pseudopanax simplex	0.70						
Nothofagus menziesii	9.20	32.94					
Weinmannia racemosa	14.00				3.17		
Elaeocarpus hookerianus	1.20				0.42		
Nothofagus solandri var. *cliffortioides*	13.10				13.62	0.71	
Metrosideros umbellata	2.90				5.63	0.41	
Pseudopanax colensoi		0.35					
Aristotelia serrata		0.47					
Hoheria glabrata		0.24					
Pseudopanax linearis					0.42		
Dicotyledons Shrubs							
Neomyrtus pedunculata	2.00						
Coprosma pseudocuneata	0.30	6.41					
Myrsine divaricata	0.30	0.53			0.75		
Coprosma foetidissima	1.50				0.33		
C. colensoi	1.10						
C. astonii		2.53					
Brachyglottis bennettii		1.12					
Coprosma propinqua		1.29					
C. ciliata		1.12					
Archeria traversii		0.94					
Coprosma rugosa		0.24					
Olearia lacunosa		0.29					
Gaultheria depressa		0.65	1.27				
Coprosma cheesemanii			1.32	0.50			
C. pumila			0.59	2.25			
Dracophyllum uniflorum			1.86		0.29	4.18	2.31
Myrsine nummularia			1.59			0.59	0.94
Pernettya alpina				3.30			
Dracophyllum longifolium					1.04		
Leptospermum scoparium					4.38	11.18	
Cyathodes juniperina					1.92	0.71	
Pimelea oreopila						0.24	
Melicytus alpinus						0.29	0.30
Muehlenbeckia axillaris							0.63
Hebe odora							0.33
Cyathodes fraseri							0.27
Dicotyledons Herbs							
Lagenifera petiolata		1.71					
Acaena hirsutula		0.35					
Celmisia armstrongii			1.41				
C. petriei			3.32				

Table 7. Continued

	Schist parent material				Ultramafic parent material		
	(i)	(ii)	(iii)	(iv)	(i)	(ii)	(iii)
Dicotyledons Herbs (Continued)							
Caltha novae-zelandiae			0.68				
Anisotome haastii			0.55				
Celmisia walkeri			0.50				
Forstera sedifolia			0.50				
Gentiana montana			0.41				
Pratia angulata			0.47				
Celmisia sessiliflora				2.58			
Anisotome aromatica				1.17			
Raoulia grandiflora				1.00			
Celmisia hectorii				0.92			
Aciphylla congesta				0.75			
Drapetes sp.				0.58			
Celmisia durietzii				0.25			
Brachyglottis bellidioides						0.29	
Helichrysum bellidioides							0.31
Wahlenbergia albomarginata							0.28
Monocotyledons Grasses							
Microlaena avenacea	0.30						
Chionochloa cf. *rigida*		1.24	24.82			3.18	6.10
Poa colensoi		0.41	1.91	2.50		0.53	3.95
Chionochloa crassiuscula			2.14	1.67			
C. pallens			0.41	1.67			
C. oreophila				4.83			
Microlaena colensoi				0.33			
Rytidosperma setifolium						1.47	3.06
Monocotyledons Herbs							
Astelia nervosa	0.30	0.65	3.14				
Uncinia filiformis		2.71					
Phormium cookianum		0.71	0.27				0.44
Oreobolus pectinatus			0.50				
Aporostylis bifolia			0.32				
Astelia linearis			2.00	1.00			
Carpha alpina			1.77	1.82			
Schoenus pauciflorus			6.77			4.12	3.20
Oreobolus impar				1.00			
Luzula crinita var. *petriana*				0.42			
Gahnia procera					3.29	3.06	
Pteridophytes							
Blechnum discolor	1.30						
B. 'blackspot'	0.90						
Cyathea smithii	0.70						
Blechnum procerum	7.00				0.88		
Hypolepis millefolium		2.06					
Polystichum vestitum		2.76					
Lycopodium fastigiatum			0.82				

Most of the surface on southern ultramafites is distinctively coloured, uncolonised skeletal soil, rock outcrops or mobile scree slopes, features perpetuated by the glacially oversteepened slopes and rapid rock weathering. Erosion on the northern ultramafites is on a more limited scale due to the maturity of the ultramafic landscape. Eroding areas are each only a few metres in extent amongst an extensive plant cover.

The older northern South Island ultramafic habitats are floristically richer (440 species) than those in the south (350 species), though the rela-

Table 8. Comparison of major plant groups and environmental factors in an altitudinal vegetation sequence on ultramafic and adjoining schist parent material, Red Mountain, north-west Otago, New Zealand

	Schist parent material				Ultramafic parent material		
	Lowland mixed beech podocarp forest	Montane-subalpine *Nothofagus menziesii* forest	Low-alpine tall tussock grassland	High alpine short tussock grassland	Lowland-montane *Nothofagus solandri* var. *cliffortioides* mixed podocarp forest	Montane mixed grass shrubland	Subalpine tall tussock grassland-fellfield
Species with a cover value >0.3%							
Gymnosperms							
no.	2	0	1	0	4	4	3
% cover	16.2	0	0.91	0	22.3	15.5	4.1
Dicotyledons trees							
no.	8	4	0	0	5	2	0
% cover	46.0	34.0	0	0	23.2	1.1	0
no.	5	20	5	3	7	6	6
% cover	5.2	15.1	6.6	6.1	9.0	17.2	4.8
no.	0	2	8	7	0	1	2
% cover	0	2.1	7.8	7.3	0	0.3	0.6
Monocotyledons grasses							
no.	1	2	4	5	0	3	3
% cover	0.3	1.6	29.6	11.0	0	5.2	13.1
Monocotyledons herbs							
no.	1	3	7	4	1	2	2
% cover	0.3	4.1	14.8	4.2	3.3	7.2	3.6
Pteridophytes no.	4	2	1	0	1	0	0
% cover	9.9	4.8	0.8	0	0.9	0	0
Total no. of species	21	23	26	19	18	18	16
% cover	77.9	81.9	59.6	28.6	58.7	46.5	26.2
Mean canopy ht (m)	15.8	6.7	0.2	0.02	4.5	0.3	0.07
Slope mean°	13.7	23.8	16.9	25.6	23.1	15.3	4.6
Altitude (m)							
mean	384	1,035	1,086	1,463	588	775	1,018
range	368–411	975–1,066	1,021–1,127	1,410–1,483	365–701	640–902	883–1,127
Mean no. spp. per quadrat	22.6	23.4	30.1	28.7	24.8	15.7	12,8
% bare ground	3.5	2.4	5.4	35.2	6.7	33.5	55.8

tive contributions of the major plant groups to the total flora is similar (Table 4). Regional differences in endemism have been discussed.

Variation between ultramafic outcrops in the composition and development of forest primarily reflects current climatic differences. The gymnosperms and dicotyledenous trees, important forest-forming species on north-west Otago and Fiordland ultramafites, are sparse on drier areas in northern Southland, and on the Nelson–Marlborough outcrops, where they form minor components of more widespread *Leptospermum scoparium* shrubland. Although the soils are similar chemically, a drier climate reduces the ultramafic tolerance of most tree species.

Primary succession

Rapid and extensive erosion on ultramafic outcrops in north-west Otago and Fiordland repeatedly creates new surfaces for plant colonization

Table 9. Floristic comparison of vegetation of ultramafic and adjoining non-ultramafic soils at similar elevation at Lake Ronald, Anita Bay ultramafities, Fiordland. The number of plots was 10 on non-ultramafic soil derived from gneiss and 70 on the ultramafites. Species with a mean percentage cover greater than 0.3% are listed. Major community types: – Non ultramafic community on gneiss rock: A. Nothofagus menziesii forest; – Ultramafic communities: B. Transitional forest along ecotone; C. (i) ultramafic heath forest, C. (ii) shrubland, C. (iii) mixed grass-shrubland

	A	B	C		
			(i)	(ii)	(iii)
Gymnosperms					
Podocarpus hallii		2.94			
Phyllocladus aspleniifolius var. *alpinus*		11.43	4.38		
Lepidothamnus intermedius		17.24	34.53	29.83	5.00
L. laxifolius			3.91		
Halocarpus biformis		3.24	9.50	4.46	2.75
Dicotyledons Trees					
Pseudopanax crassifolius	1.00				
P. simplex	3.25	3.62			
Nothofagus menziesii	61.70	16.82	0.31		
Griselinia littoralis	15.51	4.36	0.62		
Weinmannia racemosa	52.30	46.10	17.34	0.88	
Pseudopanax colensoi	4.26	4.61	1.72	0.14	
Metrosideros umbellata	13.50	35.71	20.16	5.92	0.50
Nothofagus solandri var. *cliffortioides*		36.80	26.25	10.63	0.50
Pseudopanax linearis			0.94	0.20	
Dicotyledons Shrubs					
Pseudowintera colorata	8.00				
Aristotelia fruticosa	0.26				
Rubus cissoides	0.25				
Coprosma astonii	0.75	0.42			
Myrsine divaricata	0.54	0.83			
Coprosma foetidissima	8.25	7.93	0.31		
C. colensoi	0.56	7.15	2.34	0.14	0.25
Dracophyllum longifolium		2.10		2.00	0.50
Archeria traversii		2.15	0.16		
Coprosma pseudocuneata		0.42	0.16		
Leptospermum scoparium		2.51	4.22	22.63	25.00
Cyathodes juniperina		1.00	5.94	4.93	2.25
Hebe odora				0.20	
Melicytus alpinus				0.27	
Dracophyllum uniflorum				0.92	4.81

and spatial differences in composition between plant communities can be interpreted as seral stages in the re-establishment of forest.

Early successional species vary according to the coarse fraction particle size of the available surface. Trees and shrubs are the first plants to grow on coarse scree.

The establishment of pioneer species on finer material is hindered by the inability of many of the early successional indigenous lichens, bryophytes, herbs and woody species to grow on ultramafic soils. Presumably early primary successional species have high nutrient requirements, particularly for N and P, which cannot be met on ultramafic soil, irrespective of any toxicity factor. Mine tailings on Dun Mountain and spoil heaps at Black Ridge have remained virtually plant-free for over 30 years. Evidence for the time taken for the establishment of forest on ultramafic soils was obtained from an analysis of the age of gymnosperms in the four communities associated with the Anita Bay ultramafites in Fiordland. Trees in forests along the margins and in the centre of the ultramafic outcrop were aged between 350–450 years, and were significantly older than conspecifics in either the shrubland (150–175 years) or

Table 9. Continued

	A	B	C		
			(i)	(ii)	(iii)
Dicotyledons Herbs					
Nertera sp.	0.25				
Celmisia gracilenta				0.30	0.50
Monocotyledons					
Microlaena avenacea	2.25				
Uncinia clavata	0.50				
Astelia nervosa	0.50				
Gahnia procera		1.00	7.81	7.12	2.25
Phormium cookianum		2.10	0.16		
Chionochloa cf. *rigida*				1.51	3.25
Schoenus pauciflorus				0.32	7.51
Lepidosperma australe				0.50	2.50
Poa colensoi				0.20	0.75
Oreobolus pectinatus					1.00
Carpha alpina					0.75
Rytidosperma setifolium					0.50
Pteridophytes					
Cyathea smithii	20.50				
Blechnum discolor	25.56				
Leptopteris superba	14.81				
Asplenium bulbiferum	1.50				
Blechnum fluviatile	0.74				
B. procerum	4.78	3.24			
B. 'black spot'	1.75				
Grammitis billardierei	0.25	0.42			
Trichomanes reniforme		0.42			
Gleichenia dicarpa				0.50	
Schizaea fistulosa					0.25

mixed grass-shrubland (90–100 years) communities on ultramafic soil. The development of analogous vegetation on coarse blocky ultramafic screes on Little Red Hill is associated with an increase in total N, extractable P, cation exchange capacity and organic matter in the stony soils. Mg availability and acetic acid soluble Ni levels showed an increase in soils supporting forest but their influence is apparently ameliorated by the organic matter, particularly in the organic horizon, and the improved nutrient status in the later stages of succession.

Ecological affinities of ultramafic vegetation

Cockayne (1910) considered the vegetation on the Nelson–Marlborough ultramafites to be ecologically and physiognomically equivalent to the communities of the so called 'dry' New Zealand Mountains, a view that was shared by Bell *et al.* (1911). Oliver (1944), while in agreement with Cockayne, stressed the similarities in composition between the D'Urville Island ultramafic vegetation and the narrow-leaved shrub communities in extreme edaphic conditions elsewhere, such as sulphur-rich thermal soils, infertile clay soils, pumice lands and swamps.

Ultramafic vegetation has close floristic similarities to, and shares with, communities on glacially smoothed rock surfaces, quartzose sandstone, granite, impoverished gley podzols and volcanic ash (Burrows *et al.* 1983), features such as stunted tree growth, uneven discontinuous low forest, species poverty, paucity of herbs, preponderance of microphyllous shrubs, physiognomic

Table 10. Comparison of major plant groups and environmental factors in plant communities on ultramafic and adjoining gneiss parent material, Lake Ronald, Fiordland, New Zealand

	Nothofagus menziesii forest	Transition forest	Ultramafic communities		
			Forest	Shrubland	Mixed grass-shrubland
Species with a cover value >0.3%					
Gymnosperms					
no.	0	4	4	2	2
% cover	0	34.8	52.3	33.9	8.3
Dicotyledons trees					
no.	7	7	7	5	2
% cover	151.5	148.0	67.3	17.8	1.0
Dicotyledons Shrubs					
no.	7	9	6	7	5
% cover	18.6	24.6	13.1	31.0	32.8
Dicotyledons Herbs					
no.	1	0	0	1	1
% cover	0.3	0	0	0.3	0.5
Monocotyledons Grasses					
no.	1	0	0	2	3
%cover	2.3	0	0	3.7	4.5
Monocotyledons herbs					
no.	2	3	2	3	5
% cover	1.0	3.1	8.0	7.9	14.0
Pteridophytes					
no.	8	4	0	1	1
% cover	69.8	4.5	0	0.5	0.3
Total no. of species	26	27	19	21	19
% cover	243.5	215.0	140.0	95.1	61.4
Mean canopy ht (m)	16.0	9.0	3.1	1.1	0.3
Mean attitude (m)	515	528	524	522	491
Bare soil or rock (%)	1.0	2.1	13.9	26.4	41.0
Mean no. spp. per quadrat.	19.5	23.1	16.2	14.2	15.0

variability of the vegetation, inability of naturalised plant species to enter even disturbed areas, and the presence of a few narrowly restricted endemic taxa. However, with the exception of the morainic sites, it lacks the extensive restiad, sedge, cushion and fern communities associated with permanently wet or peaty soils. Ultramafic vegetation also includes a small group of species, such as *Coprosma propinqua*, *Melicytus alpinus*, *Aristotelia fruticosa* and *Muehlenbeckia axillaris*, that are not common on infertile soils, but can presumably tolerate the high Mg and heavy metal concentrations in ultramafic soil.

Plant response to ultramafic soil

Morphology

A range of morphological differences has been observed in species on ultramafic soil that are not found in conspecifics growing off the ultramafites. Such features include prostrate scrambling habit, extensive root system, elongated flowering stalks, smaller and fewer inflorescences, precocious flowering, darker leaf colouration, thicker and shorter leaves, and severely stunted growth.

Photo 18. Metrosideros umbellata, a tree reaching 15–20 m in many New Zealand forests, occurs on Red Mountain, north-west Otago, as a small shrub. This flowering plant is 0.3 m tall.

Growth reduction of tree species

The stunted growth of trees and shrubs is a conspicuous feature of most ultramafic outcrops (Photo 18). Trees and shrubs on normal soil become dwarf bushes on ultramafics. On Red Mountain, *Nothofagus* showed a reduction in radial increment growth of 36% (*N. menziesii*) and 64% (*N. solandri* var. *cliffortioides*) compared with trees of similar age on schist-derived soil (Table 11). Growth rates in different communities on the Anita Bay ultramafics show a similar reduction with increasing ultramafic influence even amongst species quite tolerant of ultramafic soils. However these growth rates are not excessively low for these species and are well within the ranges recorded by other workers (Wardle 1963, Wardle 1970, Herbert 1972) on edaphically extreme non-ultramafic sites.

Ecotypes

Of the fifteen species (Table 12) that have been experimentally investigated for the existence of genetically distinct ultramafic ecotypes, seven were demonstrated to have ultramafic-tolerant races with a faster relative growth rate than non-ultramafic populations on ultramafic soil (Lee *et al.* 1983). These comprise a rush, four grasses and two dicotyledonous herbs, and include four naturalised species, widely known to have genetically distinct races, that must have developed ultramafic ecotypes within the last century.

Morphological differences between ultramafic-tolerant and non-tolerant populations are few. Leaves of plants from ultramafic soil of the creeping rhizomatous herb *Neopaxia australasica* were noticeably more succulent than the leaves of non-ultramafic plants, a difference that was maintained on potting compost. Similarly, with *Leptinella pyrethrifolia* the ultramafic-tolerant population, recognised as a separate variety by Lloyd (1972), has linear leaves in contrast to the normal pinnatifid form of the non-tolerant plants.

Ultramafic ecotypes are also distinguished from non-ultramafic conspecifics by high shoot concentrations of Ca and Ni on both ultramafic soil and potting compost and a low shoot/root ratio on ultramafic soil. In *Agrostis capillaris*, root growth of ultramafic-tolerant plants is less affected by

411

Table 11. Comparison of mean radial growth rate (mm 10 yr^{-1}) of tree species on ultramafic and non-ultramafic soil. Data derived from increment borings of 10 plants per species. a) *Nothofagus menziesii* and *Nothofagus solandri* var. *cliffortioides* on ultramafic soil (Red Mountain, northwest Otago) and non-ultramafic soil. b) Major forest species in each community type on Anita Bay ultramafites. Solid line joins values not significantly different (p=0.05)

a)

	Ultramafic soil	Non-ultramafic soil
Nothofagus menziesii	6.8	8.4
Nothofagus solandri var. *cliffortioides*	5.8	16.4

b)

	Forest on non-ultramafic soil	Transition community	Anita Bay ultramafites		
			Mixed forest	Shrubland	Mixed grass shrubland
Halocarpus biformis	n.a.	2.8	1.9	0.9	0.7
Lepidothamnus intermedius	n.a.	2.0	1.2 :	1.4	0.7
Nothofagus solandri var. *cliffortioides*	n.a.	5.9	2.0	n.a.	n.a.
Nothofagus menziesii	7.0	5.9	n.a.	n.a.	n.a.

Table 12. Species tested for ultramafic–tolerant ecotypes. T=tree; S=shrub; D=dicotyledonous herb; R=rush; E=sedge; G= grass; *=naturalised species

Species with ultramafic-tolerant ecotypes		Species lacking ultramafic–tolerant ecotypes	
Neopaxia australasica	D	*Nothofagus solandri* var. *cliffortioides*	T
Leptinella pyrethifolia	D	*Griselinia littoralis*	T
Luzula crinita var. *petriana*	R	*Leptospermum scoparium*	S
Poa colensoi	G	*Rumex obtusifolius**	D
*Agrostis capillaris**	G	*Schoenus pauciflorus**	E
*Anthoxanthum adoratum**	G	*Dactylis glomerata**	G
*Lolium perenne**	G	*Holcus lanatus**	G
		Poa cita	G

single ion solution of Mg and, surprisingly, Ca, than non-tolerant plants. There is no evidence that ultramafic tolerant plants preferentially exclude the potentially toxic metal ions from their shoots. For two species with ultramafic-tolerant ecotypes (*Neopaxia australasica* and *Luzula crinita* var. *petriana*), growth of the tolerant plants on normal soil was significantly less than that of non-tolerant plants, a loss of vigour probably attributable to the physiological cost of achieving ultramafic tolerance.

Disjunct distributions

The only clear example of a latitudinal extension on ultramafic soil is *Chionochloa acicularis*, a tall tussock grass that grows on the ultramafic moraine forming the Cascade and Gorge plateaux in North-west Otago. These populations represent the northern limit for the species, and are at least 120 km from the nearest known non-ultramafic population in Fiordland. It is possible that *Chion-*

ochloa acicularis persisted in the area during the Pleistocene and has become restricted locally to the ultramafites following the retreat of ice and amelioration of the climate, through the spread of forest communities on to surrounding country.

Small-scale distribution of species

As has been mentioned in the section on endemism, the distribution of species within a community on ultramafic soil is strongly correlated with variations in soil factors. On Dun Mountain in mixed grass-shrubland on serpentinite, most trees and shrubs grow on microsites with low Mg and high P, K and Ca concentrations associated with deep soils. Ultramafic endemic taxa and small shrubs such as *Melicytus alpinus* and *Dracophyllum pronum* occupy sites with the contrasting properties: usually flat bare areas that are created by water runoff (Lee 1980).

On some ultramafic outcrops variation in soil conditions is due to the presence of different rock

412

Table 13. Growth (mean height (cm) and dry weight (g)) of *Avena sativa* cv 'Mapua' after eight weeks on several ultramafic soils and Potting Compost

Site	Height (cm)	Total	Shoot	Root	S/R
				Dry weight	
North Cape	12	0.81	0.61	0.20	3.0
Dun Mountain	17	0.70	0.57	0.13	4.4
Mine Tailings	12	0.24	0.10	0.13	0.8
Red Hills	15	0.49	0.29	0.19	1.5
Red Mountain	19	0.51	0.43	0.08	5.4
Livingstone Mountains	16	0.67	0.54	0.13	4.2
Black Ridge	9	0.28	0.18	0.09	2.0
Anita Bay	13	0.68	0.53	0.15	3.5
John Innes Potting Compost	46	1.41	0.78	0.65	1.2

types. For instance, the lower slopes of West Dome support *Nothofagus* forest and several shrubland communities that grow on physically similar but chemically distinct soils (McIntosh & Lee 1986). The typical ultramafic shrubland with the local ultramafic endemic *Celmisia spedenii*, occurs where soil Mg levels are highest, and appears to be the stable vegetation on these serpentinite soils.

Bioassays

The cause of the conspicuous vegetation on ultramafites was considered by early botanists to be due to an excess of soil magnesium (Cockayne 1910, Scott-Thompson 1935, Oliver 1944). Lyon *et al.* (1968, 1970, 1971) suggested that several factors were involved, including high concentrations of metals such as Cr, Ni and Co.

Avena sativa has been used traditionally to compare the fertility of ultramafic soils, particularly the availability of the metal ions. The growth of *Avena sativa* after eight weeks on representative ultramafic soils is shown in Table 13. Compared with that on potting compost, plant growth is reduced by at least 50% and shoot/root quotients markedly increased on ultramafic soils. The smallest plants, which had swollen stunted roots and were the only ones to produce foliar symptoms of Ni toxicity, were on the Mine Tailings soil from Dun Mountain and the outcrop at Black Ridge, northern Southland. Foliar macronutrient deficiency symptoms were widespread, particularly for the major nutrients.

Foliar element concentrations in *Avena sativa*

on the ultramafic soils (Table 14) confirm the deficiency symptoms observed in the plants, particularly demonstrating the low concentrations of P, K, Ca and N in these soils. Mg concentrations in *Avena sativa* shoots known to severely retard plant growth (Proctor 1971) are reached or approached in three ultramafic soils, namely Livingstone Mountains, Mine Tailings from Dun Mountain, and Black Ridge, which have a high Mg/Ca quotient. The results also demonstrate the widespread availability of Ni, Cr and CO ions in New Zealand ultramafic soils. A comparison of these levels with those found to limit the growth of *Avena sativa* suggests that only Ni in soils from the ultramafics in the south-west of the South Island is likely to reduce plant growth critically.

On nearly all soils tested *Avena sativa* responded to Ca, while K, P and N produced significant responses on four, five and three soils respectively (Table 15). Again it is the multiplicity of responses that is evident, with five soils showing deficiencies of three or more nutrients.

The Anita Bay and North Cape soils were atypical. The former showed a negative response to N that was exacerbated in the presence of either P or K. The North Cape soil failed to respond positively to any nutrients and in fact produced a consistently negative response to Ca. The reasons for these reactions are unclear.

Response to nutrients of tolerant and non-tolerant species

Generalisations about the nutrient factors limiting growth in ultramafic soils based on *Avena*

414

Table 14. Element content of *Avena sativa* cv. '*Mapua*' (shoot material) grown for eight weeks on New Zealand ultramafic soil

Experimental soils	N (%) DW	P (%) DW	K (mg/g) AW	K (mg/g) DW	Ca (mg/g) AW	Ca (mg/g) DW	Mg (mg/g) AW	Mg (mg/g) DW	Mg/Ca ratio	Ni (μg/g) AW	Ni (μg/g) DW	Cu (μg/g) AW	Cu (μg/g) DW	Co (μg/g) AW	Co (μg/g) DW	Cr (μg/g) AW	Cr (μg/g) DW
North Cape	0.94	0.079	180.9	17.5	28.6	2.8	41.5	4.0	1.45	519	50	139	13	100	10	583	43
Dun Mountain	0.82	0.097	185.0	14.9	18.3	1.5	80.7	6.6	4.41	554	38	256	18	39	2	308	21
Mine Tailings	1.67	0.261	95.4	8.0	19.4	1.6	129.9	10.9	6.69	919	72	248	20	61	5	168	15
Red Hill	1.15	0.090	212.4	14.8	25.0	1.7	32.2	3.3	1.28	181	12	197	14	16	2	348	24
Red Mountain	0.73	0.094	156.0	5.4	13.4	1.3	52.8	5.2	3.94	523	52	114	1	29	3	169	14
Livingstone Mts	0.97	0.053	141.2	11.6	19.3	1.6	144.5	11.8	7.48	991	82	126	10	60	5	196	13
Black Ridge	1.86	0.230	90.7	7.0	14.9	1.2	125.8	9.7	8.44	686	53	331	25	74	6	291	19
Lake Ronald	1.27	0.128	162.5	11.2	17.9	1.2	97.5	6.7	5.44	1,785	116	182	12	125	8	236	17
Potting Compost	1.41	0.403	224.4	14.9	32.7	3.9	20.6	2.4	0.63	45	5	130	16	10	2	12	2

AW = ash weight; DW = dry weight.

Table 15. Significant responses from a factorial soil amendment experiment with nitrogen ($HN_4NO_4 + NaNO_3$), phosphorus ($NaH_2PO_4\,2H_2O$), potassium ($K_2SO_4 + KCl$) and calcium ($CaCl_2$) using *Avena sativa* cv '*Mapua*' on several ultramafic soils. Data were log transformed. $P < 0.001$ +++, < 0.01 ++, < 0.05 +)

Interactions	Nth Cape	Dun Mt	Mine tailings	Red Hill	Red Mts	Liv. Mts	Black R	L Ronald
N			+++	+++	+++			---
P		+++		+++	+++		+++	++
K		++	+++		+		+++	
Ca	–	+++	+++	+++	+++	++	+++	++
NP				+++	+++	↑		–
NK								–
NCa	– –			– –				
PK					+			
PCa	–			+			++	–
KCa						NA		+
NPK	–				++			–
NPCa	– – –							
NKCa								
PCKa	–							
NPKCa						↓		

sativa bioassays are likely to be misleading because plant species vary in their nutrient requirements. *Avena sativa* shows a universal growth response to Ca on most New Zealand ultramafic soils that is not found in either tolerant or non-tolerant indigenous species (Table 16). The poor growth response of the two indigenous non-tolerant species (*Nothofagus menziesii* and *Griselinia littoralis*) to a complete fertilizer mix on ultramafic soil suggests that toxicity rather than nutrient impoverishment may be excluding these species from ultramafic soil. Amongst the species tested, ultramafic-tolerant species are distinguished by their ability to maintain growth equivalent to that on potting compost in at least one ultramafic treatment and often in more. None of these species required additional Ca; but there was a widespread response to P, and a differential response to N and K.

Element uptake in indigenous species on ultramafic soil

On Dun Mountain (Lyon *et al*. 1968, 1970, 1971) and Red Mountain (Lee 1980), the unusual chemical composition of the ultramafites is reflected in a multi-element analysis of leaves of a range of species, although there is considerable within-plant and between-plant variation. On ultramafic soil, plants in general have higher foliar concentrations of Mg, Ni, Cr and perhaps Co than conspecifics on non-ultramafic soil (Lyon *et al*. 1971, Lee 1980). For six out of the nine species studied on Red Mountain, foliar levels of N, P and K were little altered while Ca was slightly depressed on ultramafites. In some species foliar concentrations are strongly dependent on those in soil, making them useful biogeochemical indicators, e.g. *Cassinia vauvilliersii* (Cr, Co, Ni; Lyon *et al*. 1968), *Halocarpus biformis* (Ni; Lee 1980). A small group of species appears to accumulate preferentially Ni ions. The most notable are the two small shrubs *Pimelea suteri* and *Melicytus alpinus*, which maintain plant/soil ratios of Ni >1 (ash weight basis), associated with high plant concentrations of the major-nutrients (Brooks *et al*. 1974, Lee 1980). There is evidence that the ameliorating effects of individual macronutrients on the potentially toxic effects of Ni, Cu, Co and Cr are utilised by ultramafic tolerant species. The highly significant positive correlations between Mg–Ni and the negative correlations between Ca–Mg and K–Mg in foliage of species growing on ultramafites are possible examples (Lyon *et al*. 1971, Lee 1980).

Conclusions

Ultramafic soils provide a unique ensemble of chemical and physical conditions for plants in New Zealand. The vegetation and flora of these areas, while supporting numerous ultramafic endemic taxa, is broadly similar to plant life on other edaphically extreme sites. Because there are numerous soil factors that could limit plant growth on ultramafic outcrops, generalisations regarding ultramafic tolerance or avoidance are difficult to establish, and are perhaps best examined at the species level.

Acknowledgements

Tony Druce provided floristic information for the North Cape and Nelson–Marlborough ultramafic outcrops, and commented on the text. I am also grateful to Daphne Lee, Ralph Allen, Peter Wardle, Peter Williams, Warwick Harris and Allan Hewitt for their improvements to drafts of the manuscript. Alan F. Mark and J. Bastow Wilson supervised much of the work included in this chapter, as a Ph.D. thesis at the University of Otago, Dunedin, New Zealand. Finally, my thanks to the many colleagues who contributed information on different aspects of the ecology of plant life on ultramafic soils in New Zealand.

References

Allan, H. H. 1961. Flora of New Zealand, vol. 1. Government Printer, Wellington. 1085 pp.

Banks, J. 1770. Endeavour Journal of Joseph Banks, J. C. Beaglehole. (ed.), Vol. 1, p. 474, N.S.W.

Bell, J. M., E. de C. Clarke & P. Marshall. 1911. The geology of the Dun Mountain subdivision, Nelson. N.Z. Geological Survey Bulletin, n.s. 12, 71.

Betts, M. W. 1918. Notes on the autecoloty of certain plants of the peridotite belt, Nelson. Part 1 – structure of some of the plants. No. 1. Transactions of the N.Z. Institue 50: 230–242.

Betts, M. W. 1919. Notes on the autecology of certain plants of the peridotite belt, Nelson. Part 1 – structure of some of the plants. No. 2. Transactions of the N.Z. Institute 51: 136–156.

Betts, M. W. 1920. Notes on the autecology of certain plants of the peridotite belt, Nelson. Part 1 – structure of some of the plants. No. 3. Transactions of the N.Z. Institute 52: 276–314.

Bliss, L. C. & A. F. Mark. 1974. High-alpine environments and primary production on the Rock and Pillar Range, Central Otago, New Zealand. New Zealand Journal of Botany 12: 445–483.

Brooks, P. R., J. Lee & T. Jaffré. 1974. Some New Zealand and New Caledonian plant accumulators of nickel. Journal of Ecology 62: 523–529.

Burrows, C. J., D. R. McQueen, A. E. Esler & P. Wardle. 1983. New Zealand Heathlands. In: Heathlands and related shrublands of the World, A. Descriptive studies. R. L. Specht (ed.), pp. 339–364.

Cockyane, L. 1910. New Zealand Plants and their story (1st edition). Government Printer, Wellington, 190 pp.

Cockayne, L. 1922. The vegetation of a portion of the 'Mineral Belt'. N.Z. Nature Notes: 39.

Cockayne, L. 1928. The vegetation of New Zealand. 2nd edition. Engelmann, Leipzig, 456 pp.

Coleman, R. G. 1966. New Zealand serpentinites and associated metasomatic rocks. N.Z. Geological Survey Bulletin n.s. 76: 102 pp.

Connor, H. E. & E. Edgar. 1987. Name changes in the indigenous New Zealand Flora, 1960–1968 and Nomina Nova IV, 1983–1986. New Zealand Journal of Botany 25: 115–170.

Coombs, D. S., C. A. Landis, R. J. Norris, J. M. Sinton, D. J. Borns & D. Craw. 1976. The Dun Mountain Ophiolite Belt, New Zealand, its tectonic setting, constitution, and origin, with special reference to the southern portion. American Journal of Science 276: 561–603.

Coulter, J. D. 1975. II. The climate. In: Biogeography and ecology in New Zealand. G. Kuschel (ed.), Junk Publishers, The Hague, pp. 87–138.

Dansererau, P. 1964. Six problems in New Zealand vegetation. Bulletin of the Torrey Botanical Club 91: 114–140.

Druce, A. P., J. K. Bartlett & R. D. Gardner. 1979. Indigenous vascular plants of the serpentine area of Surville Cliffs and adjacent cliff tops, north-west of North Cape, New Zealand. Tane 25: 187–206.

Fisher, F. J. F. 1952. Observations on the vegetation of screes of Canterbury, New Zealand. Journal of Ecology 40: 156–157.

Herbert, J. 1972. The growth of Silver Beech in Northern Fiordland. N.Z. Journal of Forestry Science 3: 131–171.

Hochstetter, F. von. 1864. Dunit, Korniger Olivinfels vom Dun Mountain bei Nelson, New-Seeland: Deutsche geol. Gesell. Zeitschr. V. 16: 341–344.

Lee, W. G. 1980. Ultramafic plant ecology of the South Island, New Zealand. Ph.D. Thesis, University of Otago, Dunedin, New Zealand. 285 pp.

Lee, W. G. & A. H. Hewitt. 1982. Soil changes associated with development of vegetation on an ultramafic scree north-west Otago, New Zealand. Journal of the Royal Society of New Zealand 12: 229–242.

Lee, W. G., A. F. Mark & J. B. Wilson. 1983. Ecotypic differentiation in the ultramafic flora of the South Island, New Zealand. New Zealand Journal of Botany 21: 141–156.

Lee, W. G. & D. R. Given. 1984. Celmisia spedenii, G. Simpson, an ultramafic endemic, and Celmisia markii, sp. nov., from southern New Zealand. New Zealand Journal of Botany 22: 585–592.

Lee, J., R. R. Brooks, R. D. Reeves, C. R. Boswell. 1975. Soil factors controlling a New Zealand serpentine flora. Plant and Soil 42: 153–160.

Lee, W. G. & R. D. Reeves. 1989. Growth and chemical

composition of *Celmisia spedenii*, an ultramafic endemic, and *Celmisia markii* on ultramafic soil and garden loam. New Zealand Journal of Botany 27: 595–598.

Lloyd, D. G. 1972. A revision of the New Zealand, subantarctic and South American species of Cotula, section Leptinella. N.Z. Journal of Botany 10: 277–372.

Lyon, G. L., R. R. Brooks, P. J. Peterson & G. W. Butler. 1968. Trace elements in a New Zealand serpentine flora. Plant and Soil 29: 225–240.

Lyon, G. L., R. R. Brooks, P. J. Peterson and G. W. Butler. 1970. Some trace elements in plants from serpentine soils. N.Z. Journal of Science 13: 133–139.

Lyon, G. L., R. R. Brooks, P. J. Peterson & G. W. Butler. 1971. Calcium, magnesium and trace elements in a New Zealand serpentine flora. Journal of Ecology 59: 421–429.

Mark, A. F. 1977. Vegetation of Mount Aspiring National Park, New Zealand. National Parks Scientific Series Number 2. 79 pp.

McIntosh, P. & W. G. Lee. 1986. Soil factors influencing the vegetation on the Dun Mountain Ophiolite Belt at West Dome, Southland, New Zealand. New Zealand Journal of the Royal Society 16: 363–379.

Metson, A. J. 1974. Magnesium in New Zealand soils. I. Some factors governing the availability of soil magnesium: a review. N.Z. Journal of Experimental Agriculture 2: 277–319.

Moore, L. B. & E. Edgar. 1970. Flora of New Zealand, vol. II. Government Printer, Wellington. 354 pp.

N.Z. Department of Scientific and Industrial Research Soil Bureau 1968a: General survey of the soils of South Island, New Zealand. N.Z. Soil Bureau Bulletin 27: 1–403.

Ogden, J. 1970. Botany and entomology of the Red Mills. Massif 3: 43–44.

Oliver, W. R. B. 1944. The vegetation and flora of D'Urville and Stephen Islands. Records of the Dominion Museum 1: 193–227.

Park, J. 1887. On the district between the Dart and Big Bay. Geological Explorations 1885–86, pp. 121–137.

Proctor, J. 1971. The plant ecology of serpentine. The influence of a high Mg/Ca ratio and high nickel and chromium levels in some British and Swedish serpentine soils. Journal of Ecology 59: 827–842.

Scott-Thompson, J. 1935. Some aspects of the vegetation and flora of the South Island. Journal of the New Zealand Institute of Horticulture 4.

Soil Survey Staff, 1975. Soil Taxonomy. A basic system of soil classification for making and interpreting soil surveys. Soil Conservation Service, U.S. Department of Agriculture, Agriculture Handbook 436.

Swindale, L. D. 1966. A mineralogical study of soils derived from basic and ultrabasic rocks in New Zealand. N.Z. Journal of Science 9: 484–506.

Thompson, R. C., K. A. Rodgers & J. E. Braggin. 1974. The relationship of serpentine and related floras to laterite and bedrock type at North Cape. N.Z. Journal of Botany 12: 275–282.

Thompson, R. D. & K. A. Rodgers. 1977. Laterization of the ultramafic-gabbro association at North Cape, Northernmost, New Zealand. Journal of the Royal Society of New Zealand 7: 347–377.

Walcott, R. I. 1969. Geology of the Red Hills complex, Nelson, New Zealand. Transactions of the Royal Society of New Zealand. Earth Science 7: 57–58.

Walls, G. Y. & M. D. Laffan. 1986. Native vegetation and soil patterns in the Marlborough Sounds, South Island, New Zealand. N.Z. Journal of Botany 24: 293–313.

Wardle, J. 1970. The ecology of Nothofagus solandri 4. Growth and general discussion to parts 1–4. No. 2. N.Z. Journal of Botany 8: 609–646.

Wardle, J., J. Hayward & J. Herbert. 1971. Forests and shrublands of northern Fiordland. N.Z. Journal of Forestry Science 1: 80–115.

Wardle, P. 1963. Vegetation studies on Secretary Island, Fiordland Part 2: The plant communities. N.Z. Journal of Botany 1: 171–87.

Wardle, P., A. F. Mark & G. T. S. Baylis. 1973. Vegetation and landscape of the West Cape District, Fiordland, New Zealand. N.Z. Journal of Botany 11: 599–626.

Wardle, P., P. N. Johnson & R. P. Buxton. 1986. Botany of the Gorge River, South Westland. Botany Division, DSIR unpublished report.

Webb, C. J., W. R. Sykes & P. J. Garnock-Jones. 1988. Flora of New Zealand. vol. IV. Christchurch, Botany Division, Department of Scientific and Industrial Research. 1365 pp.

Wilson, J. B., W. G. Lee & A. F. Mark. 1990. Species diversity in relation to ultramafic substrate and to altitude in south Western New Zealand. Vegetatio 86: 15–20.

Wood, B. L. 1972. Metamorphosed ultramafites and associated formations near Milford Sound, New Zealand, N.Z. Journal of Geology and Geophysics 15: 88–128.

W. G. Lee
DSIR Land Resources
Private Bag
Dunedin
New Zealand

XVI

Concluding remarks

B. A. ROBERTS & J. PROCTOR

The study of the ecology of serpentinized areas has presented a great challenge to the geologists, biologists and agriculturists in many parts of the world. This text reports on many of the problems associated with such studies including the confusion from terminology, a complex geology and trying to apply a detailed plant ecological approach over a wide geographic area. It has been difficult to relate many of the ecological factors over a wide area since some vegetation though restricted to serpentinized material is restricted to a narrow habitat within it, e.g. rock crevice.

Some generalizations emerge. The importance of magnesium is great in many instances and it functions in two ways. Firstly by occupying a high proportion of cation exchange sites it helps maintain a high soil pH. Secondly it is often at toxic concentrations and is likely to be involved in interactions with calcium and nickel. The importance of nickel is apparently great in some cases but more work is required. The current emphasis on nickel accumulation (of which the ecological aspects are unknown) needs transferring to the effect of nickel, and plant response to it, on non-accumulator species (which make up the bulk of the flora).

The shortage of nutrients in ultramafic soils is well attested and can limit their colonization possibly interactively with water stress. Ultramafic soils are often drought prone, even in the wettest areas, and this factor alone can account for the physiognomic contrasts with the vegetation on adjacent rock types. The role of micro-organisms in the soils is still unknown but there are signs that they could be critical and their role in plant resistance to soil toxins should be thoroughly investigated.

Particular puzzles remain concerning speciation in ultramafics. Are the endemics of the ultramafics of Zimbabwe all palaeoendemics? Why have so few intra-specific differences been reported for taxa which occur on and off ultramafic soils in the tropics? Why does *Asplenium adiantum-nigrum* in the British Isles show parallel morphological changes in disjunct populations on ultramafic rocks of differing character? Why is the flora of the Keen of Hamar in Shetland the richest in varieties and endemics in Britain when it is so far north? Presumably there has been less time for evolution there then further south. Is there a possibility that nunataks are involved more generally in accounting for unusual high latitude floras?

B. A. Roberts and J. Proctor (eds), The ecology of areas with serpentinized rocks. A world view, 419–420.
© 1992 *Kluwer Academic Publishers.*

This text also points out clearly that there is a wide variety of soils produced from serpentinized material. Some of these soils change constantly, e.g., scree slopes, and others have two or more distinct processes happening at once, e.g., bisequal development of podzol/gleyed or podzol/regosol or other soil process combinations. The soils of serpentinized areas can be greatly affected by the different quantities and quality of soil seepage which may not be detected by analysis of dry soil at certain seasons. The role of plant water from snow beds and ground water seepage is likely to be important and contrasts with the role of drought in determining other types of ultramafic flora.

It is clear that in studying the ecology of serpentinized areas the biologist is presented with many complex problems such as the effects of toxic metals, chemical imbalances, trace element deprivation, and in a natural habitat over a very long evolutionary period of time. The ecological study of these naturally occurring habitats is very important and can reveal many answers to help solve many of the pollution problems which have become the most important concern in this decade.

accumulators 2, 101, 104, 106, 156, 217, 232, 234, 237, 265, 266, 267, 313, 325, 330, 343
acidicolous, acidophilous 75, 87, 97, 140, 153, 183, 192
adaptations to
  serpentine 67, 68, 243
aeolian, loess 80, 135, 141, 153, 285, 290, 293, 296, 346, 347, 351, 360
aereoles 19, 24, 25
aerial Photos 78, 118, 259
albite 273
Alfisols 40
alluvial 41, 42, 68, 252, 253, 272, 278, 347, 355, 356, 368, 397
all-terrain vehicles 68
aluminum in soils, in plants 85, 155
alusite 242, 273
amino acids 239
amesite 9
amphibolite 11, 24, 116, 119, 120, 122, 123, 129, 130, 192, 221, 272, 273, 344, 345, 349, 351, 355, 356, 358, 359, 361, 367, 368
analcime 175, 177, 182
andalusite 272
andesitic 19, 335
Angiosperms 325, 383
animals, *see* serpentine animals
anorthite 1
anorthosite 137
anthocyanic 32, 219
anthophyllite 178, 182
anthropogenic disturbances (burning, grazing, mining, etc.) 68, 162, 170, 178, 191, 199, 200, 205, 207, 254, 257, 282, 286, 294, 317, 334, 380, 381, 392, 397
antigorite 1, 7–9, 11, 12, 170, 171, 176, 179, 182
apatite 334, 335, 339

Archaean 314, 334, 344
aretes 379
argillite 36
Aridisols 40
Argixerolls 43
asbestos 1, 3, 4, 7, 8, 68, 208,
Asian 125
augite 18
avoidance 52, 60, 61, 64
awaruite 8, 10

bacteria 48, 57
basalt 17, 220, 225–229, 254, 334, 344, 349
basicolous taxa 75, 87, 97, 153, 154, 164, 183, 220, 222
bastite 16, 139, 170, 176
beetles and serpentine 204
biogeochemical prospecting 415
biogeographic 390
biophysical 78
biotite 277
biotype's 32, 66
bird, bird droppings 155, 161, 204
bodenvag taxa 31, 32, 48, 50, 60, 64, 67
Boreal 76, 115, 125, 128, 130, 223
Braun-Blanquet 78, 118, 143, 192, 199, 203
breccia 36, 220
bronzite 18
Brown ranker soil 141, 147
brucite 8, 10, 11, 25
Brunisolic, Brunisols 80, 118–120, 130
butterflies and serpentine 32, 65, 66

Caesalpino 1, 218
calcifuga taxa, calciphilous taxa, calcicoles 153, 154, 159, 222

calcite, calcareous rocks 18, 24, 83, 136, 140, 141, 153, 155, 183 219, 221, 225–227, 282, 313, 334, 343
calcrete 334, 336, 339, 347
Caledonian age 137, 138
Cambisols 81
Cambrian, Pre 79, 80, 141, 314
carbonatite 334–336
Caruel 2, 218
Carpophores 154
cation exchange capacity, C.E.C. 31, 41, 42, 43, 59, 82, 83, 86, 251, 383, 409
cattle droppings 283, 284
cell division 329, 330
Cenozoic 170
chalcedony 250
chamaephytes 52
check list of serpentine plants 107, 126, 127, 151, 184, 186, 212
chert 36, 220, 221
chloanthite 178
chlorite 13, 41, 42, 141, 170, 175, 177, 178, 182
chlorotic, chlorosis 32, 156, 237, 279, 282, 328, 413
chromite 11, 18, 19, 20, 23, 36, 37, 138, 170, 171, 182, 274, 313, 318
chromium, in plants, 46, 76, 104, 106, 107, 156, 188, 189, 232–234, 236, 239, 263, 325, 328, 353, 366, 415
chromium, in soils 42, 85, 104, 142, 143, 145, 236, 271, 273, 343, 351, 352, 356, 357, 360, 380–381, 412
chrysotile 1, 7–13, 15, 25, 42, 142, 170, 176, 181
cirque 98, 378–380, 398
citric acid, citrate 237, 266
clay analyses 182, 255, 272, 273, 275, 278
climate, 78, 79, 96, 116, 118, 130, 131, 135, 140, 162, 169, 170, 172, 173, 194, 200, 211, 226, 231, 242, 250, 286–288, 293, 316, 325, 334, 343, 344, 348, 355, 356, 358, 376, 379, 380, 407, 412
clinopyroxene 8, 19, 23, 155
coalingite 10
colbalt 45, 46, 50, 76, 85, 104, 106, 142, 145, 170, 182, 183, 188, 232–234, 236, 239, 263, 279, 343, 351, 352, 360, 366, 381, 381, 412, 415
colloids 41, 42
colluvial 41, 42, 75, 81, 172, 252, 253, 265, 345, 347, 351, 355, 358, 361, 381
competitive ability 53, 97, 125, 256
complexes 1, 7, 18
concluding remarks 161, 267, 330, 369, 415, 419
concretions 348
conservation 4, 32, 69, 205
copper, tolerance 47, 67, 163, 169, 182, 183, 280, 283, 325, 329, 335, 343, 347, 351–353, 357, 360, 366, 380, 381, 413, 415
cordierite 272, 273
Cordilleran 32, 125
coriaceous 254
crassism 305
Cretaceous 32, 37, 219, 220
crevice 88, 91, 150, 151, 187, 191, 192, 221, 321, 419

cronstedite 9
crop plants
    nickel content of 62, 273, 279
    nickel toxicity symptoms in 62, 237, 279–284, 328
    tolerance of to serpentine 61, 328
    yield of on fertilized serpentine soils 63
Cryochrepts 381
cryosolic 80
cryoturbation (frost churning, heaving) 2, 75, 80–82, 85, 91, 115, 116, 119, 120, 125, 128, 130, 131, 140, 163, 313, 314, 378, 380
curtans 177
cuticle 254
cyclones 344
cystine 239

debris 135, 141
Devonian 40, 80
deweylites 13
detritic 170, 220
diabase 40
dialagitle 202
dichromate 275
Dicotyledoneae 95, 383, 391, 198, 398, 403, 407, 411
diopside 11, 334, 353, 366, 367
diorite 57, 325, 345, 353
Dioscorides 1
diploid taxa 32
diptercarp 259
disjunct spp., disjunctions 87, 96, 97, 125, 152, 153, 164, 183, 218, 296, 412
dolerite 137, 138, 335, 336, 358
dolomite, as fertilizer 60, 61, 138, 139
drought 2, 57, 97, 140, 163, 169, 175, 185, 187, 229, 262, 266
dunite 8, 10, 13, 14, 18, 19, 20, 22, 23, 25, 32, 35, 37, 38, 39, 40, 80, 82, 116, 139, 141, 142, 155, 273, 378, 379, 380, 381, 395, 419
dwarfism 50, 68, 135, 218, 222, 225, 227, 231, 375, 409
Dystrochrept 381
dykes 138, 336, 337, 344, 351

earth core hummocks 116
Eastern NA spp. 96
ecotypes 87, 375, 376, 409–412
ecophysiological research 217, 231
ecologite 8
element concentrations in plants 46, 102, 105, 156, 157, 188, 232–240, 262, 264, 341, 375, 414, 415
    in soils 41, 42, 85, 103, 106, 144, 145 233–236, 262, 275
    in rocks 36, 103, 139, 174, 202, 232, 274
eluvial 119
endangered species 69
endemic, animals 204
endemism in serpentine plants
    of Balkans 201, 206
    of Britain 152
    of Great Dyke – coloured photo of late Professor Wild

of Italy 217, 218, 221, 222, 225–231, 240
of Japan 272, 296, 299, 302, 305–309
of New Caledonia 256
of New Zealand 375, 380, 391–390, 395, 407, 409, 412, 415
of Portugal 172, 183, 185, 187, 191
of Sabah 258
of Sulawesi 259
of Solomon Islands 257
of United States 31, 33, 43, 45, 50, 59, 61–64, 66, 69
of Zimbabwe 312, 322, 324–326, 330
enstatite 171, 175–178, 182
Entisols 40, 41
environmental gradients 58
enzymes 161, 164, 231, 242
epidote 366, 367
epiphytes 155, 256, 261, 262, 322, 383
erosion, eroded 43, 137, 141, 142, 149, 152, 153, 161, 162, 169, 172, 173, 176, 207, 208, 209, 211, 265, 267, 314, 347, 390, 395, 395, 398, 401, 407
Eutrochrepts 381
evolutionary considerations 66, 419
exchangeable bases, total 83, 85, 86, 125, 187, 251, 252, 280, 282, 320, 383
exclusives 124, 125
exploitation 32, 68, 254

fauna 65, 204, 205
fayelite 170
Ferralsol 253
ferritchromite 10
ferromagnesium silicate 42
fertilizers, serpentine in, fertilization 60, 61, 63, 140, 164, 283, 285, 328, 329, 334, 380, 413
fire, see anthropogenic
floras
    Balkans 203
    Britain 150
    New Caledonia 251
    New Guinea 256
    New Zealand 384
    Newfoundland 87
    Portugal 183
    Quebec 123
    United States 59
    Zimbabwe 324
flushes 94, 95, 99, 148, 395
fluvial 272
flowering 409
forsterite 170
frost see cryoturbation
fungi, micro-organisms 48, 50, 53, 54, 154, 419
    ascomycetes 54
    mycorrhizae 4, 53, 54, 164
    phycomycetes 54
    imperfecti 54
future research 4, 69, 243, 419

gabbro, meta gabbro 18–21, 23, 24, 36, 39, 40, 93, 137, 138, 143, 148, 201, 220, 225, 227–229, 254, 273, 314, 316, 324, 345, 353, 355, 370, 376, 378, 392
garnet 8
garnierite 9
genecological 48, 53, 69
genetic 96, 411
geobotanical studies 338–341
geochemistry 339–341
geology, general 1, 7–28
    Balkans 200
    Great Britain 137
    Italy 218
    Japan 272
    Newfoundland 79
    New Zealand 376
    Portugal 170
    Quebec 116
    South Africa 334
    Tropical Far East 250
    United States 34
    Western Australia 344
    Zimbabwe 314
geophytes 52
gibbsite 21
glacial, till, drift 81, 116, 135, 141, 378, 379, 397, 397, 409, 412
glaciation, Pleistocene 37, 58, 100, 101, 116, 136, 141, 170, 172, 221, 222, 225, 226, 231, 233, 236, 244, 296, 375, 378, 380, 391, 404, 412
glaucescence 32, 50, 68, 183, 187, 189, 191, 201, 229, 272, 273
Gleyed Regosols 75, 80, 81
Gleysolic, gleysols, gleyed soils 80, 118, 119, 120, 123, 130, 141, 142, 147, 152, 409, 419
glinmerite 336
glossary – geological terms 25–28
gneiss 20, 36, 80, 137, 218, 251, 272, 273, 334, 335, 347, 404, 407
goethite 21, 177, 182
gold mining 344
gophers and serpentine 66
gossan 356
granite 40, 80, 130, 137, 141, 150, 220, 272, 273, 277, 317, 319, 322, 324, 335–337, 339, 344, 347, 357, 358
granodiorite 36, 254
greenalite 9
greenschist 11
Greenstone Belts, greenstones 36, 344, 347, 348, 358, 369, 385
Gymnosphermae 95, 255, 383, 397, 403, 403, 404–409
gypsum 345–347

Haploxeralfs 43
Hapludalfs 80
hairs, hariness 150, 152, 258
halite 345–347
harrisitic cumulate 139

424

harzburgite 8, 13, 14, 17, 20, 22, 23, 36, 40, 80, 116, 155, 202, 376, 380, 381
heathlands 94, 140, 143, 144, 146, 148, 149, 150, 153, 162, 259, 290, 293
hematite 8, 177
hemicryptophytes 52
herbarium specimens, analysis of 4, 104
hillite 335
hornblende 18, 19
hornfels 353, 354, 366, 367, 370
Humisols 120, 321
hybrids 183
hydrological 313, 324, 330
hydromagnetite 10
hydromorphic soils 172, 255, 316
hygrophilious 120, 128, 254, 255
hyperaccumulators of nickel 4, 32, 33, 46, 50, 68, 75, 96, 104, 106, 135, 156, 185, 189, 191, 217, 234, 237, 266, 351
    classification of 3
hypersthenite 345, 353, 355, 370

Iapetus 20
Ice Ages 135, 376, 412
illite 141
ilmenite 19, 175
illuvial 119
infertile soils 41
Inceptisols 40, 251, 252
indicators, indicator species 2, 43, 55, 59, 61, 63, 101
indifferents 124
insectivorous 101
iron in plants 46, 105, 135, 188, 235, 236
iron in soils, iron oxides 42, 106, 131, 177–179, 182, 233, 271, 280–282, 283, 343, 348
ironstone 358, 360

jadeite 272, 373
Jurassic 35, 37, 38, 218–221

karst 204, 210
kaolinite 10, 13, 81, 141, 277, 347
klippen 79, 80
kopjes 334
krummholz growth form 94, 95, 115, 116, 119, 125, 127, 129, 130
Kruepelholz 261
Kyanite 272

landslides 4, 220, 272, 276–278, 293, 299
lateritic, soils, 176, 257, 316, 318, 322, 345–348, 353, 355, 356, 358, 360, 366, 367–370, 375, 376, 381, 392
lawsonite 272
lead 67, 154, 352, 357, 360
lherzolite 8, 13, 18, 20, 23, 36, 80, 170, 202, 221, 225–227, 376
lianes 383, 398
life form spectra of serpentine vegetation 52, 60, 183, 185,

218, 222, 227
lignicolous 54
limestone, marl 8, 80, 140, 153, 154, 183, 218–220, 257, 272, 308, 309, 335, 347
Lithic Argixerolls 43, 47
Lithic Rhodoxeralfs 43
Lithoregosol 81
Lithosols 40, 43, 172–174, 177, 178, 182, 187, 191, 192, 195, 227, 316
livestock grazing 4, 68, 162, 205
Lizzard Cornwall 16
lizardite 1, 7–9, 11–13, 16, 24, 25, 176, 182
lunettes 347
Luvisolic 80, 253

macrorhizy 183, 187, 189, 201
magma chamber 21, 137, 170
magnesite 11, 16, 68
Magnesium/Calcium Quotients 31, 41, 43, 44, 47, 75, 81–83, 85, 86, 102–104, 115, 120, 123, 130, 131, 143, 145, 155, 156, 159, 160, 162, 163, 169, 172, 173, 187, 188, 194, 203, 232, 233, 237, 238, 239, 252, 262, 263, 267, 319, 324, 328, 329, 337, 351, 360–371, 375, 381, 381, 390, 413
magnesium toxicity 2, 4, 44, 103, 135, 155, 156, 159, 163, 328, 419
magnetite 8, 10, 11, 18, 19, 23, 25, 232
malic, malonic acid, malate 237, 241, 266
manganese toxicity 249, 263–265, 267
maritime climate 33, 141, 156, 223
maquis 251, 254, 255, 262, 265
medicinal uses 1, 268
Mediterranean Climate, eiement 33, 169, 170, 172, 185, 194, 195, 200, 221, 222, 226, 228
mercury 32, 45, 46, 68
Mesozoic, Pre 20, 218–220, 272, 273
metadiabases 36
methionine 239
mica 182, 251, 277, 339
microphylls 254, 334, 375, 398, 403, 409
Millisols 40
mine talings 4, 395, 409, 413
mineral nutrition studies, macro & micro 43, 75, data 85, 100, 101, data soils 102, 103, 135, 156, 162, 163, 169, 174, 233, 236, 262, 266, 267, 280
mineral potential 380
Miocene 229, 314, 347, 391
mohorovicid (moho) discontinuity 21, 23
molybdenu.ı 41, 44, 237, 238, 271, 272, 279
Monocotyledons 95, 324, 383, 403, 407
montane 255, 375, 403
montmorillonite 42, 81, 141, 275–277, 316
moraines see glaciation
morphological see variants
muscovite 273

nanism 32, 50, 183, 187, 189, 191, 201, 254, 302, 305–307
nappe 138, 220

425

Neocene, Neogene 200
nepouite 9
Newfoundland Geology 14, 17, 79
nickle, nickle toxicity 2, 8, 44, 135, 156, 159, 160, 163, 169, 178, 189, 194, 195, 237, 243, 249, 265–267, 271, 273, 278–282, 313, 328, 329, 360
nickle citrate 4
nickle in plants 46, 104, 105, 156, 188, 232–234, 236, 239, 241, 249, 263–267, 282, 283, 325, 329, 330, 341, 348, 353, 362, 366, 415
nickle in soils 42, 44, 75, 76, 83, 85, 86, 104, 106, 107, 142, 143, 145, 172, 178, 182, 183, 233, 236, 239, 249, 271, 278, 357, 360, 381, 381, 412, 413
nickel mining 358
nickel plants (see hyperaccumulators of nickel
nodules 48, 347, 351
norite 314, 316, 319, 322, 324
nunataks, unglaciated 100, 101, 183

obligate serpentinophytes 201, 203, 209
olivine 11, 15, 18–20, 23, 24, 36, 37, 40, 139, 155, 170, 175–178, 182, 202, 334, 336
ophiolites 1, 7, 12, 18–21, 23–25, 31, 35–40, 75, 80, 218–223, 225–227, 229, 230, 232, 233, 376, 378, 379, 380, 381
Ordovician 40, 79, 80, 116
organic matter 80, 83, 85, 86, 116, 118–120, 123, 169, 174, 178, 182, 187, 191, 194, 222, 251, 320, 392, 409
Orthic Humo-Ferric Podzols 80, 119, 130
Orthic Regosols 75, 80, 81, 83
orthoclase 277
orthopyroxene 8, 14, 16, 17, 20, 23, 40
oxalic acid 142
Oxisols 251, 255, 262

Palaeozoic 20, 21, 40, 79, 218, 272, 273
Pangaea 330
parasitic 54
pans 348, 356
patterned ground 75, 81, 82, 84, 92
peatland (bog, fen & mire) 81, 88, 89, 92, 99, 101, 128, 129, 144, 147, 148, 267, 278, 296, 308, 395, 395, 401, 403, 409
pectinates 369
pectite 12
pegmatite 19, 334
pelagic sediments 19
peinomorpism 55
pentlandite 10
percoraite 9
peridotite 1, 7, 8, 10, 11, 13, 15, 16, 21–25, 31, 32, 35–40, 80, 82, 116, 137–139, 141, 155, 170, 202, 251, 267, 273, 283, 296, 298, 302, 344, 345, 351–353, 358–361, 370, 379, 380, 381, 395, 398
peridotite complexes 1, 7, 24
Permian 35, 376
pH 1, 7, 11, 31, 41, 42, 46, 82, 83, 85, 86, 104, 120, 145, 153, 172–174, 176, 187, 221, 222, 225–228, 233, 236,

251, 252, 275, 278–280, 282, 283, 317, 320, 325, 328, 339–341, 347, 375, 381, 383, 390, 390, 419
phoscorite 334–336
phosphatase 4, 159, 161, 164, 242
phosphorus, limiting natural 162, 262, 337, 366, 370
photosynethesis 53, 244
phyllite 335
physiographic 76, 77, 314, 334, 344, 346, 348
physiognomy 116, 254, 369, 375, 403, 409
phytocenoses 204
phtogeographic 227
phytosociological analysis, table, survey 78, 126, 127, 129, 222, 289–301
phytotoxicity 123
picrolite 15
pillow lava 19, 21, 23, 36
plagioclases 7, 12, 18, 23, 277
plagiogranite 19, 40
plagiotropism 32, 50, 183, 187, 189, 191, 201, 218
plant sociology of serpentine vegetation of Japan 289
platinum 285
Podzolic, Podzollation 80, 81, 118–120, 141, 176, 296, 419
polyploidy 66
poikilitic grains 23
porphyry 356, 357, 368
preadaption 67
preferentials 124, 125
prehnite 12
principal components analysis 51
principal coordinate analysis 115, 118, 120–124, 130
proline 239
Proterozoic 334
pteridophytes 383, 403, 407
pubescence 218, 222
purpurescence, erithrism 183, 201, 239
pyroaurite 10
pyrophyllite 177, 182
pyroxenes, pyroxenite 7, 10, 12, 13, 15, 18–20, 24, 36, 40, 314, 316, 317, 319, 322–324, 333–341, 345, 353, 354, 358, 359, 366, 367

quartzite, quartz 250, 258, 272, 273, 277, 335, 344, 349, 351, 368

race, edaphic, non-serpentine 43, 96, 135, 150–152, 160, 272
rain fall 97, 250, 275, 314, 315, 324, 325, 334, 344, 380
rain shadow 37
rare species 56, 87, 161
Red Mediterranean Soil 172, 174–179, 182, 191, 192, 196
Rego Gleysols 75, 80, 81, 83, 120, 130
Regosolic soils 80, 82, 94, 118–120, 130, 141, 419
relic see serpentinicolous
remote sensing and Aerial photos 51, 78, 344
residual soils 41
resin, resin bleeding 206, 207
resistance 32, 43, 48, 49, 60, 67, 68, 266, 419
restriction to serpentine 55, 115, 124

rhyolite 335
rice 282
riparian species 56
rodingites 12, 24, 25
roots 4, 155, 159, 161, 188–190, 218, 225, 239, 241, 254,
    327, 329, 409, 412, 413
rubefaction 173, 175, 177
rutile 175

salamanders and serpentine 66
salinas 347, 348
Saltational species 67
sandstone 80, 154, 220, 232, 258, 259
sap, vacuolar 4, 240
saponite 142
satalite imagery *see* remote sensing and aerial photos
saturation of Mg 82, 83, 86, 103, 280
saxonite 40
schist and micaschist 36, 80, 223, 251, 272, 273, 317, 351,
    399, 403, 403, 409
Scandinavian Spp. 96
sclerophyllous, shrubs, trees 50, 183, 194, 249, 254, 257,
    343, 344, 353, 404
scree 225, 227, 229, 289, 296, 302–305, 307–309, 381,
    383, 395, 397, 403, 409, 419
sea, sea spray, salt, halophytic 136, 143, 148, 154, 220,
    334, 347, 348, 356, 357, 368
sedentary soils 278
seeds, seed adenylate, seed ATP, ADP, AMP 188, 207,
    209–211, 225, 231, 232, 239, 242, 243, 390
seeps, seepage 56, 58, 83, 87, 99, 334, 399, 401, 419
septechlorites 9
serpentine, animals on
    earthworms, snails 4, 327
    fish and reptiles 66, 204, 327
    insects 65, 204, 327, 328
    cattle and other mammals, herbivors 4, 66, 68, 162, 204,
        266, 267, 282, 283, 380, 381, 399
serpentine syndrome 34, 40, 43
serpentinicolous, relics 48, 52, 65, 96, 106, 115, 123–125,
    128–131, 172, 183, 184, 192, 193, 201, 271, 272, 293,
    296, 299, 302, 308
serpentinifuges 124, 125
serpentinomorphism, serpentinimorphoses 55, 169, 183,
    185–187, 189, 192, 201, 272, 296
serpentinophytes 169, 183–185, 187–189, 191, 192, 199,
    201, 203, 209, 211, 212, 229, 230, 232–235, 239, 271,
    302
serpentinite 13, 24, 25, 32, 35–37, 137, 138, 141, 219–221,
    225 228, 272, 274, 275, 278, 280, 283, 285, 294, 314,
    319, 322, 324, 325, 333–341, 344, 345, 349, 351, 353,
    355–358, 360, 368–370, 378, 380, 381, 395, 395,
    399–401, 412
serpentinization process 10, 15–17, 32
serpophite 170, 176
shoots 155, 188, 189, 329, 412, 413
shale, slates 80, 92, 97, 101, 172, 273, 335, 359
sheeted diabase dykes, dykes 19, 21, 80, 138

siliceous 155, 254, 273, 316, 358, 360
sillimanite 272, 273
Silurian 80
skeletal soils 140–145, 147, 150–152, 154, 159, 161–163,
    222, 347–349, 351, 356, 375, 381, 407
smectite 41, 42, 177, 178, 182, 183, 250, 255
snow and snowbeds 82, 94, 98, 119, 122, 125, 200, 290,
    291, 293, 297, 302, 376, 378, 380
soapstone 155
soils,
    Balkans 200
    Great Britain 141
    Italy 221
    Newfoundland 80
    Portugal 172
    Quebec 118
    Tropical Far East 250
    United States 31
soil genesis 42, 80, 81, 121, 141, 172, 173
solifluction terraces 81
sols bruns eutrophes hypermagnesiens 251, 252, 254, 255,
    262, 263, 265
sols ferrallitiques 265
spinel 8, 23
staurolite 272
stenophyally, stenoendemism 68, 183, 187, 192, 201, 204,
    223, 229, 302, 305–307
stone polygons, stone sorting 75, 81, 82, 84, 91, 94, 116,
    149
stone strips 75, 81, 82, 116
stone works 1, 2, 285
stoney grounds 209
stratiform complexes 1, 7, 17, 18, 23
succulent 319, 321, 411
sulphides, sulphur 353, 360
suscession 170, 191, 194, 211, 407
syenite 334–336

talc, steatite 10, 11, 138, 141, 175, 177, 178, 182, 275, 344,
    351, 361, 370
tarns 379
termites and serpentine 327, 328
Tertiary 137, 220, 347, 376
tholeiite 19
titanium 175
tolerance 33, 43–45, 49, 53, 61, 67, 68, 76, 96, 106, 160,
    161, 163, 169, 231, 239, 240, 242, 244, 267, 280, 319,
    329, 376, 381, 383, 391, 390, 407–413, 415
toncite 11
tors 334, 378
toxic and toxicity of Ni and Mg concentrations (also *see*
    nickle toxicity 2, 3, 4, 43, 75, 103, 104, 106, 135, 142,
    278, 313, 319, 322, 324, 328, 330, 367, 409
trace elements 2, 4, 104, 106, 343
travertine 83
tree ferns 383
tremolite 11, 135, 141, 351, 353, 366, 367, 370
Triassic, Pre 220

troctolites 23, 202
tundra 115, 125, 128, 130, 131
tunnels 275, 276

ultramafic soils (*see* sperpentine soils)
  geology 1, 7–30
  Balkans 200
  Britain 137
  Italy 221
  Japan 278
  Newfoundland 79
  Portugal 170
  Quebec 116
  United States 34
uranium 347

variants, morphological 43, 87, 150, 151, 187, 189, 191, 201, 203, 205, 206, 223, 226, 243, 324, 409
vermiculite 142, 334–336
vegetation types,
  Balkans 203
  Britain 143
  Italy 221
  Japan 285, 294
  New Caledonia 251
  Newfoundland 87
  New Guinea 256
  New Zealand 390
  Philippines 257
  Portugal 191
  Quebec 125
  Solomon Islands 257
  South Africa 334
  United States 51
  Western Australia 348
  Zimbabwe 319

Vertisols 40, 251, 316

wallastonite 12
water, extracts, culture, soluble fraction 4, 12, 78, 83, 86, 120, 125, 128, 129, 239–241, 274, 313, 324, 329
  metoric and non-metoric 12, 13
  carbonated 176, 277–278
  loss, surplus 151, 154, 170, 172, 173, 175–177, 192, 194, 251, 255, 275, 314, 316, 378, 412
  sea 4, 24, 262
websterites 8, 17
weeds 65
wehrlite 8, 18, 25
wind 140

xeric, xerophytic, conditions 54, 56, 62, 87, 154, 183, 191, 218, 221, 222, 225–227, 231, 243, 290, 291, 294
xeromophrism 55, 97, 154, 316
xerophytism 32
xonolite 12, 97
xylem 324

zinc, tolerance 67, 85, 154, 182, 188, 189, 242, 271, 272, 279, 282, 284, 329, 352, 357, 360
Zurich – Montpellier (Z–M) 78, 118

Geobotany

1. J.B. Hall and M.D. Swaine (eds.): *Distribution and Ecology of Vascular Plants in a Tropical Rain Forest*. Forest Vegetation in Ghana. 1981   ISBN 90-6193-681-0
2. W. Holzner and M. Numata (eds.): *Biology and Ecology of Weeds*. 1982   ISBN 90-6193-682-9
3. N.J.M. Gremmen: *The Vegetation of the Subantarctic Islands Marion and Prince Edward*. 1982
                                                                                 ISBN 90-6193-683-7
4. R.C. Buckley (ed.): *Ant-Plant Interactions in Australia*. 1982   ISBN 90-6193-684-5
5. W. Holzner, M.J.A. Werger and I. Ikusima (eds.): *Man's Impact on Vegetation*. 1983 ISBN 90-6193-685-3
6. P. Denny (ed.): *The Ecology and Management of African Wetland Vegetation*. 1985   ISBN 90-6193-509-1
7. C. Gómez-Campo (ed.): *Plant Conservation in the Mediterranean Area*. 1985   ISBN 90-6193-523-7
8. J.B. Faliński: *Ecological Studies in Białowieza Forest*. 1986   ISBN 90-6193-534-2
9. G.A. Ellenbroek: *Ecology and Productivity of an African Wetland System*. The Kafue Flats, Zambia. 1987
                                                                                 ISBN 90-6193-638-1
10. J. van Andel, J.P. Bakker and R.W. Snaydon (eds.): *Disturbance in Grasslands*. Causes, Effects and Processes. 1987   ISBN 90-6193-640-3
11. A.H.L. Huiskes, C.W.P.M. Blom and J. Rozema (eds.): *Vegetation Between Land and Sea*. Structure and Processes. 1987   ISBN 90-6193-649-7
12. G. Orshan (ed.): *Plant Pheno-morphological Studies in Mediterranean Type Ecosystems*. 1988
                                                                                 ISBN 90-6193-656-X
13. B. Dell, J.J. Havel and N. Malajczuk (eds.): *The Jarrah Forest*. A Complex Mediterranean Ecosystem. 1988   ISBN 90-6193-658-6
14. J.P. Bakker: *Nature Management by Grazing and Cutting*. 1989   ISBN 0-7923-0068-8
15. J. Osbornová, M. Kovářová, J. Lepš and K. Prach (eds.): *Succession in Abandoned Fields*. Studies in Central Bohemia, Czechoslovakia. 1990   ISBN 0-7923-0401-2
16. B. Gopal (ed.): *Ecology and Management of Aquatic Vegetation in the Indian Subcontinent*. 1990
                                                                                 ISBN 0-7923-0666-X
17. B.A. Roberts and J. Proctor (eds.): *The Ecology of Areas with Serpentinized Rocks*. A World View. 1991.
                                                                                 ISBN 0-7923-0922-7
18. J.T.A. Verhoeven (ed.): *Fens and Bogs in the Netherlands*. Vegetation, History, Nutrient Dynamics and Conservation. 1991   ISBN 0-7923-1387-9

KLUWER ACADEMIC PUBLISHERS – DORDRECHT / BOSTON / LONDON